Insect Hemocytes

T0213459

Insect Hemocytes

Development, forms, functions, and techniques

Edited by
A. P. Gupta
Professor of Entomology
Rutgers University

Cambridge University Press

Cambridge
London   New York   Melbourne

CAMBRIDGE UNIVERSITY PRESS
Cambridge, New York, Melbourne, Madrid, Cape Town, Singapore, São Paulo, Delhi

Cambridge University Press
The Edinburgh Building, Cambridge CB2 8RU, UK

Published in the United States of America by Cambridge University Press, New York

www.cambridge.org
Information on this title: www.cambridge.org/9780521113038

First published 1979
This digitally printed version 2009

A catalogue record for this publication is available from the British Library

Library of Congress Cataloguing in Publication data
Main entry under title:
Insect hemocytes.
Includes index.
1. Insects – Cytology.   2. Blood cells.
I. Gupta, A. P., 1928-
QL494.8.I57   595.7′01′13   78-10477

ISBN 978-0-521-22364-5 hardback
ISBN 978-0-521-11303-8 paperback

Contents

Preface

Insect hematology is a difficult and controversial subject. Developing suitable techniques to study hemocytes – their origin, development, differentiation, multiplication, and functions – has presented many problems; and the existing differences of opinion among insect hematologists on almost all aspects of insect hemocytes surely indicate the controversial nature of the subject. As a consequence, the study of insect hemocytes has not been popular, and this in turn has hindered progress in this area. Disagreement over the exact number of basic hemocyte types still remains, and I doubt if a consensus on this will ever be reached because this is largely a matter of opinion. Fortunately, however, insect hematologists recognize the existence of certain morphologically distinct hemocyte types, and there is a more or less general agreement on the names of these types. This has facilitated, to some degree, comparisons of various descriptions and has made it possible to define more accurately the physiological roles of at least some hemocyte types. Much of the physiological significance of hemocytes, however, remains to be understood, particularly their role in the insect endocrine system.

The book seeks no consensus on any aspect of hemocytes, but attempts to present a critical and, I hope, a balanced account of the subject as it is presently perceived by the contributors. Each contributor has had complete freedom to develop, interpret, and present his/her views on the subject. Portions of the chapters by Drs. Arnold, J. A. Hoffmann, Jones, Shapiro, and Sohi were presented at the XV International Congress of Entomology (1976) in a Special Interest Conference on Hemocytes. I invited others later to contribute to the book.

The book is organized into four sections: (1) Development and Differentiation, (2) Forms and Structure, (3) Functions, and (4) Techniques. Some overlaps in the coverage of some topics have inevitably occurred, but all such overlaps have been kept to a minimum, and nearly all are complementary to one another. Wherever relevant, overlaps and divergences of opinion in various chapters have been cross-referenced. Most authors have pointed out the important controversies about their respective topics, and many have suggested areas for further research. The native tongue of several of the contributors is not English, and in all such cases, editing has been largely confined to removing obvious infelicities in order to retain the original script style and the contents.

A book of this nature cannot be completed without the cooperation and assistance of many individuals. I am most grateful to all the authors who responded to my invitation to contribute and thus made the book possible. The following individuals, journals, societies, and publishers gave their permission to reproduce published material: N. M. Costin (U.K.), G. Devauchelle (France), M. Florkin (Belgium), Horlant-Miewis (Belgium), A. V. Loud (U.S.A., for the late Dr. M. Hagopian), A. K. Raina (U.S.A.), G. Salt (U.K.); Academic Press, Archives de Biologie, Archives Internationales de Physiologie et la Biochimie, Entomological Society of America, Entomological Society of Canada, Pergamon Press Ltd, Rockefeller University Press, Van Nostrand–Reinhold Company.

In addition to the above acknowledgments, I must express my gratitude to my wife and children for their infinite patience and understanding in tolerating my total preoccupation with the book during its preparatory phases. Their generosity in not demanding my attention and time during that period made my task considerably easier. My wife, Dr. Goda Gupta, also provided considerable help in the arduous task of preparing the indexes. The overall stewardship of Ms. Rhona Johnson and her colleagues at the Cambridge University Press during the production phase greatly improved the book.

Finally, I hope I will not be deemed presumptuous for believing that the book fills the need for an up-to-date reference work on insect hemocytes. And I shall consider all our efforts truly rewarding if the book stimulates further research and discussion.

A. P. GUPTA

New Brunswick, N.J.
June, 1979

Contributors

H. Akai—Sericultural Experiment Station, Wada, Suginami-ku, Tokyo 166, Japan (Ch. 5)

J. W. Arnold—Biosystematics Research Institute, Research Branch, Agriculture Canada, Ottawa, Ontario K1A 0C6, Canada (Ch. 8, 18)

D. E. Ashhurst—Department of Structural Biology, St. George's Hospital Medical School, Cranmer Terrace, London, SW17 ORE, U.K. (Ch. 12, 22)

R. J. Baerwald—Department of Biological Sciences, University of New Orleans, Lakefront, New Orleans, Louisiana 70122, U.S.A. (Ch. 6, 21)

M. Brehélin—Laboratory of General Biology, Louis Pasteur University Research Group of CNRS 118, 12, rue de l'Université, 67000 Strasbourg, France (Ch. 2)

A. C. Crossley—School of Biological Sciences, University of Sydney, Sydney, N.S.W. 2006, Australia (Ch. 15)

D. Feir—Biology Department, Saint Louis University, Saint Louis, Missouri 63103, U.S.A. (Ch. 3, 14)

G. Goffinet—Laboratory of General and Comparative Biochemistry, University of Liège, 17 Place Delcour, B-4020 Liège, Belgium (Ch. 7)

Ch. Grégoire—Laboratory of General and Comparative Biochemistry, University of Liège, 17 Place Delcour, B-4020 Liège, Belgium (Ch. 7)

A. P. Gupta—Department of Entomology and Economic Zoology, Rutgers University, New Brunswick, New Jersey 08903, U.S.A. (Ch. 4, 17)

C. F. Hinks—Biosystematics Research Institute, Research Branch, Agriculture Canada, Ottawa, Ontario K1A 0C6, Canada (Ch. 18)

D. Hoffmann—Laboratory of General Biology, Louis Pasteur University Research Group of CNRS 118, 12, rue de l'Université, 67000 Strasbourg, France (Ch. 2)

J. A. Hoffmann—Laboratory of General Biology, Louis Pasteur University, Research Group of CNRS 118, 12, rue de l'Université, 67000 Strasbourg, France (Ch. 2)

J. C. Jones—Department of Entomology, University of Maryland, College Park, Maryland 20742, U.S.A. (Ch. 10)

H. Mori—Department of Natural History, Faculty of Science, Tokyo

ix

Metropolitan University, Setagaya-ku, Fukazawa, Tokyo 158, Japan (Ch. 1)

A. Porte—Laboratory of Electron Microscopy, Institute of Physiology, Louis Pasteur University, 21, rue René Descartes, 67000 Strasbourg, France (Ch. 2)

N. A. Ratcliffe—Department of Zoology, University College of Swansea, Singleton Park, Swansea, SAZ 8PP, U.K. (Ch. 13)

A. F. Rowley—Department of Zoology, University College of Swansea, Singleton Park, Swansea, SA2 8PP, U.K. (Ch. 13)

S. Sato—Faculty of Agriculture, Tokyo University of Agriculture, Sakuragaoka, Setagaya-ku, Tokyo, Japan (Ch. 5)

M. Shapiro—Gypsy Moth Methods Development Laboratory, U.S. Department of Agriculture, Otis Air Force Base, Massachusetts 02542, U.S.A. (Ch. 16, 19, 20)

S. S. Sohi—Insect Pathology Research Laboratory, Department of Fisheries and Forestry, Sault Sainte Marie, Ontario P6A 5M7, Canada (Ch. 9)

V. B. Wigglesworth—Department of Zoology, University of Cambridge, Cambridge CB2 3EJ, U.K. (Ch. 11)

D. Zachary—Laboratory of General Biology, Louis Pasteur University Research Group of CNRS 118, 12, rue de l'Université, 67000 Strasbourg, France (Ch. 2)

I. Development and differentiation

1 Embryonic hemocytes: origin and development

H. MORI

Department of Natural History, Faculty of Science, Tokyo Metropolitan University, Setagaya-ku, Fukazawa, Tokyo, 158, Japan

Contents

1.1. Introduction

Although embryonic hemocytes, or blood cells as previously termed, have been described in papers dealing with organogenesis, the available information is fragmentary. The origin of hemocytes has been more or less closely observed, but the cytological features of hemocytes have not been adequately described. Furthermore, the development of hemocytes has hardly been noted. Functions and activities of embryonic hemocytes have been neglected almost completely.

Perhaps the reason why insect embryologists have not given much attention to embryonic hemocytes is because the role of these free cells in embryogenesis is quite obscure. No direct participation of hemocytes in the formation of any of the embryonic organs seems to have been observed. Embryologists have been devoting much attention to the details of midgut epithelium formation, because this elucidates the mode of entoderm formation. The metameric segmentation of the embryos also has been closely followed to discover the number of segments in the insect head. However, the unique nature of the embryonic hemocytes seems to have completely escaped the attention of embryologists.

The embryonic hemocytes are unique among embryonic cells because these cells are single and may circulate in the embryo even during the final phase of embryogenesis; they may move by themselves or be moved by the hemolymph. As embryogenesis proceeds, morphogenetic movement of each cell becomes slow, and the topographical positions of various cells in the embryo become gradually fixed. For example, the drastic shift in the positions of cells seen at the stage of germ band formation seems to be lost when the embryo is very close to hatching.

Dohrn (1876) first described embryonic hemocytes in *Bombyx* embryos and considered them to be derived from the yolk cells. Subsequently, six theories on the origin of embryonic hemocytes were proposed. Six groups of cells were thought to give rise to embryonic hemocytes: (1) yolk cells, (2) serosal cells, (3) cells at the junction of somatic and visceral musculature, (4) wall of the heart, (5) median mesoderm cells (i.e., cells at the median part of the inner layer), and (6) suboesophageal body. The most detailed description and discussion of embryonic hemocytes may be found in Heymons's (1895) work. Later reviews on embryonic hemocytes may be found in the works of Roonwal (1937) and Johannsen and Butt (1941). It is generally agreed that embryonic hemocytes are derived from the median mesoderm (Anderson, 1972a,b). However, some authors (Bock, 1939; Striebel, 1960) consider them entodermal in origin.

In this review, theories on the origin and development of embryonic hemocytes proposed by previous authors will be evaluated.

Causes that might have brought about this controversy among hemocyte formation theories will also be discussed. Cytohistological observations on the hemocytes in the embryos of *Gerris* will be presented and certain questions regarding the origin and development of the embryonic hemocytes discussed. In addition, the prospect of embryonic hemocyte investigation for the future will be discussed, including some previously unused techniques.

1.2. Theories of hemocyte formation

Conclusions of previous authors regarding the origin of embryonic hemocytes in various insects are summarized in Table 1.1.

1.2.1. Median mesoderm theory

Most authors agree that embryonic hemocytes are derived from the median mesoderm, namely the median part of the inner layer. Korotneff (1885) studied the embryogenesis of *Gryllotalpa* and was the first to propose this theory. More detailed observations of orthopteran and dermapteran embryos led Heymons (1895) to the same conclusion. Since then, many authors have supported this theory, including Nelson (1915) and Bock (1939). Although these two authors used different materials, they came to the same conclusions. The following account is based on their works.

After the differentiation of the germ band and amnion had occurred, segregation of cells from a position on the germ band took place along the longitudinal axis of the embryo. This segregation process may be called gastrulation, and the segregated cells may be considered as the inner layer. The cells of the inner layer then spread horizontally over the dorsal (yolk) side of the germ band and formed a single layer. This longitudinally continuous cell layer then divided into several groups of cells some time after the segmentation in the germ band (ectoderm) had occurred; metameric segmentation of the embryo was thus established. The inner layer at each lateral part of the embryo then became two-layered, but it remained one-layered at its median part. The former was destined to become the coelomic sac, but the latter, termed median mesoderm (= primary median mesoderm) (Fig. 1.1A), produced the embryonic hemocytes.

Dissociation of cells in the median mesoderm, sometimes accompanied by transformation of dissociated cells, could be seen (Fig. 1.1B); it eventually resulted in the liberation of cells from the median mesoderm (Fig. 1.1B,C). The liberated cells were seen in a space between the embryo and the yolk, the space being delimited by a very thin membranous structure, the yolk boundary. This space is the epineural sinus (Fig. 1.1B–D) and would eventually become the hemocoel. The

Table 1.1. Origin of embryonic hemocytes in different groups of insects (ent = entodermal; mes = mesodermal; se = serosa; sub = suboesophageal body; y = yolk cell)

Insect	Origin[a]	References
Odonata		
Epiophlebia superstes	mes	Ando (1962)
Blattodea		
Blatta germanica	mes	Wheeler (1899); Heymons (1895)
Mantodea		
Hierodula crassa	ent	Görg (1959)
Phasmatodea		
Carausius morosus	ent	Leuzinger et al. (1926)
Orthoptera		
Gryllotalpa vulgaris	mes	Korotneff (1885)
	mes	Heymons (1895)
	ent	Nusbaum and Fuliński (1906)
Gryllus domesticus	mes	Heymons (1895)
Oecanthus niveus	se	Ayers (1884)
Locusta migratoria	mes	Roonwal (1937)
Dermaptera		
Forficula auricularia	mes	Heymons (1895)
Isoptera		
Kalotermes flavicollis	ent	Striebel (1960)
Hemiptera		
Pyrilla perpusilla	mes	Sander (1956)
Pyrrhocoris apterus	mes	Seidel (1924)
Rhodnius prolixus	mes	Mellanby (1937)
Oncopeltus fasciatus	mes	Butt (1949)
Gerris paludum insularis	mes	Mori (1969)
Neuroptera		
Chrysopa perla	ent	Bock (1939)
Coleoptera		
Euryope terminalis	ent	Paterson (1932)
Calandra oryzae	mes	Tiegs and Murray (1938)
Tenebrio molitor	mes	Jackson (1939)
Lepidoptera		
Bombyx mori	y	Dohrn (1876)
	sub	Toyama (1902); Wada (1955a,b)
Donacia crassipes	ent	Hirschler (1909)
Diacrisia virginica	mes	Johannsen (1929)
Pieris rapae	mes	Eastham (1930)
Antheraea pernyi	mes	Saito (1937)
Heliothis zea	mes	Presser and Rutschky (1957)
Chilo suppressalis	mes	Okada (1960)
Trichoptera		
Neophylax concinnus	mes	Patten (1884)
Stenopsyche griseipennis	mes	Miyakawa (1974)
Diptera		
Dacus tryoni	mes	Anderson (1963b)
Calliphora erythrocephala	mes	Starre-van der Molen (1972)
Siphonaptera		
Ctenocephalides felis	mes	Kessel (1939)
Hymenoptera		
Pteronidea ribesii	mes	Shafiq (1954)
Pimpla turionellae	mes	Bronskill (1959)
Athalia proxima	mes	Farooqi (1963)
Apis mellifera	mes	Nelson (1915)

liberated cells (free cells) in the epineural sinus have been identified as the embryonic hemocytes (Fig. 1.1*B–E*).

The embryonic hemocytes must be distinguished from cells liberated from the inner layer at earlier stages; the latter cells have been termed paracytes (Heymons, 1895) or secondary vitellophages (Bock, 1939). Whereas the paracytes and the vitellophages were thought to enter the yolk and then degenerate, the embryonic hemocytes were

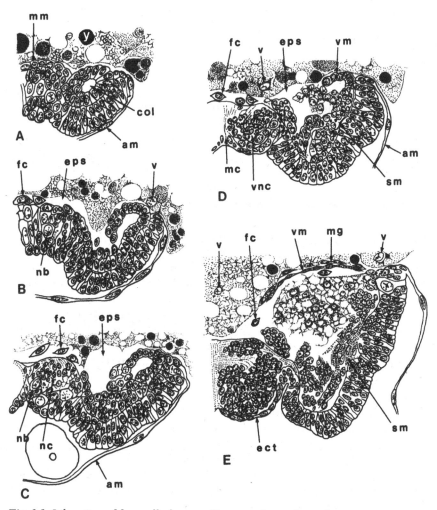

Fig. 1.1. Liberation of free cells from median mesoderm, shown by transverse sections through metathoracic segment (mesothoracic segment in **E**) of *Chrysopa* embryo. **A**, 36 hr; **B**, 42 hr; **C**, 48 hr; **D**, 54 hr; **E**, 78 hr. *am* = amnion; *col* = coelom; *ect* = ectoderm (external epithelium); *eps* = epineural sinus; *f* = fat body; *fc* = free cell; *mc* = median cord; *mg* = midgut anlage; *mm* = median mesoderm; *nb* = neuroblast; *nc* = nerve cell; *sm* = somatic musculature; *v* = vitellophage; *vm* = visceral musculature; *y* = yolk. (Redrawn from Bock, 1939)

seen to appear in the epineural sinus and never enter the yolk. Another modified method of median mesoderm formation has been described in detail by Ando (1962). According to him, the inner layer cells in the odonatan embryo were seen to disappear completely from the median part of the germ band about the time of the coelomic sac formation in the lateral parts of the embryo; the median part of the ectoderm was, however, covered later by cells that had proliferated from both lateral parts of the inner layer. Ando adopted the term "secondary median mesoderm," following Roonwal (1937), to describe this newly formed part of the inner layer. The embryonic hemocytes were concluded to be derived from this secondary median mesoderm. This method of embryonic hemocyte formation will be discussed later in this chapter, under "Observations on Embryonic Hemocytes in *Gerris.*"

At about the time when the median mesodermal cells are liberated, the prospective median cord was clearly identified in the ectoderm, and differentiation of neuroblasts and epidermal cells could be seen. Stomodaeum could be observed at the anterior part of the embryo.

The opinion of previous authors regarding the identification of the liberated cells from the median mesoderm may be divided into two categories: (1) many authors considered these cells as mesodermal because they were siblings of the somatic mesodermal cells; (2) although the method of formation of the embryonic hemocytes was identical, the hemocytes were thought to be entodermal by several authors because the same median mesoderm was believed to contribute to the formation of the midgut epithelium (Bock, 1939; Görg, 1959; Striebel, 1960). This view, however, is not convincing. Illustrations from Bock's paper (Fig. 1.1) reveal quite convincingly that the embryonic hemocytes are derived from the median mesoderm; however, the midgut epithelium formation is not illustrated in detail, and this does not allow one to understand the relationship between the median mesoderm and the midgut epithelium.

Finally, a somewhat modified method of hemocyte formation from the median mesoderm should be noted (Jackson, 1939). In *Tenebrio* embryos, cells liberated from the upper layer of the median mesoderm became micronucleocytes, whereas those from the lower layer were concluded to become macronucleocytes. The other type of hemocytes, the spherule cells, were said to have arisen from cells deep in the lateral region of the mesoderm, and finally the neural crest produced oenocytoids. Jackson further claimed that the first two types of embryonic hemocytes liberated from the median mesoderm were seen to enter the yolk first and then appeared in the definitive body cavity; however, this peculiar behavior of the embryonic hemocytes in this insect has never been reconfirmed; Ullmann (1964) made no comments on hemocyte formation in the same insect.

1.2.2. Coelomic sac theory

As early as 1884, Patten reported that hemocytes in phryganid embryos were derived from the cells of the coelomic sac, especially those
found at the junction where the two lateral tips of visceral and somatic
musculature meet. He therefore concluded that the hemocytes were
mesodermal. Later, Roonwal (1937) also suggested that some *Locusta*
hemocytes might have arisen from the coelomic sacs.

It seems that these authors proposed this theory on the ground that
the embryonic hemocytes were seen at places not far from the coelomic sacs. However, the appearance of free cells not far from the coelomic sacs may also be interpreted as a result of hemocyte movement;
a cell might have arisen from the median mesoderm and then moved
toward the coelomic sac. Pictorial evidence that might demonstrate
clearly the exact time of hemocyte formation from coelomic sac seems
to be lacking in these papers.

1.2.3. Suboesophageal body theory

The suboesophageal body is a cluster of cells observed in the trito-
cerebral segment or at a place beneath the stomodaeum. In *Bombyx*
embryos, Toyama (1902) suggested that embryonic hemocytes could
have been derived from the suboesophageal body, and later Wada
(1955a,b) published reports that confirmed Toyama's theory. More recently, however, Okada (1960) concluded that, although the hemocytes could be found more abundantly in the anterior part of the embryo, the suboesophageal body in *Chilo* could not be considered as
the site of origin of embryonic hemocytes. He concluded that they
were derived from the median mesoderm. As Okada's observation was
far more accurate than those of previous authors, hemocyte formation
from the suboesophageal body seems to be less convincing.

Liberation of cells from the suboesophageal body, however, has
been demonstrated (Johannsen, 1929). As Johannsen did not identify
these cells as embryonic hemocytes, reinvestigation of hemocyte formation in lepidopteran embryos appears to be necessary.

1.3. Causes of controversies among hemocyte formation theories

It is evident from the foregoing account that there is as yet no agreement on the origin of embryonic hemocytes. We will now consider
some of the causes that might have led to the differences of opinion.
Embryologists generally rely on histological observations of normally
developing embryos in discussing the origin and development of embryonic hemocytes. Information obtained from such observations,
however, should be interpreted carefully, since the inherent nature of

the embryonic hemocytes and the very limitations of histological technique may cause misinterpretation of facts.

1.3.1. Inherent nature of embryonic hemocytes as cause of misinterpretation

The cytological features of embryonic hemocytes (information most easily obtained by histological observations) may change quite remarkably owing to transformation of the hemocytes for a variety of reasons. In addition to the changes caused by fixatives, short-term transformations may occur because of the hemocyte's ameboid movement; long-term transformation may result from further differentiation of the hemocytes.

Ameboid movement

Embryonic hemocytes supposedly are capable of ameboid movement, since they disappear soon after being liberated from the median mesoderm and reappear in various parts in the epineural sinus. Transformation of the embryonic hemocytes attributable to their development and differentiation has been described in the embryos of *Chrysopa* (Bock, 1939). He tried to follow the developmental sequence of the hemocytes, and concluded that there were three pathways of hemocyte development (Fig. 1.2) (see also later in this chapter, under

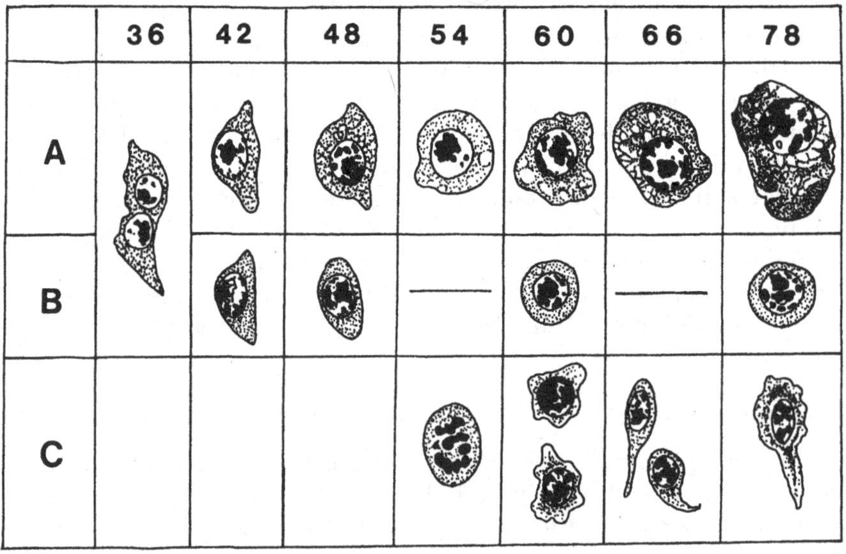

Fig. 1.2. Development of three types of free cells found liberated in epineural sinus of *Chrysopa* embryo. Numerals at top of each column indicate embryonic stages (hours after oviposition). **A**, lymphatic cell (= plasmatocyte); **B**, blood cell (= prohemocyte); **C**, large amoebocyte (= granulocyte). (Redrawn from Bock, 1939)

"Cytological Features of Embryonic Hemocytes Previously Observed").

Topographical positions of embryonic hemocytes

This information is obtained from histological observation. The topographical positions of embryonic hemocytes may change drastically because of hemocyte movements. It is necessary to remember that embryonic hemocytes are freely circulating single cells in the embryo. Motility or ameboid movement of the cells may be a cause of change in the topographical positions of the hemocytes at earlier stages, whereas at later stages a stream of hemolymph, presumably generated by pulsative movement of the embryonic body, may transport the embryonic hemocytes from this position to another; it is very likely that positions of embryonic hemocytes may change at every moment. In fact, histological observations of *Gerris* embryos at the same stage of development reveal that even though the site of observation is identical, distribution of hemocytes is not the same. As stated earlier, different interpretations may be possible from the position of a cell found in the embryo; for example, the cell might have come from somewhere to that position, or might be on its way from that spot to somewhere else.

1.3.2. Technical limits as cause of misinterpretation

Although embryogenesis may be interpreted as a complex series of events that takes place according to definite chronological and topographical sequences, a continuous trace of a specific cell throughout the embryogenesis in one individual will be impossible if we employ histological technique as means of observation.

It must also be remembered that histological technique allows us to identify only cells that have similar cytological features; there is no reason to believe that the cells which have identical cytological features are embryologically identical. Furthermore, histological observation of embryogenesis would not reveal functions and activities of embryonic hemocytes. Because of the limitations of histological techniques mentioned above, the information obtained should be interpreted very carefully.

1.4. Observations on embryonic hemocytes in *Gerris*

Although many authors have identified various types, forms, or categories of hemocytes in insects, in this review the classification of postembryonic hemocytes described by Arnold (1972, 1974), with some minor changes in terminology, will be employed.

1.4.1. Cytological features of embryonic hemocytes previously observed

The most detailed observation of cytological features of embryonic hemocytes has been published by Bock (1939). He classified free cells liberated from the median mesoderm of *Chrysopa* embryo into four types and followed their development. Of the four types of free cells, Bock termed one cell type as secondary vitellophages and concluded that they degenerated in the yolk; the other three types – (1) lymphatic cells, (2) blood cells, and (3) large amoebocytes – were identified as the embryonic hemocytes (Fig. 1.2). On the basis of their final forms, these hemocytes would correspond with (1) plasmatocytes (PLs), (2) prohemocytes (PRs), and (3) granulocytes (GRs), respectively. Fusiform cells first liberated from the median mesoderm were seen to differentiate into blood cells (PRs) and subsequently into large amoebocytes (GRs), whereas the differentiation of lymphatic cells (PLs) seemed to occur later than that of the other two types of hemocytes (Fig. 1.2).

Similar types of hemocytes found in *Tenebrio* embryos have been termed (1) micronucleocytes, (2) macronucleocytes, (3) oenocytoids (OEs), and (4) spherule cells (Jackson, 1939). These hemocytes seem to correspond with (1) PLs, (2) PRs, (3) OEs, and (4) spherulocytes (SPs), respectively. Descriptions of the embryonic hemocytes by various authors, however, are generally insufficient to extract information. It may be said that hemocytes are ameboid soon after liberation and then become spherical as embryogenesis continues. Vacuoles appearing in some of the embryonic hemocytes, however, seem to have attracted the attention of some authors. Vacuoles were reported in the hemocytes of lepidopteran (Eastham, 1930; Presser and Rutschky, 1957; Okada, 1960), trichopteran (Miyakawa, 1974), and coleopteran embryos (Tiegs and Murray, 1938). These hemocytes with vacuoles seem to correspond with the adipohemocytes (ADs) found in the prepupa of *Galleria* (Lepidoptera) (Ashhurst and Richards, 1964).

1.4.2. Hemocyte formation in *Gerris* embryos

The hemocytes in *Gerris* embryos are derived from the median mesoderm (Mori, 1969). The following description of the formation of embryonic hemocytes is based in part on my earlier works (Mori, 1969, 1976).

Liberation of free cells from the median part of the embryo

As described elsewhere (Mori, 1969), a groove appears on the dorsal side of the germ band about the time when the formation of germ band–amnion complex is complete. This groove occurs only along the median line of the protocorm, the part of the germ band destined to become the embryonic trunk. This groove, or mass of cells, forms a single, cuboidal cell layer that covers the dorsal side of the protocorm. The protocorm thus consists of two single-cell layers: the dorsal cuboidal layer and the ventral columnar layer, respectively. The former will be designated as the inner layer; the latter, as ectoderm. The inner layer is established about the time the germ band–amnion complex becomes independent from the serosa, a part of the blastoderm remaining at the surface of the egg.

Metameric segmentation in the inner layer is seen some hours later than the occurrence of segmentation in the ectoderm. The inner layer, which was so far continuous longitudinally, splits into a chain of cell clusters distributed along the median line of the embryo. This metameric segmentation in the inner layer is seen in the gnathal and thoracic segments first; the inner layer in the abdomen remains intact at this stage, which indicates very clearly that differentiation in the inner layer occurs earlier in the anterior part of the embryo.

Differentiation in the segmented inner layer cells is seen when the extreme lateral parts of the ectoderm turn dorsally, indicating the beginning of abdominal appendage formation. In the lateral parts of the embryo, the inner layer cells are rather columnar, whereas in the median part they are somewhat flattened. These flattened parts of the inner layer could be termed the median mesoderm (= primary median mesoderm) (Fig. 1.3*B*).

The inner layer then splits in two along the median line of the embryo. A gap thus appears between the two laterally separated inner layers and expands subsequently, resulting in the disappearance of the cells that so far covered the dorsal side of the ectoderm at its median part (Fig. 1.3*C*). This expansion of the gap may be seen in the embryos undergoing embryonic rotation around the longitudinal axis of the egg (Mori, 1969). At this stage the median cord is clearly visible in the median part of the ectoderm, while in the lateral parts differentiation of neuroblasts and epidermal cells is beginning to occur (Fig. 1.3*C*).

It should be noted here that a very delicate membranous structure is seen to delimit the ventral sides of the yolk granules at this stage, thus separating the yolk from the embryonic cells; this structure is the yolk boundary described previously (Mori, 1969). The space between the yolk boundary and embryonic cells is termed the "epineural sinus."

In the embryos that have finished rotation (48 hr at 20 °C after oviposition), the gap previously found in the median part of the inner layer is closed. The cells that fill the gap are identified as secondary median mesoderm cells. Although the secondary median mesoderm is a single-cell layer, the cells are not cuboidal. They are not as closely packed as they were in the median mesoderm (primary median meso-

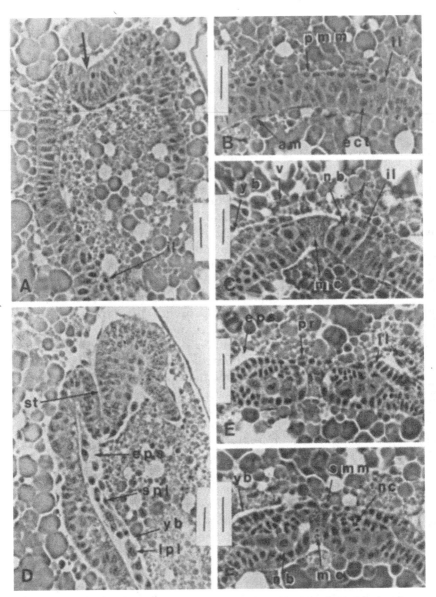

Fig. 1.3. Embryonic hemocyte formation in *Gerris*. **A.** Sagittal section of an embryo immediately after rotation (arrow indicates depression). **B.** Transverse section through prospective thorax of protocorm. **C.** Transverse section through metathoracic segment of an embryo during rotation (top of page is lateral side of egg). **D.** Sagittal section of an embryo at 9 hr (25 °C) after rotation. **E.** Transverse section through second maxillary segment of an embryo at 9 hr (25 °C) after rotation. **F.** Transverse section through metathoracic segment in an embryo of same stage as **E.** *am* = amnion; *ect* = ectoderm; *eps* = epineural sinus; *il* = inner layer; *lpl* = larger plasmatocyte; *mc* = median cord; *nb* = neuroblast; *nc* = nerve cell; *pmm* = primary median mesoderm; *pr* = prohemocyte; *smm* = secondary median mesoderm; *spl* = smaller plasmatocyte; *st* = stomodaeum; *v* = vitellophage; *yb* = yolk boundary. Bars represent 20 mμ.

derm), and each cell is rather ovoid. It is very likely that the secondary median mesoderm is established as a result of the median expansion of the inner layers, once separated laterally. Mitotic figures seen frequently at the median tips of the inner layer may explain this expansion.

Dissociation of the cells of the secondary median mesoderm – namely, the appearance of a minute gap between each two cells – can be confirmed in the embryos that develop several hours after rotation. The minute gaps then merge with the epineural sinus.

Differentiation of the cells in the secondary median mesoderm can be seen clearly in the embryos about 9 hr (25 °C) after rotation. Some of the cells are spherical and occasionally undergo mitosis (Fig. 1.3E,F), whereas others are fusiform or ameboid (Fig. 1.3D). Differences in size of cells should also be noted (Fig. 1.3D,F). Differentiation of the cells seems to have occurred earlier or more rapidly in the gnathal segments than in the thoracic ones (Fig. 1.3D–F). Expansion of the epineural sinus in gnathal segments is also far greater than that in the thoracic segments. No such differentiation of inner layer cells has been observed in the abdominal segments, where the epineural sinus was still very narrow. It seems, therefore, that there might be some correlation between the differentiation of the cells in the secondary median mesoderm and the expansion of the epineural sinus. Furthermore, it is possible to say that the liberated cells from the secondary median mesoderm begin to disperse as the epineural sinus expands (Fig. 1.3D). These liberated cells can be identified as the embryonic hemocytes because they are independent from other tissues of the embryo. The cells of the secondary median mesoderm at the beginning of liberation are equal in size and are seen to undergo mitosis (Fig. 1.3E). These cells are identified as PRs – namely, the first generation of embryonic hemocytes.

In the embryos in which liberation of the hemocytes took place, a chain of nerve cells proliferated from the neuroblasts, which demonstrates that the ventral nerve ganglion formation was under way. The ectoderm is therefore differentiated into the ventral nerve ganglia and the ventral epidermis at both the lateral parts as well as the bilaterally flat cells of the median cord in the median part (Fig. 1.3E,F).

Types of embryonic hemocytes of *Gerris*

Embryonic hemocytes observed in this insect may be divided into three types on the basis of their cytological features: (1) PRs, (2) PLs, and (3) GRs. Neither SPs nor OEs have been confirmed. The key characters employed to classify the embryonic hemocytes are (1) size and form of the cell, (2) size and form of the nucleus, (3) topographical position of the nuclei in the cells, (4) nature of the chromatin granules in the nuclei, (5) nature of the cytoplasm, and (6) cytoplasmic inclusions and vacuoles.

Prohemocytes (PRs). These are rather small, spherical cells with a single nucleus (Fig. 1.3*E*). The thin cytoplasm that surrounds the nucleus, which has densely packed chromatin granules, is neutrophilic and without cytoplasmic inclusions.

The PRs are found throughout embryogenesis at various places in the embryo; however, they are not numerous. Mitotic figures of the PRs were difficult to find, although two daughter cells after mitosis were seen more frequently (Figs. 1.4*E*, 1.5*D*). Before katatrepsis, the PRs are found at places not far from the dorsal side of the ventral nerve cord, whereas in the last phase of embryogenesis they are frequently seen in the lumen of the dorsal vessel (Fig. 1.5*B*). However, no concentration of PRs in any specific part of the embryo was observed.

Although no further development of daughter cells after cleavage of the PRs could be confirmed histologically, it is very likely that transformation of the daughter cells into other types of embryonic hemocytes takes place; this is deduced from an impression that the number of PRs seemed to change slightly.

Plasmatocytes (PLs). PLs may be identified on the basis of their forms; many of them are ameboid, others spindle-shaped or fusiform (Figs. 1.3*D*, 1.4*A*–*D*, 1.5*E*,*F*). The PLs can be divided into two subtypes: larger PLs and smaller PLs. The larger PLs are rather ovoid and sometimes attenuated (Figs. 1.3*D*, 1.4*B*); the smaller ones are fusiform (Figs. 1.3*D*; 1.4*C*,*E*,*F*; 1.5*E*,*F*). In the former subtypes of PLs, the nuclei may be found anywhere in the cell, whereas in the latter, they are centrally located and have more densely packed chromatin granules. Vacuoles may be found in both subtypes, but no cytoplasmic inclusions are seen. The cytoplasm is neutrophilic in both subtypes.

PLs are derived from the cells of the secondary median mesoderm at the very early phase of liberation of the hemocytes. The cells become ameboid as soon as they dissociate from other cells. It is reasonable to suggest that the PLs are probably second-generation embryonic hemocytes derived from the first generation, the PRs.

Both the larger and smaller PLs are seen in the epineural sinus of the embryos 9 hr (at 25 °C) after rotation (Fig. 1.3*D*), which indicates that this differentiation in the embryonic hemocytes is a mere reflection of their conditions. In the later stages, the PLs may be found not only in places on the dorsal side of the ventral nerve cord, but also in the epineural sinus at both lateral parts of the embryo (Fig. 1.4*B*); this may be attributable to the movement of the PLs. The PLs are then found to appear in various parts of the embryo; however, the larger PLs are no longer seen at this stage (Figs. 1.4*E*,*F*; 1.5*E*,*F*), which might indicate that they represent younger stages of PLs. After katatrepsis, the PLs may be found everywhere in the embryo, but aggregation of these cells was never observed. In the lumen of dorsal vessel,

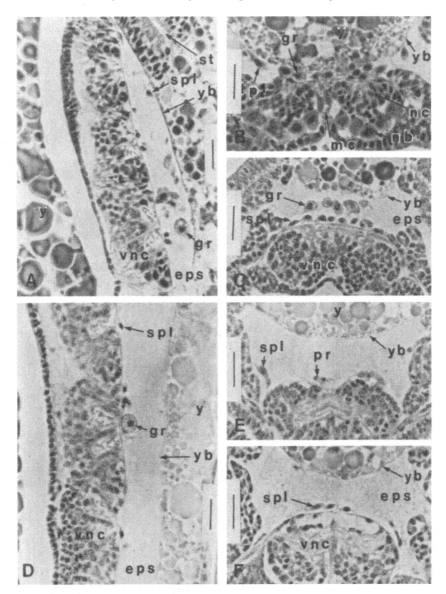

Fig. 1.4. Embryonic hemocyte formation in *Gerris*. **A.** Sagittal section of an embryo at about 48 hr (25 °C) after rotation. **B.** Transverse section through metathoracic segment of an embryo at 24 hr (20 °C) after rotation. **C.** Transverse section through second maxillary segment of an embryo at 48 hr (25 °C) after rotation. **D.** Sagittal section of an embryo about 72 hr (20 °C) after rotation. **E.** Transverse section through metathoracic segment at same stage as **D. F.** Transverse section through metathoracic segment of an embryo at same stage as **D** (different individual). *eps* = epineural sinus; *gr* = granulocyte; *lpl* = larger plasmatocyte; *mc* = median cord; *nb* = neuroblast; *nc* = nerve cell; *pr* = prohemocyte; *st* = stomodaeum; *spl* = smaller plasmatocyte; *vnc* = ventral nerve cord; *y* = yolk; *yb* = yolk boundary. Bars represent 20 mμ.

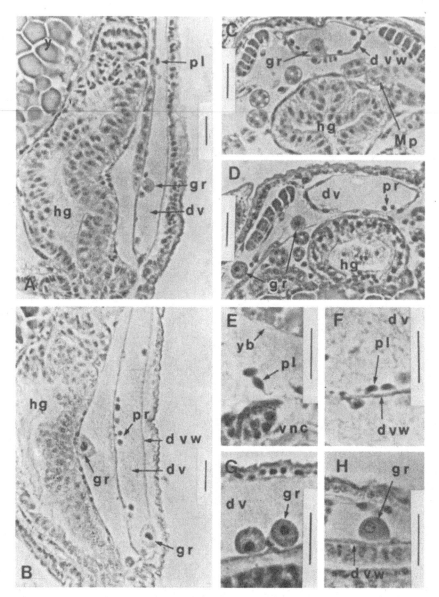

Fig. 1.5. Embryonic hemocyte in *Gerris* after katatrepsis. **A.** Sagittal section of caudal part of an embryo about 4 days (20 °C) after katatrepsis. **B.** Sagittal section of caudal part of an embryo immediately before hatching (twelfth day at 20 °C after oviposition). **C.** Transverse section through caudal part of an embryo about 11 days (20 °C) after katatrepsis. **D.** Transverse section through caudal part of an embryo at about same stage as **C. E.** Plasmatocytes in an embryo about 72 hr (20 °C) after rotation. **F.** Plasmatocyte in lumen of dorsal vessel of an embryo immediately before hatching. **G.** Granulocytes in lumen of dorsal vessel of an embryo about 11 days after oviposition (20 °C). **H.** A type of granulocyte in lumen of dorsal vessel of an embryo at same stage as **G.** *dv* = dorsal vessel; *dvw* = dorsal vessel wall; *gr* = granulocyte; *hg* = hindgut; *Mp* = malpighian tubule; *pl* = plasmatocyte; *pr* = prohemocyte; *vnc* = ventral nerve cord; *y* = yolk; *yb* = yolk boundary. Bars represent 20 mμ.

sometimes groups of PLs were seen; however, this was never taken to indicate formation of a hemopoietic tissue by the PLs. The PLs may be the hemocytes most frequently observed; however, no mitotic figures were observed.

Granulocytes (GRs). GRs are the largest of the three embryonic hemocyte types (Fig. 1.4D). They are spherical and rarely attenuated. The diameter of GRs varies only slightly, and the nucleus is centrally located. Chromatin granules in the nucleus are less densely packed than those in the PRs; the eosinophilic granules in the neutrophilic cytoplasm may be found distributed evenly and abundantly.

At the beginning of hemocyte liberation, the GRs were hardly observed. They were found to appear in the embryos some 24 hr (at 20 °C) after rotation (Fig. 1.4B). In some of the embryos at this stage, the GRs of smaller sizes could be found in the epineural sinus, not far from the dorsal side of the ventral nerve cord; this may be interpreted as proof of GR formation from PRs.

The GRs may be found at places rather far from the dorsal side of the ventral nerve cord. In one of the embryos at 48 hr (at 25 °C) after rotation, some of the GRs were seen in the epineural sinus far from the ventral nerve cord, whereas closer positions were occupied by a group of PLs (Fig. 1.4C). In contrast to the PLs, the GRs could not be seen in places close to the coelomic sacs. After katatrepsis, the GRs were found more abundantly in the caudal part of the embryo. Some of the GRs were also seen to enter the lumen of the dorsal vessel (Fig. 1.5A – C,G).

Large, rather spherical cells, whose size is roughly equal to that of the GRs, may also be found in some of the embryos on the eleventh day after oviposition (Fig. 1.5H). No granular inclusions are found in these cells; however, the cytoplasm is eosinophilic. This type of free cell may also be found in other places in the embryo at this stage. Absence of cytoplasmic inclusions should not be considered an artifact attributable to some histological technique, because the typical GRs may be found in the same embryo. These types of free cells also may be classified as GRs because of their size and eosinophilic cytoplasm. Throughout the course of embryogenesis, no mitosis of GRs was observed.

Origin of embryonic hemocytes in *Gerris*

It is concluded from observations described in the foregoing paragraphs that the embryonic hemocytes of *Gerris* are directly derived from the secondary median mesoderm. The midgut epithelium, on the other hand, is derived from vitellophages and the yolk boundary (Mori, 1976); the vitellophages distributed on the inner surface of the yolk boundary directly become the nuclei of the midgut epithelium.

The secondary median mesoderm never contributes to the formation of the midgut epithelium, as none of the cells in this cell layer becomes part of the midgut epithelium. The cells of the secondary median mesoderm may be considered as sibling cells of the coelomic sac, because the former were formed as a result of the median expansion of the latter. Therefore, the embryonic hemocytes of *Gerris* are mesodermal.

There are several reasons that do not support hemocyte formation from sources other than the secondary median mesoderm:

1. No liberation of cells from the coelomic sac has been observed. None of the cells is seen leaving the inner layer in the lateral parts of the embryo before and after liberation of cells from the secondary median mesoderm takes place. Furthermore, the free cells, which appeared in the epineural sinus after the beginning of the cell liberation from the secondary median mesoderm, were always found closer to the ventral nerve cord than to the coelomic sac. One cannot deny the possibility that cells of the coelomic sac might become independent and appear in the epineural sinus; however, I failed to observe this.

2. No hemocyte is known to be derived from the suboesophageal body, as no such tissue is found in *Gerris*. In the embryos immediately after rotation, a slight depression may be seen (Fig. 1.3A); however, no aggregation of cells is found beneath this depression.

   A mound of the inner layer cells found on the anteriormost part of the gnathal segments cannot be considered as a suboesophageal body because these cells, a part of the segmented inner layer, are always separated from the stomodaeum.

3. Formation of hemocytes from the cells of the dorsal vessel was never observed. The cells observed in the dorsal vessel lumen were either the cells present from the beginning of dorsal vessel formation or those that entered the lumen during the very late phase of embryogenesis.

4. No cells in the yolk were seen to enter the epineural sinus. In this insect, "entry" of the inner layer cells into the yolk could be observed before or at about the time when the inner layer cells undergo metameric segmentation. Some of the inner layer cells were seen to be separated from the main body of cells; however, it seems that these cells were left behind in the yolk as a horizontal flattening occurred in the germ band–amnion complex before embryonic rotation. These isolated inner layer cells were seen to degenerate subsequently. The vitellophages were seen only in the yolk, and these cells have never been seen in the epineural sinus.

1.5. Prospects of hemocyte investigation for the future

1.5.1. Unanswered questions regarding the origin and development of embryonic hemocytes

Many of the histological observations of the embryonic hemocytes seem to justify the conclusion that the first generation of embryonic hemocytes is derived from the median mesodermal cells. However, several unanswered questions regarding the origin and development of the embryonic hemocytes remain:

1. No direct proof has been obtained that would indicate that the embryonic hemocytes are the direct precursors of postembryonic hemocytes. Observations of hemocytes throughout the course of development from the embryo to adult in the same insect has been done by Jackson (1939). Unfortunately, however, his observations in *Tenebrio* need reconfirmation, particularly with regard to the method of embryonic hemocyte formation. More recently, observations in embryos close to hatching revealed that the embryonic hemocytes are very similar to postembryonic hemocytes (Anderson, 1963a; Arnold and Salkeld, 1967). However, information on this subject is not good enough to reveal a relationship between embryonic hemocytes and postembryonic hemocytes.

2. The question whether the embryonic hemocytes are derived from a single source is not yet conclusively settled. It is certain, however, that the first-generation embryonic hemocytes are derived from the median mesoderm; no free cell formation from any other source in the embryo seems to occur when the cells are liberated from the median mesoderm. However, cells might be liberated from another source at a later stage of development. However, free cells thus formed, if any, could also be identified as embryonic hemocytes.

3. Although several types of embryonic hemocytes can be identified in a single insect, no evidence that would conclusively disprove the existence of only one type of embryonic hemocyte has been obtained. Although I have identified three types of embryonic hemocytes in *Gerris*, it is likely that the GRs and PLs are modified PRs. PRs are known as germinal forms of postembryonic hemocytes, from which other types of hemocytes are produced (Wigglesworth, 1959, 1972; Arnold, 1974) (see also Chapter 4).

4. Detailed observations of hemocyte formation in *Gerris* embryos revealed that liberation of cells from the secondary median mesoderm produced embryonic hemocytes; however, the mechanism of cell liberation is obscure. Furthermore, hemocyte liberation seems to have been initiated from the cells in the anterior part of the gnathal segments. Whether or not every cell of the secondary median mesoderm is capable of initiating hemocyte development remains unknown.

1.5.2. Techniques for investigation of embryonic hemocytes

Insect embryologists have so far employed histological and histochemical techniques exclusively to investigate embryonic hemocytes. However, as stated earlier in this chapter, histological techniques have limitations. I will now suggest certain techniques that would be useful in demonstrating the characteristics and activities of embryonic hemocytes.

Time-lapse photography. For an analysis of cell behavior, it is necessary to study many histological preparations, and there always exists
the possibility of misinterpreting the histological details. Time-lapse
photography, on the other hand, will assure continuous observation of
a single cell in the living embryo. By this method, the topographical
relationship between the hemocytes and other embryonic tissues may
be observed directly. Although time-lapse photography may not be a
good tool for an analysis of organogenesis at the cellular level, it is
useful in revealing the dynamic behavior of the cells. Comparison of
results obtained by both time-lapse photography and conventional
histology would be very useful in demonstrating whether or not the
embryonic hemocytes arise from a single source.

Electron microscopy. Light microscopic observation may reveal
only a vague topographical relationship between the free hemocytes
and the embryonic tissues; electron microscopy may be employed to
determine if the embryonic hemocytes are really independent from
other embryonic tissues. Furthermore, differentiation that occurs in
the cells of the median mesoderm after liberation may also be revealed by electron microscopy.

In vitro embryonic hemocyte culture. Postembryonic hemocyte culture has already revealed that PLs are derived from PRs (Sohi, 1971).
As far as is known, in vitro culturing of embryonic hemocytes has
never been tried. Observations of these cells in vitro may reveal (1)
the character of PRs as germinal types of embryonic hemocytes, (2)
the causes of transformation into PLs, and (3) the nature of granular
inclusions found in the GRs. Activities of the embryonic hemocytes,
which have so far been neglected, may also become clearer. In vitro
behavior of embryonic hemocytes against other tissues would reveal
possible phagocytic activities of embryonic hemocytes. In vitro culture of embryos has already been done (Wolf-Neis, 1973), and has
proved a very effective method for studying embryology, and could be
used to study the behavior of embryonic hemocytes.

Immunohistology. Immunohistological technique was recently
used in insect embryology (Wolf-Neis et al., 1976). As early as 1963,
the appearance of a substance showing reactions similar to those of
yolk granules was demonstrated in the hemocytes of *Dacus* embryos
(Anderson, 1963b); these reactions supposedly disappear at a later
stage. On the basis of Anderson's observation, it is reasonable to suggest that embryonic hemocytes transport yolk to various parts of the
embryo. A positive reaction of the antibody of *Carausius* yolk in the
ameboid cells in *Carausius* (Wolf-Neis et al., 1976) may be considered
to indicate that the above suggestion could be confirmed by an im-

munohistological method; the nature of the ameboid cells of *Carausius* was not identified, but they very likely are embryonic hemocytes.

Whereas the positive reaction of yolk antibody in the ameboid cells may indicate their involvement in the intermediary metabolism of the embryo, immunohistology may be used to identify the types of embryonic hemocytes more accurately.

Microcauterization. Among the several techniques that have been used by experimental embryologists, microcauterization may be considered as the most promising method for investigating the origin of embryonic hemocytes. Application of a microcauterizer on the embryo at a rather advanced stage of embryogenesis is never fatal (Mori, 1975, 1977). Damage from cauterization can be restricted to a minimum, and hence embryogenesis of defective embryos may be allowed to continue smoothly, at least for a period needed to observe hemocyte formation. Cauterization experiments for revealing the origin and development of embryonic hemocytes have not been done. My observations of cauterized *Gerris* embryos indicated that this method could be employed for that purpose. Hemocyte formation in the defective embryos was found to occur quite normally (Fig. 1.6A,B). Because hemocyte formation in this insect takes place within several hours after embryonic rotation (first in the gnathal segments), a microcauterizer was applied on the gnathal segments of the embryo immediately after rotation; successive applications of the microcauterizer on gnathal as well as abdominal segments of the single embryo of the same stage as above were also carried out. Embryonic hemocytes formed in both groups of defective embryos. Therefore, any secondary median mesoderm cells in the posterior part of gnathal segments as well as those in the thoracic segments can play a role in initiating hemocyte formation.

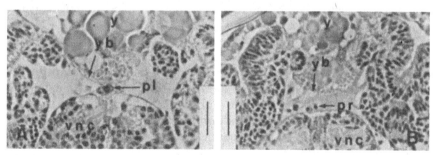

Fig. 1.6. Embryonic hemocyte formation in defective embryos of *Gerris*. **A.** Result of gnathal segment cauterization. Transverse section through metathoracic segment of a defective embryo about 72 hr (20 °C) after cauterization. **B.** Result of gnathal as well as abdominal segment cauterization. Transverse section through metathoracic segment of a defective embryo about 48 hr (20 °C) after cauterization. *pl* = plasmatocyte; *pr* = prohemocyte; *vnc* = ventral nerve cord; *y* = yolk; *yb* = yolk boundary. Bars represent 20 mμ.

The results also indicate that the cauterization technique could be applied to determine if one of the activities of the suboesophageal body is embryonic hemocyte formation.

Microinjection. Microinjection of alien bodies into insect bodies has already been proved successful in bringing about encapsulation, one of the major functions of postembryonic hemocytes (Wigglesworth, 1972; Arnold, 1974). Assuming that the embryonic hemocytes have functions similar to those of postembryonic hemocytes, microinjection of alien bodies into the embryo may bring about encapsulation and may shed some light on embryonic hemocyte multiplication.

1.6. Summary

Embryonic hemocytes may be defined as single cells that freely circulate in the hemocoel. In contrast to the majority of embryonic cells, embryonic hemocytes are seen liberated within the embryo at a relatively early stage and remain independent during the rest of embryogenesis.

Several theories have been proposed to explain the exact origin of embryonic hemocytes. However, contemporary authors seem to support the median mesoderm theory, which concludes that the embryonic hemocytes are derived from the median mesoderm. This theory has been proved convincingly. The majority of authors considers the embryonic hemocytes as mesodermal, but several authors regard them as entodermal on the ground that the median mesoderm contributed in the formation of midgut epithelium. I believe the controversy regarding modes of hemocyte formation is attributable largely to the misinterpretation of the information obtained histologically. Causes of misinterpretation may be found in the inherent nature of the embryonic hemocytes as well as in the limitations of histological techniques.

Although the origin of embryonic hemocytes has been described by many authors, cytological features of the hemocytes have been insufficiently observed. In this review I have given a detailed description of the hemocytes in *Gerris* (Hemiptera: Gerridae) embryos, together with the method of formation. The embryonic hemocytes were seen to be derived from the median mesoderm (secondary median mesoderm) and differentiated into three types: (1) prohemocytes, (2) plasmatocytes, and (3) granulocytes.

The following techniques are suggested for future investigation of embryonic hemocytes: time-lapse photography, electron microscopy, in vitro embryonic hemocyte culture, immunohistology, microcauterization, and microinjection.

Acknowledgments

I thank Professor D. T. Anderson of Sydney University for his valuable suggestions and critical reading of the manuscript.

References

(An asterisk indicates that the original reference was not seen.)

Anderson, D. T. 1963a. The embryology of *Dacus tryoni.* 2. Development of imaginal discs in the embryo. *J. Embryol. Exp. Morphol. 11*(2):339–51.

Anderson, D. T. 1963b. The larval development of *Dacus tryoni* (Frogg.) (Diptera: Trypetidae). I. Larval instars, imaginal discs and hemocytes. *Aust. J. Zool. 11*(2):202–18.

Anderson, D. T. 1972a. The development of hemimetabolous insects, pp. 95–163. *In* S. J. Counce and C. H. Waddington (eds.). *Developmental Systems: Insects,* Vol. 1. Academic Press, New York.

Anderson, D. T. 1972b. The development of holometabolous insects, pp. 165–242. *In* S. J. Counce and C. H. Waddington (eds.). *Developmental Systems: Insects,* Vol. 1. Academic Press, New York.

Ando, H. 1962. *The Comparative Embryology of Odonata with Special Reference to a Relic Dragonfly Epiophlebia superstes Selys.* Japanese Society for the Promotion of Science, Tokyo.

Arnold, J. W. 1972. A comparative study of the hemocytes (blood cells) of cockroaches (Insecta: Dictyoptera: Blattaria), with a view of their significance in taxonomy. *Can. Entomol. 104*:309–48.

Arnold, J. W. 1974. The hemocytes of insects, pp. 201–54. *In* M. Rockstein (ed.). *The Physiology of Insecta,* Vol. 5, 2nd ed. Academic Press, New York.

Arnold, J. W., and E. H. Salkeld. 1967. Morphology of the hemocytes of the giant cockroach, *Blaberus giganteus,* with histochemical tests. *Can. Entomol. 99*(11):1138–45.

Ashhurst, D. E., and A. G. Richards. 1964. Some histochemical observations on the blood cells of the wax moth, *Galleria mellonella* L. *J. Morphol. 114*(2):247–54.

*Ayers, H. 1884. On the development of *Oecanthus niveus* and its parasite, *Teleas. Mem. Boston Soc. Nat. Hist. 3*:225–81.

Bock, E. 1939. Bildung and Differenzierung der Keimblätter bei *Chrysopa perla* (L.) *Z. Morphol. Oekol. Tiere 27*:615–702.

Bronskill, J. F. 1959. The embryology of *Pimpla turionellae* (L.) (Hymenoptera, Ichneumonidae). *Can. J. Zool. 34*:655–88.

Butt, F. H. 1949. Embryology of the milkweed bug, *Oncopeltus fasciatus* (Hemiptera). *Mem. Cornell Univ. Agric. Exp. Stn. 283*:1–43.

*Dohrn, A. 1876. Notizen zur Kenntnis der Insektenentwicklung. *Z. Wiss. Zool. 26*:112–38.

Eastham, L. E. S. 1930. The embryology of *Pieris rapae:* Organogeny. *Philos. Trans. R. Soc. Lond. B. Biol. Sci. 219*:1–50.

Farooqi, M. M. 1963. The embryology of the mustard sawfly *Athalia proxima* Klug. (Tenthredinidae, Hymenoptera). *Aligarh Muslim Univ. Publ. Zool. Ser. 6*:1–68.

Görg, I. 1959. Untersuchungen am Keim von *Hierodula* (*Rhombodera*) *crassa* Giglio Tos, ein Beitrag zur Embryologie des Mantiden (Mantodea). *Dtsch. Entomol. Z. 6*:389–450.

Heymons, R. 1895. *Die Embryonalentwicklung von Dermapteren und Orthopteren unter besonderer Berücksichtigung der Keimblätterbildung.* G. Fischer, Jena.

Hirschler, J. 1909. Die Embryonalentwicklung von *Donacia crassipes* L. *Z. Wiss. Zool.* 92:627–744.

Jackson, H. W. 1939. The morphology and histogenesis of the blood of the mealworm (*Tenebrio molitor* L.) with observations on its embryology. Ph.D. thesis, Cornell University, Ithaca, New York.

Johannsen, O. A. 1929. Some phases in the embryonic development of *Diacrisia virginica* Fabr. (Lepidoptera). *J. Morphol.* 48(2):493–541.

Johannsen, O. A., and F. H. Butt. 1941. *Embryology of Insects and Myriapods.* McGraw-Hill, New York.

Kessel, E. L. 1939. The embryology of fleas. *Smithson. Misc. Collect.* 98:1–78.

Korotneff, A. 1885. Die Embryologie der *Gryllotalpa. Z. Wiss. Zool.* 41:507–604.

Leuzinger, H., R. Wiesmann, and F. H. Lehmann. 1926. *Zur Kenntnis der Anatomie und Entwicklungsgeschichte der Stabheuschrecke Carausius morosus* Br. G. Fischer, Jena.

Mellanby, H. 1937. The later embryology of *Rhodnius prolixus. Q. J. Microsc. Sci.* 79:1–42.

Miyakawa, K. 1974. The embryology of the caddisfly *Stenopsyche griseipennis* McLachlan (Trichoptera: Stenopsychidae). IV. Organogenesis: Mesodermal derivatives. *Kontyû* 42(4):451–66.

Mori, H. 1969. Normal embryogenesis of the waterstrider, *Gerris paludum insularis* Motschulsky, with special reference to midgut formation. *Jpn. J. Zool.* 16(1):53–67.

Mori, H. 1975. Everted embryos obtained after cauterization of eggs of the waterstrider, *Gerris paludum insularis* Motschulsky. *Annot. Zool. Jpn. 48(4):252–61.*

Mori, H. 1976. Formation of the visceral musculature and origin of the midgut epithelium in the embryos of *Gerris paludum insularis* Motschulsky (Hemiptera: Gerridae). *Int. J. Insect Morphol. Embryol.* 5(2):117–25.

Mori, H. 1977. Inductive role of the visceral musculature in formation of the midgut epithelium in embryos of the waterstrider, *Gerris paludum insularis* Motschulsky. *Annot. Zool. Jpn.* 50(1):22–30.

Nelson, J. A. 1915. *The Embryology of the Honey Bee.* Princeton University Press, Princeton.

Nusbaum, J., and B. Fuliński. 1906. Zur Entwicklungsgeschichte des Darmdrüsenblattes bei *Gryllotalpa vulgaris. Z. Wiss. Zool.* 93:306–48.

Okada, M. 1960. Embryonic development of the rice stem-borer, *Chilo suppressalis. Sci. Rep. Tokyo Kyoiku Daigaku Sect. B* 9:243–96.

Paterson, N. F. 1932. A contribution to the embryological development of *Euryope terminalis* Baly (Coleoptera, Phytophaga, Chrysomelidae). II. Organogeny. *S. Afr. J. Sci.* 29:414–48.

Patten, W. 1884. The development of phryganids, with a preliminary note on the development of *Blatta germanica. Q. J. Microsc. Sci.* 24:549–602.

Presser, B. D., and C. W. Rutschky, 1957. The embryonic development of the corn earworm *Heliothis zea* (Bodie) (Lepidoptera, Phalaenidae). *Ann. Entomol. Soc. Amer.* 50:133–64.

Roonwal, M. L. 1937. Studies on the embryology of the African migratory locust, *Locusta migratoria migratorioides.* II. Organogeny. *Philos. Trans. R. Soc. Lond. B. Biol. Sci.* 227:175–244.

Saito, S. 1937. On the development of the tusser, *Antheraea pernyi. J. Coll. Agric. Hokkaido Imp. Univ.* 33:35–109.

Sander, K. 1956. The early embryology of *Pyrilla perpusilla* Walker (Homoptera), including some observations on later development. *Aligarh Muslim Univ. Publ. Zool. Ser.* (Indian Insect Types) 4:1–61.

Seidel, F. 1924. Die Geschlechtsorgane in der embryonalen Entwicklung von *Pyrrhocoris apterus. Z. Morphol. Oekol. Tiere* 1:429–506.

Shafiq, S. A. 1954. A study of the embryonic development of the gooseberry sawfly *Pteronidea ribesii*. *Q. J. Microsc. Sci.* 95(1):93–114.

Sohi, S. S. 1971. *In vitro* cultivation of hemocytes of *Malacosoma disstria* Hübner (Lepidoptera: Lasiocampidae). *Can. J. Zool.* 49:1355–8.

Starre-van der Molen, L. G. van der. 1972. Embryogenesis of *Calliphora erythrocephala* Meigen. I. Morphology. *Neth. J. Zool.* 22(2):119–82.

Striebel, H. 1960. Zur Entwicklung der Termiten. *Acta Trop.* 13:193–260.

Tiegs, Ò. W., and F. V. Murray. 1938. The embryonic development of *Calandra oryzae*. *Q. J. Microsc. Sci.* 80:159–284.

Toyama, K. 1902. Contribution to the study of silkworm. I. On the embryology of the silkworm. *Bull. Coll. Agric. Tokyo Imp. Univ.* 5:73–118.

Ullmann, S. L. 1964. The origin and structure of the mesoderm and the formation of the coelomic sacs in *Tenebrio molitor* L. (Insecta, Coleoptera). *Philos. Trans. R. Soc. Lond. B. Biol. Sci.* 248:245–77.

Wada, S. 1955a. Zur Kenntnis der Keimblätterherkunft des Subösophagealkörpers am Embryo des Seidenraupe, *Bombyx mori* L. *J. Seric. Sci. Jpn.* 24(2):114–17. (in Japanese, German summary.)

Wada. S. 1955b. Zur Frage des Subösophagealkörpers als hämopoietische Gewebes am Embryo des Seidenraupe, *Bombyx mori* L. *J. Seric. Sci. Jpn.* 24(5/6):311–13. (In Japanese, German summary.)

Wheeler, W. M. 1889. The embryology of *Blatta germanica* and *Doryphora decemlineata*. *J. Morphol.* 3:291–386.

Wigglesworth, V. B. 1959. Insect blood cells. *Annu. Rev. Entomol.* 4:1–16.

Wigglesworth, V. B. 1972. *The Principles of Insect Physiology*, 7th ed. Chapman & Hall, London.

Wolf-Neis, R. 1973. Differenzierungsverlauf von *Carausius*-Embryonen *in ovo* und *in vitro*. *Zool. Jahrb. Abt. Anat. Ontog. Tiere* 91:1–18, 158–200.

Wolf-Neis, R., C. Kirchner, P. Koch, and K. A. Seitz. 1976. Immunohistological studies on the distribution of yolk proteins in the stick insect (*Carausius morosus*). *J. Insect Physiol.* 22:865–9.

2 Postembryonic development and differentiation: hemopoietic tissues and their functions in some insects

J. A. HOFFMANN, D. ZACHARY, D. HOFFMANN, M. BREHÉLIN,

Laboratory of General Biology, Louis Pasteur University, Research Group of CNRS #118, 12 rue de l'Université, 67000. Strasbourg, France

A. PORTE

Laboratory of Electron Microscopy, Institute of Physiology, Louis Pasteur University, 21 Rue René Descartes, 67000 Strasbourg, France

Contents

2.1. Introduction

The blood cells of many higher animals are formed in specialized tissues called hemocytopoietic tissues (blood-cell-making tissues; *poiesis* = "making") or hemopoietic tissues. Although less accurate, the latter term is generally used to mean the same thing as hemocytopoietic tissues and will be preferred in this chapter.

Hemopoietic tissues are concerned with producing different lines of blood cells and delivering them into the blood; in some animals they may also function to remove dying or dead cells and histolytic debris from circulation. This function is assumed by phagocytic cells that originate from a particular cell lineage, generally called reticular or reticuloendothelial cells; it is this cell type that gives to hemopoietic tissues a characteristic reticular arrangement. It must be stressed that a third major function of hemopoietic tissues in higher animals may be immunologic, involving specialized cells such as lymphocytes and plasmocytes.

That in at least some insects definite hemopoietic tissues are present was reported near the end of the last century (Kowalevsky, 1889). However, discrepancies obviously exist between the organizational level of hemopoietic tissues (or their equivalents) in different insect orders; in addition, this organization may change during the development in the same species, as will be explained for *Calliphora*.

These differences are the major source of the controversies about the postembryonic origin of hemocytes that existed between Kowalevsky (1889) and Cuénot (1896). Another source is the lack of quantitative data.

Jones (1970) published an exhaustive report on these controversies (250 references), and it is unnecessary to duplicate such a review here (see also Chapter 3).

Our objective is different; we will give a detailed description of a highly organized hemopoietic organ in an insect where it has been studied with up-to-date techniques and stress the analogies with hemopoietic tissues in higher animals. We will point out the variations in the organization of hemopoietic tissues of several insects and propose a general scheme for blood cell production in insects; this should explain most of the controversial aspects of this topic. Challenging lines of future research in the field of hemopoietic tissues and their functions in insects will be outlined at the end of the chapter (see also Chapters 5, 8).

2.2. Hemopoietic organ of *Gryllus bimaculatus:* a model

Of all the hemopoietic tissues described so far in insects, that of *Gryllus* is undoubtedly the most typical in its features and corresponds closely to the hemopoietic organs in higher animal groups, including vertebrates.

On both sides of the dorsal vessel of *Gryllus*, in the second and third abdominal segments, exist two pairs of clearly distinguishable cell accumulations that are connected to the blood vessel itself. Their outlines are regular (Fig. 2.1). The periphery is dense and the center shows a "lumen" of variable size. This "lumen" is irregularly filled with freely floating blood cells (Fig. 2.2). These cell accumulations were described near the end of the last century under the generic name "phagocytic tissues" (Cuénot, 1896). As we will see, this term, although correct, refers to only one of the properties of the main cell types of the accumulations. The hemopoietic character of the "phagocytic tissues" of *Gryllus* becomes evident when these centers are carefully examined by light and electron microscopy; it can be easily demonstrated by inducing severe hemorrhages, which result in a dramatic stimulation of these tissues.

The hemopoietic organs of *Gryllus* are surrounded by an irregular layer of cells and basement material (see Fig. 2.4). Occasionally, the cell layer is reduced to basal membranes; this is especially common when the hemopoietic tissue is in contact with fat body. Sometimes the cells form a continuous single- or double-stranded layer, and the basement material is either tenuous or locally absent. The cells of the capsule surrounding the hemopoietic tissue are clearly of single type: They show characteristic features of fibroblasts; as a rule, they appear digitated, and their cytoplasmic membranes are often seen in close contact with thin collagen-like fibrils, which are predominant in the basement material (Fig. 2.3). The periodicity of these fibrils is not always distinct (approx. 250 Å); they resemble immature collagen fibrils

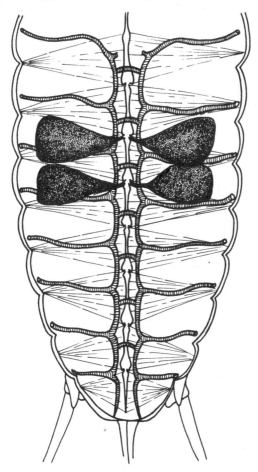

Fig. 2.1. Hemopoietic organs of G. *bimaculatus,* ventral view with dorsal vessel, aliform muscles, and tracheal trunks. Shaded hemopoietic organs are in communication with dorsal vessel.

of vertebrates. The cytoplasm of the fibroblasts shows well-developed, rough-surfaced endoplasmic reticulum (RER), numerous Golgi complexes, and often lysosomal inclusions; the microtubules are abundant in these cells, which closely resemble the exothelial cells of the dorsal vessel (Hoffmann and Lévi, 1965), with which they are continuous.

The cortical zone of the hemopoietic organs shows a striking reticular arrangement (Fig. 2.4): Thick and extremely ramified fibers form a complex network together with highly polymorphic cells that send out long, thin processes that in turn are in close contact with the fibers (Fig. 2.5). The processes of different cells frequently contact desmosome-like junctions, as shown in Fig. 2.6. The highly polymorphic cells, which we called reticular cells, are similar in many respects (Fig. 2.7). In addition to their long processes, they show numerous deep indentations at their periphery. The free ribosomes are relatively numerous in these cells, whereas the RER is not very conspicuous; the Golgi complexes are often scarce.

The reticular cells show many pinocytotic vesicles, dense bodies, and heterogeneous vacuoles, all of which are signs of active resorption of foreign material and breakdown of this material in the cells. Sometimes they contain larger inclusions, such as necrotic hemocytes.

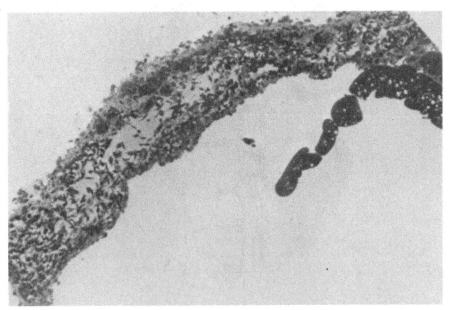

Fig. 2.2. Semithin section of hemopoietic organ of *G. bimaculatus* (fuchsin stain). Peripheral zone (cortex) is rich in germinal centers; central "lumen" contains predominantly differentiated hemocytes. Dorsal vessel (not seen in this section) is on upper right. × 140. (From Hoffmann, 1970)

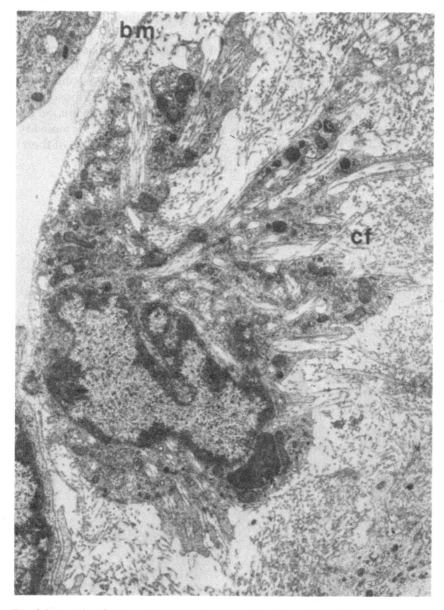

Fig. 2.3. Peripheral zone of hemopoietic organ of *G. bimaculatus:* fibroblast of capsule. Note numerous cell digitations and abundance of collagen-like fibers in close contact with this cell. *bm* = basement material; *cf* = collagen-like fibers. × 20,400.

They have the macrophagic functions of the reticuloendothelial cells in the hemopoietic tissues of vertebrates.

The reticular cells undergo frequent mitotic divisions. Some of the daughter cells differentiate into hemopoietic stem cells. Their outline appears regular in this case, and these cells show no sign of phagocytosis. Their cytoplasm becomes rapidly enriched with free ribosomes, all the other organelles and inclusions being still poorly developed or very scarce (Fig. 2.8). These cells can be considered as hemocytoblasts. They are the true stem cells of hemocytes; at this stage of their differentiation they cannot be classified.

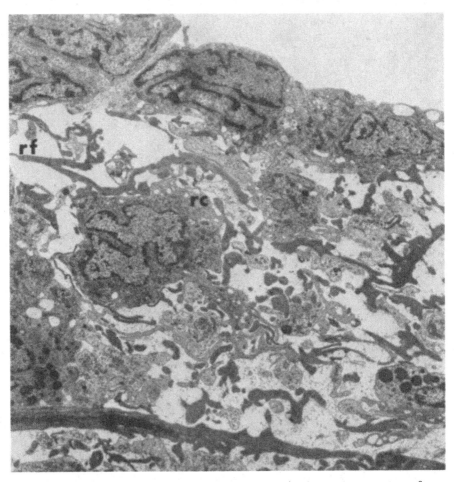

Fig. 2.4. Peripheral zone of hemopoietic organ of *G. bimaculatus*. Note presence of capsule and reticular organization of cortex. Reticular cells (*rc*) and reticular fibers (*rf*) are easily identified, as well as two granulocytes. Central axis of large reticular fiber (bottom) appears very dense. ×3,500. (From Hoffmann, 1970)

The hemocytoblasts in the cortical zone of the hemopoietic tissues of *Gryllus* differentiate rapidly: Their RER appears in the form of short cisternae with a moderately dense content; the Golgi complexes become more numerous and especially rich in small vesicles and dense granules. At this stage, the different types of granules, which characterize some of the hemocytes, are observed in the differentiating cells (Fig. 2.9).

During their differentiation, the cells remain in close contact with the reticular fibers and the neighboring reticular cells. They undergo repeated cell divisions during their process of maturation (Figs. 2.10, 2.11) and appear, for this reason, typically as clusters of cells of the

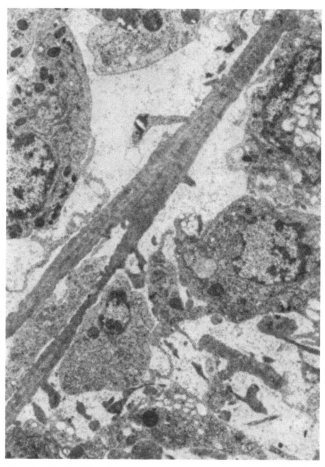

Fig. 2.5. Details of reticular fiber in cortex of hemopoietic organs of *Gryllus*. Note presence of differentiating plasmatocytes and granulocytes. ×4,200. (From Hoffmann, 1970)

same hemocyte type, at the same stage of differentiation (isogenic cell islets) (Fig. 2.12). This differentiation in isogenic islets reminds one of the process of differentiation observed in vertebrates.

In the case of *G. bimaculatus*, it is easy to monitor the differentiation of granulocytes (GRs), coagulocytes (COs), and plasmatocytes (PLs) in the hemopoietic tissue, the PL being obviously very close to the reticular cell (as are the monocytes in mammals).

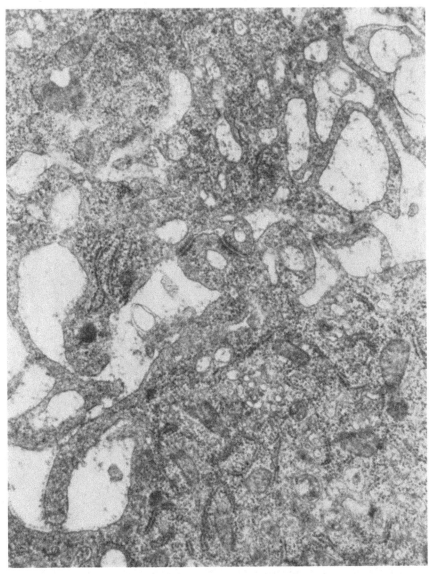

Fig. 2.6. Desmosome-like junctions between processes of two reticular cells in periphery of hemopoietic organ of *Gryllus*. ×23,000.

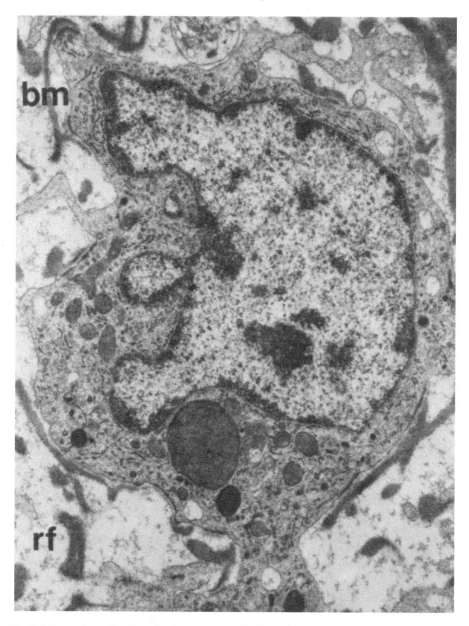

Fig. 2.7. Reticular cell of hemopoietic organ of *Gryllus*. Note presence of dense inclusion of resorptive significance. In this figure basal material (*bm*) is seen in close contact with reticular cell (hemidesmosomes are observed at points of contact), and this same basement material is in continuity with larger reticular fibers (*rf*). × 16,000. (From Hoffmann, 1970)

Once the hemocytes have achieved their differentiation, they accumulate in the irregular lumen of the hemopoietic tissue (see Fig. 2.15). Some of them immediately enter the bloodstream through the lumen of the dorsal vessel with which the hemopoietic tissue is in direct contact.

If *G. bimaculatus* is subjected to severe hemorrhage, the hemopoietic organs show important changes, resulting from a dramatic intensification of hemopoiesis (Fig. 2.13). The tissues appear more compact and voluminous. The "germinal centers" of the cortical zone tend to congregate and form a mass in which cell divisions are numerous. The

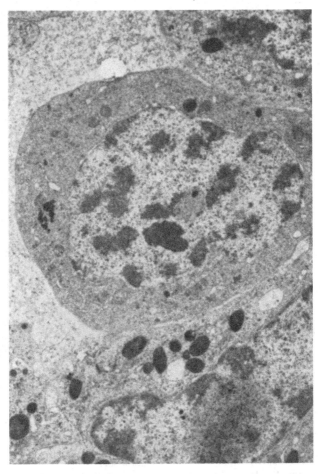

Fig. 2.8 Hemocytoblast of *Gryllus* hemopoietic organ. Cell is rich in free ribosomes, but relatively poor in conventional cytoplasmic organelles. Nucleus is large compared with cytoplasm; no signs of phagocytosis are observed in these cells. ×8,500. (From Hoffmann, 1970)

central lumen appears obstructed owing to the accumulation of differentiated hemocytes, whose elongated shape and orientation indicate their movement toward the dorsal vessel.

In electron microscopy, the reticular cells and the differentiating young hemocytes appear densely packed in the cortical zone (Fig. 2.14), the reticular fibers being shifted toward the periphery of new hemopoietic centers. Numerous mitotic figures are seen in the reticular cells, as well as in the young hemocytes. The medullary zone is filled with differentiated hemocytes that have no contact anymore with the reticular cells or reticular fibers (Fig. 2.15). Figure 2.16 summarizes the observations on the hemopoietic tissues of *G. bimaculatus*.

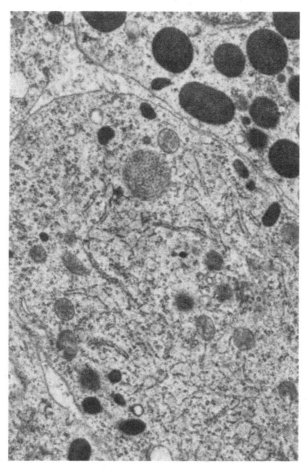

Fig. 2.9. Young differentiating coagulocyte with typical multitubular granule in cortex of hemopoietic organ of *Gryllus*. × 11,000. (From Hoffmann, 1970)

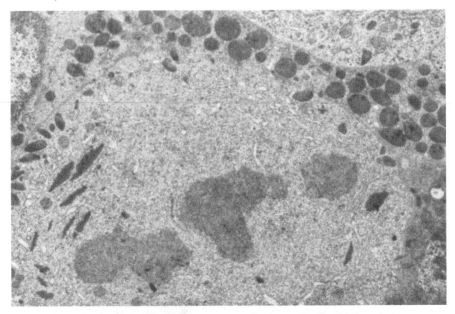

Fig. 2.10. Mitotic division of differentiating coagulocyte in *Gryllus* hemopoietic organ (cortex). × 9,500. (From Hoffmann, 1970)

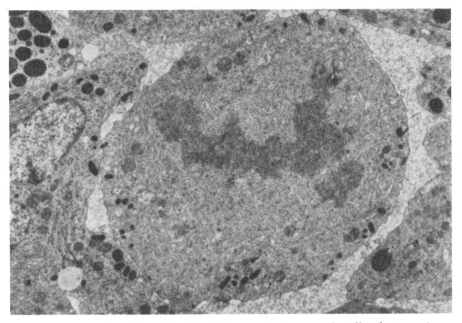

Fig. 2.11. Mitosis of granulocyte in cortex of hemopoietic organ of *Gryllus* during differentiation. Divisions of maturing hemocytes in hemopoietic organs explain formation of isogenic cell clusters. × 6,000. (From Hoffmann, 1970)

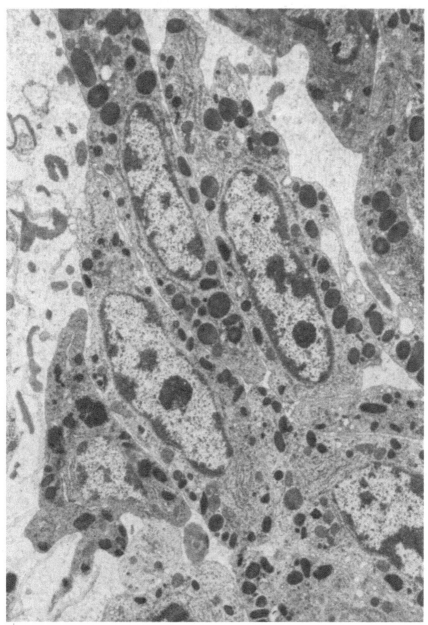

Fig. 2.12. Isogenic islets of granulocytes and coagulocytes in cortex of hemopoietic organ of *Gryllus*. ×4,500. (From Hoffmann, 1970)

Injected inert powders, such as iron saccharate, are taken up massively by the reticular cells. The interrelationship between the dual potentialities (hemocytoblastic and macrophagic) of the reticular cell is evidenced by the effect on the hemogram of injections of substances that typically induce a reduced hemopoiesis and, as a consequence, a depression of the number of circulating hemocytes (Brehélin and Hoffmann, 1971).

We have undertaken a comparative study of hemopoiesis in *Gryllus* and a few other selected insect species. The results, given below, will throw some light on the causes of the controversies. They will also explain that there is a good deal of truth in most of the controversial reports in the literature. The insect species chosen for the comparative studies were another gryllid, *Gryllotalpa gryllotalpa;* an orthopteran, *Locusta migratoria;* the coleopteran, *Melolontha melolontha;* and the dipteran, *Calliphora erythrocephala.*

The results of our study in *G. gryllotalpa* were exactly the same as those in *Gryllus* and will not be discussed here. *L. migratoria* and larvae of *M. melolontha* were found to have hemopoietic tissues with strong similarities; *C. erythrocephala,* although rather similar to other species, shows interesting pecularities in its hemopoiesis. These data will now be given in some detail.

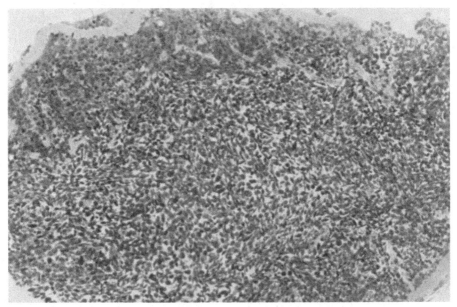

Fig. 2.13. Hemopoietic organ of *Gryllus* after severe bleeding (2 days). Compare with Fig. 2.2. Semithin section, fuchsin stain. × 140. (From Hoffmann, 1970)

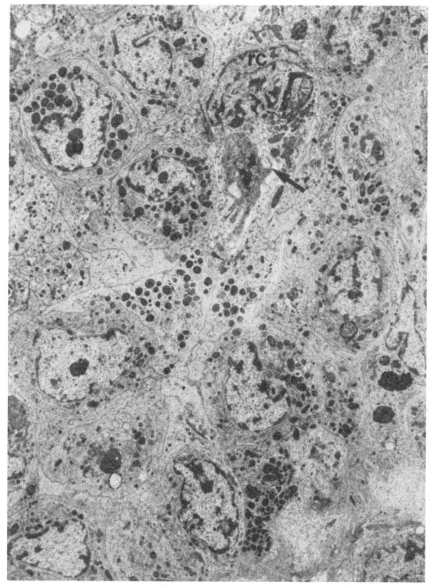

Fig. 2.14. Stimulation of hemopoiesis after severe bleeding. Massive proliferation of maturing hemocytes in cortex of hemopoietic organ of *Gryllus* two days after a hemorrhage. Note presence of cell debris (arrow) in reticular cell (*rc*). ×4,690. (From Hoffmann, 1970)

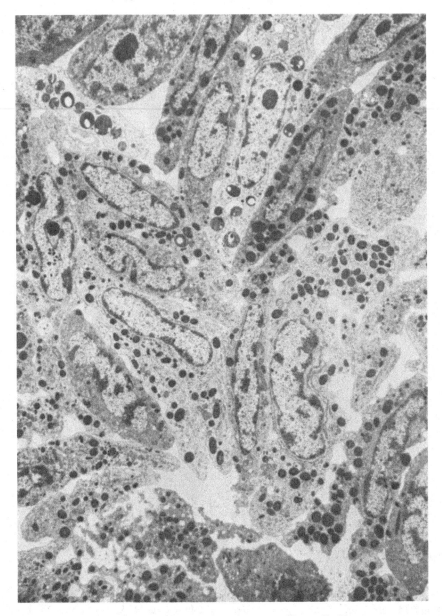

Fig. 2.15. Accumulation of differentiated hemocytes in "lumen" of hemopoietic organ of *Gryllus* two days after a severe hemorrhage. ×4,020. (From Hoffmann, 1970)

2.3. Hemopoietic tissues of *Locusta migratoria* and *Melolontha melolontha*

On the upper surface of the dorsal diaphragm of *L. migratoria,* an ir-regular accumulation of cells stretches out all along the dorsal vessel. No limiting membrane surrounds this cell accumulation, which is not connected with the dorsal vessel (Fig. 2.17). In transverse sections, the accumulations appear thinner near the contact with the heart ves-sel and on both sides in the vicinity of the tergites. These irregular cell accumulations stretch out between the pericardial cells and the fat body of the pericardial sinus. No differentiation into a cortical zone and a central lumen is seen in *L. migratoria* (Fig. 2.18).

Ultrastructural study shows the presence of several cell types in the dorsal accumulations: reticular cells, young differentiating hemo-cytes, and fully differentiated hemocytes of the different cell cate-gories normally found in circulation in this species (Fig. 2.19). *Lo-*

Fig. 2.16. Hemopoietic organ of *G. bimaculatus.* Mesothelial cell layer surrounds organ (upper left); certain cells are very digitated and in close contact with collagen-like fibers (capsule). Cortex has regular organization, regular cells (dotted) and fibers forming a dense network; isogenic islets of differentiating hemocytes are seen in this network; note numerous cell divisions. Medulla or lumen (lower part of figure) shows accumulation of differentiated hemocytes of different cell types present in he-molymph; medulla of organ is in communication with dorsal vessel. Each insect pos-sesses four hemopoietic organs (two pairs) in this species.

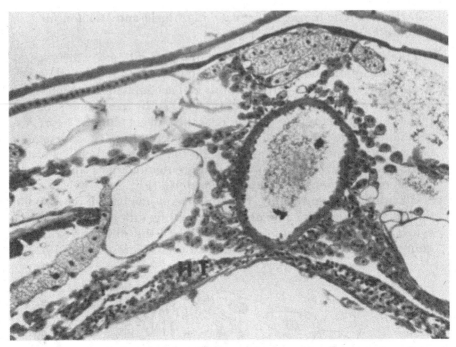

Fig. 2.17. Hemopoietic tissue (*HT*) of *L. migratoria* is situated on upper surface of dorsal diaphragm in first five abdominal segments. × 150. (From Hoffmann, 1973)

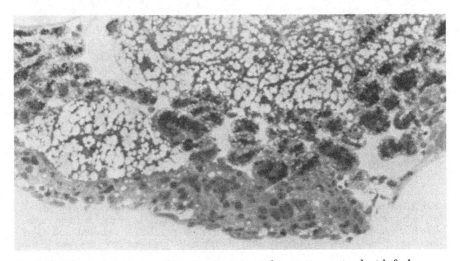

Fig. 2.18. Hemopoietic tissue of *L. migratoria* (semithin section stained with fuchsin) in contact with fat body and pericardial cells. No differentiation of cortex or "lumen" is observed in this species. × 240. (From Hoffmann, 1970)

custa does not have a network of clearly distinctive reticular fibers as has *Gryllus;* instead, a complex reticulum of fine lamellae of basement material is present throughout the cell accumulation and surrounds it loosely.

The reticular cells of *L. migratoria* resemble in every respect those of *Gryllus;* they often show signs of phagocytosis and readily take up

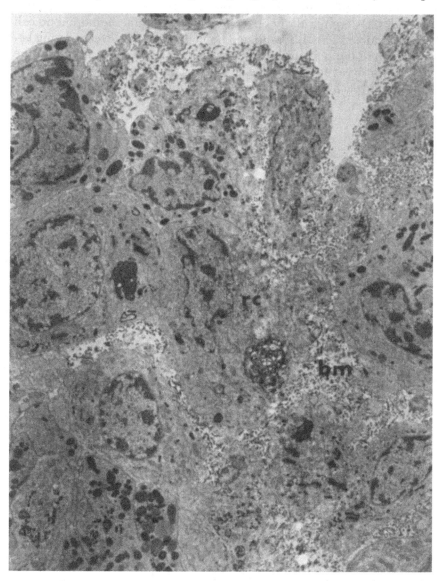

Fig. 2.19. Hemopoietic tissues of *L. migratoria*. Reticular cells (*rc*) and differentiating hemocytes are observed together with fibrillar basement material (*bm*). No capsule surrounds hemopoietic tissue (upper right). × 7,000. (From Hoffmann, 1970)

injected substances such as iron saccharate or bacteria. These cells undergo numerous divisions and frequently transform into hemopoietic stem cells that differentiate in situ; as in *Gryllus*, the differentiating hemocytes also divide abundantly, which explains why they usually appear as clusters of isogenic hemocytes (Fig. 2.20).

The differentiating hemocytes represent in the hemopoietic tissue the different cell lineages in *L. migratoria*: Isogenic clusters of PLs, GRs, or COs are regularly observed throughout the hemopoietic cell accumulations. The differentiated hemocytes enter the circulating hemolymph of the pericardial sinus by setting themselves free from the network of basement lamellae.

Without reaching the same degree of organization as that of *G. bi-*

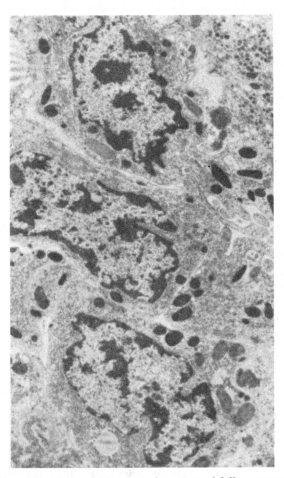

Fig. 2.20. Isogenic islet of coagulocytes at late stage of differentiation in hemopoietic tissue of *L. migratoria*. ×4,500. (From Hoffmann, 1970)

maculatus, the hemopoietic tissue of *Locusta* shows obviously an outline of an organ (Fig. 2.21). The disposition of the reticular cells is identical in both cases; the PAS-positive basement material, although it is never transformed into thick fibers, corresponds, as does that of *Gryllus*, to the reticulin fiber network of the hemopoietic tissues of vertebrates and has the same functional significance.

As in *G. bimaculatus*, the reticular cells of *Locusta* have double

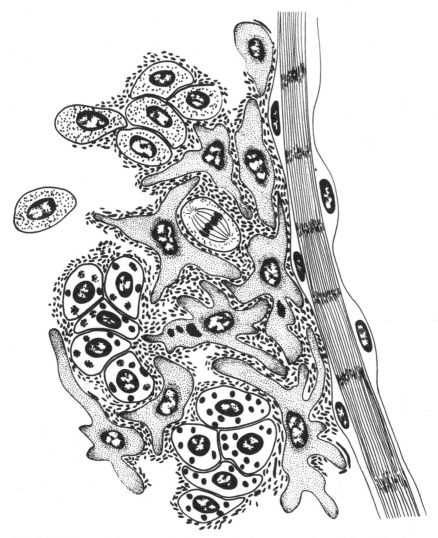

Fig. 2.21. Hemopoietic tissue of *L. migratoria*. On upper surface of dorsal diaphragm are observed reticular cells (dotted), isogenic cell islets with maturing hemocytes of different types, and differentiated hemocytes entering bloodstream. Between cells, fine lamella of basement material is present.

functions: hemopoietic or macrophagic. The interrelationship between these two functions is illustrated by the injection of iron saccharate, which stimulates the phagocytotic function of the reticular cells and, as a result, hampers the hemopoietic differentiation for several days; as a consequence, a marked decrease in the number of circulating hemocytes is observed (Fig. 2.22).

Interestingly, severe hemorrhages in *L. migratoria* expel the reticular cells into circulation, as monitored by ultrastructural studies. While in circulation, these cells are quite capable of differentiating and forming new hemocytes. This experimental situation is not observed under normal conditions in *L. migratoria*, but may exist among other insects at certain developmental stages. It shows a certain flexibility in the hemopoietic organization.

In the larvae of *M. melolontha*, the hemopoietic tissue also consists of cell accumulations that stretch out between the large pericardial cells in the dorsal part of the abdomen; the accumulations are not surrounded by a connective sheath. The arrangement of the cells in the hemopoietic tissue is finely reticular: The highly polymorphic reticu-

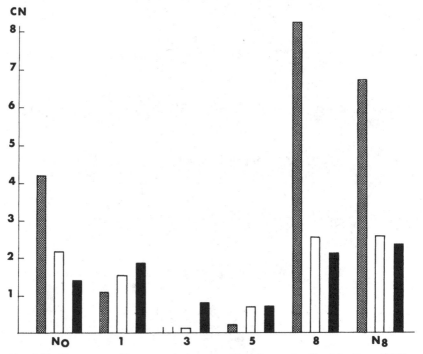

Fig. 2.22. Hemogram of last nymphal instar of *L. migratoria* after injection of 250 μg iron saccharate per insect on first day of instar. Ordinate shows number of hemocytes per cubic millimeter in thousands; abscissa shows age in instar; coagulocyte (dotted) plasmatocyte (white), and granulocyte (black) values are means of determinations from 18 insects per experiment.

lar cells in this tissue are very similar to those of *Gryllus* and are in close contact with a well-developed basal material, which often appears in the form of very thick fibers, as in *Gryllus* (Fig. 2.23). The hemopoietic tissue of *Melolontha* does not show any specific organization, and the reticular cells are mixed at random with differentiating or differentiated hemocytes of the different cell types found in *Melolontha*.

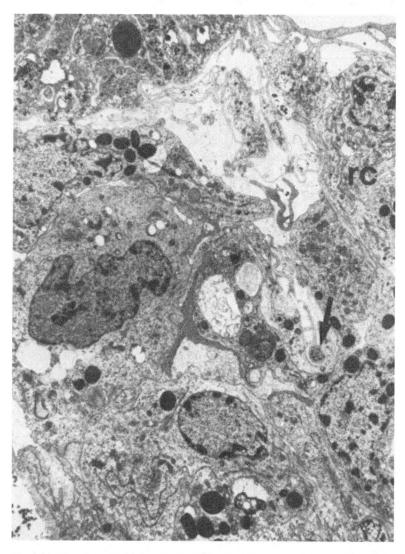

Fig. 2.23. Hemopoietic tissue of *M. melolontha*. Note reticular organization with typical reticular cells (*rc*) and reticular fibers, which sometimes have a dense core (arrow). Granulocytes are also seen in this section. × 6,000. (From Bréhélin, 1973)

2.4. Hemopoiesis in *Calliphora erythrocephala*

In *C. erythrocephala*, the hemopoietic centers appear in the last three abdominal segments of the blow fly larva; these cell accumulations are found on both sides of the dorsal vessel and extend to the two laterodorsal trunks (Fig. 2.24). They are irregular in outline and are not surrounded by any connective tissue sheath. No structural organization is recognized.

Ultrastructural study shows that the accumulations contain both

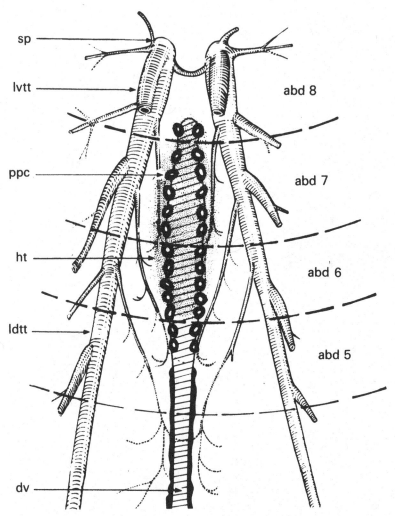

Fig. 2.24. Ventral view of hemopoietic tissue of *C. erythrocephala* larva. Tissue extends on both sides of dorsal vessel in posterior abdominal segments (abd. 5 to abd. 8). *dv* = dorsal vessel; *ht* = hemopoietic tissue (dotted); *ldtt* = laterodorsal tracheal trunk; *lvtt* = lateroventral trunk; *pcc* = posterior pericardial cells; *sp* = spiracles.

highly polymorphous cells resembling in every respect the reticular cells of the hemopoietic organs of G. *bimaculatus* and hemocytes at various stages of maturation.

The thin and often very long cytoplasmic extensions of the polymorphic reticular cells are tightly interwoven (Fig. 2.25), with frequent desmosomal contacts, and this gives the hemopoietic tissue of *Calli-*

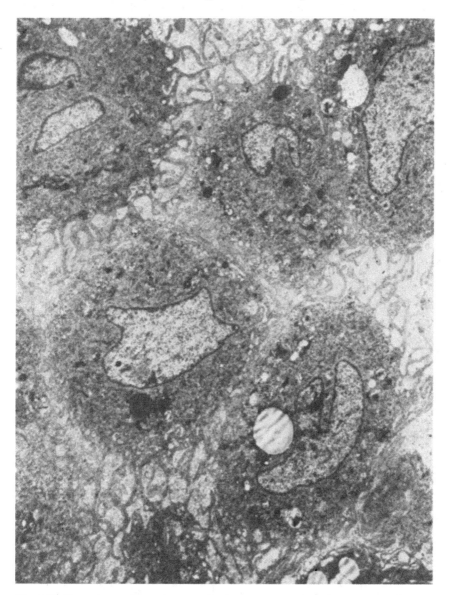

Fig. 2.25. Typical reticular organization in hemopoietic tissue of *C. erythrocephala*. Some reticular cells contain inclusions of resorptive nature. × 6,000.

phora the "reticular" organization found in *Gryllus, Locusta,* and *Melolontha.* In *Calliphora,* conspicuous basement material, some-times forming long cords, is generally found among the reticular cells. Signs of macrophagy are evident in any of these cells. As in all hemo-poietic organs, the reticular cells of *Calliphora* hemopoietic organs di-vide abundantly, and some of them initiate hemopoietic differentia-tion; they show regular profiles, and their cytoplasm becomes

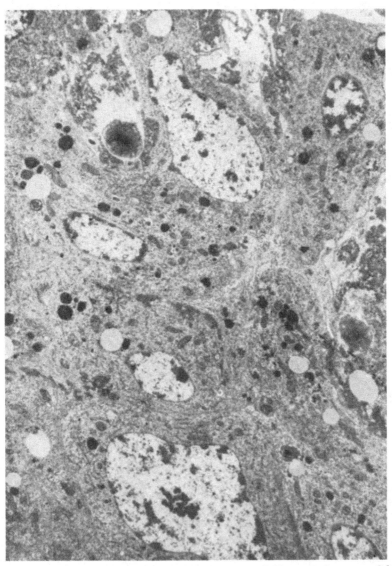

Fig. 2.26. Detail of hemopoietic tissue of larva of *Calliphora.* Early stages of dif-ferentiation of plasmatocytes. × 7,000.

remarkably enriched with free ribosomes. In the hemopoietic tissue itself, differentiation of these hemopoietic stem cells (hemocytoblasts) is normally easy to follow: The RER cisternae rapidly become more conspicuous and sometimes considerably elongated, while their content appears more dense. The Golgi complexes become more numerous, and the size of the mitochondria increases (Fig. 2.26).

An interesting feature of *Calliphora* hemopoiesis is that although the hemocyte production by the hemopoietic tissue results in the release of single hemocytes, whole clusters of maturing hemocytes, at different stages of differentiation, enter the bloodstream (Fig. 2.27). This process is not observed under normal conditions in the three foregoing species. It is especially important in *Calliphora* at the end of larval life.

Prior to pupation, the hemopoietic tissue desegregates, and clusters of differentiating hemocytes are extremely abundant in the hemolymph. In *Calliphora*, differentiation can occur in both the hemopoietic tissue and the circulating hemolymph. The latter situation becomes prevalent at the end of larval life and is the rule in the adult.

The study of the hemopoietic organs of *Gryllus* and comparisons with the corresponding tissues of *Locusta, Melolontha,* and *Calli-*

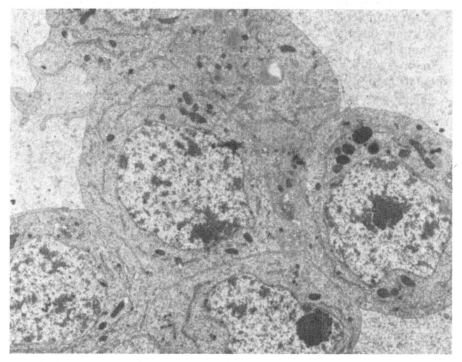

Fig. 2.27. Circulating group of poorly differentiated hemocytes in blood of *Calliphora*. These cells will mature in circulation. × 7,000.

phora lead one to conclude that there exist structural and functional similarities in all the four species. The predominant cell type is the reticular cell; it has a primordial functional importance. It is responsible for hemocyte production by giving rise to hemocytoblasts in the organ or tissue itself, and it plays a major role in the uptake and destruction of unwanted particles as well as degenerating hemocytes. Its functional capabilities appear more diverse in view of some recent experiments. Through their extremely thin processes, the reticular cells build up a complex network, together with a more or less well-developed system of basement lamellae, which sometimes can appear as a complete elaborate mesh of thick, ramified fibers. This so-called reticular organization (in analogy with hemopoietic tissues from vertebrates) is typically present in all four species investigated. However, its complexity is more or less different from one species to another.

It is of interest that the hemopoietic cells are quite capable of differentiating into hemocytes when they are set free from the hemopoietic tissue; in *Locusta*, this can be demonstrated by induced hemorrhages; in *Calliphora*, not only differentiated hemocytes, but also clusters of immature cells enter the bloodstream and differentiate, while in circulation, under normal conditions.

A series of transitions exists apparently in the organizational level of the hemopoietic cells in the insects investigated. In *Gryllus*, these cells are integrated into a highly organized hemopoietic organ, surrounded by a capsule, and show cortical germinal centers, separated from a medulla in which differentiated hemocytes accumulate prior to entering the hemolymph. Hemocyte differentiation occurs in the hemopoietic organs, which are present throughout the entire life cycle; in *Locusta* and *Melolontha*, the hemopoietic cells are located in a diffuse tissue, which shows no sign of true organization; hemopoiesis occurs under normal conditions exclusively in the hemopoietic tissues, which are present during the entire life of *Locusta* (no investigation in adult *Melolontha* has been undertaken). In *Calliphora*, the degree of organization is poor in the hemopoietic tissue, but interestingly, differentiating hemocyte clusters enter the circulation prior to their final maturation. Lastly, the hemopoietic cells can differentiate in the circulating blood in insects that possess a definite (although diffuse) hemopoietic tissue if they are set free from this tissue, as seen under experimental conditions in *L. migratoria*.

In all cases investigated, the fundamental scheme of hemopoiesis is similar and basically identical to that reported from other animal groups, including vertebrates. What differs from one insect group to another is the modality of the organization of the hemopoietic cells. We are convinced that the differences in the models of organization are the source of the numerous controversies that exist in the literature. *Calliphora* illustrates the transitions that can exist in the same species between a localized hemopoietic tissue in the young larva and circulating clusters of differentiating cells, which, at metamorphosis, become the main centers of hemopoiesis; in the adult, blood cell production is ensured by these differentiating cell clusters in circulation, the hemopoietic tissue having completely dissociated at metamorphosis. This situation in the adult corresponds exactly to the analysis by Cuénot (1896), who stated that in insects, "amibocytes" of class I (poorly differentiated circulating blood cells) in circulation represent, in dissociated form, the hemopoietic tissue of decapod crustaceans. This analysis is, in effect, correct for the particular developmental stage in *Calliphora;* however, it should be understood that it is too restricted a view of hemopoiesis in insects. Figure 2.28 summarizes the conclusions of this study.

The production of differentiated hemocytes in insects such as *Gryllus* and *Locusta* can be investigated on a quantitative basis by different types of experiments: irradiations of hemopoietic tissues, blockage of hemopoiesis by injection of massive doses of iron saccharate (which induces macrophagy in reticular cells and consequently slows the hemopoietic differentiation of these cells for some time), severe hemorrhages.

These experiments confirmed the conclusions of the ultrastructural studies on the hemopoietic role of these organs or tissues (see, for instance, Hoffmann, 1969, 1972; Brehélin and Hoffmann, 1971; Zachary and Hoffmann, 1973). Moreover, they showed that the hemopoietic

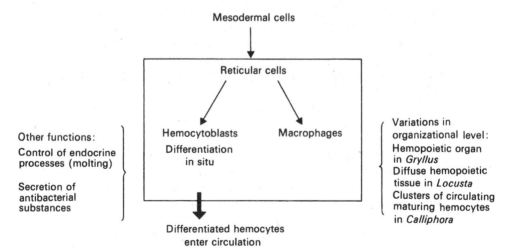

Fig. 2.28. General outlines of hemopoiesis in insects according to comparative study in four species: *G. bimaculatus, L. migratoria, M. melolontha,* and *C. erythrocephala.*

tissues play a major role in several important physiological events, such as molting, and in certain defense reactions. This aspect is at present under intensive investigation in our laboratory; we shall outline some of the data which have been obtained in the hope of stimulating future research in this field in other insects.

2.5. Hemopoietc tissues and regulation of molting

The anatomical situation of the hemopoietic tissue in *Locusta migratoria* makes it most suitable for selective X-ray treatment. If such treatment is performed during nymphal development, a complete blockage of molting is observed, provided the irradiation takes place at the beginning of the instar (regardless of the instar) (Hoffmann, 1971). The nymphs survive for months and die without ever initiating apolysis or ecdysis. If sham irradiations are performed in this insect in a ventral abdominal region, for instance (Fig. 2.29), molting is not inhibited.

This result is probably very similar to that reported by Wigglesworth (1955) in *Rhodnius* after injection of iron saccharate. Molting was considerably delayed if the inert powder was injected at the beginning of the instar. Wigglesworth's interpretation at that time was that the circulating hemocytes which take up the saccharate lose their capacity to transport brain hormone to the prothoracic glands in the blood. Indeed, as we have explained in detail, the injection of iron saccharate results in a blockage of the normal functioning of the hemopoietic tissue, and it is very likely that the two types of experiments are basically identical. Selective irradiations of the hemopoietic tissue and massive injections of iron saccharate both depress the normal functioning of the pool of reticular cells (one by blocking mitoses, which are particularly abundant in these cells and a prerequisite for the differentiation; the other by inducing macrophagy in all reticular cells).

What is the mechanism of the blockage of the molting process?

We presently have data indicating that after selective irradiation of the hemopoietic tissue, the prothoracic glands no longer produce ecdysone (although these glands are not touched by the X-rays during the experiment). Does irradiation result in an inhibition of the prothoracic glands? If these glands are extirpated at various time intervals following the X-ray treatment of the hemopoietic tissue and incubated in an appropriate medium, they do not synthesize ecdysone, whereas under the same conditions, glands from normal nymphs do produce ecdysone in fair amounts. However, experiments show that the prothoracic glands of irradiated insects retain their capacity to produce ecdysone: If they are transplanted into prothoracectomized nymphs, in which the molting process is blocked, they resume ecdysone pro-

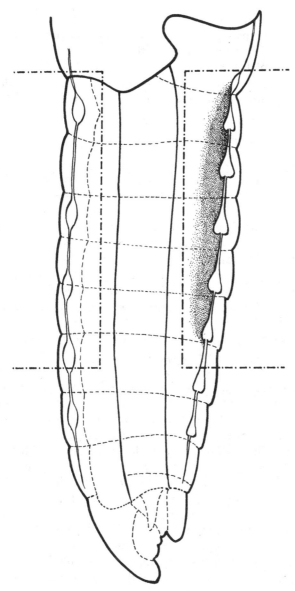

Fig. 2.29. Selective X-ray treatment of hemopoietic tissue of *L. migratoria:* lateral view of abdomen, showing dorsal vessel and diffuse hemopoietic tissue (dotted), digestive tract, and ventral nerve cord. Dashes outline regions exposed to X-radiations (dorsal hemopoietic irradiations and ventral sham irradiations). Direction of X-ray is perpendicular.

duction and restore molting in the receiver. It appears likely that in the X-ray treated nymphs, the prothoracic glands are not stimulated by the brain hormone. Our working hypothesis at present is that the radiosensitive cells of the hemopoietic tissue produce a factor that is necessary for the secretion of the brain hormone or the stimulation of the prothoracic glands by this hormone (see also Chapter 11).

2.6. Hemopoietic tissues and defense reactions

The hemopoietic tissues are involved in defense reactions against pathogens. Cuénot (1896) reported that bacteria are taken up in the "phagocytic tissues" of orthopteroid insects. This observation is confirmed by recent studies. However, the exact role of the hemopoietic tissue in the defense reaction appears more and more complex, as studies with *Locusta migratoria* have shown.

In this species, circulating hemocytes do not play a major part in taking up bacteria that have been injected into the hemolymph; instead, the bacteria are engulfed massively by the reticular cells (Fig. 2.30) of the hemopoietic tissue. Frequently, as a result of this massive uptake, reticular cells undergo necrosis, which in turn induces encapsulation by the surrounding and circulating GRs. Numerous (blackish) nodules are formed as a consequence of injections of pathogens in the hemopoietic tissue (Figs. 2.31, 2.32).

This process has an important effect on the hemogram. The number of circulating cells (of all types) rapidly decreases and remains low for 24–48 hr; after that period, however, a marked increase is seen in the hemocyte count, which parallels a noticeable proliferation in the hemopoietic tissue: The number of reticular cells and of differentiating hemocytes is increased. In fact, the first contact with pathogens induces the phagocytic and hemopoietic processes of the reticular cells, which partly explains why such insects are more resistant to a second injection. Stimulation of the phagocytic process is not the only explanation for induced protection in immunized insects. It was shown in *Locusta* (for other insects, see Whitcomb et al., 1974) that injection of a low dose of *Bacillus thuringiensis* induced rapidly (6–12 hr) an in vivo protection against a lethal dose of the same pathogen. It was interesting to note that selective irradiation of the hemopoietic tissue or massive injections of iron saccharate prior to immunization rendered the insects incapable of building up their protection against lethal doses. However, if the irradiation or the injection of iron saccharate was performed after the protection had been induced, normally lethal doses of pathogens did not kill these insects. This strongly suggested to us that the hemopoietic tissue could be responsible for the production of an antibacterial factor that was secreted into the hemolymph. The chemical nature of this factor is still uncertain,

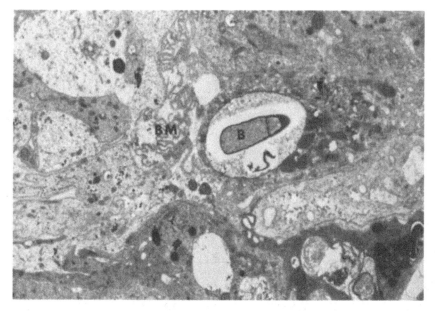

Fig. 2.30. Phagocytosis of *B. thuringiensis* by a reticular cell of hemopoietic tissue of *L. migratoria* 6 hr after injection into hemolymph. Note presence of characteristic basal material between cells of hemopoietic tissue. *B* = bacterium; *BM* = basal material. ×7,150. (From Hoffmann et al., 1974)

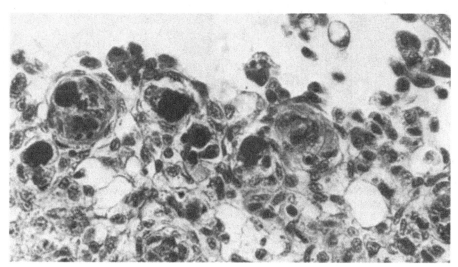

Fig. 2.31. Black nodules in hemopoietic tissue of *Locusta* after injection of *B. thuringiensis.* × 140.

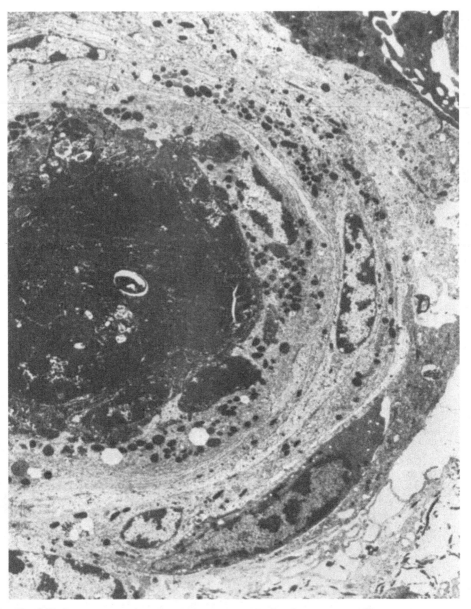

Fig. 2.32. Reactive capsule in hemopoietic tissue 24 hr after injection of live *B. thuringiensis.* Necrotic central zone is encapsulated by granulocytes. × 6,000. (From Hoffmann et al., 1974)

but we have some evidence that the reticular cells are activated by injection of an immunizing dose of pathogens to produce large amounts of a proteinaceous substance. Indeed, about 12 hr after such an injection in these cells, we observed an increase in the number and volume of mitochondria, as well as a strong development of the RER. The cisternae are at first short, but become rapidly elongated, and between 20 and 30 hr following the injection, they show a strong tendency for dilation. At that time, they contain a moderately dense material, with some zones of condensation. Eventually, this material becomes very dense and often appears as crystalloid inclusions inside the ergastoplasmic cisternae (Fig. 2.33).

Stimulation of protein synthesis in the reticular cells of the hemopoietic tissue by injection of an immunizing dose of a pathogen also induces the transitional cells between reticular elements and plasma cells in vertebrates during antigenic stimulation. The accumulation of proteinaceous material in crystalline or amorphous form is particularly common in the plasmocyte cell lineage during that process. These observations lead to the working hypothesis that differentiation of certain reticular cells into immunocompetent cells occurs in the case of an infection. This differentiation is, of course, inhibited in the case of selective irradiations of the hemopoietic tissue, as well as in the case of a massive stimulation of the macrophagic processes of all the reticular cells by a massive injection of iron saccharate prior to the infection. The possibility of a secretory differentiation of certain reticular cells in relation to the production of soluble antibacterial substances would make the hemopoietic tissue of some insects appear more similar to the hemopoietic tissues in vertebrates than was initially suspected.

2.7. Summary

Microscopic observations of the normal "phagocytic tissues" of the dorsal diaphragm in the two orthopterans, *Gryllus bimaculatus* and *Locusta migratoria*, unequivocally demonstrated the hemopoietic nature of these cell accumulations. In these species, the hemopoietic elements develop from a large number of so-called reticular cells of mesodermic origin, which resemble closely the reticular cells of the hemopoietic organs of vertebrates. As is the case in vertebrates, the differentiation of the hemopoietic elements into mature blood cells occurs in the two orthopteran species in isogenic cell islets. The phagocytic activity of the reticular cells explains why these organs were generally considered simple phagocytic organs.

The hemopoietic differentiation of the reticular cells can occur either in a poorly organized, loose tissue located along the dorsal vessel, as is the case in *Locusta*, or in a group of highly organized hemopoi-

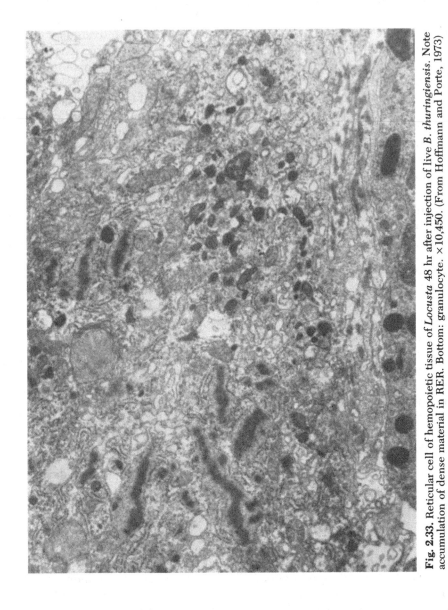

Fig. 2.33. Reticular cell of hemopoietic tissue of *Locusta* 48 hr after injection of live *B. thuringiensis*. Note accumulation of dense material in RER. Bottom: granulocyte. ×10,450. (From Hoffmann and Porte, 1973)

etic organs, as in *Gryllus;* the latter resemble far more the classic hemopoietic structures of vertebrates. We give a detailed description of both types of organs, especially of the subdivision of the hemopoietic organs of *Gryllus* into a cortex, where the hemocytes differentiate, and a medulla, where they can accumulate.

Hemopoietic tissues with the same basic organization are described in *Melolontha melolontha* and *Calliphora erythrocephala.* An interesting feature of *Calliphora* hemopoietic processes is the entrance into the bloodstream of isogenic hemocyte clusters at a very early stage of differentiation (an observation not made in orthopterans).

The normal functioning of the hemopoietic tissues and their reactions under abnormal and experimental conditions are both remarkably similar to those of the vertebrate hemopoietic organs. While their role in the continuous production of mature hemocytes is essential, the macrophagic capacity of these tissues enables them to play an important role also in eliminating wornout hemocytes and various debris and especially in contributing to defense reactions and to resistance to bacterial infection. Selective X-ray-induced lesions of this tissue affect molting and ovarian maturation. These different functional aspects are at present under investigation.

References

Brehélin, M. 1973. Présence d'un tissu hématopoïétique chez le Coléoptère *Melolontha melolontha. Experientia 29:*1539–40.

Brehélin, M., and J. A. Hoffmann. 1971. Effets de l'injection d'une poudre inerte (saccharate de fer) sur l'hémogramme et la coagulation de l'hémolymphe chez un Insecte Orthoptère *Locusta migratoria. C. R. Acad. Sci. D 272:*1409–12.

Cuénot, L. 1896. Etudes physiologiques sur les Orthoptères. *Arch. Biol. 14:*293–341.

Hoffmann, D., M. Brehélin, and J. A. Hoffmann. 1974. Modifications of the hemogram and of the hemocytopoietic tissue of male adults of *Locusta migratoria* (Orthoptera) after injection of *Bacillus thuringiensis. J. Invertebr. Pathol. 24:*238–47.

Hoffmann, D., and A. Porte. 1973. Sur la différentiation sécrétoire de cellules réticulaires de l'organe hématopoïétique de *Locusta migratoria* immunisée contre *Bacillus thuringiensis:* Etude au microscope électronique. *C. R. Acad. Sci. D 276:* 677–80.

Hoffmann, J. A. 1969. Etude de la récupération hémocytaire après hémorragies expérimentales chez l'Orthoptère *Locusta migratoria. J. Insect Physiol. 15:*1375–84.

Hoffmann, J. A. 1970. Les organes hématopoïétique de deux Insectes Orthoptères: *Locusta migratoria* et *Gryllus bimaculatus. Z. Zellforsch. Mikrosk. Anat. 106:*451–72.

Hoffmann, J. A. 1971. Obtention de larves permanentes par irradiation selective du tissu hématopoïétique de jeunes larves de stade V de *Locusta migratoria. C. R. Acad. Sci. D 273:*2568–71.

Hoffmann, J. A. 1972, Modifications of the haemogramme of larval and adult *Locusta migratoria* after selective X-irradiations of the haemocytopoietic tissue. *J. Insect Physiol. 18*(9):1639–52.

Hoffmann, J. A. 1973. Blood-forming tissues in orthopteran insects: An analogue to vertebrate hemopoietic organs. *Experientia 29:*50–1.

Hoffmann, J. A., and C. Lévi. 1965. Etude au microscope électronique du vaisseau dorsal de *Locusta migratoria. C. R. Acad. Sci. D 260*:6988–90.

Jones, J. C. 1970. Hemocytopoiesis in insects, pp. 7–65. *In* A. S. Gordon (ed.). *Regulation of Hematopoiesis.* Appleton, New York.

Kowalevsky, A. 1889. Ein Beitrag zur Kenntnis der Exkretionsorgane. *Biol. Zentralbl. 9*:33–47, 65–76, 127–8.

Whitcomb, R. F., M. Shapiro, and R. R. Granados. 1974. Insect defense mechanisms against microorganisms and parasitoids, pp. 447–536. *In* M. Rockstein (ed.). *The Physiology of Insecta*, Vol. 5, 2nd ed. Academic Press, New York.

Wigglesworth, V. B. 1955. The role of the haemocytes in the growth and moulting of an insect, *Rhodnius prolixus* (Hemiptera). *Exp. Biol. 32*:649–63.

Zachary, D., and J. A. Hoffmann. 1973. The haemocytes of *Calliphora erythrocephala* (Meig.) (Diptera). *Z. Zellforsch. Mikrosk. Anat. 141*:55–73.

3 Multiplication of hemocytes

D. FEIR

Biology Department, Saint Louis University, Saint Louis, Missouri, 63103, U.S.A.

Contents

3.1. Introduction

The postembryonic origin and multiplication of insect hemocytes continue to provide food for thought and material for research projects for many entomologists, comparative physiologists, and comparative pathologists. Review articles discussing insect hemocytes, total and differential cell counts, and hemopoiesis include Jones (1962, 1970), Wigglesworth (1959), Arnold (1974), and Crossley (1975). One should consult these articles for the historical development of the topic as well as for an extensive review of the literature.

Because the terms "hemopoiesis," "hematopoiesis," and "hemocytopoiesis" seemed to be used interchangeably in the literature, I decided to do a little checking before using one term or the other. According to Webster's *Third New International Dictionary* (1965) "hemopoiesis" and "hematopoiesis" are equivalent in meaning and in derivation and therefore can be used interchangeably. Both mean formation of blood or blood cells within the body. Webster does not list "hemocytopoiesis," but does list "hemocytogenesis" as being the part of hematopoiesis concerned with the formation of blood cells. I will use "hemopoiesis" because it is the shortest of the terms and includes the formation of blood cells in its definition (see also Chapter 10).

The process of hemopoiesis includes both cell proliferation and cell differentiation. As cell differentiation proceeds, the capacity for cell proliferation decreases. Mature cells are incapable of proliferation (Best and Taylor, 1966). Even though I will be principally concerned with cell proliferation and multiplication in this chapter, I will also cover some aspects of cell differentiation because it bears an important relationship to the amount of cell proliferation in the normal animal.

Although I have tried to emphasize articles not covered in the previously mentioned reviews, there is, of course, some overlap. I have tried to cover most of the pertinent articles published in the last few years.

3.2. Circulating hemocytes

3.2.1. Types of hemocytes

The classification of hemocytes still presents many problems in the literature. The hemocytes are highly pleomorphic, and the particular form they present at any one time seems to depend on the age, developmental stage, nutritional state, and species of insect as well as on the methods of collection and examination used by the investigator (Jones, 1962; Lai-Fook and Neuwirth, 1972).

Most authors seem able to agree on the definitions and nomenclature of the prohemocytes (PRs) and the plasmatocytes (PLs). Granulocytes (GRs) or granular hemocytes are also quite commonly and consistently used in the literature. Other hemocyte names that have been used by investigators include spherule cells (SPs) (Boiteau and Perron, 1976); oenocytoids (OEs) (Akai and Sato, 1973; Arnold and Sohi, 1974); adipohemocytes (ADs) (McLaughlin and Allen, 1965; Korecka, 1974); granulo-adipohemocytes (Murray, 1972); cystocytes (Price and Ratcliffe, 1974); podocytes (POs) (Raina and Bell, 1974); coagulocytes (COs) (Costin, 1975; Breugnon and Le Berre, 1976); lamellocytes (LAs) (Nappi, 1970); macronucleocytes, micronucleocytes, and oenocytoids (Misselunene, 1975); and the unique classification for insect cells by Gilliam and Shimanuki (1971), which included proleucocyte, neutrophil, eosinophil, basophil, normal leucocyte, pycnonucleocyte, and hyalinocyte. With the exception of the last group of terms, many investigators have used the other hemocyte terms for years (Jones, 1962) (see also Chapter 4). Gupta (1969) compared the various classifications of hemocytes in an attempt to coordinate and simplify usage. This long list of terms helps to emphasize the rather large number of types of cells that have been described and the difficulties inherent in trying to show relationships.

Is each type of cell derived from a different stem cell? This question represents the polyphyletic theory of the origin of hemocytes. If this theory is to be ultimately accepted, it is necessary to demonstrate mitosis in each of these types or mitosis in different types of cells, perhaps noncirculating cells, which ultimately differentiate to form the types described in circulation. The monophyletic theory of hemocyte origin says that there is one stem cell that differentiates into the other types under appropriate conditions. In this theory the vast majority of mitoses would be seen in the stem cell. Some of the other cell types could divide also, but the level of mitosis would be too low to account for the numbers of these types of cells present. In both these theories, it is necessary to demonstrate the steps in differentiation from one stage (or type) of hemocyte to the next stage and to investigate the conditions of differentiation (see also Chapters 4, 8).

3.2.2. Mitotic activity

Mitotic activity has been most consistently reported in the PR (Jones, 1962; Arnold and Sohi, 1974; Korecka, 1974; Breugnon and Le Berre, 1976). Jones and Liu (1968) found that just the small PL-like cells showed mitoses in unfixed smears of *Galleria mellonella*. In this same species of insect, Shapiro (1968) reported mitoses in PRs, PLs, and GRs. Arnold and Hinks (1976) reported that the PRs divided the most frequently, the GRs and SPs divided quite frequently, PLs rarely divided, and OEs were never found in mitosis.

There have only been a few reports on dividing hemocytes in culture conditions, but these types of studies promise to produce significant results in the future. In 1970, Kurtti and Brooks showed that mitoses were abundant in PRs, but not in other types of hemocytes in primary culture. Arnold and Sohi (1974) maintained hemocytes in vitro by serial passage up to 148 times. Mitoses were observed in PRs, PLs, and GRs. The ability to maintain hemocytes in vitro and to observe mitoses will certainly aid in answering some questions about types of dividing cells, conditions necessary for division, and differentiation of cells (Landureau and Grellet, 1975) (see also Chapter 9).

The levels of mitotic activity in circulating hemocytes have been reported to be less than 1% in almost all cases. Most investigators have not attempted to classify the cells undergoing mitosis. Tauber (1936) considered a count of 2,000 cells necessary in order to determine an accurate percentage of dividing cells in *Blatta orientalis*. Most of his mitotic counts were from 0.116% to 0.250% of the hemocytes. He thought that all types of hemocytes divided, but this is not the general consensus today. Lea and Gilbert (1966) determined the average mitotic index (MI) of *Hyalophora cercropia* to be from 0.38% to 0.76%. The highest MI they found was 1.79% in the fourth-instar larva. Feir and McClain (1968a) did mitotic counts at hourly intervals throughout the fifth stadium of *Oncopeltus fasciatus*. They found that the MI was very low immediately after ecdysis into the fifth instar. It began to rise at the twenty-third hour postecdysis, reached its peak (4.06%) in the 30-hr group, remained high until the seventy-fourth hour, and then declined during the remainder of the stadium. There was no consistent diurnal periodicity of mitotic activity.

Feir and O'Connor (1965) injected tritiated thymidine into different groups of *Oncopeltus* on each day of the fifth stage. At 24-hr intervals after the injection and for the remainder of the instar, they bled and sacrificed groups of insects. They made blood smears, radioautographed them, and counted the mitotic activity and the percentage of cells showing tritiated thymidine incorporation. DNA synthesis normally precedes mitosis and therefore tritiated thymidine incorporation would indicate the proportion of cells preparing for mitosis or having undergone mitosis within the time period available. Although the MIs were generally very low, one time period did show an MI of 15%, emphasizing the great variability of these determinations. Perhaps these MIs should be compared with MIs after an injury (injection of tritiated thymidine), which we will discuss later, rather than with MIs of the normal or unstimulated insect. The tritiated thymidine incorporation counts were zero fairly often; however, some groups of insects showed 66%, 78%, and 86% of their cells in a smear to have incorporated tritiated thymidine. Assuming that these are valid figures, these data indicate a high rate of turnover or a great increase of circulating cells at certain times of the instar. The lack of correlation between MI and tritiated thymidine incorporation was interpreted as indicating that much of the mitotic activity occurred in hemopoietic organs or cells temporarily sequestered from circulation.

If living insects are plunged into liquid nitrogen, they are killed very quickly, and the assumption was made by Feir and O'Connor (1969) that this method of killing would prevent any clumping or localized accumulations of circulating hemocytes in the insect. Fifth-instar milkweed bugs were fixed in this manner on each day of the instar and tissue sections made. Total hemocyte counts (THCs) and MIs of the circulating hemocytes were made on these tissue sections. The THCs were considerably different from the THCs obtained from blood collected from severed appendages and counted in hemocytometers. The MI was always below 0.7%, which was quite comparable with the MI obtained on smears of hemolymph. The authors concluded that there were no permanent hemopoietic sites in the milkweed bug.

There is very little information on the length of the cell cycle in insect hemocytes. Crossley (1975) cites several references indicating that epidermal cells of insects probably have a total cell cycle time of

24 hr. This is fairly comparable to results of mammalian studies. There is a definite need for more information on times of the various stages of mitosis in insect hemocytes.

3.2.3. Total hemocyte counts (THCs) and differential hemocyte counts (DHCs)

THCs have usually been determined on blood collected from a severed appendage or some other wound and really represent the number of hemocytes in a particular known volume of hemolymph. THCs done in this way have been used as indirect evidence for multiplication of hemocytes and changes in hemocyte populations in general under various experimental conditions. In some cases DHCs were also done to show the changes in proportions of hemocyte types in relation to THCs. Some examples of THCs reported in the literature are as follows: 15,000–23,000/mm^3 in sixth-larva armyworm (Wittig, 1966), 15,000–275,000 cells/mm^3 in the field cricket (Tauber and Yeager, 1934), 17,000–45,000/mm^3 in sixth-stage armyworm (Rosenberger and Jones, 1960), a range of 200–3600 cells/mm^3 with a mean of 1,336 cells/mm^3 in the large milkweed bug (Feir, 1964).

THCs vary with ecdysis and with developmental stage. In *Hyalophora cecropia*, Lea (1964) found that the counts were high in fourth-instar and later larval stages, that they decreased at pupation, and that they remained relatively low during diapause and adult development. In newly emerged adults the counts were high, but decreased in older adults. The PL group predominated in late fifth-instar, during adult development (pupa), and in the adult. GRs predominated in the fourth, early fifth, and diapausing pupa stages.

Patton and Flint (1959) found a reduction in the THC in *Periplaneta americana* at the time of ecdysis and a very rapid buildup of cells within 24 hr of ecdysis. Feir (1964) found an increase in THCs after ecdysis to the fourth, fifth, and adult stages of the large milkweed bug and a drop in the counts after 48 hr postecdysis. The bug *Halys dentata* (Bahadur and Pathak, 1971) showed an abrupt fall in THC after ecdysis, a rise during the middle part of the stadium, and a slight fall before the next ecdysis. The THC increased throughout the life cycle, with the adults showing the highest counts.

Gilliam and Shimanuki (1966) reported that "total hemocyte counts were essentially identical in worker and drone larvae and pupae of the same age" and that the counts were the same in honeybee larvae 5–8 days of age and in pupae 12 or 15 days old. In the adult bee, on the other hand, Kostecki (1965) concluded that the wide variations in the number of hemocytes in workers of various ages precluded the use of hemocyte determinations for disease diagnosis.

According to Clark and Chadbourne (1960), there are no statistical differences in the THCs of diapausing and nondiapausing larvae and the pupae of the pink bollworm.

As explained in a preceding paragraph, these studies on THCs represent total counts per unit, usually cubic millimeters, of hemolymph. Therefore any loss or gain in volume of water in the hemolymph would greatly affect the THC per cubic millimeter, but it might not have any effect on the absolute number of hemocytes in the hemocoel. There have been very few studies on THCs and simultaneous volume determinations. Wheeler (1963) measured the hemolymph volume and the number of hemocytes per cubic millimeter in *Periplaneta americana*. From these data he calculated the absolute total hemocyte count in the roach. The absolute number of circulating hemocytes in the entire roach did not increase prior to ecdysis, even though the number of hemocytes per cubic millimeter did increase because there was a decrease in hemolymph volume. At ecdysis the hemolymph volume increased, which gave the impression of a decrease in the THC per cubic millimeter, but the absolute number of hemocytes did not decline. The absolute number of hemocytes significantly decreased 24 hr after ecdysis, but there was no change in the THC per cubic millimeter because there was a decrease in hemolymph volume. This is a rather sobering set of data in view of the many studies on THC. However, because few studies have been as intensive as Wheeler's, sweeping conclusions cannot yet be drawn. Perhaps not all insects show so many volume changes. Nevertheless, the lack of volume determinations does raise many questions and might explain the wide variations in results reported in the preceding paragraphs (see also Chapter 16).

DHCs have also been used to give indirect evidence of hemopoiesis. Differential counts are not affected by changes in cell volume because they represent proportions of the cells counted. However, they can be affected by changes in cell adhesiveness or sequestering phenomena.

In a study of the hemocytes of an aphid, Boiteau and Perron (1976) found that the PRs were at low levels all through the larval and adults stages; the PLs were the highest of all groups until the end of the third instar, and then they declined somewhat; the GRs were low until the end of the third instar, and then they rose to be the highest in numbers of all types; the SPs and OEs remained at approximately the same levels throughout the study.

In *Rhodnius*, Jones (1967b) found that the PLs increased and the GRs decreased during the fasting period following each ecdysis. After the fourth and fifth instars took a blood meal, the PLs decreased and the GRs increased. In the adults, following feeding, the PLs increased and GRs decreased.

Diapause has been shown to alter hemocyte counts. In the pink bollworm there were significant reductions in all cell types during diapause compared with the nondiapause state. The ADs were three times higher in pharate pupae developed from diapausing larvae than in those developing from nondiapausing larvae (Raina and Bell, 1974).

3.2.4. Stimuli for hemopoiesis

Although THCs (in the sense of hemocytes per cubic millimeter), DHCs, and MIs have been used as indirect indications of hemopoietic activity, it is possible that the normally developing insect may not show enough mitotic activity to demonstrate its full hemopoietic potential. Hemopoiesis may not occur without a suitable stimulus.

Elbadry (1964) tried low doses of gamma radiation on the potato tuberworm, but found only a few morphological changes and no changes in the DHC. There were no drastic changes in the hemograms of larval mealworms even after a series of 12 hemorrhages over a period of 16 days (Jones, 1954). Ligating the thorax or abdomen of *Galleria mellonella* larvae caused a decrease in the THC posterior to the ligature (Jones and Liu, 1969). This experiment might be interpreted to indicate a need for the endocrines of the anterior portion for multiplication or release of hemocytes. Gilliam and Shimanuki (1970) showed an increase in THC with increasing elevation in honeybee larvae. The highest elevation used was 7,200 ft. Was the increase attributable to low atmospheric pressure? low O_2?

Using *G. mellonella*, Shapiro (1968) caused an increase in round PLs, GRs, PRs, and SPs and a decrease in ADs 24 hr after wounding. These differences from the normal counts disappeared 120 hr after wounding. He also demonstrated an increase in the THC, with the peak count at 72 hr after wounding. He accounted for the increase in the THC in part by mitosis of circulating hemocytes and in part by the release of hemocytes from tissue surfaces. He showed that the MI increased sixfold in the 24 hr after wounding. He concluded that the changes in THCs were attributable to changes in hemolymph volume. There is a large number of good references in this article.

Feir and McClain (1968b) measured MIs in the fifth-instar milkweed bug at 24-hr intervals after various treatments for the duration of the fifth instar. They found that bleeding, injury, and the injection of adrenaline increased mitotic activity during the 48- to 72-hour period of the fifth instar. Cobalt radiation blocked mitosis at metaphase and therefore increased the MI. The carcinogenic agents 20-methylcholanthrene and 9,10-dimethyl-1,2-benzanthracene did not affect the MI but did cause morphological changes (see also Chapter 16).

3.2.5. Hemocyte transformations

We have talked about changing proportions of hemocyte types and the interest in determining the stem cell or cells. There have been several interesting reports on hemocyte transformations that are the beginning of solving these types of problems. Whitten (1964) did not have much evidence, but did suggest that the "spherules" in their later stages may represent the ADs of some authors. Using unfixed and unstained hemocytes in hanging drops, Nappi (1970) suggested that PLs change into POs and that POs change into LAs. Price and Ratcliffe (1974) studied hemocytes from 15 orders of insects and concluded that the numerous intermediates of hemocyte types may represent different developmental and/or functional stages of one basic cell type. Beaulaton and Monpeyssin (1976) did an ultrastructural and cytochemical study and concluded that PRs are stem cells and after several intermediate stages change into PLs. SPs have been reported to differentiate from PLs (Breugnon and Le Berre, 1976), and young PLs have been said to develop gradually from SPs (Gupta and Sutherland, 1967). Using telobiotic pairs, Rizki (1962) observed PLs transforming into LAs and POs. Gupta and Sutherland (1966) reported in vitro transformations of a PL into a GR either by fragmentation of portions of the cytoplasm or by a gradual rounding off of the fusiform PL, the differentiation of the SP and the cystocyte (= CO) from a PL or from a GR, the changing of a GR into an AD, the gradual transformation of fusiform PLs into POs and vermiform cells (VEs), and the formation of OEs from fusiform PLs by cytoplasmic fragmentation. These authors suggested that the PL, not the PR, is the basic hemocyte type (see also Chapter 4).

Arnold and Hinks (1976) reported that the PRs of a noctuid represent 6% of the circulating hemocytes and that the PRs divide frequently enough to produce up to 1,000 cells/day. They concluded that the PRs differentiate exclusively into PLs (about 32.5% of the hemocytes), which rarely divide themselves but which increases in number throughout larval life. The PLs continued to increase in number even after the fourth instar, when the mitotic index of the PRs declined, and this suggested to the authors that the PLs had a secondary source. The GRs and the SPs were not involved in transformations. The OEs did not divide, but there was no indication of which cell might be their source (see also Chapter 8).

3.2.6. Amitosis

Amitosis is briefly discussed in Arnold (1974) and Jones (1970). Gilliam and Shimanuki (1971) cite Kostecki as saying that PRs divide by mitosis in young honeybees, but the older cells divide by amitosis as the PRs disappear with age. In general, discussions of amitosis and its possible occurrence in insects are in the older studies and have not been reported in recent years. This might be a result of changes in techniques as well as of increased understanding of cell divisions.

3.2.7. Fragmentation

Arnold (1972) (cited in Arnold, 1974) described an asymmetric or heteromorphic division of hemocytes that probably involved nuclear fragmentation. Gupta and Sutherland (1966) reported that OEs formed from fusiform PLs by cytoplasmic fragmentation.

A supposedly new type of hemocyte, the thrombocytoid, was described by Zachary and Hoffmann (1973). They described this hemocyte as disintegrating to form fragments and "naked nuclei." The fragments agglutinated to form a meshwork. The authors felt that this cell type resembled the vertebrate megakaryocyte.

There have been so few reports on fragmentation processes that it is impossible to draw any generalized conclusions. Perhaps if more investigators keep in mind the possibility of this process in insect hemolymph, they will be more alert in their observations on hemocytes, particularly in culture situations.

3.3. Hemopoietic organs

The rather low MI of circulating hemocytes in most insects has led some investigators to suggest the presence of special hemopoietic tissue (Taylor, 1935; Hrdý, 1959; Arnold and Hinks, 1976). Crossley (1964) gives a good historical summary of hemopoietic studies, and Jones (1970) gives an exceptionally thorough discussion of hemopoietic tissues and organs in the various orders of insects. Hemopoietic organs are undoubtedly the same as the leucopoietic organs of Arvy (1952) and the phagocytic organs of Cuénot (1896) (see also Chapters 2, 8).

There has been a fair amount of discussion in the literature concerning whether groups of hemocytes seen in tissue sections are simply aggregates or clumps of cells resulting from the method of fixation or whether they are specific and definite hemopoietic organs (Shapiro, 1968).

O'Connor and Feir (1968) suggested several criteria to be applied in determining whether groups of hemocytes indicated hemopoietic sites or were simply accidental accumulations. The criteria included the demonstration of higher levels of DNA synthesis and/or mitotic activity in the groups as compared with circulating cells, a more consistent number of cells in the groups at the same times in the life cycle as compared with random accumulations, a definite location within the insect body regardless of the method of fixation, and the presence of a membrane or some clear organ morphology.

Jones (1970) listed seven criteria for a hemopoietic tissue: (1) histological discovery of a compact tissue of hemocytes, (2) the occurrence and (3) the quantitation of mitotic divisions in the tissue and/or (4) the presence of various stages of differentiation of various types of cells within such tissue, (5) specific correlations between the numbers of cells being produced in and (6) being released from such organs, and (7) statistically valid increases in the number of circulating cells. Admittedly, the whole set of criteria is very difficult to satisfy. However, since the criteria have been published, the descriptions of new hemopoietic tissues have been more complete and include at least some of the criteria.

Several investigators, using morphological criteria, have described hemopoietic organs in insects, some of which were not in Jones's review (1970). Klein and Coppel (1969) identified dense clusters of hemocytes enclosed by a thin cellular sheath in the head, thorax, and abdomen of larvae from the second instar through prepupal stages in the introduced pine sawfly. There were numerous mitotic figures in these clusters, although no comparisons with circulating hemocyte mitotic activity were given. The most complex grouping of these hemopoietic clusters was found ventrally. From one to seven clusters of cells were found in a ventral fold of tissue at the base of the legs in all three thoracic segments and all but the ninth abdominal segment. Lateral organs were just ventral and posterior to the spiracles or corresponding areas in the thorax and abdomen. Clusters were also found in the head and dorsally along the entire length of each larva. The largest organ measured was 174 × 100 μm. Over 7,000 cells were counted in one organ. The authors have some good pictures to corroborate their statements.

A very detailed description of a hemopoietic mass in *Gryllus* was given by Hoffmann et al. (1969). The hemopoietic mass had a membrane or capsule, a peripheral zone with a compact cellular arrangement, and a central area in which the circulating hemocytes differentiated. The authors described a reticular network similar to that seen in vertebrate hemopoietic organs. The reticular cells were seen to undergo many mitoses after hemorrhage, and all types of hemocytes were said to differentiate from the reticular cells. *Locusta migratoria* showed a less highly developed hemopoietic tissue. There were just irregular accumulations of cells on top of the dorsal diaphragm. There was no limiting membrane and no organization into cortex and medulla. Instead of a reticular meshwork, there was a complex network covering the whole hemopoietic tissue in a loose manner without consitituting a true capsule. The reticular cells looked just like those of *Gryllus*, according to the authors, and they functioned in destroying old hemocytes as well as transforming into hemopoietic elements that differentiated into isogenic islets of various hemocyte lines. The mature hemocytes were described as entering the circulating hemolymph by going through the basement membrane network. At 24 hours after hemorrhage, there was a decrease in cells in the hemopoietic sites and an increase in undifferentiated cells in the circulating hemolymph. The authors concluded that these undifferentiated cells were set free from the hemopoietic tissue (for more details see Chapter 2).

Akai and Sato (1971) have some very good pictures of the hemopoietic organs of *Bombyx mori*. These organs are located near the imaginal wing disks in the thorax. The size of the organs increased during the early stages of each instar, and the number of organs increased during each apolysis to ecdysis. The volume of all mature hemopoietic organs decreased during apolysis to ecdysis, and the authors concluded that this showed that the hemocytes were released from the hemopoietic organ into the hemocoel. Openings in the acellular sheath surrounding the hemopoietic organ allowed the hemocytes to enter the circulating blood. The hemopoietic organs measured a maximum of 60×80 μm, and all the organs disappeared early in the pupal stage.

Phase and electron microscopy as well as histochemical techniques were used to describe the hemopoietic organ in *Thermobia domestica* (François, 1975). The organ is similar to that described for Orthoptera, Coleoptera, and Diptera. It is an irregular accumulation of cells lying along the dorsal blood vessel between the pericardial cells and the dorsal diaphragm. It was said to consist of star-shaped reticular cells, fibroblasts, and cellular islets formed by groups of hemocytes of the same type and same stage of differentiation. The hemocytes originated from the reticular cells and then differentiated within the islets. A very interesting point is that the hemopoietic organ remained in the adult and did not degenerate, as reported in the adults of other insects. The author suggested that this was because Thysanura continue to molt as long as they live.

The preceding identifications of hemopoietic sites have been principally on the basis of morphological evidence. Additional evidence for hemopoietic sites and activity in insects comes from responses of these sites to various experimental conditions. Selective irradiation of the hemopoietic tissue of larval and adult locusts caused the THC of circulating hemocytes to fall by 50% from the initial figure. It took 5–6 days for return to the initial level of THC (Hoffmann, 1972). Crossley (1964) showed a marked increase in division rate in hemopoietic areas 18 hr after hemorrhage. Treatment with X-rays caused a 50% increase in hemocyte number from day 6 to day 8 in *Calliphora erythrocephala* when the normal increase in cell count was 20% (Hoffmann 1973; Zachary and Hoffmann, 1973). These same authors ligated the posterior part of the abdomen between days 6 and 7 of the larval stage and found a rapid fall in the hemocyte count in the anterior and an increase in the posterior part comparable to that found in the unligated controls. The hemopoietic sites were described from the posterior region of the *Calliphora* larva (see also Chapter 2).

Mitsuhashi (1972) has cultured the hemopoietic tissue of *Papilio xuthus*. Unfortunately, the cells that migrated from the hemopoietic tissue were rather short-lived. The successful culturing of hemopoietic tissue would be a valuable tool for many studies. Hemocytes could be labeled with a radioactive isotope, for example; injected into the same species as the hemopoietic organ was taken from; and much needed information on turnover times might be obtained.

The evidence for definite hemopoietic tissue, if not discrete organs, in insects is certainly increasing.

3.4. Role of endocrines in hemocyte multiplication

It has been reported frequently that the THC varies with the developmental stage and physiological condition of the insect. Synchronized changes in hemocyte MI and population size are circumstantial indications of humoral controls (Crossley, 1975). Crossley gives quite a complete discussion of the endocrine regulation of hemopoiesis in insects, and I could not find any articles on the subject more recent than the ones he covered. I will cite just a few examples to emphasize the fact that there probably is endocrine regulation but that we do not have a great deal of direct evidence for it.

Jones (1967a) repeatedly bled *Rhodnius* and concluded that the amount of hemolymph available from an appendage, as well as the types and numbers of hemocytes, were regulated, at least in part, by the insect's hormones.

Hoffmann (1970) presented some good evidence for endocrine regulation of hemopoiesis. He concluded from his studies that the corpora allata stimulated the production and differentiation of hemocytes in adult locusts and that the prothoracic glands stimulated the production and differentiation of the hemocytes in the last nymphal instar. Cardiacectomy slowed the rate of differentiation of the COs and the GRs. If the locusts are not fed, the production of these two types of cells stops almost completely.

Ligating the thorax or abdomen of *Galleria mellonella* caused a decrease in the THC posterior to the ligature (Jones and Liu, 1969). As I stated earlier, this might be interpreted to indicate a need for the endocrines of the anterior portion for multiplication or release of hemocytes. Not all experiments are consistent with this interpretation.

3.5. Summary and conclusions

According to various authors, insect hemocytes multiply only in the hemolymph or in part in the hemolymph and in part in hemopoietic tissues or sites. In some cases the evidence for these conclusions is very good. The control of the multiplication of hemocytes seems just beginning to receive some attention, and there are indications that it may, at least in part, be hormonal. The story of insect hemocyte multiplication is obviously far from complete.

Several problems need to be investigated in this area:

1. The length of the various stages of mitosis and the total length of the mitotic cycle are important to know to fully interpret the data on mitotic indexes.
2. The turnover time of each cell or each cell type needs to be known; it might not be a firm time but might vary with applied stimuli or stress situations and the control of multiplication itself. In vitro techniques will certainly aid in these studies, but sooner or later that information has to be related back to the in vivo situation.
3. In order to have significance, THCs in an animal with an open circulatory system need to be related to hemolymph volume; in cases where hemolymph volume has been shown to be directly proportional to live weight, this will not be so time-consuming to do.

Most of these problems are technically rather difficult and time-consuming. But perhaps we should not feel too badly about the amount of information still needed to complete the insect hemocyte multiplication story. After all, it is still not known for certain how long granulocytes and agranulocytes live in the mammalian circulation, and a great many investigators have been trying to answer this question for a very long time.

References

Akai, H., and S. Sato. 1971. An ultrastructure study of the haemopoietic organs of the silkworm, *Bombyx mori. J. Insect Physiol.* 17:1665–76.

Akai, H., and S. Sato. 1973. Ultrastructure of the larval hemocytes of the silkworm, *Bombyx mori* L. (Lepidoptera: Bombycidae). *Int. J. Insect Morphol. Embryol.* 2(3):207–31.

Arnold, J. W. 1974. The hemocytes of insects, pp. 201–54. *In* M. Rockstein (ed.). *The Physiology of Insecta,* Vol. 5, 2nd ed. Academic Press, New York.

Arnold, J. W., and C. F. Hinks. 1976. Haemopoiesis in Lepidoptera. I. The multiplication of circulating haemocytes. *Can. J. Zool.* 54(6):1003–12.

Arnold, J. W., and S. S. Sohi. 1974. Hemocytes of *Malacosoma disstria* Huebner (Lepidoptera: Lasiocampidae): Morphology of the cells in fresh blood and after cultivation *in vitro. Can. J. Zool.* 52(4):481–5.

Arvy, L. 1952. Particularités histologiques des centres leucopoïetiques thoraciques chez quelques Lepidoptères. *C. R. Acad. Sci. Paris* 235:1539–41.

Bahadur, J., and J. P. N. Pathak. 1971. Changes in the total haemocyte counts of the bug, *Halys dentata*, under certain specific conditions. *J. Insect Physiol. 17*(2):329–34.

Beaulaton, J., and M. Monpeyssin. 1976. Ultrastructure and cytochemistry of the hemocytes of *Antheraea pernyi* Guer. (Lepidoptera, Attacidae) during the 5th larval stage. I. Prohemocytes, plasmatocytes, and granulocytes. *J. Ultrastruct. Res. 55*(2):143–56. (In French, English summary.)

Best, C. H., and N. B. Taylor. 1966. *The Physiological Basis of Medical Practice*, 8th ed. Williams & Wilkins, Baltimore.

Boiteau, G., and J. M. Perron. 1976. Study of hemocytes of *Macrosiphum euphorbiae* (Thomas) (Homoptera Aphididae). *Can. J. Zool. 54*:228–34. (In French, English summary.)

Breugnon, M. and J. R. Le Berre. 1976. Fluctuation of the haemocyte formula and haemolymph volume in the caterpillar *Pieris brassicae* L. *Ann. Zool. Ecol. Anim. 8*(1):1–12. (In French, English summary.)

Clark, E., and D. S. Chadbourne. 1960. Haemocytes of nondiapause and diapause larvae and pupae of the pink bollworm. *Ann. Entomol. Soc. Amer. 53*:682–5.

Costin, N. M. 1975. Histochemical observations of the haemocytes of *Locusta migratoria*. *Histochem. J. 7*(1):21–43.

Crossley, A. C. 1964. An experimental analysis of the origin and physiology of haemocytes in the blue-fly *Calliphora erythrocephala*. *J. Exp. Zool. 157*:375–98.

Crossley, A. C. 1975. The cytophysiology of insect blood. *Adv. Insect Physiol. 11*:117–221.

Cuénot, L. 1896. Etudes physiologiques sur les Orthoptères. *Arch. Biol. 14*:293–341.

Elbadry, E. 1964. The effect of gamma irradiation on the hemocyte counts of larvae of the potato tuberworm *Gnorimoschema operculella* (Zeller). *J. Insect Pathol. 6*:327–30.

Feir, D. 1964. Haemocyte counts on the large milkweed bug, *Oncopeltus fasciatus. Nature (Lond.) 202*:1136–7.

Feir, D., and E. McClain. 1968a. Mitotic activity of the circulating hemocytes of the large milkweed bug, *Oncopeltus fasciatus. Ann. Entomol. Soc. Amer. 61*(2):413–16.

Feir, D., and E. McClain. 1968b. Induced changes in the mitotic activity of hemocytes of the large milkweed bug, *Oncopeltus fasciatus. Ann. Entomol. Soc. Amer. 61*(2):416–21.

Feir, D., and G. M. O'Connor, Jr. 1965. Mitotic activity in the hemocytes of *Oncopeltus fasciatus* (Doll.). *Exp. Cell Res. 39*:637–42.

Feir, D., and G. M. O'Connor, Jr. 1969. Liquid nitrogen fixation: A new method for hemocyte counts and mitotic indices in tissue sections. *Ann. Entomol. Soc. Amer. 62*:246–7.

François, J. 1975. Hemocytes and the hematopoietic organ of *Thermobia domestica* (Packard) (Thysanura: Lepismatidae). *Int. J. Insect Morphol. Embryol. 4*(6):477–94. (In French.)

Gilliam, M., and H. Shimanuki. 1966. Total hemocyte counts in hemolymph of immature honey bees. *Amer. Bee J. 106*(10):376. (Abstract only.)

Gilliam, M., and H. Shimanuki. 1970. Total hemocyte counts of honey bee larvae (*Apis melifera*) from various elevations. *Experientia 26*:1006.

Gilliam, M., and H. Shimanuki. 1971. Blood cells of the worker honey bee. *J. Apic. Res. 10*(2):79–85.

Gupta, A. P. 1969. Studies of the blood of Meloidae. I. The hemocytes of *Epicauta cinerea* and a synonymy of haemocyte terminologies. *Cytologia 34*(2):300–44.

Gupta, A. P., and D. J. Sutherland. 1966. *In vitro* transformation of the insect plasmatocyte in some insects. *J. Insect Physiol. 12*:1369–75.

Gupta, A. P., and D. J. Sutherland. 1967. Phase contrast and histochemical studies of spherule cells in cockroaches (Dictyoptera). *Ann. Entomol. Soc. Amer. 60*(3):557–65.

Hoffmann, J. A. 1970. Endocrine regulation of the production and differentiation of hemocytes in an orthopteran insect: *Locusta migratoria migratoroides. Gen. Comp. Endocrinol. 15*(2):198–219. (In French.)

Hoffmann, J. A. 1972. Modifications of the haemogramme of larval and adult *Locusta migratoria* after selective X-irradiations of the haemocytopoietic tissue. *J. Insect Physiol. 18*(9):1639–52.

Hoffmann, J. A. 1973. Blood-forming tissues in orthopteran insects: An analogue to vertebrate hemopoietic organs. *Experientia 29*:50–1.

Hoffmann, J. A., A. Porte, and P. Joly. 1969. L'hématopoïèse chez les insectes Orthoptères. *C. R. Séances Soc. Biol. 163*:2701–3.

Hrdý, I. 1959. Development of the blood picture of the cricket *Acheta domesticus* L., pp. 106–10. *In* I. Hrdý (ed.). *Ontogeny of Insects*. Academic Press, New York.

Jones, J. C. 1954. A study of mealworm hemocytes with phase contrast microscopy. *Ann. Entomol. Soc. Amer. 47*(2):308–15.

Jones, J. C. 1962. Current concepts concerning insect hemocytes. *Amer. Zool. 2*:209–46.

Jones, J. C. 1967a. Effects of repeated haemolymph withdrawals and of ligaturing the head on differential haemocyte counts of *Rhodnius prolixus* Stal. *J. Insect Physiol. 13*(9):1351–60.

Jones, J. C. 1967b. Normal differential counts of haemocytes in relation to ecdysis and feeding in *Rhodnius. J. Insect Physiol. 13*(8):1133–41.

Jones, J. C. 1970. Hemocytopoiesis in insects, pp. 7–65. *In* A. S. Gordon (ed.). *Regulation of Hemopoiesis*, Vol. 1. Appleton, New York.

Jones, J. C., and D. P. Liu. 1968. A quantitative study of mitotic divisions of haemocytes of *Galleria mellonella* larvae. *J. Insect Physiol. 14*(8):1055–61.

Jones, J. C., and D. P. Liu. 1969. The effect of ligaturing *Galleria mellonella* larvae on total hemocyte counts and on mitotic indices among haemocytes. *J. Insect Physiol. 15*:1703–8.

Klein, M. G., and H. C. Coppel. 1969. Hemocytopoietic organs in larvae of the introduced pine sawfly, *Diprion similis. Ann. Entomol. Soc. Amer. 62*(6):1259–61.

Korecka, T. 1974. The haemolymph in *Quadraspidiotus ostreaeformis* (Curt.) (Homoptera, Cocoidea, Aspidiotini). *Acta Biol. Cracov. (Ser. Zool.) 17*(1):85–93.

Kostecki, R. 1965. Investigation on the hemocytes and hemolymph of honeybees. *J. Apic. Res. 4*(1):49–54.

Kurtti, T. J., and M. Brooks. 1970. Growth of lepidopteran epithelial cells and hemocytes in primary cultures. *J. Invertebr. Pathol. 15*:341–50.

Lai-Fook, J., and M. Neuwirth. 1972. Importance of methods of fixation in the study of insect blood cells. *Can. J. Zool. 50*:1011–13.

Landureau, J. C., and P. Grellet. 1975. New permanent cell lines from cockroach hemocytes: Physiological and ultrastructural characteristics. *J. Insect Physiol. 21*(1):137–52. (In French.)

Lea, M. S. 1964. A study of the hemocytes of the silkworm *Hyalophora cecropia*. Ph.D. dissertation, Northwestern University, Evanston, Illinois.

Lea, M. S., and L. I. Gilbert. 1966. The hemocytes of *Hyalophora cecropia* (Lepidoptera). *J. Morphol. 118*(2):197–216.

McLaughlin, R., and G. Allen. 1965. Description of the hemocytes and the coagulation process in the boll weevil, *Anthonomus grandis* Bohemian. *Biol. Bull. (Woods Hole) 128*(1):112–24.

Misselunene, I. 1975. Morphology of haemolymph cells in caterpillars of cabbage butterfly (*Pieris brassicae* L.). *Tsitologiia 17*(6):647–52.

Mitsuhashi, J. 1972. Primary culture of the haemocytopoietic tissue of *Papilio xuthus* Linne. *Appl. Entomol. Zool. 7*(1):39–41. (Abstract only.)

Murray, V. I. E. 1972. The haemocytes of *Hypoderma* (Diptera: Oestridae). *Proc. Ento-mol. Soc. Ont. 102:*46–63. (Abstract only.)

Nappi, A. J. 1970. Hemocytes of larvae of *Drosophila euronotus* (Diptera: Drosophili-dae). *Ann. Entomol. Soc. Amer. 63*(5):1217–24.

O'Connor, G. M., Jr., and D. Feir. 1968. Application of quantitative criteria for haemo-poietic activity to insect haemocytes. *J. Insect Physiol. 14*(12):1779–84.

Patton, R., and R. Flint. 1959. The variation in the blood cell count of *Periplaneta americana* (L.) during a molt. *Ann. Entomol. Soc. Amer. 52:*240–2.

Price, C. D., and N. A. Ratcliffe. 1974. A reappraisal of insect haemocyte classification by the examination of blood from fifteen insect orders. *Z. Zellforsch. Mikrosk. Anat. 147*(3):313–24.

Raina, A. K., and R. A. Bell. 1974. Haemocytes of the pink bollworm *Pectinophora gos-sypiella,* during larval development and diapause. *J. Insect Physiol. 20*(11):2171–80.

Rizki, T. M. 1962. Experimental analysis of hemocyte morphology in insects. *Amer. Zool. 2*(2):247–56.

Rosenberger, C., and J. C. Jones. 1960. Studies on total blood cell counts of the southern armyworm larva, *Prodenia ridania* (Lepidoptera). *Ann. Entomol. Soc. Amer. 53:*351–5.

Shapiro, M. 1968. Changes in the haemocyte population of the wax moth, *Galleria mel-lonella,* during wound healing. *J. Insect Physiol. 14*(12):1725–33.

Tauber, O. E. 1936. Mitosis of circulating cells in the hemolymph of the roach, *Blatta orientalis. Iowa State Coll. J. Sci. 10:*373–81.

Tauber, O. E., and J. F. Yeager. 1934. On the total blood (hemolymph) cell count of the field cricket, *Gryllus assimilis pennsylvanicus* Burn. *Iowa State Coll. J. Sci. 9*(1):13–24.

Taylor, A. 1935. Experimentally induced changes in the cell complex of the blood of *Periplaneta americana. Ann. Entomol. Soc. Amer. 28:*135–45.

Webster's Third New International Dictionary of the English Language, Unabridged. 1965. Merriam, Springfield, Massachusetts.

Wheeler, R. 1963. Studies on the total haemocyte count and haemolymph volume in *Periplaneta americana* (L.) with special reference to the last moulting cyle. *J. In-sect Physiol. 9:*223–35.

Whitten, J. M. 1964. Haemocytes and the metamorphosing tissues in *Sarcophaga bul-lata, Drosophila melanogaster,* and other cyclorrhaphous Diptera. *J. Insect Physiol. 10:*447–69.

Wigglesworth, V. B. 1959. Insect blood cells. *Annu. Rev. Entomol. 4:*1–16.

Wittig, G. 1966. Phagocytosis by blood cells in healthy and diseased caterpillars. II. A consideration of the method of making hemocyte counts. *J. Invertebr. Pathol. 8*(4):461–77.

Zachary, D., and J. A. Hoffmann. 1973. The haemocytes of *Calliphora erythrocephala* (Meig.) (Diptera). *Z. Zellforsch. Mikrosk. Anat. 141*(1):55–73.

II. Forms and structure

4   Hemocyte types: their structures, synonymies, interrelationships, and taxonomic significance

A. P. GUPTA

Department of Entomology and Economic Zoology, Rutgers University, New Brunswick, New Jersey 08903, U.S.A.

Contents

4.1. Introduction

Hemocytes of arthropods and several other invertebrate groups have been studied (see Gupta, 1979). Among arthropods, they have been most extensively studied in insects, followed by crustaceans, arachnids, and myriapods. Hemocytes of a few onychophorans also have been described. It is not surprising, therefore, that the need for a reliable, uniform classification of various hemocyte types has been felt more keenly by insect hematologists than by those of other arthropod groups. Fortunately, a generally acceptable hemocyte classification in insects, based largely on morphological characteristics, now exists.

Hemocyte classifications both in insects and other arthropods have been variously based on morphology, functions, and staining or histochemical reactions of hemocytes. Thus, it is not unusual to find the same hemocyte type or its various forms being referred to by different names in various arthropods, by different authors – a situation that has inevitably resulted in a confusing mass of terminology. Consequently, it becomes very difficult to compare hemocytes of one species with those of others. This has particularly hindered any phylogenetic consideration of the evolution of hemocyte types in various arthropod groups and the Onychophora. Clearly, there is a need for a uniform hemocyte classification for insects as well as other arthropod groups. The insect hemocyte classification that is generally used has evolved over more than half a century. According to Millara (1947), Cuénot (1896) was the first to classify insect hemocytes into four categories and was later followed in this attempt by Hollande (1909, 1911) and others. Wigglesworth (1939) summarized most of the earlier classifications and presented a classification that was widely accepted. He modified it later (Wigglesworth, 1959). On the American side, Yeager's (1945) work stimulated considerable interest in the study of hemocytes. Jones (1962) revised and greatly improved Yeager's classification.

In order to adopt a uniform hemocyte classification for discussing hemocytes and their physiological significance in various insects, it is necessary to homologize terminologies used by different authors on the bases of descriptions, observed functions, line drawings, and mi-

Table 4.1. Summary of hemocyte types in various taxa (orders), based on published and unpublished information and personal observation

Taxa	Prohemocyte (PR)	Plasmatocyte (PL)	Granulocyte (GR)	Spherulocyte (SP)	Adipohemocyte (AD)	Coagulocyte (CO)	Oenocytoid (OE)
Collembola	—	—	GR	—	—	—	—
Thysanura	—	PL	GR	SP	—	CO	—
Ephemeroptera	PR	PL	GR	—	—	—	OE
Odonata	—	PL	GR	—	—	—	—
Orthoptera (= Cursoria)	PR	PL	GR	SP^a	—	—	—
Dermaptera	PR	PL	GR	SP	—	CO	—
Blattaria	PR	PL	GR	SP	AD	CO	OE
Mantodea	PR	PL	GR	SP	—	—	—
Plecoptera	PR	PL	GR	SP	—	CO	—
Hemiptera	PR	PL	GR	—	AD	—	OE
Hymenoptera	PR	PL	GR	—	AD	—	OE
Coleoptera	PR	PL	GR	SP	AD	CO	OE
Megaloptera	PR	PL^a	GR	—	—	CO	OE
Neuroptera	PR	PL	GR^a	SP	AD	—	OE
Trichoptera	PR	PL	GR	—	—	—	—
Lepidoptera	PR	PL	GR	SP	AD	CO	OE
Diptera	PR	PL	GR	SP	AD	CO	OE

^{a} These terms were not used by the original authors, but have been adopted by me after scrutiny of the original micrographs and figures.

crographs of hemocytes studied by those authors. A summary of the seven main hemocyte types in various insect orders is presented in Table 4.1. The three terms indicated by the table's footnote were not used by the original authors, but have been adopted by me after scrutinizing original descriptions and figures. Hemocytes categorized as amoebocytes and/or phagocytes by the original authors have been assigned mostly to the category of plasmatocyte (PL), although they could be included in granulocyte (GR), spherulocyte (SP), and/or adipohemocyte (AD), inasmuch as these last three forms also are supposedly phagocytic in certain insects.

4.2. Main hemocyte types

There is disagreement among insect hematologists about the number of hemocyte types in various insects. From one or a few to as many as nine or more types have been described, particularly by light microscopy. Ultrastructurally, however, only seven types have so far been identified in various insects: prohemocyte (PR), plasmatocyte (PL), granulocyte (GR), spherulocyte (SP), adipohemocyte (AD), oenocytoid (OE), and coagulocyte (CO). Of these seven, AD has been reported only by Devauchelle (1971) and CO by Goffinet and Grégoire (1975) and Ratcliffe and Price (1974). Podocyte (PO) and vermicyte (VE) have not been recognized as distinct types in electron microscopic studies so far, primarily because ultrastructurally they appear similar to PLs (Devauchelle, 1971). A general description of various hemocyte types, based on both light and electron microscopic studies, their synonymies, and interrelationships is presented below. I must emphasize that although I am including PO and VE in the following description, I do not regard them as distinct types. Furthermore, I believe that description of "new" hemocyte types, based on superficial dissimilarities, should be avoided (see also Chapters 8, 10).

4.2.1. *Prohemocyte* (PR)

Structure

PRs are small round, oval, or elliptical cells with variable sizes (6–10 μm wide and 6–14 μm long). The plasma membrane is generally smooth (Fig. 4.1A), but may show vesiculation (Fig. 4.3A). The nucleus is large, centrally located, and almost filling the cell; nuclear size is variable (3.6–12 μm) in various insects; several nuclei and nucleoli may be present. A thin or dense, homogeneous and intensely basophilic layer of cytoplasm surrounds the nucleus, the nucleocytoplasmic ratio being 0.5–1.9 or more. The cytoplasm may contain granules, droplets, or vacuoles (Fig. 4.3A).

The laminar nature of the plasma and nuclear membranes may not

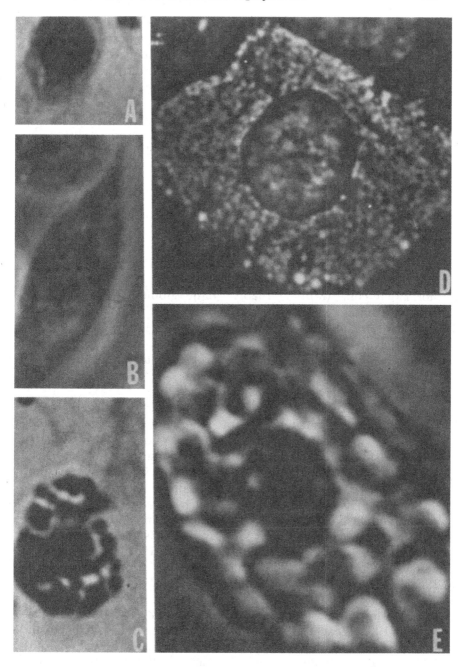

Fig. 4.1. A. Prohemocyte of *Periplaneta americana*. Ca. × 8,000. B. Plasmatocyte of *P. americana*. Ca. × 10,000. C. Spherulocyte of *Nauphoeta cinerea*. Ca. × 8,500. D. Granulocyte of *Locusta migratoria*. E. Spherulocyte of *N. cinerea*. Ca. × 25,000. (C and E from Gupta and Sutherland, 1967; D from Costin, 1975)

be evident. The cytoplasm generally contains a low concentration of endoplasmic reticulum (ER), mitochondria, and Golgi bodies. However, free ribosomes, rough endoplasmic reticulum (RER), and even mitochondria may be numerous. Centrioles – indicating the mitotic nature of PRs – and microtubules have been observed in the cytoplasm.

PRs are generally found in groups and appear indistinguishable from young or small PLs. They may be numerous, rare, or absent, depending on the developmental and physiological state of the insect at the time of observation. PRs are seldom seen in vivo.

Synonymies

The term that has survived to date with little or no change since its adoption by Hollande (1911) is "proleucocyte." Yeager (1945) used the term "proleucocytoid" and Jones and Tauber (1954) "prohemocytoid." I believe Arnold (1952) was the first to use the term "prohaemocyte." Other synonyms for PR are "macronucleocyte" (Paillot, 1919); "formative cell" (Müller, 1925); "jeune globule" (Bruntz, 1908); "smooth-contour chromophilic cell" (Yeager, 1945); "jeune leucocyte" (Millara, 1947); "plasmatocytelike cell" (Jones, 1959); "young plasmatocyte" (Gupta and Sutherland, 1966; Gupta, 1969); "young granulocyte" (François, 1974); and "proleucocyte" (many authors).

Interrelationship with other types

The controversial questions often raised regarding PRs are: (1) are they the stem cells that transform into other hemocytes? and (2) if they are, are they the main postembryonic source of hemocytes? Although there are substantiating reports that PRs do transform into at least a few other hemocyte types, evidence on their being the main postembryonic source is inconclusive. The term "prohemocyte" suggests that these cells give rise to other types, but it has not yet been demonstrated conclusively that all hemocyte types are derived from PRs. The most generally accepted view is that PRs transform into PLs (Yeager, 1945; Arnold, 1952, 1970, 1974; Jones, 1954, 1956, 1959; Shrivastava and Richards, 1965; Mitsuhashi, 1966; Wille and Vecchi, 1966; Beaulaton, 1968; Devauchelle, 1971; Lai-Fook, 1973; Beaulaton and Monpeyssin, 1976, 1977). Several authors have suggested that PRs transform into other types as well (Muttkowski, 1924; Bogojavlensky, 1932; Yeager, 1945; Arvy and Lhoste, 1946; Ashhurst and Richards, 1964). Arnold (1952) stated that "haemocytes, with the possible exception of the Oenocytoids, apparently develop originally from a common source, the prohaemocytes," but has now changed his mind. Wille and Vecchi (1966), however, suggested that PR can give rise to OE. Arnold (1970), in *Diploptera punctata*, stated that PRs are likely stem cells for PLs, GRs, and SPs, but the direction of differentia-

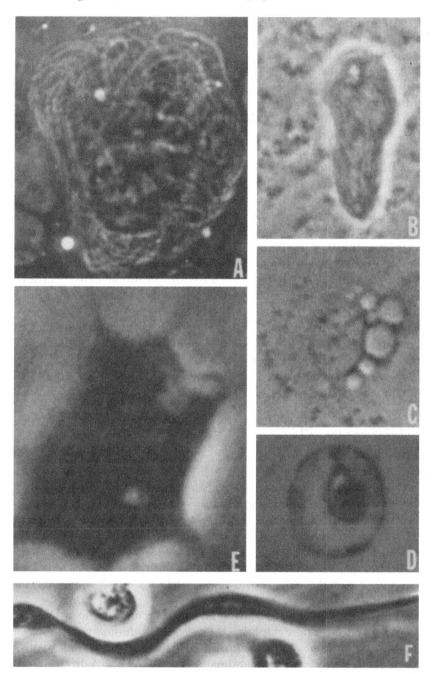

Fig. 4.2. A. Oenocytoid of *L. migratoria*, showing cytoplasmic filaments. **B.** Oenocytoid of *P. americana*. Ca. × 6,800. **C.** Adipohemocyte of *P. americana* nymph. Ca. × 7,225. **D.** Coagulocyte of *Epicauta cinerea*. Ca. × 6,800. **E.** Podocyte of *P. americana* nymph. Ca. × 7,225. **F.** Vermicyte of *P. americana*. Ca. × 1,190. Costin, 1975; **C** from Gupta and Sutherland, 1966; **D**–**F** from Gupta, 1969)

tion is determined early, perhaps in the hemopoietic tissue. Yeager (1945) and Jones (1959) reported that PRs can give rise to SPs and ADs. Devauchelle (1971) reported that PRs, PLs, GRs, and ADs are derived from each other. François (1974) found that PRs transform into GRs. Recently, Sohi (1971) indicated by subculturing that PRs are the germinal cells from which other categories develop, while Arnold and Sohi (1974) indicated two cell lines in subculture. Some authors (Gupta and Sutherland, 1966; Hoffmann et al., 1968; Akai, 1969; Gupta, 1969; Zachary and Hoffmann, 1973) did not recognize the existence of PRs (see also Chapters 2, 8).

As I stated earlier, the evidence on whether PRs constitute the main postembryonic source of hemocytes is inconclusive. There is growing evidence that PRs reside in the hemopoietic organs (Hoffmann et al., 1968; Arnold, 1970; Akai and Sato, 1973; Zachary and Hoffmann, 1973; François, 1974; Hinks and Arnold, 1977) and differentiate into other hemocyte types. Hoffmann (1967), Arnold (1974), and Beaulaton and Monpeyssin (1976) stated that PRs are germinal cells. Earlier (1970), Arnold stated that PRs appear in the hemolymph only intermittently and often in groups, suggestive of their release from hemopoietic tissue. Wille and Vecchi (1966) reported that PRs are abundant in newly emerged bees, but rare in old ones. Gupta and Sutherland (1968) reported an increase in PLs, GRs, SPs, and COs (= CYs) in *Periplaneta americana* following treatment with sublethal doses of chlordane.

4.2.2. *Plasmatocyte* (PL)

Structure

PLs are small to large, polymorphic cells with variable sizes (3.3–5 μm wide and 3.3–40 μm long). The plasma membrane may have micropapillae, filopodia, or other irregular processes, as well as pinocytotic or vesicular invaginations (Figs. 4.1B; 4.3B,C). The nucleus may be round or elongate and is generally centrally located. It may be lobate (Fig. 4.3C), vary in size (3–9 μm wide and 4–10 μm long) in various insects, and appear punctate. Scattered chromatin masses may be present along with the nucleolus (Fig. 4.3C). Occasionally, binucleate PLs may be found.

The laminar nature of the plasma and nuclear membranes may or may not be visible. The cytoplasm is generally abundant and may be granular or agranular; it is basophilic and rich in organelles. Generally, there is well-developed and extensive RER (Fig. 4.4B), which may form greatly distended cisternae or a vacuolar system. Golgi bodies (= dictyosomes = golgiosomes or internal reticular apparatus) (Fig. 4.4A) and lysosomes (membrane-bounded, electron-dense bodies, 0.1–1.30 μm in size) may be numerous; lysosomes can be

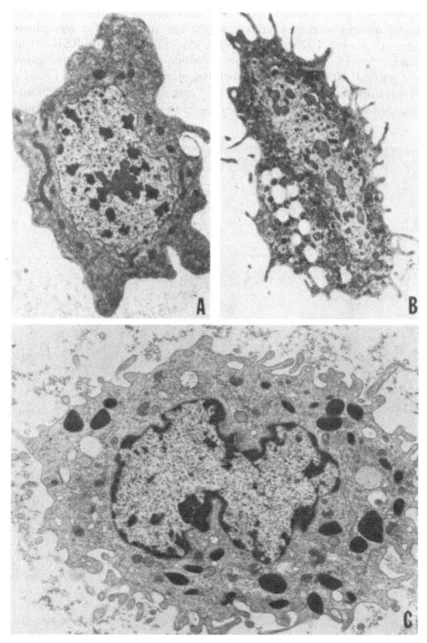

Fig. 4.3. A. Prohemocyte of *Pectinophora gossypiella.* Ca. × 8,760. **B.** Plasmatocyte of *P. gossypiella,* showing micropapillae. Ca. × 10,950. **C.** Plasmatocyte of *Carausius morosus,* showing micropapillae and lobate nucleus. × 6,570. (**A** and **B** from Raina, 1976; **C** from Goffinet and Grégoire, 1975)

identified by the presence in them of the reaction products of the hydrolytic marker enzymes, acid phosphatase and thiamine pyrophosphatase (Scharrer, 1972), and are often associated with the RER or the vacuolar system. The Golgi bodies produce the electron-dense granules (generally 0.5 μm in diameter) that one observes in the PLs. Microsomes and cisternae of the ER (= "ergastoplasme" of French authors) may be present. Free ribosomes (polysomes or polyribosomes)

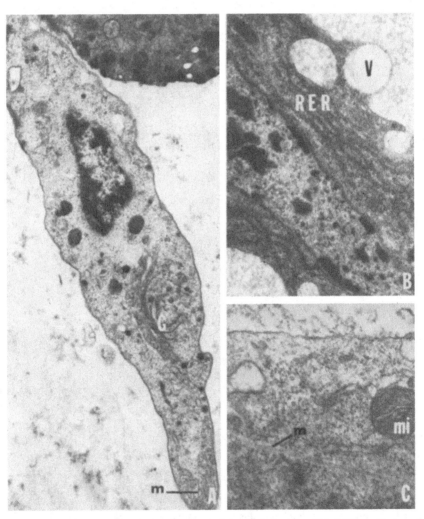

Fig. 4.4. A. Plasmatocyte of *Melolontha melolontha*, showing Golgi (*G*) and intracytoplasmic microtubules (*m*). × 12,600. **B.** Portion of plasmatocyte of *P. gossypiella*, showing rough endoplasmic reticulum (*RER*) and vacuoles (*V*). Ca. × 21,350. **C.** Portion of plasmatocyte of *Ephestia kühniella*, showing mitochondrion (*mi*) and intracytoplasmic microtubule (*m*). Ca. × 44,100. (**A** reinterpreted from Devauchelle, 1971; **B** from Raina, 1976; **C** reinterpreted from Grimstone et al., 1967)

or those attached to microsomes or RER may be present; intracytoplasmic microtubules are present, sometimes arranged in bundles
(Fig. 4.4A,C).

PLs are generally abundant and in some insects may be indistinguishable from PRs and GRs. Several types (most often the transitional forms) of PLs have been described on the bases of their sizes
and shapes.

Synonymies
Yeager and Munson (1941) first introduced the term "plasmatocyte." Some of the commonly used synonyms of PL are "leucocyte"
(Kollman, 1908; Metalnikov, 1908); "micronucleocyte" (Paillot, 1919);
"phagocyte," "amoebocyte," and "lymphocyte" (many authors), "podocyte" (Devauchelle, 1971); and "vermiform cell" (Lea and Gilbert,
1966; Devauchelle, 1971). PLs also include the "lamellocyte" of some
authors and the "nematocyte" of Rizki (1953).

Interrelationship with other types
The first real problem one encounters with PLs is that of distinguishing them, particularly the so-called young PLs, from the PRs.
This situation is further complicated by the presence of many transitional forms between these two types. The distinction between PRs
and PLs is generally based on the relative cell and nuclear sizes, intensity of cytoplasmic basophilia, and the extent and development of
the intracellular organelles.

The question that is often raised regarding PLs is whether they are
the primary cells that give rise to other forms by secondary transformation. Taylor (1935) claimed that amoebocytes (= mostly PLs), and not
chromophils (= PRs), are the basic types. Gupta and Sutherland
(1966) and Gupta (1969) supported Taylor's claim and considered PRs
as young PLs. Direct transformation of PLs into GRs (Yeager, 1945;
Jones, 1956; Gupta and Sutherland 1966; Hoffmann, 1967; Devauchelle, 1971; Beaulaton and Monpeyssin, 1977), SPs (Devauchelle,
1971; Breugnon and Le Berre, 1976; Beaulaton and Monpeyssin,
1977), ADs (Yeager, 1945; Shrivastava and Richards, 1965; Gupta and
Sutherland, 1966; Raina, 1976), COs (Gupta and Sutherland, 1966;
Devauchelle, 1971), OEs (Gupta and Sutherland, 1966; Hoffmann,
1967; Beaulaton and Monpeyssin, 1977), VEs (Tuzet and Manier,
1959; Gupta and Sutherland, 1966; Lea and Gilbert, 1966; Devauchelle, 1971; François, 1974, 1975), and POs (Gupta and Sutherland,
1966; Nappi, 1970; Devauchelle, 1971; François, 1974, 1975) has been
reported, but not substantiated. Devauchelle (1971) considered VEs
and POs ultrastructurally similar to PLs. That it is the PL which trans-
forms into other types is indicated also by the corresponding decrease
of PLs and increase of other types in differential hemocyte counts. For

example, in *Prodenia,* when PLs fall in number, spheroidocytes
(= ADs) increase (Yeager, 1945); in *Drosophila melanogaster,* when
POs increase, PLs decrease (Rizki, 1962); and in *P. americana,* within 4
hr of antennal hemorrhage, GRs increase, while PLs decrease (pers.
observ.).

It has also been suggested (Gupta and Sutherland, 1966; Moran,
1971; Scharrer, 1972; Price and Ratcliffe, 1974; Beaulaton and Monpeyssin, 1976) that insects have only one basic type of hemocyte and
that the commonly recognized types of hemocytes are merely different physiological manifestations of the same type, depending on the
physiological needs of the insect at different times. Although the PL
has been regarded as the primary type in insects, a survey of the hemocyte types in other arthropod groups reveals that the GR, not the
PL, is the basic type (Gupta, 1979) (see also Chapter 8).

4.2.3. Granulocyte (GR)

Structure

GRs are small to large, spherical or oval cells (Figs. 4.1*D;* 4.5*A,B;*
4.6*A,B*) with variable sizes (10–45 μm long and 4–32 μm wide, rarely
larger). The plasma membrane may or may not have micropapillae, filopodia, or other irregular processes. The nucleus may be relatively
small (compared with that in the PL), round or elongate, and is generally centrally located. Nuclear size is variable (2–8 μm long and 2–7
μm wide).

The laminar nature of the plasma and nuclear membranes may not
be visible. The cytoplasm is characteristically granular (Figs. 4.1*D;*
4.5*A,B;* 4.6*A*). Several types of membrane-bounded granules have
been described in the GRs of various insects (Figs. 4.6*A,B;* 4.7*A–C;*
4.8*A,B*). Recently, Goffinet and Grégoire (1975) summarized and synonymized various types of granules into three categories, based on
their observations in *Carausius morosus.* The following summary and
synonymy of granules are based on these and other works:

1. Structureless, electron-dense granules: = unstructured inclusions (type
   1) of Baerwald and Boush (1970); melanosome-like granules of Hagopian
   (1971); opaque body of Moran (1971); type 2 bodies of Scharrer (1972);
   and electron-dense granules of Raina (1976) and others
2. Structureless, thinly granular bodies: = type 3 of Scharrer (1972); and
   electron-lucent granules of Raina (1976)
3. Structured granules: = "globules" or "granules multibullaires" of Beaulaton (1968); "grains denses structures" in the AD of Devauchelle (1971);
   and Landureau and Grellet (1975); "corpus fibrillaires" of Hoffmann et
   al. (1968, 1970); cylinder inclusions (type 2), regular-packed inclusions
   (type 3), and inclusions with bandlike units (type 4) of Baerwald and
   Boush (1970); "Granula mit tubularer Binnenstruktur" of Stang-Voss

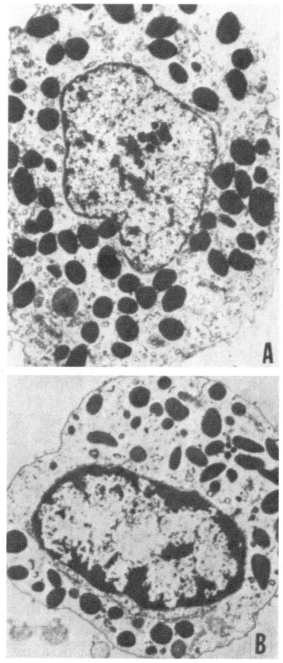

Fig. 4.5. A. Granulocyte of *M. melolontha. N* = nucleus. × 6,300. **B.** Granulocyte of *Thermobia domestica.* × 9,400. (**A** courtesy of Dr. G. Devauchelle; **B** from François, 1975)

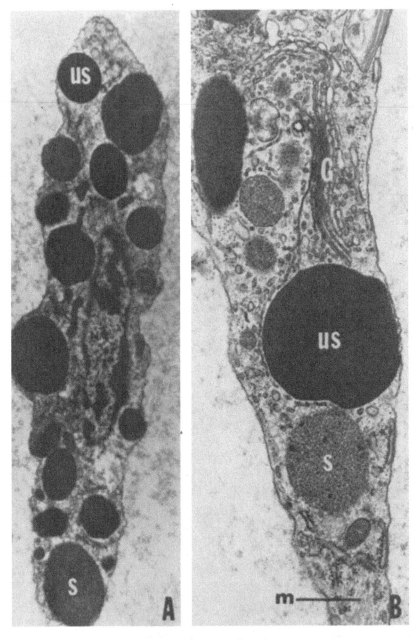

Fig. 4.6. A. Granulocyte of *P. americana,* showing structured (*s*) and unstructured (*us*) granules. Ca. × 16,000. **B.** Portion of granulocyte of *Leucophaea maderae,* showing derivation of structured (= premelanosome-like) granule (*s*) from Golgi (*G*), structureless or unstructured (= melanin-like) granule (*us*), and intracytoplasmic microtubules (*m*). × 3,000. (**A** reinterpreted from Baerwald and Boush, 1970; **B** reinterpreted from Hagopian, 1971)

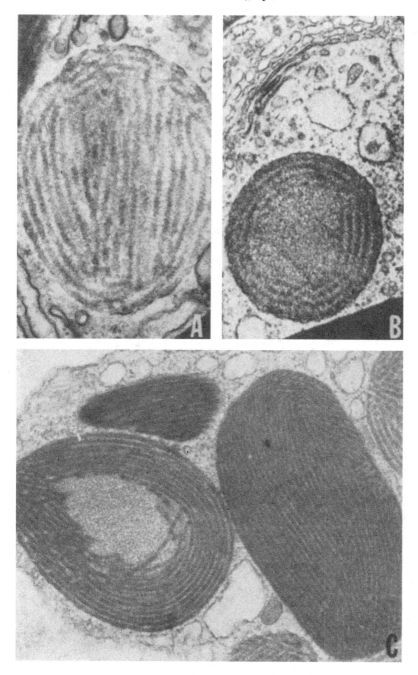

Fig. 4.7. **A.** Structured granule from granulocyte of *L. maderae,* showing internal microtubules. × 40,000. **B.** An earlier stage of development of internal microtubules. × 50,000. **C.** Section of a structured granule showing concentric arrangement of internal microtubules. × 38,000. (**A–C** from Hagopian, 1971)

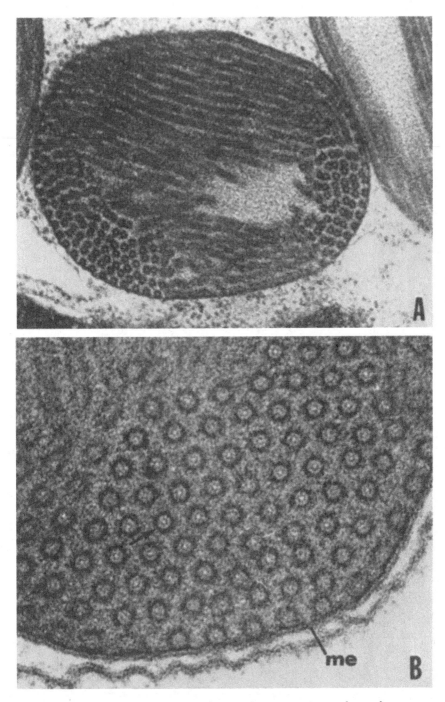

Fig. 4.8. A. Section of structured granule of a granulocyte of *L. maderae*, showing arrangement of microtubules about 25 nm in diameter. × 63,000. **B.** Highly magnified view of microtubules of structured granules. Note micro-microtubules (5 nm in diameter, arrow) within microtubules and limiting membrane (*me*) of granule. × 240,000. (**A** and **B** from Hagopian, 1971)

(1970); premelanosome-like granules of Hagopian (1971); tubule-containing bodies or TCB of Moran (1971); type 1 of Scharrer (1972); and granules with a microtubular structure of Ratcliffe and Price (1974).

The length or the diameter of the structureless granules varies from 0.15 to 3 μm or more in various insects, while that of the structured granules varies from 0.5 to 2 μm. The shape of the granules may be spherical, ovoid, elongate, or irregularly polygonal (Figs. 4.5A,B; 4.6A,B; 4.7B,C; 4.8A). The diameter of the microtubules within the structured granules varies from 15 to 80 nm) in various insects. Internally, the microtubules may show micro-microtubules about 5 nm in diameter (Hagopian, 1971) (Fig. 4.8B). Akai and Sato (1973) also have described "subunits of fibrils" in their so-called secretory vesicles. The number of microtubules per granule may vary from 9 to 80. From the accounts provided by Hagopian (1971), Scharrer (1972), Akai and Sato (1973), and François (1975), it appears that the granules are derived from the Golgi bodies (Fig. 4.6B), the microtubules developing during the later stages of morphogenesis. It is conceivable that the structureless, electron-dense granules represent the final stage of development of these granules in which the structured nature becomes obliterated. Supposedly, the granules are eventually released into the hemolymph. Histochemical analysis shows that most of the granules contain sulfated, periodate-reactive sialomucin and other glycoproteins or neutral mucopolysaccharides (François, 1974, 1975; Costin, 1975). Occasionally, lipid droplets may be present, especially in older GRs.

In addition to the structureless and structured granules, the cytoplasm is rich in free ribosomes (polysomes), Golgi bodies, both ER and RER, and lysosomes. Mitochondria are generally few in number. Marginal bundles of intracytoplasmic microtubules are also present (Fig. 4.6B).

Synonymies

Jones (1846) first established the category of granular cells, and later Cuénot (1896) mentioned "amoebocytes" with finely granular cytoplasm. Shrivastava and Richards (1965) and Lea and Gilbert (1966) treated GRs as ADs. The so-called "pycnoleucocytes" of Wille and Vecchi (1966) are probably GRs. Recently, Devauchelle (1971) synonymized GRs with cystocytes (= COs), and François (1975) did with SPs. GRs have been referred to as "phagocytes," "amoebocytes," and "hyaline cells."

Interrelationship with other types

GRs have been widely misinterpreted and confused with PLs, SPs, ADs, and COs (= cystocytes). François (1975) considered the SPs, described by Gupta and Sutherland (1966) and Price and Ratcliffe

(1974), as GRs. Goffinet and Grégoire (1975) reported separate categories of GRs and COs in *Carausius morosus*. As a matter of fact, the separate existence of the GR (not to be confused with PL, SP, AD, and CO) is now recognized by most authors, although Devauchelle (1971) has included both GR and CO in his type III.

How are GRs formed? Are they derived from PRs or PLs? Both sources of origin have been suggested (Arnold, 1974). Gupta and Sutherland (1966) indicated that the derivation of GR from PL is a short step. Takada and Kitano (1971) reported that GRs showed a trend to increase and PLs to decrease with time. Are GRs capable of transforming into other types of hemocytes? There are reports that indicate that GRs do indeed give rise to SPs, ADs, and COs (Gupta and Sutherland, 1966). Arnold (1974) has suggested that GRs "might be considered basic units from which more precisely structured and functioning classes of cells have developed." This is supported by my survey of hemocyte types in Arthropoda (Gupta, 1979). Hinks and Arnold (1977), however, have suggested separate origins of GRs and SPs.

The presence of microtubules in the granules and in the cytoplasm of the GR also has caused debate. According to Crossley (1975), the microtubules of the granules do not have the "dimensions of typical cytoplasmic microtubules (24–27 nm diameter), nor have been demonstrated to be sensitive to colchicine or vinblastine . . . and therefore they should not be called microtubules." According to him, only in *Leucophaea* are the dimensions of the inclusion tubules (25 nm) comparable to those of the intracytoplasmic microtubules.

The intracytoplasmic microtubules have been described in several insects (Grimstone et al., 1967; Baerwald and Boush, 1970, 1971; Devauchelle, 1971; Hagopian, 1971; Scharrer, 1972; to mention a few investigators) and are generally narrower in diameter than the microtubules of the granules. These intracytoplasmic microtubules may be arranged into marginal bundles (Hagopian, 1971) or may be randomly distributed in the cytoplasm (Devauchelle, 1971) and supposedly are found in all hemocyte types, except OEs (Devauchelle, 1971), although Raina (1976) has described them in OEs.

4.2.4. Spherulocyte (SP)

Structure

SPs are ovoid or round cells (Figs. 4.1C,E; 4.9A,B) with variable sizes (9–25 μm long and 5–10 μm wide) and usually larger than GRs. The plasma membrane may or may not have micropapillae, filopodia, or other irregular processes. The nucleus is generally small (5–9 μm long and 2.5–6 μm wide), central or eccentric, rich in chromatin bodies, and generally obscured by the membrane-bounded, electron-dense, intracytoplasmic spherules that are characteristic of these cells.

The number of the spherules may vary from few to many, and the diameter from 1.5 to 5 µm. The spherules contain granular, fine-textured, filamentous, or flocculent material (Raina, 1976). The granules within the spherules may vary from 15 to 17 nm in diameter (Akai and Sato, 1973). In addition to the spherules, the cytoplasm contains polyribosomes (Fig. 4.10C), Golgi bodies (moderately to well developed) (Fig. 4.10A), membrane-bounded vacuoles (= lysosomes) (Fig. 4.9C), numerous randomly distributed microtubules, elongated mito-

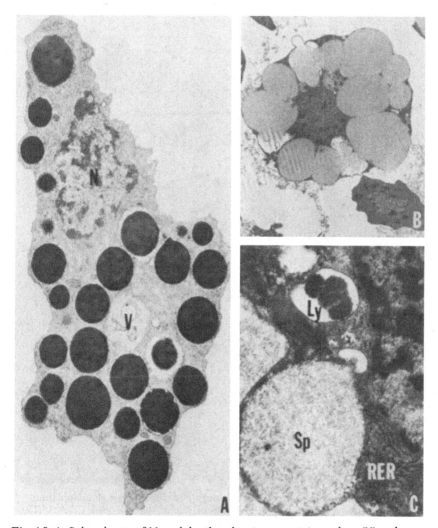

Fig. 4.9. A. Spherulocyte of *M. melolontha*, showing eccentric nucleus (*N*) and nu-numerous spherules and a vacuole (*V*). × 9,450. **B.** Spherulocyte of *Bombyx mori*. Ca. × 7,000. **C.** Portion of spherulocyte of *P. gossypiella*, showing rough endoplasmic reticulum (*RER*), spherule (*Sp*) with granular contents, and lysosome (*Ly*). Ca. × 21,000. (**A** from Devauchelle, 1971; **B** from Akai and Sato, 1973; **C** from Raina, 1976)

chondria, and RER (Figs. 4.9*C*, 4.10*A,C*). Devauchelle (1971) has also described a more or less loose network of fibrils in the cytoplasm (Fig. 4.10*B*). SPs release the material in their spherules into the hemolymph by exocytosis.

Histochemically, the spherules have been reported to contain neu-

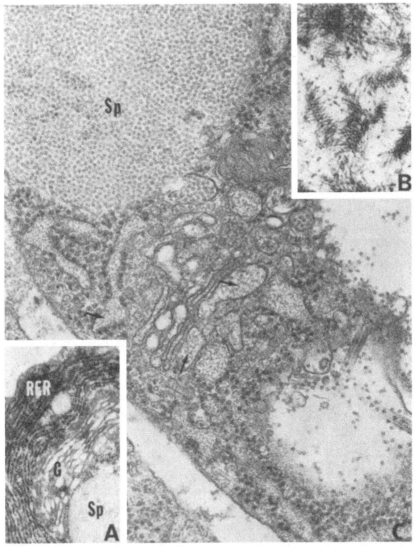

Fig. 4.10. A. Portion of spherulocyte of *P. gossypiella*, showing rough endoplasmic reticulum (*RER*) and Golgi (*G*) involved in formation of spherule (*Sp*). Ca. × 15,750. **B.** Portion of spherulocyte of *M. melolontha*, showing loose network of intracytoplasmic fibrils. **C.** Portion of spherulocyte of *B. mori*, showing spherule (*Sp*) with fine granules, rough endoplasmic reticulum containing fibrous material in its cisternae (arrows), and ribosomes. × 50,400. (**A** from Raina, 1976; **B** from Devauchelle, 1971; **C** from Akai and Sato, 1973)

tral or acid mucopolysaccharide and glycomucoproteins by several authors (Vercauteren and Aerts, 1958; Nittono, 1960; Ashhurst and Richards, 1964; Gupta and Sutherland, 1967; Costin, 1975; Beaulaton and Monpeyssin, 1977). Much earlier, Hollande (1909) reported that the spherules contain "lipochrome" (a kind of carotenoid lipid). The presence of tyrosinase has been reported by Dennell (1947), Jones (1956), and Rizki and Rizki (1959). Most recently, Costin (1975) reported the presence of nonsulfated sialomucin, in addition to glycoproteins and neutral mucopolysacchrides.

Synonymies

Hollande (1909) was the first to use the term "spherule cells," which is now generally used by most workers. Other terms that have been used by various authors are "cellules sphéruleuses" or "cellules à sphérules" (Paillot, 1919; Paillot and Noel, 1928); "spherocytes" (Bogojavlensky, 1932); "eruptive cells" (Yeager, 1945); "oenocytoids" (Dennell, 1947); "rhegmatocytes" (?)(Hrdý, 1957); and "hyaline cells" (Whitten, 1964, her Fig. 1R). Harpaz et al. (1969) classified SPs as ADs.

Interrelationship with other types

The main controversies about SPs concern the transformation of these cells into other types, formation of the spherules, and the functions of these cells. The transformation of SPs into other types is not well documented. Gupta and Sutherland (1966) suggested that SPs are capable of transforming into ADs and COs (= cystocytes) and that SPs are themselves derived from GRs. Millara (1947) and Arnold and Salkeld (1967) also considered the SP as a phase in the life of a GR. Later, Arnold (1974) stated that "they seem to be another specialized cell within the granular hemocyte complex." Beaulaton (1968) has suggested that SPs are degenerated OEs. Hinks and Arnold (1977) consider SPs as separate types with mitotic capabilities.

Little information is available on the formation of the spherules. According to Akai and Sato (1973), the material in the spherules is first observed in enlarged cisternae of the RER, then transferred into the Golgi complex, where it is packaged into the membrane-bounded spherules.

The role of SPs is highly controversial. Hollande (1909) considered these cells respiratory in function because of the presence of the so-called lipochrome. It has been demonstrated by Åkesson (1945), Ashhurst and Richards (1964), Arnold and Salkeld (1967), Gupta and Sutherland (1967), and Costin (1975) that the material contained in the spherules is neutral or acid mucopolysaccharide, not a carotenoid lipid. Hollande (1909) stated also that these cells contained an oxidase. Dennell (1947), Jones (1959), and Rizki and Rizki (1959) reported tyrosinase in the spherules of various Diptera, but Gupta and Sutherland (1967) found no tyrosinase in the SPs of cockroaches. Gupta and Sutherland (1965) were supposedly the first to report SPs in cockroaches. Whitten (1964) suggested that SPs (= her hyaline cells) may play a role in the darkening of the puparium in some cyclorrhaphous Diptera. Pérez (1910) reported that SPs took part in histolysis. Although this has been disputed by Åkesson (1945), the histolytic role of SPs should not be surprising, considering the fact that before and after molting several SPs are observed to congregate on histolyzing tissue. Gupta (1970) has suggested the probable histolytic or phagocytic functions of SPs. It is probable that SPs both histolyze and phagocytize tissues in at least a few insects. The phagocytic function was reported by Kollman (1908), Cameron (1934), and Åkesson (1945), but further work is needed to demonstrate clearly the histolytic role of SPs. Raina (1976) found no evidence of their role in phagocytosis. Metalnikov and Chorine (1929) and Metalnikov (1934) found that the SPs in *Galleria* are related to bacterial immunity. Nittono (1960) stated that strains of silkworm larvae that lacked SPs completely or incompletely tended to produce relatively smaller quantities of silk. Wigglesworth (1959) suggested that SPs are involved in the uptake and transport of other substances, such as hormones. Akai and Sato (1973) suggested that SPs are sources of some blood proteins.

4.2.5. *Adipohemocyte* (AD)

Structure

ADs are small to large, spherical or oval cells (Fig. 4.2C) with variable sizes (7–45 μm in diameter). The plasma membrane may or may not have micropapillae, filopodia, or other irregular processes. The nucleus is relatively small (compared with that in a PL or SP), round or slightly elongate, and centrally or eccentrically located. Nuclear size is variable (4–10 μm in diameter). The nucleus may appear concave, biconvex, punctate, or lobate.

The laminar nature of the plasma and nuclear membranes may not be visible. The cytoplasm contains characteristic small to very large refringent fat droplets (0.5–15 μm in diameter) and other nonlipid granules (0.5–9 μm in diameter) and vacuoles, which, according to Arnold (1974), become filled with lipids under certain conditions. In addition, the cytoplasm contains well-developed Golgi bodies, mitochondria, and polyribosomes.

Histochemically, ADs are reported to contain PAS-positive substance in the granules (Ashhurst and Richards, 1964; Lea and Gilbert, 1966). Costin (1975) did not recognize ADs as a type in her study.

Synonymies

Hollande (1911) first introduced the term "adipoleucocyte," although Kollman (1908) had earlier used the term "adipo-spherule cell" for some hemocytes of invertebrates. Other terms used for ADs are "spheroidocytes" (Yeager, 1945; Arnold, 1952; Rizki, 1953; Jones, 1959); "later stages of spherules" (Whitten, 1964); and "adipocytes" (Wigglesworth, 1965) (see also Chapter 8).

Interrelationship with other types

The main controversy about ADs concerns their identity as a distinct category of hemocytes. Scrutiny of the literature leads one to believe that they are not a distinct type. Several authors have reported that it is difficult to distinguish them from GRs (Jones, 1970; Arnold, 1974), and many others did not recognize the category of ADs in their studies (Wittig, 1968; Akai and Sato, 1973; Costin, 1975; François, 1975; Goffinet and Grégoire, 1975; Boiteau and Perron, 1976; Raina, 1976; Beaulaton and Monpeyssin, 1977; to mention a few recent ones). Raina (1976) noted a progressive accumulation of lipid drops in GRs and on that basis considered ADs as functional stages of GRs. Gupta and Sutherland (1966) also have reported the transformation of GRs into ADs. Only one of the ultrastructural studies (Devauchelle, 1971) of hemocytes includes ADs as separate category. However, his micrographs (his Figs. 20, 21, 24) are strikingly similar to the GRs (cf. Fig. 4.11 with Fig. 4.6B) described by Hagopian (1971) and other authors.

On the basis of the histochemical nature of these cells also it is difficult to justify the term, and hence the category, of ADs. According to Crossley (1975), "ultrastructural studies of so-called adipohaemocytes include cells which contain no reported lipid (Devauchelle, 1971, Fig. 18–22), lipid of doubtful authenticity (Pipa and Woolever, 1965) or material believed to be mucoprotein or mucopolysaccharide (Beaulaton, 1968, Fig. 7)." Costin (1975) also did not recognize the AD on the basis of the histochemistry of hemocytes in *Locusta migratoria*.

The resemblance of ADs to fat body cells is also confusing and adds to the difficulty of identifying ADs in fresh hemolymph samples. For example, one may find all gradations between small ADs and fat body cells (Wigglesworth, 1955). Jones (1965) suggested that hemocytes "with excentric nucleus and many brilliant fatlike droplets should be termed ADs only when they can be clearly distinguished from fat body cells." According to him (Jones, 1975), ADs are at least 10 times smaller than fat body cells.

Gupta and Sutherland (1966) suggested that under certain conditions, such as chilling, starvation, and diapause (periods of nonfeeding and reduced metabolic rate), the PLs respond by changing into ADs with "lipid" droplets. This was based on their observation that meal-worm larvae, when chilled for 20–24 hr at 5 °C, showed numerous "lipid" droplets and ADs, other types of hemocytes being rare. Such larvae subsequently recovered. Ludwig and Wugmeister (1953) noted an increased amount of free fats in the hemolymph of the starving Japanese beetle, *Popillia japonica*, and Clark and Chadbourne (1960) reported a greater number of ADs (= their spheroidocytes) in diapausing larvae of the pink bollworm, *Pectinophora gossypiella*. Thus, it

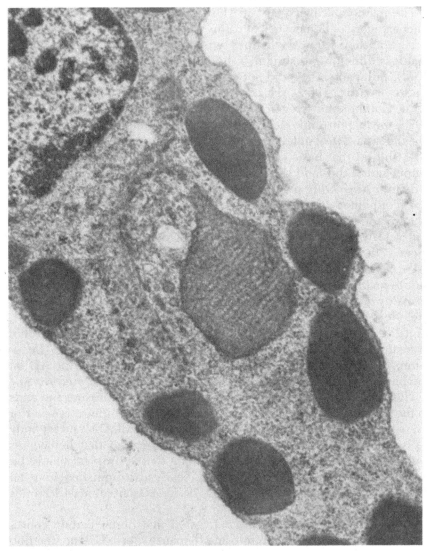

Fig. 4.11. Supposed adipohemocyte of *M. melolontha*. But note that it looks like a granulocyte. × 41,500. (From Devauchelle, 1971)

seems very likely that the appearance (or transformation from PLs) of ADs in the hemolymph at certain times is governed by the physiological state of the insect.

4.2.6. *Oenocytoid* (OE)

Structure

OEs are small to large, thick, oval, spherical, or elongate cells (Figs. 4.2*A,B;* 4.12*A,B*) with widely variable sizes (16–54 μm or more) and shapes. The plasma membrane is generally without micropapillae, filopodia, or other irregular processes. The nucleus is generally small, round or elongate, and generally eccentrically located (Figs. 4.2*B,* 4.12*A*). Nuclear size may vary (3–15 μm). Occasionally, two nuclei may be present.

The laminar nature of the plasma and nuclear membranes may not be visible. The cytoplasm is generally thick and homogeneous and has several kinds of plate-, rod-, or needlelike inclusions. According to Costin (1975), the OE is distinguished by an elaborate system of filaments that fills the cytoplasm (Figs. 4.2*A;* 4.12*C,D*) and is visible under a phase-contrast microscope. Histochemically, the filaments resemble the cytoplasm and hence are not visible in stained preparations. Hoffmann (1966) and Hoffmann et al. (1968) also have reported such filaments. In addition to the above intracytoplasmic inclusions and filaments, a few electron-dense spherules may be present in the cell periphery (Devauchelle, 1971). With the exception of polyribosomes and the abundant, large mitochondria, which are conspicuous, other organelles, such as ER and Golgi, are poorly developed. Supposedly, lysosomes are absent.

Histochemically, OEs are reported to contain tyrosinase (Dennell, 1947), protein (Akai and Sato, 1973), and PAS-positive-only granules, indicating the presence of glycoproteins or neutral mucopolysaccharides and sulfated, periodate-reactive sialomucin (Costin, 1975).

One peculiarity of OEs seems to be their highly labile nature. They are particularly fragile in vitro and lyse quickly, ejecting material in the hemolymph. They are nonphagocytic.

Synonymies

No two terms have caused as much confusion as "oenocytes" and "oenocytoids." Oenocytes differ from oenocytoids in that they are ectodermal in origin, usually segmentally arranged, yellow in color, and not hemocytes. Poyarkoff (1910) first introduced the term OE, followed by Hollande (1911). In order to avoid any confusion between oenocytes (proposed by Wielowiejsky, 1866) and oenocytoids, Hollande (1914) proposed to replace the term oenocyte by "cerodecytes." It is not surprising that several earlier authors mistook oenocytes for

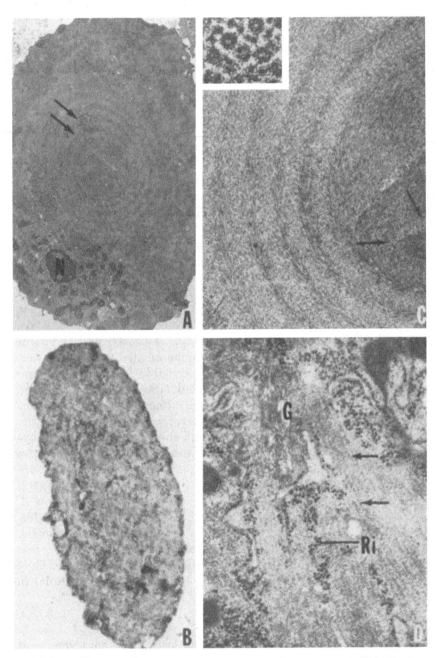

Fig. 4.12. A. Oenocytoid of *B. mori*, showing concentric arrangement of intracytoplasmic fibrils (arrows) and eccentric nucleus (*N*). × 3,000. **B.** Oenocytoid of *P. gossypiella* × 6,600. **C.** Portion of oenocytoid of *B. mori*, showing highly magnified view of concentric rings of intracytoplasmic fibrils. Note also unoriented fibrils (arrows). Ca. × 30,000. **Inset** (ca. × 225,000) shows fibrils in cross section. **D.** Portion of oenocytoid of *P. gossypiella*, showing longitudinally arranged intracytoplasmic microtubules (arrows), Golgi (*G*), and ribosomes (*Ri*). × 35,475. (**A** and **C** from Akai and Sato, 1973; **B** and **D** from Raina, 1976)

OEs. Even after Hollande's (1920) detailed description of OEs, several authors (Metalnikov and Gaschen, 1922; Müller, 1925; Tateiwa, 1928; Metalnikov and Chorine, 1929; Bogojavlensky, 1932; and Cameron, 1934) used the term oenocyte instead of OE in their respective works.

Other synonyms used for OEs are "oenocyte-like cells" (Yeager, 1945); large "non-granular spindle cells" and "non-phagocytic giant hemocytes" (Wigglesworth, 1933, 1955; see discussion in Jones, 1965); "crystalloid" and "dark hyaline hemocytes" (Selman, 1962); "crystal cells" (Rizki, 1953, 1962; Nappi, 1970); and COs (Hoffmann and Stoekel, 1968).

Interrelationship with other types

The main controversy about OEs concerns their identity as a separate category, particularly their distinction from COs. The view that OEs are part of the CO complex has received support owing to the observation by some authors (Lea and Gilbert, 1966) that OEs are unstable in vitro and that they undergo rapid and drastic transformation into hyaline cells (= COs). These authors report that in *Hyalophora cecropia*, OEs begin to transform within 15–30 sec, and fully transformed OEs are found within 15 min. Jones (1959) and Nittono (1960) also have reported such transformation of OEs in *Prodenia* and *Bombyx*, respectively. Coupled with these observations are the reports by many authors that OEs are either found in very small numbers or are absent. This may partly explain why several authors either have not reported OEs in their studies or do not recognize OEs as a distinct category. Crossley (1975), however, stated that ultrastructurally OEs and COs are different, and indeed some authors (Hoffmann et al., 1968) have described both OEs and COs in their ultrastructural studies. The ultrastructural identity of OEs is also supported by the fact that although these cells eject material into the hemolymph as COs do, this does not result in plasma gelation (Arnold, 1974).

The origin or derivation of these cells is also controversial. Gupta and Sutherland (1966) and Beaulaton and Monpeyssin (1977) suggested that OEs are differentiated from PLs. Devauchelle (1971) has indicated that OEs might be derived from PRs. Arnold (1974) stated that "the cells seem to be allied with the complex of granular cells, but their origins and relationships are not understood." Hinks and Arnold (1977) found them originating in the hemopoietic tissue.

4.2.7. *Coagulocyte* (CO)

Structure

COs are generally small to large (3–30 μm long), spherical, hyaline, fragile, and unstable cells, combining the features of GRs and OEs

(Arnold, 1974). The plasma membrane is generally without any micro-papillae, filopodia, or other irregular processes. The nucleus is relatively small (5–11 μm long), generally eccentric, oval, sharply outlined, and under phase-contrast may appear cartwheel-like owing to the arrangement of the chromatin in that fashion (Fig. 4.2D). According to Goffinet and Grégoire (1975), there is a pronounced perinuclear cisterna (Fig. 4.13A), which, together with microruptures in the plasma membrane, supposedly distinguishes these cells from other types.

The laminar nature of the plasma and nuclear membranes may not be visible. The plasma membrane may show microruptures. The cytoplasm is hyaline and rich in polyribosomes, but has fewer mitochondria and moderately developed ER. In addition, the cytoplasm has some spherical or elongate granular inclusions, about 1 μm in diameter (Fig. 4.13A). François (1975) has described four types of such granules in *Thermobia domestica:* (1) electron-dense, homogeneous granules, generally resembling those in GRs; (2) moderately electron-dense, homogeneous granules; (3) heterogeneous granules with a central or lateral dense zone, the remaining portion being homogeneously granular; and (4) structured granules, with internal microtubules (15 nm in diameter), arranged in a parallel fashion and 40 nm apart. Goffinet and Grégoire (1975) also described structural granules in the COs of *C. morosus.* It is obvious that there is a very close resemblance between GRs and COs.

Histochemically, COs are clearly distinguishable from PLs and GRs according to the periodic acid–Schiff (PAS) test (Costin, 1975). According to her, "compared with the cytoplasm of the other types of blood cells, that of coagulocyte has much reduced basophilia." It is very weakly PAS-positive.

Synonymies

Grégoire and Florkin (1950) for the first time introduced the term "coagulocyte" or "unstable hyaline hemocyte" in *Gryllulus* and *Carausius.* Earlier, Yeager (1945) for the first time used the term "cystocyte" for cells with cystlike inclusions. Jones (1950) used that term for "coarsely granular haemocytes," and later (Jones, 1962) suggested that the "term coagulocyte for these cells may be preferred to cysto-cyte because these cells are only identified by their function." Wigglesworth (Chapter 11) synonymized "thrombocytoids" of Zachary and Hoffmann (1973) with COs.

Interrelationship with other types

The main controversies about COs concern their identity, function, and origin. It is still debatable whether the COs are ultrastructurally different from GRs. Devauchelle (1971) found them indistinguishable from GRs and synonymized them with the latter. Moran (1971) found a type of cell in *Blaberus discoidalis* (frequently in newly molted, untanned adults) with membrane-bounded, tubule-containing bodies (TCB) filled with rows of 34-nm tubules, which are quite different from the intracytoplasmic microtubules. He suggested that these cells are equivalent to COs (cystocytes). Ratcliffe and Price (1974), however, have identified COs (their cystocyte) in their work. Most recently, Goffinet and Grégoire (1975) and Grégoire and Goffinet (Chapter 7) claimed a separate identity for them. According to them, the perinuclear cisternae of the COs are much more pronounced than those of PRs, PLs, and GRs, and their plasma membrane is ruptured during coagulation, whereas those in PRs, PLs, and GRs remain intact.

The role of the COs in hemolymph coagulation is generally accepted and has been recently reconfirmed by Grégoire (1974), François (1975), Goffinet and Grégoire (1975), and Grégoire and Goffinet (Chapter 7). Gupta and Sutherland (1966), however, have suggested that COs are the effect rather than the cause of coagulation on the basis of their observation that as soon as coagulation starts, several PLs transform into COs. This view, however, is not accepted by Grégoire. Supposedly, COs also contain phenol-oxidizing enzymes (Crossley, 1975). It must be mentioned here that in several arthropod groups coagulation of the hemolymph is caused by the GR (Gupta, 1979), and evidence is accumulating that this is also true in some insects (Rowley and Ratcliffe, 1976; Rowley, 1977).

The origin of COs is still debatable. Grégoire and Goffinet (Chapter 7) and Hinks and Arnold (1977) have suggested that these cells originate in the hemopoietic organs. However, if we accept the premise that hemocytes respond to bodily injury in the insect, it is conceivable that either the injury itself would induce the production of COs or some other type of hemocyte would produce them by transformation. For more details on the structure, functions, and origin of COs, the reader is referred to Chapter 7.

4.3. Other hemocyte types

4.3.1. *Podocyte* (PO)

Structure

These hemocytes have not been recognized as a separate category in any ultrastructural study and are not ordinarily observed in the hemocyte samples under the light microscope. They should be regarded as a variant form of PL. According to Arnold (1974), they have been correctly identified only in *Prodenia* (Yeager, 1945; Jones, 1959). However, I (Gupta, 1969) have observed them in *P. americana* nymphs. More often than not, radiate PLs with pseudopodia are mistaken for POs.

These hemocytes are very large (Fig. 4.2E), extremely flattened PL-like cells with several cytoplasmic extensions (Fig. 4.13B). The nucleus is generally large and centrally located and may appear punctate.

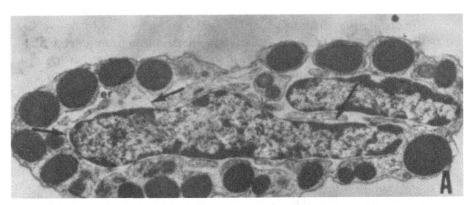

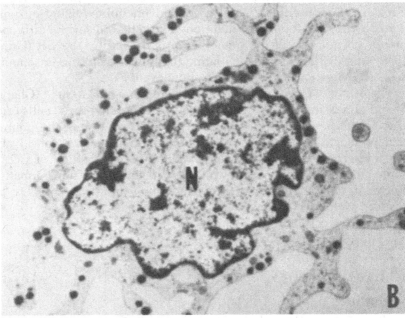

Fig. 4.13. A. Coagulocyte, showing perinuclear cisternae (arrows) and those of endoplasmic reticulum (*er*). Note presence of electron-dense (structureless) granules such as are found in granulocytes. × 20,000. **B.** Podocyte, showing pseudopodia. Note resemblance to prohemocyte or young plasmatocyte. *N* = nucleus. × 9,500. (**A** and **B** courtesy Dr. G. Devauchelle)

Synonymies

Yeager (1945) introduced the term "podocyte." Graber's (1871) "star-shaped amoebocytes" and Lutz's (1895) radiate cells may have included POs.

Interrelationship with other types

POs are derived from PLs. Gupta and Sutherland (1966) have suggested their transformation from PLs. This seems to be supported by Rizki's (1962) observation that in *D. melanogaster* when POs increase in differential counts, PLs decrease. Rizki's (1953) POs appear to be PLs. Whitten (1964) questioned the concept of POs, and Devauchelle (1971) considered them variant forms of PLs.

4.3.2. *Vermicyte* (VE)

Structure

This form is generally called vermiform cell and should not be regarded as a separate category. As the name suggests, these are extremely elongated cells with slightly granular or agranular cytoplasm. The nucleus may be located centrally or eccentrically (Fig. 4.2F).

Synonymies

Yeager (1945) introduced the term "vermiform cell," but the term "vermicyte" is more appropriate. Tuzet and Manier (1959) used the term "giant fusiform cells" for VEs.

Interrelationship with other types

The origin of VEs is unknown. However, it is conceivable that they are derived from PLs, as has been suggested by Gupta and Sutherland (1966). Lea and Gilbert (1966) considered them a variant form of PLs. According to Arnold (1974), "they seem to occur mainly just prior to pupation, but never in large numbers."

4.3.3. Additional miscellaneous hemocyte types

In addition to the above nine hemocyte types, several authors have, from time to time, reported hemocytes, many or all of which have not been generally accepted – for example: "haemocytoblast" of Bogojavlensky (1932); "leucoblast" of Arvy and Gabe (1946) and Arvy (1954); "proleucocytoid" and "prohaemocytoid" of Yeager (1945) and Jones (1950), respectively. Yeager also introduced the term "nematocyte." Rizki (1962) in his works used the terms "lamellocyte" and "crystal cell." The latter was adopted also by Whitten (1964). According to Arnold (1974), crystal cells and lamellocytes are considered variants of

OEs and PLs, respectively. Gupta (1969) also suggested that the crystal cell is probably an OE. Terms such as "seleniform cell" (Poyarkoff, 1910); "miocyte" (Tillyard, 1917); "splanchnocyte" (Muttkowski, 1924); "teratocyte" (Hollande, 1920); "pycnonucleocyte" (Morganthaler, 1953; Wille and Vecchi, 1966); "nucleocyte" and "rhegmatocyte" (Hrdy, 1957) are rarely encountered in the literature; most likely several of these cells are not even hemocytes. Jones (1965) introduced the term "granulocytophagous" cell in his work on *Rhodnius prolixus*, and Ritter (1965) and Scharrer (1965) "anucleate crescent body" and "crescent cell," respectively, in the cockroach, *Gromphadorhina portentosa*. Zachary and Hoffmann (1973) described the hemocyte "thrombocytoid" that takes part in encapsulation in *Calliphora erythrocephala* (Zachary et al., 1975).

Several years ago, I (Gupta, 1969) reported that the hemocytes in several orders of insects have not been studied. For example, as of that year, among Apterygota, only Thysanura had been studied. Among the orthopteroid groups, Isoptera and Embioptera awaited studies. In the hemipteroid complex, no account of hemocytes was available in Zoraptera, Phthiraptera, Corrodentia, and Thysanoptera; and finally in the neuropteroid group, hemocytes were yet to be studied in Raphidioidea, Mecoptera, and Siphonaptera. It seems that situation has changed very little since then, for hemocytes in most of the above groups are still awaiting studies.

4.4. Hemocyte types in various orders

In terms of the number of species studied in various orders, Lepidoptera, Hymenoptera, Coleoptera, and Diptera appear to be the most extensively studied groups. In addition to the Heteroptera, Homoptera, and Odonata, of which only a few species have so far been studied, Dermaptera, Plecoptera, Trichoptera, and Thysanoptera are the most poorly studied groups. Hemocytes of several insect orders are unknown.

With the exception of the GR and possibly also PLs, all other types of hemocytes are not present in all insect orders (see Table 4.1). According to Arnold (1974), all hemocyte types have been reported only in *Prodenia* (Yeager, 1945; Jones, 1959). Most insects seem to possess PRs, PLs, and GRs (see also Chapter 8).

4.5. Plesiomorphic hemocyte and its differentiation into other types

I have reported elsewhere (Gupta, 1979) that the granulocyte (GR) is the plesiomorphic hemocyte, and it is the only hemocyte type that has been reported in all major arthropod groups, including all studied insect orders, and the Onychophora. The following account of the presence of GR in Insecta is based on Arnold's (1974) reinterpretation of various works, my own (Gupta, 1969) survey of the hemocyte literature in many insects, and the most recent transmission electron microscopic (TEM) studies of hemocytes. As far as is known, only Bruntz (1908), Millara (1947), Barra (1969), Gupta (1969), and François (1974, 1975) have worked on the hemocytes of the Apterygota; and on the basis of these works, both Collembola (it is controversial whether they should be included in Apterygota) and Thysanura possess GRs. All higher orders of insects possess GRs (see Table 4.1). Some of the most recent TEM studies (Hoffmann et al., 1968, 1970; Baerwald and Boush, 1970; Hagopian, 1971; Moran 1971; Scharrer, 1972; Ratcliffe and Price, 1974; Goffinet and Grégoire, 1975; Beaulaton and Monpeyssin, 1976; Brehélin et al., 1976; Ratcliffe et al., 1976; Rowley and Ratcliffe, 1976; Rowley, 1977; Schmit and Ratcliffe, 1977) have reported (or can be interpreted to show) GRs in various insects. The most highly specialized neuropteroid orders also possess GRs.

Since it has been reported by several authors that one hemocyte type can and does differentiate into another type, it is conceivable that during evolution the plesiomorphic GR differentiated into other hemocyte types. It can be postulated also that the GR originates from the so-called prohemocyte (PR) or stem cell and goes through the plasmatocyte (PL) stage before becoming a distinct GR type. In taxa in which only GRs have been observed (e.g., Xiphosura), the PR and PL are merely evanescent stages and have not achieved distinctness as types. In taxa that are reported to possess other types besides PR, PL, and, GR, the last perhaps further differentiated into SP, AD, CO, and OE, not necessarily in that order. The post-GR differentiation is generally accompanied by distinct PRs and PLs. Furthermore, in the more highly evolved taxa any of the types may be suppressed. The main differentiation pathways as postulated above may be represented as shown in the diagram.

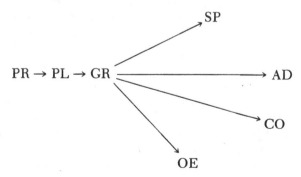

4.6. Phylogenetic significance of hemocyte types

The prospect of using variations in hemocyte types in various insect orders for phylogenetic considerations is severely limited owing to (1) lack of uniform terminology, and hence the difficulty of establishing comparisons, and (2) paucity of comprehensive studies of hemocyte types in large numbers of species within various orders to enable one to draw meaningful conclusions on the basis of important variations. However, hemocyte types and their numerical variations in many insects orders are known. Does the number of hemocyte types in various orders have any phylogenetic significance? Arnold (1972a,b, 1976) and Arnold and Hinks (1975) have used hemocytes in insect taxonomy, and several authors have suggested (see Gupta, 1979) that some phylogenetic relationship among various arthropod groups can be demonstrated on the basis of the occurrence of the GR and other hemocyte types.

A review of the insect hemocyte literature indicates some phylogenetic trends in the diversity of the hemocyte types in Insecta as a group as the evolutionary ladder is ascended (Fig. 4.14). This was observed also by Arnold (1974), although he attributed this diversity more to "shifts in the emphasis on certain functions or in the assignment of functions to different tissues" than to phylogeny. And at least in certain instances he may be right.

As far as is known, only Bruntz (1908), Millara (1947), Barra (1969), Gupta (1969), and François (1974, 1975) have worked on the hemocytes of the Apterygota; and on the basis of these works, Collembola possess only the plesiomorphic GR, whereas the Thysanura (Lepismatidae) seem to possess PL, GR, SP, and CO. Assuming that the Thysanura originated from Symphyla-like ancestors, and that the latter had hemocyte types comparable to those of *Scutigerella* (Gupta, 1968), we find that a reduction from six (PR, PL, GR, SP, AD, and CO) in the symphylan ancestor to four in the Thysanura has occurred. I have no evidence to suggest whether this is a secondary suppression and/or reduction or, as Arnold (1974) suggested, attributable to shifts in functions. It is also possible that future studies will reveal more types than are presently known.

According to Carpenter (1976), the derivation of the Pterygota from the apterygote Thysanura is almost universally accepted. It is also generally believed that the pterygotes evolved along four evolutionary lines: paleopteroid, orthopteroid, hemipteroid, and neuropteroid. It is interesting to note (Fig. 4.14) that in the Palaeoptera, although the number of hemocyte types has not increased from the ancestral thysanuran number, the OE has already made its appearance in the very beginning of the evolution of winged insects. In addition, the PR has achieved distinctness, and the CO is either suppressed or its function is taken over by the GR, which is generally the case in some other arthropods (aquatic Chelicerata and some Crustacea).

It is also evident from Fig. 4.14 that beyond the paleopteroid line the number of hemocyte types increased, and all the six or seven types were realized in the orthopteroid, hemipteroid, and neuropteroid lines. It seems that by the time the orthopteroid line evolved, the plesiomorphic GR had already differentiated into all the distinct types presently known in pterygote insects, and that no further evolution in the hemocyte types occurred beyond that point. It is doubtful whether within the orthopteroid group any phylogenetic significance of the hemocyte types exists. The number of hemocyte types reported in various orders of this group (see Gupta, 1969; Arnold, 1974) is so variable that it is very difficult to discern any phylogenetic trends. On the basis

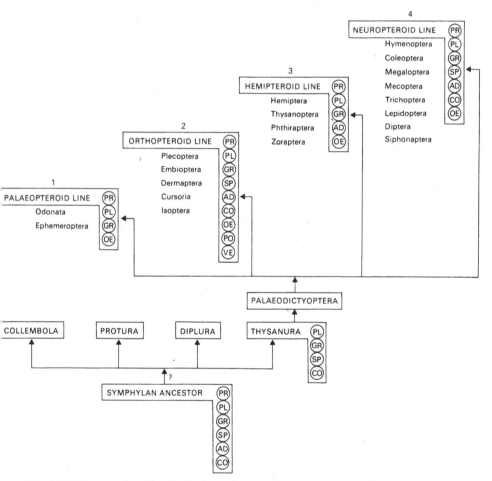

Fig. 4.14. Diagram showing distribution of various hemocyte types in four evolutionary lines (groups) and in symphylan ancestor of Insecta. Note that listed orders under each evolutionary line are only some examples of representative orders, not necessarily those in which hemocyte types designated under each evolutionary line are present. *AD* = adipohemocyte; *CO* = coagulocyte; *GR* = granulocyte; *OE* = oenocytoid; *PL* = plasmatocyte; *PO* = podocyte; *PR* = prohemocyte; *SP* = spherulocyte; *VE* × vermicyte. (From Gupta, 1979)

of the information presently available, the hemocyte types vary from three to eight or nine by light microscope, and seven (PR, PL, GR, SP, CO, AD, OE) types have been demonstrated by TEM. Unfortunately, hemocytes of several hemipteroid orders have not been studied (Gupta, 1969; Arnold, 1974), most of the work being confined to a few species in the order Hemiptera (Poisson, 1924; Hamilton, 1931; Khanna, 1964; Wigglesworth, 1955, 1956; Jones, 1965; Lai-Fook, 1970; Zaidi and Khan, 1975). Five (PR, PL, GR, AD, OE) types by light microscopy and four (PR, PL, GR, OE) by TEM studies have been identified. The apparent absence of SP and CO in the hemipteroid group as a whole is probably attributable to lack of enough studied species. It is difficult to imagine that these two types have been suppressed in the hemipteroid orders.

In the neuropteroid group, we find the seven major types (PR, PL, GR, SP, AD, CO, OE), although not all these types have been reported from each of the orders in this group (Gupta, 1969), and not all are recognized as types by all authors. As a matter of fact, the types of hemocytes reported vary widely even within an order in this group. For example, two to seven types have been reported in Lepidoptera, Coleoptera, and Diptera; five types in Hymenoptera, Neuroptera, and Megaloptera; and only three (PR, PL, GR) in Trichoptera (see Gupta, 1969, for various listings; Arnold, 1974). Since all the orders are highly evolved, it is quite conceivable that most or all major types would be found in all the orders as more studies become available.

4.7. Summary

Among arthropods, hemocytes have been most extensively studied in insects. A uniform hemocyte classification is still lacking for insects as well as for other arthropod groups. There is disagreement among insect hematologists about the number of hemocyte types in various insects. From one or a few to as many as nine or more types have been described, particularly by light microscopy. Ultrastructurally, however, only seven types have so far been identified in various insects: prohemocyte (PR), plasmatocyte (PL), granulocyte (GR), spherulocyte (SP), adipohemocyte (AD), oenocytoid (OE), and coagulocyte (CO). Of these seven, AD and CO have been described by only a few authors. All seven types have been described under various other names, and these synonymies are mentioned. It is suggested that of the seven types of hemocytes, the GR is the plesiomorphic hemocyte type and has evolved into other hemocyte types. I have postulated that the GR originates from the PR and, goes through the PL stage before becoming a distinct GR type. In taxa in which only GRs have been found, the PR and PL are merely evanescent stages and have not achieved distinctness as types. In taxa that are reported to possess other types besides PR, PL, and GR, the last further differentiates into SP, AD, and OE, not necessarily in that order. This post-GR differentiation is generally accompanied by distinct PRs and PLs. Furthermore, in more highly evolved taxa, as well as in some lower ones, any of the types may be suppressed.

A review of the insect hemocyte literature indicates some phylogenetic trends in the diversity of the hemocyte types as the evolutionary ladder is ascended in Insecta.

Acknowledgments

The material presented in this chapter is adopted and modified from Gupta (1979). Without the published illustrations I have included, this review would not have been possible. For providing me with prints or negatives and allowing me to reproduce their published illustrations, I am most indebted to Drs. H. Akai, R. J. Baerwald, N. M. Costin, G. Devauchelle, J. François, G. Goffinet, Ch. Grégoire, A. V. Loud (for the late Dr. M. Hagopian), A. K. Raina, G. Salt, and S. Sato. I am grateful to Dr. J. W. Arnold for his comments and suggestions. The entire credit and my deep appreciation for preparing all the illustrations of this review go to Dr. Y. T. Das and Mr. S. B. Ramaswamy. I sincerely appreciate the secretarial assistance of Mrs. Joan Gross.

References

Akai, H. 1969. Ultrastructure of haemocytes observed on the fat-body cells in *Philosamia* during metamorphosis. *Jpn. J. Appl. Entomol. Zool. 13:*17–21.

Akai, H., and S. Sato. 1973. Ultrastructure of the larval hemocytes of the silkworm, *Bombyx mori* L. (Lepidoptera: Bombycidae). *Int. J. Insect Morphol. Embryol. 2*(3):207–31.

Åkesson, B. 1945. Observations on the haemocytes during the metamorphosis of *Calliphora erythrocephala* (Meig.). *Ark. Zool. 6*(12):203–11.

Arnold, J. W. 1952. The haemocytes of the Mediterranean flour moth, *Ephestia kühniella* Zell. (Lepidoptera: Pyralididae). *Can. J. Zool. 30:*352–64.

Arnold, J. W. 1970. Haemocytes of the Pacific beetle cockroach, *Diploptera punctata. Can. Entomol. 102*(7):830–5.

Arnold, J. W. 1972a. Haemocytology in insect biosystematics: The prospect. *Can. Entomol. 104:*655–9.

Arnold, J. W. 1972b. A comparative study of the haemocytes (blood cells) of cockroaches (Insecta: Dictyoptera: Blattaria), with a view of their significance in taxonomy. *Can. Entomol. 104:*309–48.

Arnold, J. W. 1974. The hemocytes of insects, pp. 201–54. *In* M. Rockstein (ed.). *The Physiology of Insecta*, Vol. 5, 2nd ed. Academic Press, New York.

Arnold, J. W. 1976. Biosystematics of the genus *Euxoa* (Lepidoptera: Noctuidae). VIII. The hemocytological position of *E. rockburnei* in the "declarata group." *Can. Entomol. 108:*1387–90.

Arnold, J. W., and C. F. Hinks. 1975. Biosystematics of the genus *Euxoa* (Lepidoptera: Noctuidae). III. Hemocytological distinctions between two closely related species, *E. campestris* and *E. declarata. Can. Entomol. 107;*1095–1100.

Arnold, J. W., and E. H. Salkeld. 1967. Morphology of the haemocytes of the giant cockroach, *Blaberus giganteus*, with histochemical tests. *Can. Entomol. 99:*1138–45.

Arnold, J. W., and S. S. Sohi. 1974. Hemocytes of *Malacosoma disstria* Hübner (Lepidoptera: Lasiocampidae): Morphology of the cells in fresh blood and after cultivation *in vitro. Can. J. Zool. 52*(4):481–5.

122   A. P. Gupta

Arvy, L. 1954. Présentation de documents sur la leucopoïèse chez *Peripatopsis capensis* Grube. *Bull. Soc. Zool. Fr.* 79:13.

Arvy, L., and M. Gabe. 1946. Identification des diastrases sanguines chez quelques insectes. *C. R. Soc. Biol. (Paris)* 140:757–8.

Arvy, L., and J. Lhoste. 1946. Les variations du leucogramme au cours de la métamorphose chez *Forficula auricularia* L. *Bull. Soc. Zool. Fr.* 70:114–48.

Ashhurst, D. E., and A. G. Richards. 1964. Some histochemical observations on the blood cells of the wax moth, *Galleria mellonella* L. *J. Morphol.* 114:247–53.

Baerwald, R. J., and G. M. Boush. 1970. Fine structure of the hemocytes of *Periplaneta americana* (Orthoptera: Blattidae) with particular reference to marginal bundles. *J. Ultrastruct. Res.* 31:151–61.

Baerwald, R. J., and G. M. Boush. 1971. Vinblastine-induced disruption of microtubules in cockroach hemocytes. *Tissue Cell* 3(2):251–60.

Barra, J. A. 1969. Tégument des Collemboles: Présence d'hémocytes à granules dans le liquide exuvial au cours de la mue (Insectes, Collemboles). *C. R. Acad. Sci. Paris* 269D:902–3.

Beaulaton, J. 1968. Etude ultrastructurale et cytochimique des glandes prothoraciques de vers à soie aux quatrième et cinquième âges larvaires. I. La *tunica propria* et ses relations avec les fibres conjonctives et les hémocytes. *J. Ultrastruct. Res.* 23:474–98.

Beaulaton, J., and M. Monpeyssin. 1976. Ultrastructure et cytochimie des hémocytes d'*Antheraea pernyi* Guér. (Lepidoptera, Attacidae) au cours du cinquième âge larvaire. I. Prohémocytes, plasmatocytes et granulocytes. *J. Ultrastruct. Res.* 55(2):143–56.

Beaulaton, J., and M. Monpeyssin. 1977. Ultrastructure et cytochimie des hémocytes d'*Antheraea pernyi* Guér. (Lepidoptera, Attacidae). II. Cellules à sphérules et oenocytoïdes. *Biol. Cell.* 28(1):13–8.

Bogojavlensky, K. S. 1932. The formed elements of the blood of insects. *Arch. Russ. Anat. Hist. Embryol.* 11:361–86. (In Russian.)

Boiteau, G., and J. M. Perron. 1976. Etude des hémocytes de 'Macrosiphum euphorbiae (Thomas) (Homoptera: Aphidiidae). *Can. J. Zool.* 54(2):228–34.

Brehélin, M., D. Zachary, and J. A. Hoffmann. 1976. Fonctions des granulocytes typiques dans la cicatrisation chez l'orthoptère *Locusta migratoria* L. *J. Microsc. Biol. Cell.* 25(2):133–6.

Breugnon, M., and J. R. Le Berre. 1976. Fluctuation of the hemocyte formula and hemolymph volume in the caterpillar *Pieris brassicae. Ann. Zool. Ecol. Anim.* 8(1):1–12. (In French.)

Bruntz, L. 1908. Nouvelles recherches sur l'excrétion et la phagocytose chez les Thysanoures. *Arch. Zool. Exp. Gén.* 38:471–88.

Cameron, G. R. 1934. Inflammation in the caterpillars of Lepidoptera. *J. Pathol. Bacteriol.* 38:441–66.

Carpenter, F. M. 1976. Geological history and evolution of the insect. *Proc. 15th Int. Congr. Entomol.* 1976:63–70.

Clark, E. W., and D. S. Chadbourne. 1960. The hemocytes of non-diapause and diapause larvae and pupae of the pink bollworm. *Ann. Entomol. Soc. Amer.* 53:682–5.

Costin, N. M. 1975. Histochemical observations of the haemocytes of *Locusta migratoria. Histochem. J.* 7:21–43.

Crossley, A. C. 1975. The cytophysiology of insect blood. *Adv. Insect Physiol.* 11:117–221.

Cuénot, L. 1896. Etudes physiologiques sur les Orthoptères. *Arch. Biol.* 14:293–341.

Dennell, R. 1947. A study of an insect cuticle. *Proc. R. Soc. Lond. (B)* 134:79–110.

Devauchelle, G. 1971. Etude ultrastructurale des hémocytes du Coléoptère *Melolontha melolontha* (L.). *J. Ultrastruct. Res.* 34:492–516.

François, J. 1974. Etude ultrastructurale des hémocytes du Thysanoure *Thermobia domestica* (Insecte, Aptérygote). *Pedobiologia* 14:157–62.

François, J. 1975. Hémocyte et organe hématopoïètique de *Thermobia domestica* (Packard) (Thysanura: Lepismatidae). *Int. J. Insect Morphol. Embryol.* 4(6):477–94.

Goffinet, G., and Ch. Grégoire. 1975. Coagulocyte alterations in clotting hemolymph of *Carausius morosus* L. *Arch. Int. Physiol. Biochim.* 83(4):707–22.

Graber, V. 1871. Ueber die Blutkorperchen der Insekten. *Sitzb. Akad. Math.-Nat. Wiss. Wien* 64:9–44.

Grégoire, Ch. 1974. Hemolymph coagulation, pp. 309–60. *In* M. Rockstein (ed.). *The Physiology of Insecta*, Vol. 5, 2nd ed. Academic Press, New York.

Grégoire, Ch., and M. Florkin. 1950. Blood coagulation in arthropods. I. The coagulation of insect blood, as studied with the phase contrast microscope. *Physiol. Comp. Oecol.* 2(2):126–39.

Grimstone, A. V., S. Rotheram, and G. Salt. 1967. An electron-microscope study of capsule formation by insect blood cells. *J. Cell Sci.* 2:281–92.

Gupta, A. P. 1968. Hemocytes of *Scutigerella immaculata* and the ancestry of Insecta. *Ann. Entomol. Soc. Amer.* 61(4):1028–9.

Gupta, A. P. 1969. Studies of the blood of Meloidae (Coleoptera). I. The haemocytes of *Epicauta cinerea* (Forster), and a synonymy of haemocyte terminologies. *Cytologia* 34(2):300–44.

Gupta, A. P. 1970. Midgut lesions in *Epicauta cinerea* (Coleoptera: Meloidae). *Ann. Entomol. Soc. Amer.* 63:1786–8.

Gupta, A. P. 1979. Arthropod hemocytes and phylogeny, pp. 669–735. *In* A. P. Gupta (ed.). *Arthropod Phylogeny*. Van Nostrand Reinhold, New York.

Gupta, A. P., and D. J. Sutherland. 1965. Observations on the spherule cells in some Blattaria (Orthoptera). *Bull. Entomol. Soc. Amer.* 11:161.

Gupta, A. P., and D. J. Sutherland. 1966. *In vitro* transformations of the insect plasmatocyte in certain insects. *J. Insect Physiol.* 12:1369–75.

Gupta, A. P., and D. J. Sutherland. 1967. Phase contrast and histochemical studies of spherule cells in cockroaches. *Ann. Entomol. Soc. Amer.* 60(3):557–65.

Gupta, A. P., and D. J. Sutherland. 1968. Effects of sublethal doses of chlordane on the hemocytes and midgut epithelium of *Periplaneta americana*. *Ann. Entomol. Soc. Amer.* 61(4):910–18.

Hagopian, M. 1971. Unique structures in the insect granular hemocytes. *J. Ultrastruct. Res.* 36:646–58.

Hamilton, M. A. 1931. The morphology of the water scorpion, *Nepa cinerea* Linn. *Proc. Zool. Soc. Lond.* 193:1067–1136.

Harpaz, F., N. Kislev, and A. Zelcer. 1969. Electron-microscopic studies on hemocytes of the Egyptian cottonworm, *Spodoptera littoralis* (Boisduval) infected with a nuclear-polyhedrosis virus, as compared to noninfected hemocytes. I. Noninfected hemocytes. *J. Invertebr. Pathol.* 14:175–85.

Hinks, C. F., and J. W. Arnold. 1977. Haemopoiesis in Lepidoptera. II. The role of the haemopoietic organs. *Can. J. Zool.* 55(10):1740–55.

Hoffmann, J. A. 1966. Etude des oenocytoïdes chez *Locusta migratoria* (Orthoptère). *J. Microsc.* (Paris) 5:269–72.

Hoffmann, J. A. 1967. Etude des hémocytes de *Locusta migratoria* L. (Orthoptère). *Arch. Zool. Exp. Gén.* 108:251–91.

Hoffmann, J. A., A. Porte, and P. Joly. 1970. On the localization of phenoloxidase activity in coagulation of *Locusta migratoria* (L.) (Orthoptera). *C. R. Hebd. Séances Acad. Sci. Paris* 270D:629–31.

Hoffmann, J. A., and M. E. Stoekel. 1968. Sur les modifications ultrastructurales des coagulocytes au cours de la coagulation de l'hémolymphe chez un insecte Orthopteroïde: *Locusta migratoria* C. R. *Séances Soc. Biol. Strasbourg* 162:2257–9.

Hoffmann, J. A., M. E. Stoekel, A. Porte, and P. Joly. 1968. Ultrastructure des hémocytes de *Locusta migratoria* (Orthoptère). *C. R. Hebd. Séances Acad. Sci. Paris* 266:503–5.

124   A. P. Gupta

Hollande, A. C. 1909. Contribution á l'étude du sang des Coléoptères. *Arch. Zool. Exp. Gén. (Ser. 5) 21:*271–94.

Hollande, A. C. 1911. Etude histologiques comparée du sang des insectes à hémorrhée et des insectes sans hémorrhée. *Arch. Zool. Exp. Gén. (Ser. 5) 6:*283–323.

Hollande, A. C. 1914. Le cerodecytes ou "oenocytes" des insectes considères au point de vue biochimique. *Arch. Anat. Microsc. 16:*1–66.

Hollande, A. C. 1920. Oenocytoïdes et teratocytes du sang des chenilles. *C. R. Acad. Sci. Paris 170:*1341–4.

Hrdy, I. 1957. Comparison of preparation and staining techniques of insect blood cells. *Acta Soc. Entomol. Cech. 54:*305–11.

Jones, J. C. 1950. The normal hemocyte picture of the yellow mealworm *Tenebrio molitor* Linnaeus. *Iowa State Coll. J. Sci. 24:*356–61.

Jones, J. C. 1954. A study of mealworm hemocytes with phase contrast microscopy. *Ann. Entomol. Soc. Amer. 47:*308–15.

Jones, J. C. 1956. The hemocytes of *Sarcophaga bullata* Parker. *J. Morphol. 99*(2):233–57.

Jones, J. C. 1959. A phase contrast study of the blood cells in *Prodenia* larvae (Order Lepidoptera). *Q. J. Microsc. Sci. 100*(1):17–23.

Jones, J. C. 1962. Current concepts concerning insect hemocytes. *Amer. Zool. 2:*209–46.

Jones, J. C. 1965. The hemocytes of *Rhodnius prolixus* Stål. *Biol. Bull. (Woods Hole) 129:*282–94.

Jones, J. C. 1970. Hematopoiesis in insects, pp. 7–65. *In* A. S. Gordon (ed.). *Regulation of Hematopoiesis*, Vol. 1. Appleton, New York.

Jones, J. C. 1975. Forms and functions of insect hemocytes. pp. 119–128. *In* K. Maramorosch and R. E. Shope (eds.). *Invertebrate Immunity.* Academic Press, New York.

Jones, J. C., and O. E. Tauber. 1954. Abnormal hemocytes in mealworms (*Tenebrio molitor* L.). *Ann. Entomol. Soc. Amer. 47*(3):428–44.

Jones, T. W. 1846. The blood corpuscle considered in its different phases of development in the animal series. *Philos. Trans. R. Soc. 136:*1–106.

Khanna, S. 1964. The circulatory system of *Dysdercus koenig* F. (Hemiptera: Pyrrhocoridae). *Indian J. Entomol. 26*(4):404–10.

Kollmann, M. 1908. Recherches sur les leucocytes et le tissu lymphoïde des invertèbres. *Ann. Sci. Nat. Zool. 9:*1–238.

Lai-Fook, J. 1970. Haemocytes in the repair of wounds in an insect (*Rhodnius prolixus*). *J. Morphol. 130:*297–314.

Lai-Fook, J. 1973. The structure of the haemocytes of *Calpodes ethlius* (Lepidoptera). *J. Morphol. 139:*79–104.

Landureau, J. C., and P. Grellet. 1975. Obtentions de lignées permanentes d'hémocytes de Blatte: Caractéristiques physiologiques et ultrastructurales. *J. Insect Physiol. 21:*137–51.

Lea, M. S., and L. I. Gilbert. 1966. The hemocytes of *Hyalophora cecropia* (Lepidoptera). *J. Morphol. 118*(2):197–216.

Ludwig, D., and M. Wugmeister. 1953. Effects of starvation on the blood of Japanese beetle (*Popillia japonica* Newman) larvae. *Physiol. Zool. 26:*254–9.

Lutz, K. G. 1895. Das Bluten der Coccinelliden. *Zool. Anz. 18:*244–5.

Metalnikov, S. 1908. Recherches expérimentales sur les Chenilles de *Galleria mellonella. Arch. Zool. Exp. (Ser. 4) 8:*489–588.

Metalnikov, S. 1934. *Rôle du système nerveux et des facteurs biologiques et psychiques dans l'immunité.* Monograph Pasteur Institute. Masson, Paris.

Metalnikov, S., and V. Chorine. 1929. On the natural and acquired immunity of *Pyrausta nubilalis* Hb. *Int. Corn Borer Invest. (Chicago) 2:*22–38.

Metalnikov, S., and H. Gaschen. 1922. Immunité cellulaire et humoral chez la chenille. *Ann. Inst. Pasteur 36:*233–52.

Millara, P. 1947. Contributions à l'étude cytologique et physiologique des leucocytes d'Insectes. *Bull. Biol. Fr. Belg. 81*:129–53.

Mitsuhashi, J. 1966. Tissue culture of the rice stem borer, *Chilo suppressalis* Walker (Lepidoptera: Pyralidae). II. Morphology and *in vitro* cultivation of hemocytes. *Appl. Entomol. Zool. 1*:5–20.

Moran, D. T. 1971. The fine structure of cockroach blood cells. *Tissue Cell 3*:413–22.

Morganthaler, P. W. 1953. Blutuntersuchungen bei Bienen. *Mitt. Schweiz. Entomol. Ges. 26*:247–57.

Müller, K. 1925. Ueber die korpuskularen Elemente der Blutflüssigkeit bei der erwaschsenen Honigbiene *(Apis mellifica)*. *Erlanger Jahr. Blenenkunde 3*:5–27.

Muttkowski, R. A. 1924. Studies on the blood of insects. II. The structural elements of the blood. *Bull. Brooklyn Entomol. Soc. 19*:4–19.

Nappi, A. J. 1970. Hemocytes of larvae of *Drosophila euronotus* (Diptera: Drosophilidae). *Ann. Entomol. Soc. Amer. 63*(53):1217–24.

Nittono, Y. 1960. Studies on the blood cells in the silkworm, *Bombyx mori* L. *Bull. Seric. Exp. Stn. Tokyo 16*:171–266 (In Japanese, English summary.).

Paillot, A. 1919. Le karyocinetose, nouvelle réaction d'immunité naturelle observé chez les chenilles de macrolépidoptères. *C. R. Acad. Sci. Paris 169*:396–8.

Paillot, A., and R. Noel. 1928. Recherches histophysiologiques sur les cellules péricardiales et les éléments du sang des larves d'insectes *(Bombyx mori* et *Pieris brassicae)*. *Bull. Hist. Appl. Physiol. Pathol. 5*:105–28.

Pérez, C. 1910. Recherches histologiques sur la métamorphose des muscides *(Calliphora erythrocephala* Mg.). *Arch. Zool. Exp. Gén. 4*:1–274.

Pipa, R. L., and P. S. Woolever. 1965. Insect neurometamorphosis. II. The fine structure of perineural connective tissue, adipohemocytes, and the shortening ventral nerve cord of a moth *Galleria mellonella*. *Z. Zellforsch. Mikrosk. Anat. 68*: 80–101.

Poisson, R. 1924. Contribution à l'étude des Hémiptères aquatiques. *Bull. Biol. Fr. Belg. 58*:49–305.

Poyarkoff, E. 1910. Recherches histologiques sur la métamorphose d'un Coléoptère (La Galéruque de l'Orme). *Arch. Anat. Microsc. 12*:333–474.

Price, C. D., and N. A. Ratcliffe. 1974. A reappraisal of insect haemocyte classification by the examination of blood from fifteen insect orders. *Z. Zellforsch. Mikrosk. Anat. 147*:537–49.

Raina, A. K. 1976. Ultrastructure of the larval hemocytes of the pink bollworm, *Pectinophora gossypiella* (Saunders) (Lepidoptera: Gelechiidae). *Int. J. Insect Morphol. Embryol. 5*(3):187–95.

Ratcliffe, N. A., S. J. Gagen, A. F. Rowley, and A. Schmit. 1976. The role of granular hemocytes in the cellular defense reactions of the wax moth, *Galleria mellonella*. *Proc. 6th European Congr. Electron Microsc. 1976*:295–97.

Ratcliffe, N. A., and C. D. Price. 1974. Correlation of light and electron microscope hemocyte structure in the Dictyoptera. *J. Morphol. 144*:485–97.

Ritter, H., Jr. 1965. Blood of a cockroach: Unusual cellular behavior. *Science (Wash., D.C.) 147*:518–19.

Rizki, M. T. M. 1953. The larval blood cells of *Drosophila willistoni. J. Exp. Zool. 123*:397–411.

Rizki, M. T. M. 1962. Experimental analysis of hemocyte morphology in insects. *Amer. Zool. 2*:247–56.

Rizki, M. T. M., and R. M. Rizki. 1959. Functional significance of the crystal cells in the larva of *Drosophila melanogaster. J. Biophys. Biochem. Cytol. 5*:235–40.

Rowley, A. F. 1977. The role of the haemocytes of *Clitumnus extradentatus* in haemolymph coagulation. *Cell Tissue Res. 182*(4):513–24.

Rowley, A. F., and N. A. Ratcliffe. 1976. The granular cells of *Galleria mellonella* during clotting and phagocytic reactions *in vitro*. *Tissue Cell 8*:437–46.

Scharrer, B. 1965. The fine structure of an unusual hemocyte in the insect *Gromphado-rhina portentosa*. *Life Sci. 4*:1741–4.

Scharrer, B. 1972. Cytophysiological features of hemocytes in cockroaches. *Z. Zellforsch. Mikrosk. Anat. 129*:301–13.

Schmit, A. R., and N. A. Ratcliffe. 1977. The encapsulation of foreign tissue implants in *Galleria mellonella* larvae. *J. Insect Physiol. 23*:175–84.

Selman, J. 1962. The fate of the blood cells during the life history of *Sialis lutaria* L. *J. Insect Physiol. 8*:209–14.

Shrivastava, S. C., and A. G. Richards. 1965. An autoradiographic study of the relation between haemocytes and connective tissue in the wax moth, *Galleria mellonella*. *Biol. Bull. (Woods Hole) 128*:337–45.

Sohi, S. S. 1971. *In vitro* cultivation of hemocytes of *Malacosoma disstria* Hübner (Lepidoptera: Lasiocampidae). *Can. J. Zool. 49*:1355–8.

Stang-Voss, C. 1970. Zur Ultrastruktur der Blutzellen wirbelloser Tiere. I. Ueber die Haemocyten der Larve des Mehlkäfers *Tenebrio molitor*. *Z. Zellforsch. Mikrosk. Anat. 103*:589–605.

Takada, M., and H. Kitano. 1971. Studies on the larval hemocytes of *Pieris rapae crucivora* Boisduval with special reference to hemocyte classification, phagocytic activity and encapsulation capacity (Lep. Pieridae). *Kontyû 39*:385–94.

Tateiwa, J. 1928. Le formule leucocytaire du sang des chenilles normales et immunisées de *Galleria mellonella*. *Ann. Inst. Pasteur 42*:791–806.

Taylor, A. 1935. Experimentally induced changes in cell complex of the blood of *Periplaneta americana* (Blattidae: Orthoptera). *Ann. Entomol. Soc. Amer. 28*:135–45.

Tillyard, R. J. 1917. *The Biology of Dragonflies (Odonata or Paraneuroptera)*. Cambridge University Press, Cambridge.

Tuzet, O., and F. F. Manier. 1959. Recherches sur les hémocytes fusiformes géants (= "vermiform cells" de Yeager) de hémolymphe d'insectes holométaboles et hétérométaboles. *Ann. Sci. Nat. Zool. Biol. Anim. 12*(1):81–9.

Vercauteren, R. E., and F. Aerts. 1958. On the cytochemistry of the hemocytes of *Galleria mellonella* with special reference to polyphenoloxidase. *Enzymologia 20*:167–72.

Whitten, J. M. 1964. Haemocytes and the metamorphosing tissues in *Sarcophaga bullata*, *Drosophila melanogaster*, and other cyclorrhaphous Diptera. *J. Insect Physiol. 10*:447–69.

Wielowiejsky, H. 1886. Ueber das Blutgewebe der Insekten. *Z. Wiss. Zool. 43*:512–36.

Wigglesworth, V. B. 1933. The physiology of the cuticle and of ecdysis in *Rhodnius prolixus* (Triatomidae, Hemiptera), with special reference to the function of the oenocytes and of the dermal glands. *Q. J. Microsc. Sci. 76*:269–319.

Wigglesworth, V. B. 1939. *The Principles of Insect Physiology*, 1st ed. Methuen, London.

Wigglesworth, V. B. 1955. The role of haemocytes in the growth and moulting of an insect, *Rhodnius prolixus* (Hemiptera). *J. Exp. Biol. 32*:649–63.

Wigglesworth, V. B. 1956. The function of the amoebocytes during molting in *Rhodnius. Ann. Sci. Nat. Zool. (Ser. 11) 18*:139–44.

Wigglesworth, V. B. 1959. Insect blood cells. *Annu. Rev. Entomol. 4*:1–16.

Wigglesworth, V. B. 1965. *The Principles of Insect Physiology*, 6th ed. Methuen, London.

Wille, H., and M. A. Vecchi. 1966. Etude sur l'hémolymphe de l'abeille (*Apis mellifica* L.). I. Les frottis de sang de l'abeille adelle adulte d'été. *Mitt. Schweiz. Entomol. Ges. 34*:69–97.

Wittig, G. 1968. Electron microscopic characterization of insect hemocytes. *Proc. 26th Annu. Meeting EMSA. 1968*:68–9.

Yeager, J. F. 1945. The blood picture of the southern armyworm (*Prodenia eridania*). *J. Agric. Res. 71*:1–40.

Yeager, J. F., and S. C. Munson. 1941. Histochemical detection of glycogen in blood cells of the southern armyworm (*Prodenia eridania*) and in other tissues, especially midgut epithelium. *J. Agric. Res. 63*(5):257–94.

Zachary, D., M. Brehélin, and J. A. Hoffmann. 1975. Role of the "Thrombocytoids" in the capsule formation in the dipteran *Calliphora erythrocephala*. *Cell Tissue Res. 162:*(3):343–48.

Zachary, D., and J. A. Hoffmann. 1973. The haemocytes of *Calliphora erythrocephala* Meig. (Diptera). *Z. Zellforsch. Mikrosk. Anat. 141:*55–7.

Zaidi, Z. S., and M. A. Khan. 1975. Changes in the total and differential hemocyte counts of *Dysdercus cingulatus* (Hemiptera, Pyrrhocoridae) related to metamorphosis and reproduction. *J. Anim. Morphol. Physiol. 22*(2):110–19.

5 Surface and internal ultrastructure of hemocytes of some insects

H. AKAI

Sericultural Experiment Station, Wada, Suginami-ku, Tokyo 166, Japan

S. SATO

Faculty of Agriculture, Tokyo University of Agriculture, Sakuragaoka, Setagaya-ku, Tokyo, Japan

Contents

5.1. Introduction

Numerous light microscopic observations concerning the classification of insect hemocytes have been published (Yeager, 1945; Wigglesworth, 1959; Nittono, 1960; Jones, 1962, 1964; Gupta, 1969; Arnold, 1972). However, there are considerable differences of opinion concerning hemocyte classifications and terminologies. There are also some remarkable differences between the hemocyte types in various insect orders. For example, in the larva of *Prodenia eridania*, Yeager (1945) recognized 10 classes, containing 32 types of hemocytes; in *Bombyx* larvae, Nittono (1960) reported five classes of hemocytes: prohemocytes (PRs), plasmatocytes (PLs), granulocytes (GRs), spherulocytes (SPs), and oenocytoids (OEs). Gupta (1969) classified insect hemocytes into eight types: PLs, GRs, OEs, SPs, adipohemocytes (ADs), podocytes (POs), vermicytes (VEs), and cystocytes (= coagulocytes, COs). Some authors recognize only three basic classes: PRs, PLs, and GRs. It also appears that there is greater differentiation of hemocytes in higher insect orders than in lower ones (Jones, 1962, 1964) (see also Chapter 4). Some authors consider amoebocytes, lamellocytes, POs, and VEs as various forms of PLs; COs and VEs as various forms of PLs; and COs and ADs as belonging to the category of GRs.

Recent electron microscopic studies have proved helpful in distinguishing various hemocyte types. For example, in *Bombyx mori*, the larval PLs and GRs, which were frequently indistinguishable under the light microscope, were clearly identifiable by their ultrastructures (Akai and Sato, 1973; Akai, 1976). Scanning electron microscopic (SEM) studies of the insect hemocytes also are necessary to elucidate their characteristic surface structures. Such studies are likely to supplement the information obtained by light and transmission electron microscopic (TEM) observations. For example, in *Bombyx* larvae, much morphological and functional information, which was difficult to obtain by TEM, was found by SEM (Akai and Sato, 1973, 1976).

Insect hemocytes have been classified also on the basis of their functions, such as phagocytosis, encapsulation, and wound healing. It appears that the same function is performed by different hemocyte types in various orders. For example, the GR is the only type of hemocyte that performs the phagocytic function in Lepidoptera, but it is the PL that is phagocytic in other orders.

Hemopoietic organs are very important in the consideration of hemocyte classification and differentiation. There organs are known to produce several hemocyte types.

In this chapter, we will describe by SEM and TEM the ultrastructure of the hemocyte types of five species, representing five orders.

5.2. Surface and internal ultrastructures

The insects used in this study are *Bombyx mori* (Lepidoptera), *Panesthia angustipennis* (Dictyoptera), *Locusta migratoria* (Orthoptera), *Holotrichia kiotoensis* (Coleoptera), and *Lucilia illustris* (Diptera).

5.2.1. *Bombyx mori* (Lepidoptera)

Ultrastructural studies of *B. mori* hemocytes have been reported by Akai and Sato (1971, 1973, 1976) and Sato and Akai (1977). There are five types of hemocytes.

Prohemocytes

In *Bombyx* larvae, PRs are usually round or oval in shape and small in size. They are characterized by a low concentration of intracellular organelles, especially the endoplasmic reticulum (ER), Golgi complex, and cytoplasmic inclusions (Fig. 5.9). By SEM, typical PRs are clearly distinguishable from other hemocytes by their surface structure, shape, and size (Fig. 5.1), but they are frequently confused with PLs.

Plasmatocytes

PLs are usually spindle-shaped (10–20 μm in length) or round (about 10 μm in diameter) (Figs. 5.1, 5.3, 5.4, 5.8). The ER and Golgi complex are well developed. The PL is one of the most common hemocytes and is found throughout the larval and adult stages. PLs have no phagocytic function, but do participate in encapsulation and wound healing.

By SEM, *Bombyx* PLs are recognized as spindle-shaped (Fig. 5.3), leaf-shaped (Fig. 5.4), sickle-shaped, and round (Fig. 5.1) cells, showing their active movement in circulating blood. PLs change their shapes several minutes after withdrawal, as pointed out by Nittono (1960).

Granulocytes

The GR also is one of the most common hemocytes. GRs are found throughout the larval and adult stages. Under the light microscope, GRs are recognized as spherical or oval cells that vary considerably in

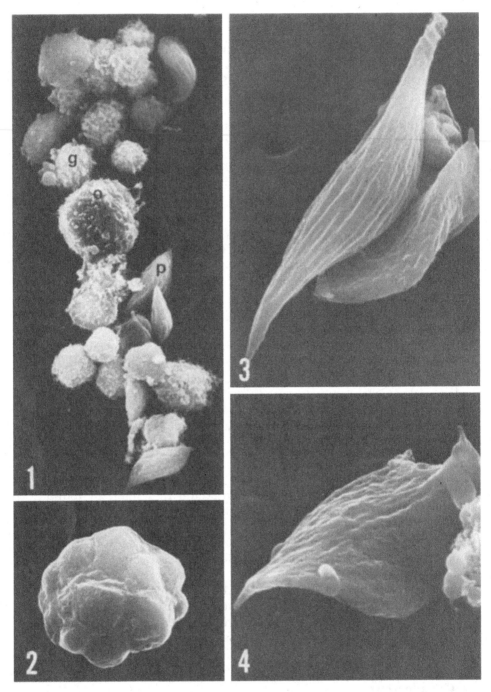

Figs. 5.1–5.4 SEM micrographs of larval hemocytes of *B. mori*. **5.1** Low-magnification picture of an aggregation of several types of larval hemocytes: plasmatocytes (*p*), granulocytes (*g*), and oenocytoids (*o*). × 1,600. **5.2** Typical spherulocyte. × 5,000. **5.3.** Spindle-shaped plasmatocyte near a leaf-shaped plasmatocyte. × 5,200. **5.4.** Leaf-shaped plasmatocyte: top (right upper corner) and tail (left lower corner) of cell. × 5,000. (**5.1–5.4** from Akai and Sato, 1976)

size. The cytoplasm contains numerous cytoplasmic inclusions, which are considered to be mucopolysaccharides (Nittono, 1960). Under TEM, the GRs are commonly characterized by specific secretory granules, which are packed with fibrous materials (Fig. 5.8). At higher magnification, the individual fibers are found to be 200–250 Å in diameter, each consisting of subunits (Akai and Sato, 1973). The GRs usually contain numerous enlarged granular cisternae packed with fine fibers. These cisternae frequently fuse with Golgi complexes, and the materials in the cisternae seem to be transferred into Golgi complexes and concentrated into secretory droplets, which are released into the hemolymph (Akai and Sato, 1973). These secretory materials seem similar to the materials of basement membranes in various tissues (Wigglesworth, 1973).

In *Bombyx* larvae, the GRs actively phagocytize India ink particles or hyphal bodies (Fig. 5.15). The ink particles are phagocytized into the cytoplasm and digested through lysosomes; however, it is not known whether the hyphal bodies of *Botrytis bassiana* are phagocytized and digested in the GRs or whether the GRs are killed by them. In the case of a larger foreign body (e.g., a nylon thread), GRs react actively and encapsulate it (Figs. 5.6, 5.7). In this case, the sheath is composed of two layers (Sato et al., 1976). The outer layer consists of only PLs (Fig. 5.13) and the inner one of only GRs, which release secretory materials between the cells in the tissue (Fig. 5.14). Melanization also occurs in the sheath (Fig. 5.14).

By SEM, the GRs are clearly distinguishable from other cells. They extend many cytoplasmic projections, which are usually covered with fibrous materials (Fig. 5.5).

Spherulocytes

SPs are clearly distinguishable from other hemocytes. In *Bombyx* larvae these hemocytes are oval in shape and contain many characteristic spherules in their cytoplasm (Figs. 5.2, 5.10). The spherules give strong reactions for neutral mucopolysaccharides and mucoproteins (Nittono, 1960). Under TEM, the spherules are usually 2–5 μm in diameter and contain homogeneous granular materials. Mature SPs release the fine granular materials into the hemolymph by exocytosis. At higher magnifications, the fine fibrous materials are detected in enlarged cisternae of rough endoplasmic reticulum (RER) and appear to be transferred into the Golgi complex. The Golgi complex concentrates the granular materials within a limiting membrane, and the mature spherule results. The fine granules in the mature spherule are uniformly 150–170 Å in diameter (Akai and Sato, 1973). By SEM, the SPs are clearly distinguishable by their characteristic surface structures (Fig. 5.2).

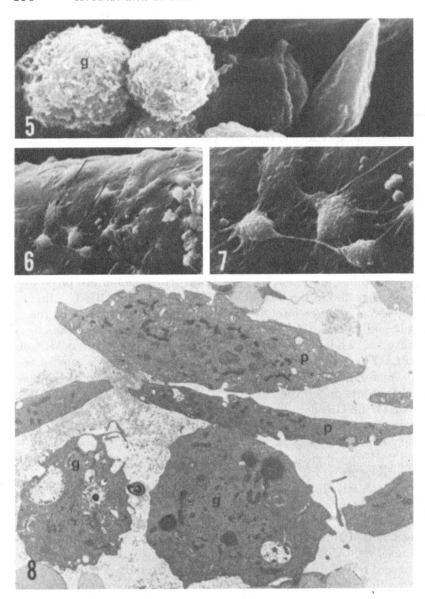

Figs. 5.5–5.7. SEM micrographs of larval hemocytes of *B. mori*. **5.5.** Granulocytes (*g*) are covered with fibrous materials. ×4,590. **5.6.** Adhered hemocytes (arrows) on surface of nylon thread inserted into body cavity of *Bombyx* larvae. ×1,275. **5.7.** Enlarged picture of hemocytes shown in Fig. 5.6. Cells contact each other by their cytoplasmic processes. ×3,315. (**5.5** from Akai and Sato, 1976; **5.6** and **5.7** from Sato et al., 1976)

Figs. 5.8–5.12. TEM micrographs of larval hemocytes of *B. mori*. **5.8.** A thicker and two slender plasmatocytes (*p*) in upper part and a large and a medium-size granulocyte (*g*) in lower part. ×3,145. **5.9.** Typical prohemocyte, showing several mitochon-

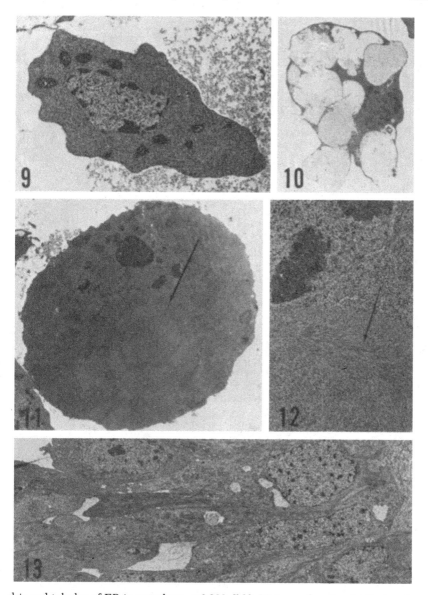

dria and tubules of ER in cytoplasm. ×6,800. **5.10.** Mature spherulocyte discharging materials from spherule. ×3,570. **5.11.** Mature oenocytoid, showing nucleus and large cytoplasmic inclusion (arrow). ×2,550. **5.12.** Part of cytoplasmic inclusion in mature oenocytoid, showing oriented fibers (arrow) near nucleus. ×34,000. (**5.8–5.10** from Akai and Sato, 1973)

Fig. 5.13. Part of outer layer of capsule tissue around nylon thread inserted into body cavity of *B. mori* larva, showing accumulation of plasmatocytes. ×3,400. (From Sato et al., 1976)

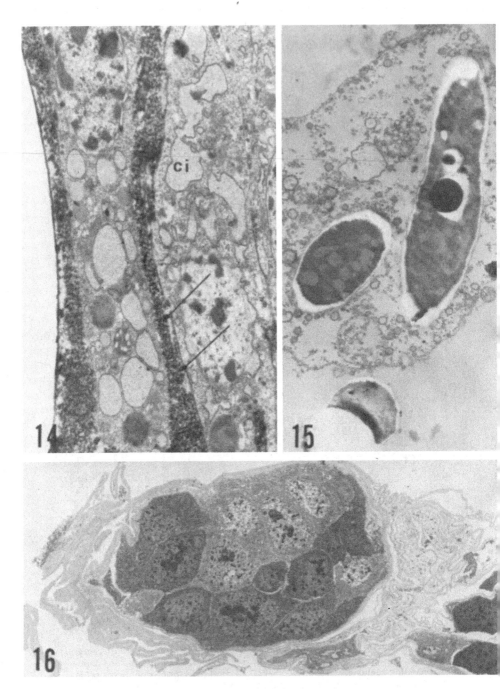

Fig. 5.14. Part of inner layer of capsule tissue around nylon thread inserted into body cavity of *B. mori* larva, showing several characteristic ultrastructural changes: enlarged cisternae of granular ER (*ci*), swelled nuclear membrane, and accumulated melanized materials (arrows). ×8,000. (From Sato et al., 1976)

Fig. 5.15. Phagocytized hyphal bodies in granulocyte of *B. mori* larva. ×8,000.

Fig. 5.16. Hemopoietic organ in fourth instar of *Bombyx* larva containing many young hemocytes. ×3,000. (From Sato and Akai, 1977)

Oenocytoids

In *Bombyx* larvae, OEs are large, spherical cells characterized by large cytoplasmic inclusions (Figs. 5.1, 5.11, 5.12). These cells range in diameter from 8 to 20 μm. In mature OEs, the inclusions appear to be made up of clusters of fine fibers that are arranged in concentric circles (Figs. 5.11, 5.12). The surrounding cytoplasm is filled with numerous ribosomes and bundles of fine fibers, but the ER is poorly developed at this stage. At higher magnifications, sectioned fibers are seen to have a diameter of about 300 Å, and each is composed of several fibrils (Akai and Sato, 1973).

5.2.2. *Panesthia angustipennis* (Dictyoptera)

By light and electron microscopic studies, five types of hemocytes (PRs, PLs, GRs, SPs, and COs) are observed in several species of dictyopteran insects. They are *Periplaneta americana* (Baerwald and Boush, 1970; Baerwald, 1975); *Leucophaea maderae* (Hagopian, 1971); *Blaberus cranifer, Byrsotria fumigata,* and *Gromphadorhina portentosa* (Scharrer, 1972); and *Blattella germanica, Pycnocellus surinamensis,* and *Sphodromantis bioculata* (Price and Ratcliffe, 1974; Ratcliffe and Price, 1974). We found four types in *Panesthia angustipennis.*

Prohemocytes

The typical PR is a small round or elongate cell, with a large nucleus. The cytoplasm is undifferentiated and has numerous free ribosomes, a few mitochondria, and electron-dense granules.

Plasmatocytes

PLs are round or spindle-shaped cells with irregular profiles (Figs. 5.20, 5.21). The cytoplasm contains a small amount of RER, large numbers of free ribosomes, poorly differentiated Golgi complexes, and a variable number of electron-dense granules. Most of the granules in the PLs are homogeneous and electron-dense. By SEM, the PLs are seen as flat and round figures. Also, they are found in higher ratio than other hemocytes (Fig. 5.18).

Granulocytes

GRs are easily identifiable under the light microscope. They are highly refractive, nonameboid, round or oval cells, 12–16 μm in diameter, with a central nucleus that is often masked by large numbers of granules, 1–1.5 μm in diameter (Fig. 5.22).

By TEM, it is quite difficult to distinguish the GRs from the granular PLs. At higher magnification, the cytoplasm shows several types of

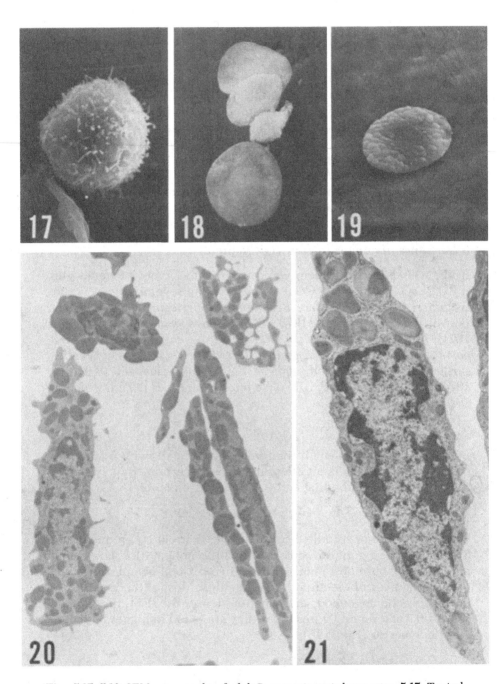

Figs. 5.17–5.19. SEM micrographs of adult *P. angustipennis* hemocytes. **5.17.** Typical granulocyte characterized by numerous fine cytoplasmic processes. × 3,000. **5.18.** Three flat plasmatocytes and a small cell, which may be a prohemocyte. × 1,200. **5.19.** Flat spherulocyte. Sculptured patterns are visible on cell surface. × 1,500.
Figs. 5.20, 5.21. TEM micrographs of adult *P. angustipennis* hemocytes. **5.20.** Medium-sized plasmatocyte (lower left side) and small flat plasmatocytes (lower right side). × 4,000. **5.21.** Spindle-shaped plasmatocyte containing a large nucleus. × 8,000.

138

granules that are distinct in size, electron density (Fig. 5.24), and internal structure (Fig. 5.25).

In typical GRs, several cytoplasmic processes are resolved by TEM (Fig. 5.22) and numerous processes in surface structure by SEM (Fig. 5.17). Cytoplasmic processes of the GRs may characterize this type of hemocyte.

Spherulocytes

SPs are usually round, 14–20 μm in diameter, with a small, often eccentric nucleus that is usually masked by a large number of spherules, 1.5–3 μm in diameter (Fig. 5.23). In many spherules, a single electron-dense core is centrally located and surrounded by a granular region bounded by a unit membrane (Fig. 5.26). Occasionally, the SPs discharge the amorphous materials in the spherules into the hemolymph (Fig. 5.23).

By SEM, SPs are easily distinguished by their surface structure and are usually round and flat, with a characteristic surface pattern attributable to the presence of the spherules inside, as shown in Fig. 5.19.

5.2.3. *Locusta migratoria* (Orthoptera)

On the basis of electron microscopic observations, four types of hemocytes have been reported in *Locusta migratoria* (Hoffmann et al., 1969). We found five types and reticular cells.

Prohemocytes

The PRs are usually small, round to oval cells, 6–13 μm in diameter, with a thin rim of cytoplasm. By SEM, they appear as small and round cells (Fig. 5.27).

Plasmatocytes

Usually, a PL is spindle-shaped or round with a large nucleus (Figs. 5.31, 5.32). In the cytoplasm, a small amount of filamentous RER and a few electron-dense granules (0.3–0.6 μm in diameter) are scattered. Several cytoplasmic projections are detected on their cell surfaces. By SEM, PLs appear as smooth-surfaced, spindle-shaped cells (Fig. 5.28).

Granulocytes

The GRs are round and oval, with irregular surface structure and numerous electron-dense granules (0.2–0.8 μm in diameter) in the cytoplasm (Fig. 5.29). Although some tubular ER is observed in the cytoplasm, the Golgi complexes and mitochondria are scarcely detected. The cell surface seems to undulate, along with some processes. GRs

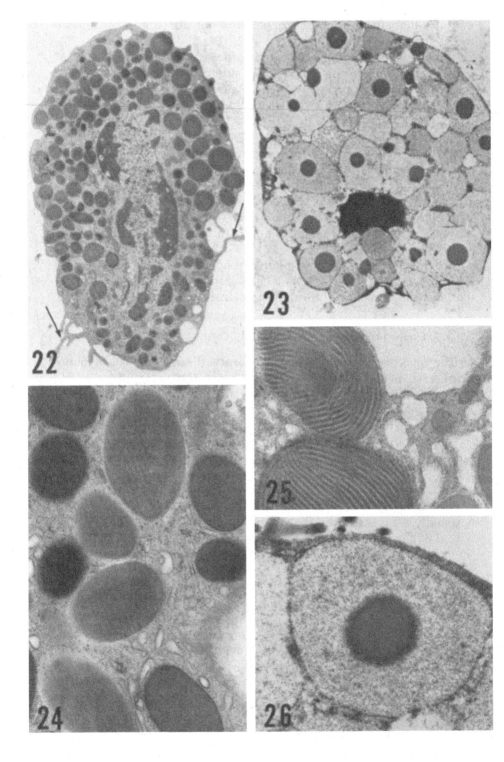

participate in encapsulation of foreign bodies and in wound healing (Brehélin et al., 1976).

By SEM, their characteristic undulated surface shows many processes (Fig. 5.27).

Oenocytoids

The OEs are large cells, showing characteristic spindle- or leaf-shaped profiles (Figs. 5.27, 5.30). The nucleus is large, with a large amount of nuclear material (Fig. 5.30). Around the nucleus, numerous characteristic tubular structures in parallel distributions (Fig. 5.30) are observed. These features have already been described by Cassier and Fain-Maurel (1968) and Hoffmann (1970). In the cytoplasm, small numbers of electron-dense granules are scattered (Fig. 5.30). Small amounts of tubular ER and mitochondria are distributed. Scattered polysomes are seen at high magnification. By SEM, their profiles are leaf-shaped rather than spindle-shaped, with several cytoplasmic projections (Fig. 5.27).

Coagulocytes

Hoffmann et al. (1970) observed the COs in *L. migratoria* and reported well-developed ER, electron-dense granules, and fibrous materials in the cytoplasm. Our observations support this description. Brehélin et al. (1975) reported that the COs show phagocytosis.

Reticular cells

In addition to the above hemocytes, a small number of large reticular cells (25–30 μm in diameter) are sometimes seen among the hemocytes (Fig. 5.33). The nucleus in these cells is centrally located. The cytoplasm shows reticular figures and numerous vacuoles and cytoplasmic processes (Fig. 5.33). At high magnification, large amounts of tubular ER and numerous well-developed Golgi complexes are seen.

5.2.4. *Holotrichia kiotoensis* (Coleoptera)

Devauchelle (1971) described six types of hemocytes: PRs, PLs, GRs, ADs, SPs, and OEs. He considered pòdocytes (POs) and vermicytes (VEs) as PLs, and found COs barely distinguishable from GRs. In addition, many intermediate stages among PRs, PLs, GRs, and ADs were

---

Figs. 5.22–5.26. TEM micrographs of adult *P. angustipennis* hemocytes. **5.22.** Typical granulocyte containing numerous dense granules in cytoplasm. Several cytoplasmic processes (arrows) can be seen. × 6,500. **5.23.** Spherulocyte characterized by many large spherules, each containing a single dense core. × 4,500. **5.24.** Enlarged dense granules in cytoplasm of granulocyte. × 35,000. **5.25.** Enlarged cytoplasmic granules showing premelanosome-like structure in granulocyte. × 23,000. **5.26.** Enlarged spherule containing a single dense core. × 23,000.

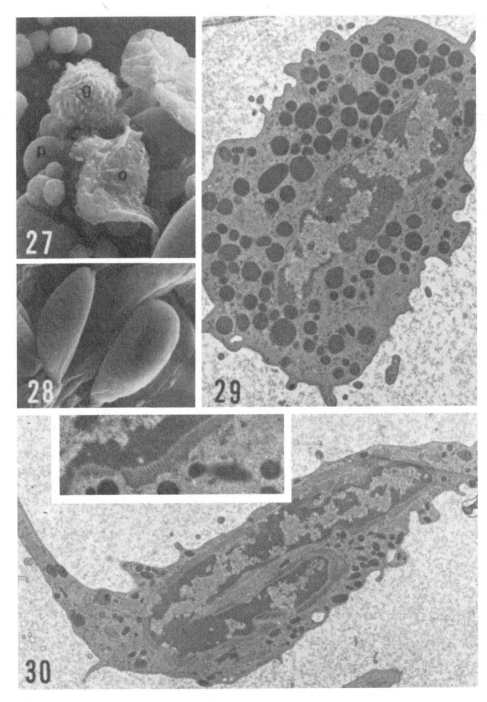

observed. We found the same six hemocytes in *H. kiotoensis* by SEM and TEM.

Prohemocytes
The PRs are round or spindle-shaped. Small numbers of electron-dense granules are detected. Both ER and Golgi complexes are usually undeveloped, but in some cells well-developed Golgi complexes were seen.

Plasmatocytes
The PLs are spindle-shaped (Fig. 5.37), round, ameboid, and vermiform cells. These PLs are characterized by the small number and size of electron-dense granules in the cytoplasm. Small amounts of ER and mitochondria are also detected in the cytoplasm. Intermediate forms between PLs and GRs are frequently observed. They contain more electron-dense granules in the cytoplasm. By SEM, PLs appear as round and flat cells (Fig. 5.34).

Granulocytes
The GR is one of the most common hemocytes in Coleoptera. GRs are usually round or barrel-shaped, containing electron-dense granules (0.3–0.5 μm in diameter) in the cytoplasm (Fig. 5.40). Well-developed tubular ER and Golgi complexes are present. Several vacuoles are detected. Intermediate cells between the PLs and GRs are frequently observed (Fig. 5.39). By SEM, GRs appear as barrel-shaped granular cells (Fig. 5.36). They are easily distinguished by the characteristic surface patterns of the cells.

Adipohemocytes
The ADs are characterized by many electron-dense granules with tubular structures. Well-developed Golgi complexes are frequently detected in the cytoplasm.

Spherulocytes
The SP is a common hemocyte type in Coleoptera. The cells may be rod-shaped, spindle-shaped, and of various other shapes. They are characterized by large electron-dense spherules (Fig. 5.38). The nu-

Figs. 5.27, 5.28. SEM micrographs of *L. migratoria* hemocytes. **5.27.** Large granulocyte (*g*), leaf-shaped oenocytoid (*o*), and small prohemocytes (*p*). × 1,500. **5.28.** Two spindle-shaped plasmatocytes. × 1,500.
Fig. 5.29, 5.30. TEM micrographs of *L. migratoria* hemocytes. **5.29.** Granulocyte containing numerous electron-dense granules in cytoplasm. × 7,500. **5.30.** Spindle-shaped oenocytoid. Characteristic tubular structures are distributed around nucleus. × 5,300. **Inset:** Enlarged picture of tubular structures in cytoplasm. × 18,000.

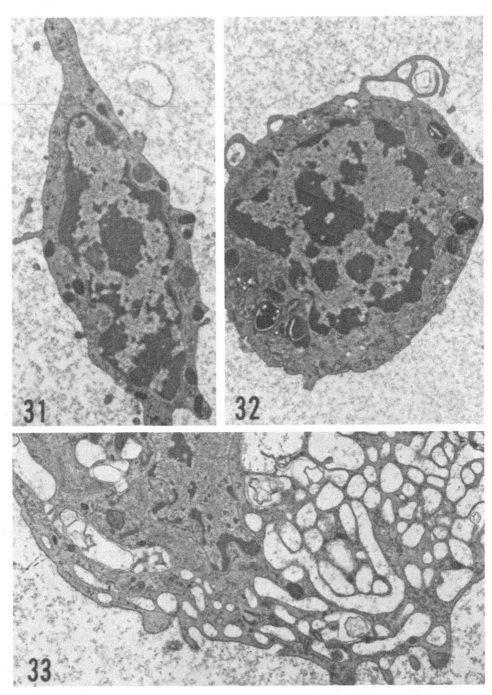

Figs. 5.31–5.33. TEM micrographs of *L. migratoria* hemocytes. **5.31.** Spindle-shaped plasmatocyte containing a large nucleus. ×10,000. **5.3^.** Round plasmatocyte containing a large nucleus. ×7,500. **5.33.** Part of a large reticular cell showing vacuolated cytoplasm. ×9,000.

cleus is usually small. Small numbers of Golgi complexes are clearly detected. Intermediate cells between the PLs and SPs are found. By SEM, SPs are easily distinguished by the undulated surface patterns attributable to the presence of spherules (Fig. 5.35).

Oenocytoids

The OEs are giant cells, about 25 μm in diameter. The nucleus is small. In the cytoplasm, there are numerous ribosomes, polysomes, and a small number of mitochondria. Small numbers of lytic vacuoles are also detected.

5.2.5. *Lucilia illustris* (Diptera)

We studied the larval hemocytes of *L. illustris* by TEM and SEM (Akai and Sato, 1977) and found four types of hemocytes: PRs, PLs, GRs, and ADs. In a few cases, oenocytoid-like large cells were recognized by SEM.

Prohemocytes

The PRs are present in small numbers during the larval stages. Typical PRs are small (about 5 μm in diameter) and are characterized by undeveloped intracellular organelles (Fig. 5.42). The nucleus is comparatively large, containing a big nucleolus. Small amounts of RER and mitochondria are scattered in the cytoplasm. By SEM, PRs frequently appear as round cells that correspond to PLs (Fig. 5.41).

Plasmatocytes

PLs are fairly common hemocytes in the larval stages. The typical PL is about 10 μm in diameter. The cell surface is smooth without any pseudopod-like projections (Fig. 5.43). There is a large nucleus containing a massive nucleolus. In the cytoplasm, tubular ER increases as compared with the PRs. Sometimes, small vacuoles and lytic vacuoles appear.

Granulocytes

The GRs are the most common hemocytes in the larval stages. They vary (6–11 μm in diameter) in size and are characterized by several long, pseudopod-like cytoplasmic projections (Figs. 5.44, 5.45). In the cytoplasm, well-developed granular ER and Golgi complexes are found. Lysosomes and lytic vacuoles are also located in the cytoplasm, indicating active phagocytosis. In some large GRs, phagocytized materials are frequently seen. The GRs contain many droplets with dense and intermediate electron density. The relative number of these droplets may vary in various cells. By SEM, GRs appear as rod- and spindle-shaped hemocytes (Fig. 5.41).

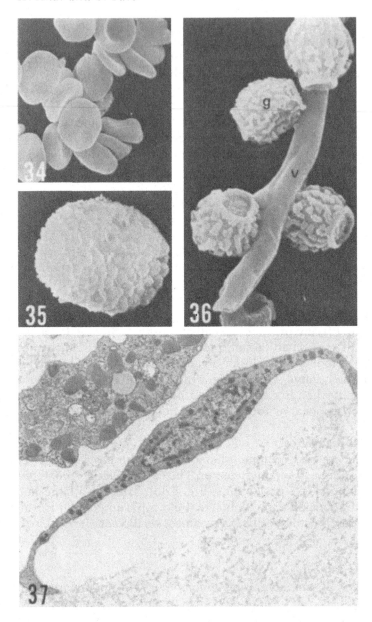

Figs. 5.34–5.36. SEM micrographs of *H. kiotoensis* hemocytes. **5.34.** Aggregated plasmatocytes. ×2,295. **5.35.** Typical spherulocyte characterized by sculptured surface patterns. ×3,400. **5.36.** Barrel-shaped granulocytes (g) and vermiform plasmatocyte (v). Characteristic cytoplasmic projections are visible. ×6,715.

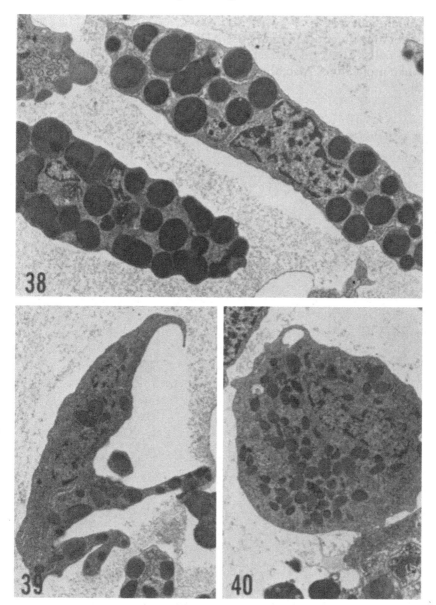

Figs. 5.37–5.40. TEM micrographs of *H. kiotoenis* hemocytes. **5.37.** Spindle-shaped plasmatocyte. ×6,800. **5.38.** Two spherulocytes containing numerous electron-dense spherules. ×8,700. **5.39.** An irregular-shaped cell (intermediate cell between plasmatocyte and granulocyte). ×6,525. **5.40.** Granulocyte containing a small nucleus and numerous dense granules in cytoplasm. ×6,525.

Adipohemocytes

Under the electron microscope, ADs appear as rod- or spindle-shaped cells. They are characterized by the absence of any pseudopod-like projections. Also, they contain many dense lytic bodies (Fig. 5.46).

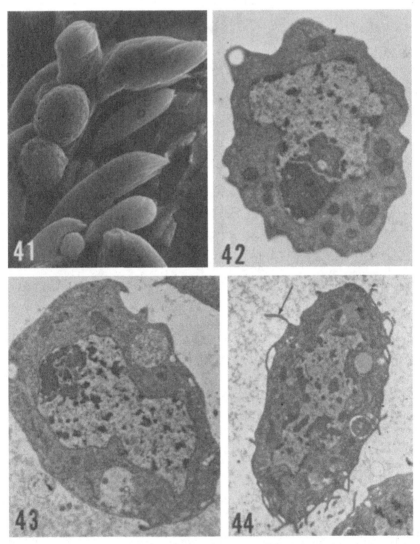

Fig. 5.41. SEM micrograph of *L. illustris* hemocytes. Plasmatocytes (*p*) and granulocytes (*g*) can be seen. ×1,914.
Figs. 5.42–5.46. TEM micrographs of *L. illustris* hemocytes. **5.42.** Typical prohemocyte, showing a large nucleus and undifferentiated cytoplasm. ×11,310. **5.43.** Plasmatocyte, showing a large nucleus and some large vacuoles in cytoplasm. ×6,960. **5.44.**

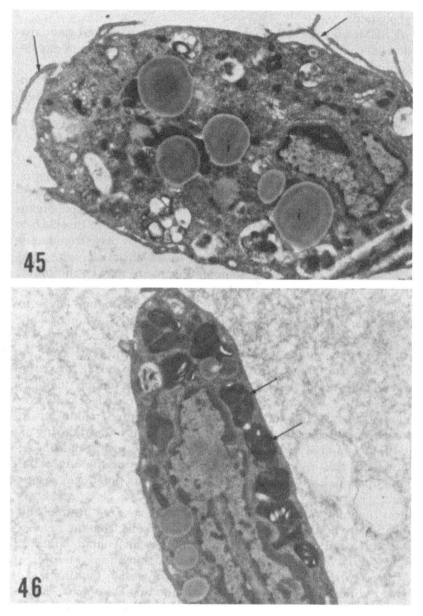

Granulocyte characterized by several pseudopod-like cytoplasmic projections (arrow). ×8,000. **5.45.** Enlarged picture of a granulocyte. Several pseudopod-like cytoplasmic projections (arrows), cytoplasmic inclusions (*i*), and lytic vacuoles are visible. ×11,310. **5.46.** Adipohemocyte containing many dense granules (arrows) in cytoplasm. ×11,310.

Oenocytoids

With SEM, a few large hemocytes were detected in aggregated materials from centrifuged blood. Such large cells, however, were not detected under TEM.

In *Drosophila*, various figures of crystalline materials were observed in the cytoplasm of crystal cells (= OEs) under TEM (Gateff, 1977, pers. comm.).

5.3. Hemopoietic organs and hemocytic differentiation

Several kinds of hemopoietic tissues – ranging from freely circulating blood cells, cells loosely adherent to the tissues, or cells forming transient aggregations, to collections of cells that form distinct organs more or less constant in size and position – are known in insects (see also Chapter 2). In Lepidoptera, the hemocytes may arise from hemopoietic organs located near or attached to the imaginal wing disks (Arvy, 1952). In *B. mori*, these hemopoietic organs may be the source of PRs and PLs (Nittono et al., 1964).

Recently, we studied the hemopoietic organs of *Bombyx* larvae by light and electron microscopy (Akai and Sato, 1971). The hemopoietic organs increase in size and number during larval development. The size of the organs increases only during the early stages of each of the instars, and the number of the organs increases only during the period between apolysis and ecdysis. The volume of all hemopoietic organs decreases during the same period, owing to the release of cells from hemopoietic organs into the hemolymph. During the larval instars incorporation of tritiated thymidine into the organs is highest during the period between apolysis and ecdysis. In the fifth instar, there is extremely heavy incorporation of the thymidine during the 3 days preceding pupation. Early in the pupal stage, the hemopoietic organs disappear.

Ultrastructurally, the hemopoietic organs are covered by an acellular sheath with openings that allow passage of cells from the organ to the blood. During the intermolt stages, the organs contain spherical cysts made up of electron-opaque cells and numerous undifferentiated hemocytes (Fig. 5.16). However, during each molting stage, fairly well-differentiated hemocytes are found in these organs. For example, during the larval-pupal molting stage, PLs, GRs, SPs, and OEs are recognized in the organs. This clearly indicates that in some Lepidoptera the hemocytes are already differentiated into various types, with characteristic structures in the hemopoietic organs, and that probably differentiation of one type into another does not occur in the hemolymph. In *Bombyx* larvae, however, dividing PRs, GRs, and SPs are found in the blood (Nittono, 1960).

5.4. Discussion

We studied the surface and the internal ultrastructures of hemocytes in five species and found considerable differences among these insects. Even in the same species, the hemocytes may differ in different strains or at different developmental stages. In lepidopterous and dictyopterous insects, for example, the SPs are absent in some species and strains. In addition, cells intermediate between types of hemocytes were detected. Therefore, it is generally difficult to distinguish a distinct type from an intermediate form.

In the case of *Bombyx* larvae, the hemocytes are mainly of five types (Nittono, 1960; Akai and Sato, 1973). Among these hemocytes, OEs, SPs, and GRs are clearly distinguishable from other hemocytes by their characteristic ultrastructures. However, the distinction between the PRs and PLs is occasionally unclear, although typical cells have the characteristic profiles. When we observed the PLs by SEM, we detected several characteristic shapes, such as a typical spindle shape, a long spindle shape, leaf shape, sickle shape, and pear shape. Spindle- and leaf-shaped PLs are able to move, but not so the sickle- and pear-shaped ones. The internal ultrastructures are fairly similar in all types of PLs. In addition to their ultrastructures, we compared the functions of the hemocytes in five orders from published papers (Table 5.1).

In *B. mori*, functions of hemocytes have been studied by light microscope (Iwasaki, 1930; Nittono, 1960), transmission electron microscope (Akai and Sato, 1971, 1973; Akai, 1976; Sato et al., 1976), and scanning electron microscope (Akai and Sato, 1976; Sato et al., 1976). Both the GRs and the PLs function in defense against larger foreign

Table 5.1. Various hemocytes and some of their functions

Insect	Phagocytosis	Encapsulation	Wound healing	Trephocytosis
Bombyx mori (Lepidoptera)	Granulocytes	Granulocytes Plasmatocytes	Plasmatocytes	Granulocytes Spherulocytes Oenocytoids
Leucophaea maderae (Dictyoptera)	?	Granulocytes Plasmatocytes	?	Spherulocytes
Locusta migratoria (Orthoptera)	Plasmatocytes Coagulocytes	Granulocytes	Granulocytes	Oenocytoids
Calliphora erythrocephala (Diptera)	Plasmatocytes	Thrombocytoids	?	Spherulocytes Oenocytoids (= crystal cells)
Rhodnius prolixus (Hemiptera)	Plasmatocytes	?	Plasmatocytes	Granulocytes Oenocytoids

bodies in *Bombyx* larvae. However, both hemocyte types make different layers and have different roles in the encapsulated tissue. In Dictyoptera, the same types of hemocytes participate in encapsulation
(Table 5.1).

In wound healing, the PLs are involved in *B. mori* and the GRs in *L. migratoria*. SPs and OEs are typical trephocytes without any characteristic behavior. They release the cytoplasmic secretory materials
into the blood in *Bombyx* larvae. The larval SPs disappear in the early
pupal stage, and then the adult SPs appear. The OEs disappear during
the early pupal stage. The function of these two types of hemocytes
seems to be related to the nutritional situation in the blood. On the
other hand, the OEs of *P. angustipennis* are spindle-shaped hemocytes with active behavior in profile (Fig. 5.30). It seems that the OEs
in these two species differ considerably in their functions. Phagocytosis and encapsulation are most useful tools for analyzing hemocyte
functions. Hemopoietic organs also are very important in this respect.
In *Bombyx* larvae, all types of hemocytes are produced in these
organs, located around the wing disks; they release these hemocytes
only during each molting stage and larval-pupal metamorphosis (Akai
and Sato, 1971; Sato and Akai, 1977). In Orthoptera, the hemopoietic
organs are functional in adults (Hoffmann, 1970). Information on hemopoietic organs and their role in hemocyte differentiation is not yet
enough. Although cell division of the circulating hemocytes may
occur in the hemolymph, differentiation of hemocytes seems to occur
in the hemopoietic organs.

5.5. Summary

Using scanning and transmission electron microscopy, we studied
comparative ultrastructures of hemocyte types in five species, representing five orders: *Bombyx mori* (Lepidoptera), *Panesthia angustipennis* (Dictyoptera), *Locusta migratoria* (Orthoptera), *Holotrichia
kiotoensis* (Coleoptera), and *Lucilia illustris* (Diptera).

In *B. mori*, five types of hemocytes (prohemocytes, plasmatocytes,
granulocytes, spherulocytes, and oenocytoids) were observed on the
basis of their characteristic ultrastructures and functions. Hemopoietic organs attached to the imaginal wing disks produce some types of
young hemocytes during the molting stages and the larval-pupal metamorphosis. In *P. angustipennis*, four types of hemocytes (prohemocytes, plasmatocytes, granulocytes, and spherulocytes) were recognized. In *L. migratoria*, five types of hemocytes (prohemocytes,
plasmatocytes, granulocytes, oenocytoids, and coagulocytes) and reticular cells were seen. In *H. kiotoensis*, six types of hemocytes (prohemocytes, plasmatocytes, granulocytes, adipohemocytes, spherulocytes, and oenocytoids) were observed. In *L. illustris*, four types of

hemocytes (prohemocytes, plasmatocytes, granulocytes, and adipohemocytes) were recognized.

It is suggested that comparative hemocyte morphology should be studied by scanning and transmission electron microscopy. Also, ultrastructures, related to hemocyte functions in many more species, should be studied to resolve the question of hemocytic differentiation.

References

Akai, H. 1976. Hemocytes, pp. 183–236. *In* H. Akai (ed.). *Ultrastructural Morphology in Insects*. University of Tokyo Press, Tokyo.

Akai, H., and S. Sato. 1971. An ultrastructural study of the hemopoietic organs of the silkworm, *Bombyx mori*. *J. Insect Physiol.* 17:1665–76.

Akai, H., and S. Sato. 1973. Ultrastructure of the larval hemocytes of the silkworm, *Bombyx mori* L. (Lepidoptera: Bombycidae). *Int. J. Insect Morphol. Embryol.* 2:207–31.

Akai, H., and S. Sato. 1976. Surface ultrastructure of the larval hemocytes of the silkworm, *Bombyx mori* L. (Lepidoptera: Bombycidae). *Int. J. Insect Morphol. Embryol.* 5:17–21.

Akai, H., and S. Sato. 1977. Surface and internal ultrastructure of the hemocytes in *Lucilia illustris* (in preparation).

Arnold, J. W. 1972. A comparative study of the hemocytes (blood cells) of cockroaches (Insecta: Dictyoptera: Blattaria), with a view of their significance in taxonomy. *Can. Entomol.* 104:309–48.

Arvy, L. 1952. Particularités histologiques des centres leucopoïétiques thoraciques chez quelques Lépidoptères. *C. R. Acad. Sci.* 23:1539–41.

Baerwald, R. J. 1975. Inverted gap and other cell junctions in cockroach hemocyte capsules: A thin section and freeze-fracture study. *Tissue Cell* 7:575–85.

Baerwald, R. J., and G. M. Boush. 1970. Fine structure of the hemocytes of *Periplaneta americana* (Orthoptera: Blattidae) with particular reference to marginal bundles. *J. Ultrastruct. Res.* 31:151–61.

Brehélin, M., J. A. Hoffmann, G. Matz, and A. Porte. 1975. Encapsulation of implanted foreign bodies by hemocytes in *Locusta migratoria* and *Melolontha melolontha*. *Cell Tissue Res.* 160:283–9.

Brehélin, M., D. Zachary, and J. A. Hoffmann. 1976. Functions of typical granulocytes in scar formation in *Locusta migratoria* L. *J. Microsc. Biol. Cell* 25:133–6.

Cassier, M. P., and M. A. Fain-Maurel. 1968. Sur la présence de microtubules dans l'ergastoplasme et l'espace perinucléaire des oenocytoïdes du criquet migrateur, *Locusta migratoria migratorioides* (R. et F.). *C. R. Acad. Sci. Paris* 266:686–9.

Devauchelle, G. 1971. Etude ultrastructurale des hémocytes du Coléoptère *Melolontha melolontha* (L.). *J. Ultrastruct. Res.* 34:492–516.

Gupta, A. P. 1969. The hemocytes of *Epicauta cinerea*, and a synonymy of hemocyte terminologies. *Cytologia* 34(2):300–44.

Hagopian, M. 1971. Unique structures in the insect granular hemocytes. *J. Ultrastruct. Res.* 36:646–58.

Hoffmann, J. A. 1970. Haemopoietic organ in Orthoptera. *Z. Zellforsch. Mikrosk. Anat.* 106:451–72.

Hoffmann, J. A., A. Porte, and P. Joly. 1970. Sur la localisation d'une activité phenol-oxydasique dans les coagulocytes de *Locusta migratoria* L. (Orthoptère). *C. R. Acad. Sci. Paris* 270:629–31.

Hoffmann, J. A., M. E. Stoekel, A. Porte, and P. Joly. 1969. Ultrastructure des hémocytes de *Locusta migratoria* (Orthoptère). *C. R. Acad. Sci. Paris* 266:503–5.

Iwasaki, Y. 1930. Researches on the larval blood corpuscles of *Bombyx mori* and nine other Lepidoptera. *Bull. Kagoshima Imp. Coll. Agric. Forest.* 8:172–284.

Jones, J. C. 1962. Current concepts concerning insect hemocytes. *Amer. Zool.* 2:209–46.

Jones, J. C. 1964. The circulatory system of insects, pp. 2–94. *In* M. Rockstein (ed.). *The Physiology of Insecta*, 1st ed. Academic Press, New York.

Nittono, Y. 1960. Studies on the blood cells in the silkworm, *Bombyx mori* L. *Bull. Seric. Exp. Stn.* 16:171–266.

Nittono, Y., S. Tomabechi, and N. Onodera. 1964. Formation of hemocytes near the imaginal wing disc in the silkworm, *Bombyx mori* L. *J. Seric. Sci. Jpn.* 33:43–5.

Price, C. D., and N. A. Ratcliffe. 1974. A reappraisal of insect haemocyte classification by the examination of blood from fifteen insect orders. *Z. Zellforsch. Mikrosk. Anat.* 147:537–49.

Ratcliffe, N. A., and C. D. Price. 1974. Correlation of light and electron microscopic hemocyte structure in the Dictyoptera. *J. Morphol.* 144:485–98.

Sato, S., and H. Akai. 1977. Development of the hemopoietic organs of the silkworm, *Bombyx mori* L. *J. Seric. Sci. Jpn.* 46:397–403.

Sato, S., H. Akai, and H. Sawada. 1976. An ultrastructural study of capsule formation by *Bombyx mori*. *Annot. Zool. Jpn.* 49:177–88.

Scharrer, B. 1972. Cytophysiological features of hemocytes in cockroaches. *Z. Zellforsch. Mikrosk. Anat.* 129:301–19.

Wigglesworth, V. B. 1959. Insect blood cells. *Annu. Rev. Entomol.* 4:1–16.

Wigglesworth, V. B. 1973. Haemocytes and basement membrane formation in *Rhodnius*. *J. Insect Physiol.* 19:831–44.

Yeager, J. F. 1945. The blood picture of the southern armyworm (*Prodenia eridania*). *J., Agric. Res.* 71(1):1–40.

6   Fine structure of hemocyte membranes and intercellular junctions formed during hemocyte encapsulation

R. J. BAERWALD

Department of Biological Sciences, University of New Orleans,
Lakefront, New Orleans, Louisiana, 70122, U.S.A.

Contents

6.1 Introduction

Hemocyte activities involving the cell membrane – including in vitro cell migrations, general membrane substructure, and encapsulation – will be considered in this chapter in terms of the now widely accepted fluid mosaic model of the biological membrane (Singer and Nicolson, 1972). Singer's model (Fig. 6.1) is widely accepted.

The fluid mosaic concept of the living biological membrane suggests that phospolipids form a bilayered matrix or backbone of the

155

"unit" membrane. The hydrophilic phosphate groups are oriented toward both surfaces of the membrane, with the hydrocarbon tails of the phospholipids sequestered in the middle of the membrane, excluding water. A very important part of the model suggests that in the living membrane the phospholipids are free to move by changing places with each other in the plane of the membrane, but ordinarily do not change sides; hence, the "fluid" nature of the membrane. Hydrophobic bonding holds the acyl chains together within the thickness of the

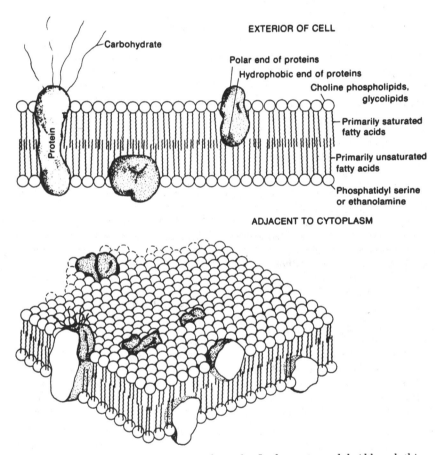

Fig. 6.1. Diagrammatic representation of popular fluid mosaic model. Although this model has since been modified, it remains essentially as depicted here. Note three types of integral proteins or protein assemblies embedded in phospholipid bilayer: (1) ectoproteins protruding into extracellular space, (2) endoproteins in contact with cytoplasm, and (3) protein complexes that extend completely through unit membrane. Integral proteins are seen as IM (intercalated membrane) particles by freeze-fracture method. Nonintegral surface proteins may also be loosely associated with membrane as well as cytoplasmic microtubules or microfilaments, depending on membrane or cell type. (From Ville and Dethier, 1976)

membrane so that ordinarily 180° rotations or flip-flops of the phospholipids with respect to the two-dimensional plane of the membrane would not occur.

The "mosaic" portion of the model describes the arrangement of protein components embedded in the membrane. These "integral" proteins (Nicolson, 1974) are also free to move within the fluid plane of the membrane and are of three types: (1) ectoproteins, (2) endoproteins, and (3) protein assemblies traversing the phospholipid bilayer (Fig. 6.1). Proteins completely embedded within the phospholipid bilayer have not been demonstrated (Rothman and Lenard, 1977). Other proteins and glycoproteins may also be peripherally associated with this complex, loosely bound to the surface of the membrane by weak electrostatic forces. One recent elaboration on the 1972 fluid mosaic model suggests that contractile microtubule-microfilament assemblies may be associated with the endoproteins by weak electrostatic forces (Nicolson, 1974). Many lines of experimental evidence now support, or are at least consistent with, these concepts, including the hemocyte membrane substructural details produced by freeze-fracture technique considered in this chapter.

6.2. Factors influencing hemocyte migrations

6.2.1 In vivo observations

It is now clear that insect hemocytes engage in very active ameboid-type movements involving the plasma membrane, especially in vitro. These activities are similar to those observed with mammalian macrophages and lymphocytes. Such migrations are of potential use to the insect in wound healing, phagocytosis, and capsule formation. In vivo observations of single hemocytes have been observed through transparent wing veins (Arnold, 1959, 1972) by phase-contrast light microscopy. This approach has the advantage of observing the cells under "natural conditions." However, because the cells must be viewed through a cylindrical wing vein, resolution is somewhat limited, although a remarkable amount of cellular detail can still be seen in many cockroach species (Arnold, 1972), but not all.

6.2.2. In vitro observations

In vitro observations (Baerwald and Boush, 1971a), using polished glass plane surfaces, as expected, yielded superior resolution when compared with images obtained in vivo through a cylindrical wing vein, although the experimental condition has the disadvantage of being "unnatural." This condition stimulates hemocyte attachment to

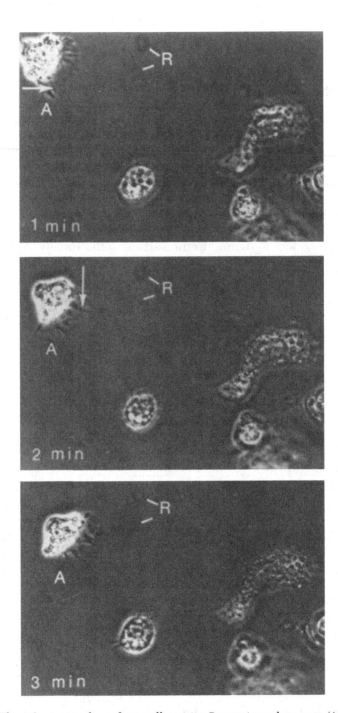

Fig. 6.2. Three 1-min time-lapse frames illustrating *P. americana* hemocyte (A) migration *in vitro*. As a point of reference, note position on frames, of unidentified material (R), which did not move during this period. Note fan shape (arrows) associated with migrating hemocyte compared with other hemocytes (in pictures) that were not migrating and assumed a variety of random shapes. ×1,800. (From Baerwald and Boush, 1971a)

the substrate and allows a detailed description of the overall cell shape during migrations. Hemocytes migrating in vitro displayed an overall fan shape, with a flattened leading edge and a rounded-up trailing edge pulling the nucleus along behind (Fig. 6.2). Hemocyte migration speed averaged 1.7 μm/min and varied from 0.6 to 3.0 μm/min. The overall shape and speed of migrating hemocytes in vitro were similar to those of migrating mammalian fibroblasts in vitro (Ingram, 1969; Abercrombie et al., 1970).

6.2.3. Effects of mitotic poisons on hemocytes

The in vitro hemocyte study (Baerwald and Boush, 1971a) also showed that very high concentrations (hemocoel injections) of vinblastine (10^{-3} M) resulted in an accumulation of hemocytes in early metaphase, but had no effect on cell migrations. The mitotic index increased from a normal of 0.1% to a high of 10% after 48 hr. This concentration was eventually fatal (48–72 hr postinjection) to adult *Periplaneta americana*. However, lower concentrations (10^{-4} M or 10^{-5} M) achieved similar effects without death of the insect (pers. observ.). Accumulation of mitotic figures, all in early metaphase, indicates that the microtubule mitotic spindle was disrupted under these conditions. In addition, by standard sectioning techniques and electron microscopy, Baerwald and Boush (1971b) demonstrated an expected complete loss of microtubules making up the marginal bundle. A microtubule marginal bundle from *P. americana* hemocytes is shown in cross section in Fig. 6.3. These data, coupled with observations made by mammalian cell researchers (Borisy and Taylor, 1967a,b) studying the effects of vinblastine and other alkaloids (colchicine) that depolymerize microtubules, suggest that microtubules are not involved in active plasma membrane movements associated with the cell migration. However, it has been shown that the mold extract cytochalasin B (Spooner et al., 1971) affects microfilaments and that certain microfilaments are apparently involved in the type of hemocyte migrations observed in vitro and in vivo. Although it is clear that cytochalasin B can indeed interfere with certain types of cell movements, the mechanism of action of this drug on microfilaments and indeed the exact role microfilaments play in ameboid-type cell movements remain controversial.

Nevertheless, useful experiments could be conducted to determine if cytochalasin B and other related drugs will reversibly affect hemocyte migrations both in vitro and in vivo. If they do, similar experiments could be designed to determine possible reversible effects on encapsulation and other related hemocyte activities in the presence of these drugs.

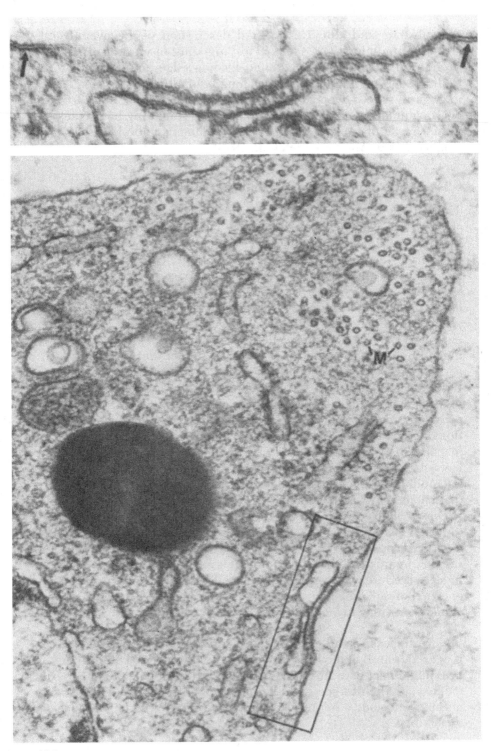

6.3. External hemocyte membrane

6.3.1. Standard electron microscopic method

The plasma membrane (Fig. 6.3) and internal membranes (Fig. 6.4) are revealed by standard electron microscopy as a typical "trilaminar" unit membrane. En bloc staining with uranyl acetate enhances membrane profiles. The typical "unit membrane" (Robertson, 1960) can be resolved in high-resolution photomicrographs as having two $\simeq$25-Å electron-dense laminae separated by a $\simeq$30-Å electron-lucent lamina. This unit membrane cross-section profile is thought by at least some researchers to be an artifact rather than to reflect the true nature of the membrane (see review by Staehelin, 1974); others feel that some useful information can be gleaned from this profile (see review by McNutt and Weinstein, 1973). These latter authors stated: "The structure represented by this trilaminar appearance does accurately reflect the spatial position of the main permeability barrier at membranes." In any case, standard thin-section techniques have revealed very few useful details with respect to the true macromolecular arrangement of the unit membrane or membrane specializations.

6.3.2. Freeze-fracture method

The freeze-fracture or freeze-cleave method introduced by Moor and Muhlethaler (1963) has without doubt contributed more to our understanding of cell membrane substructure than any other electron microscope technique. There are several important reasons for the current popularity of the freeze-fracture method among membranologists. The method exposes large areas of biological membranes in three-dimensional relief with resolutions comparable to those obtainable with standard thin-section techniques. Another important feature of this method is that the cleavage plane splits the membrane down the middle, generating two novel fracture faces (see McNutt and Weinstein, 1973, for review on the fracturing process).

The juxtacytoplasmic half of the membrane viewed by this technique was originally termed "fracture face A," and the half membrane closest to the extracellular space has been called "fracture face B" (Satir and Gilula, 1973). These two fracture faces have recently been renamed "fracture face P" (protoplasmic) and "fracture face E" (extracellular) (Branton et al., 1975). The terms P and E fracture faces will

Fig. 6.3. *P. americana* hemocyte illustrating microtubule (*M*) marginal bundle and limiting trilaminar plasma membrane (box and **inset** with arrows). En bloc staining with uranyl acetate enhances membrane profiles. **Main figure** × 100,000; **inset** × 257,000.

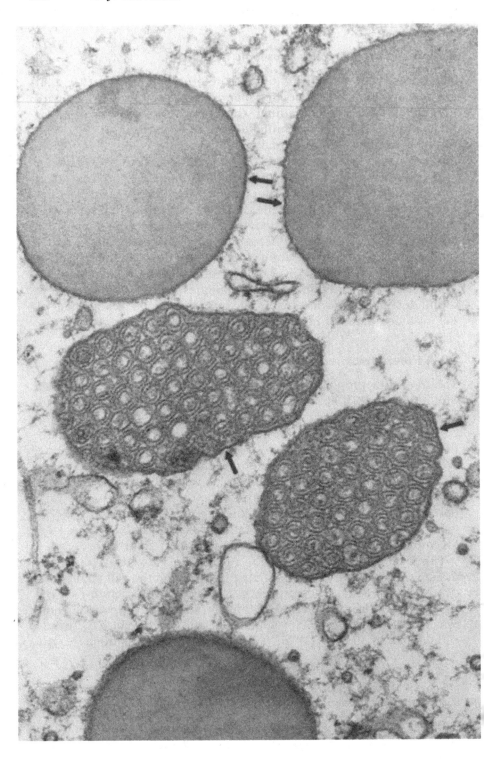

be used in this chapter. A diagrammatic cross-section profile of a hypothetical freeze-fractured hemocyte has been provided in Fig. 6.5 to assist in interpretation. In addition, the reader is referred to Chapter 21 in this book for more detailed descriptions of general freeze-fracture technique.

In addition to revealing novel membrane fracture faces, this technique also usually allows direct visualization of ≈80-Å bumps or intercalated membrane (IM) particles. These particles have been shown to represent protein assemblies (Branton and Deamer, 1972), with the P face usually bearing higher concentrations of IM particles than the E face.

Finally, small areas of the true surface of a membrane can be exposed by an optional ancillary procedure called freeze etching, if glycerol is avoided during the freezing step. Freeze etching removes some of the ice surrounding the cell by sublimation, exposing the external unfractured "true" surface of the plasma membrane. Because true freeze etching is a completely different process from freeze fracturing, these terms should not be used interchangeably.

Many of the important physiological activities of the hemocyte – including phagocytosis, wound repair, cell migrations, release of cell

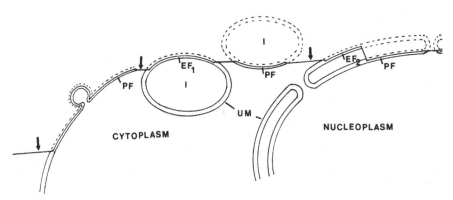

Fig. 6.5 Diagrammatic representation of a hypothetical freeze-fractured hemocyte including two cytoplasmic inclusion bodies (*I*). Labeling is based on current freeze-cleave nomenclature (Branton et al., 1975). Drawing is not to scale. Unfractured unit membranes are represented by wide parallel solid lines (*UM*). Narrow parallel solid lines represent remaining freeze-fractured half membrane (*PF* = protoplasmic face; *EF*₁ = exoplasmic face; *EF*₂ = endoplasmic face). Broken lines indicate cellular material removed during fracturing process; jagged lines (arrows) indicate cross-fractured hemolymph or cytoplasm.

(facing page)
Fig. 6.4. *P. americana* hemocyte showing membrane-bounded structured and unstructured inclusion bodies. En bloc staining with uranyl acetate greatly enhances visualization of the unit membrane (arrows). Freeze-fractured profiles of membrane surrounding unstructured inclusion bodies are seen in Fig. 6.9A,B. × 90,000.

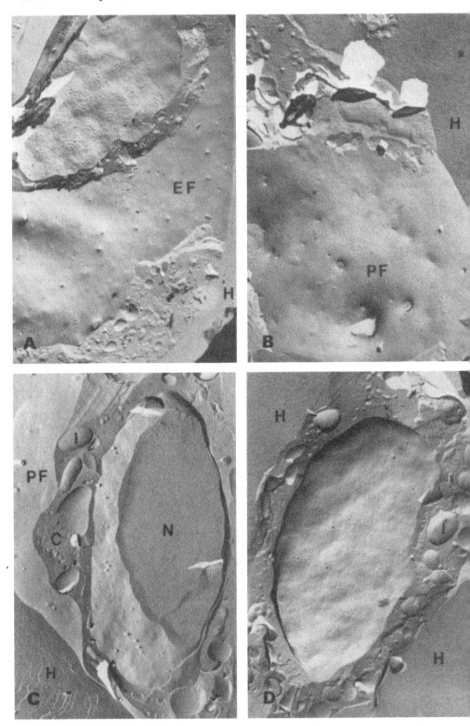

inclusions, encapsulation, and certain aspects of hemolymph coagulation–involve biological membranes. In view of this, it is disappointing that this method has not been utilized more frequently in the study of hemocyte structure and function.

Freeze-fractured hemocyte plasmalemma

The hemocyte plasma membrane not involved in capsule formation was examined by the freeze-fracture method and found to be similar to the plasma membranes of mammalian cells in having an asymmetric distribution of randomly scattered IM particles ($\approx$2,500/μm²; P face; $\approx$800/μm², E face) (Baerwald, 1974). The sizes of the largest E face particles were somewhat unusual (150 Å in diameter) compared with mammalian IM particles (60–120 Å in diameter) (McNutt and Weinstein, 1973). This is noteworthy because the E-type gap junctions are apparently formed during capsule formation from these rather large IM particles (Baerwald, 1975). Four selected freeze-fractured hemocytes are presented in Fig. 6.6 at low magnification to illustrate some of the membrane profiles revealed by this method.

Cytoplasmic blebbing was observed by this method to good advantage (Fig. 6.7) and probably represents a normal activity such as exocytosis of the cell, although the possibility that the blebbing may be produced as a fixation artifact cannot be ruled out at this time. In any event, the blebs are of interest in that they were found to be particle-free on both P and E faces (arrows, Fig. 6.7) even though they were apparently formed from or passed through the normal particle-studded plasma membranes (Fig. 6.7) (Baerwald, 1975).

The freeze-fracture method offers exciting potential for studying several interesting aspects of hemocyte lysis and gross membrane blebbing typically associated with coagulation in insects. Further, comparative studies using this method may reveal some fundamental structural differences in hemocyte membranes between species in which hemolymph coagulation is apparently accomplished in different ways, as can be determined by light microscopy (see Grégoire, 1974, for review on coagulation).

Fig. 6.6. Four low-magnification micrographs illustrating some of various ways hemocytes can be freeze-fractured. Each view is from a different *P. americana* hemocyte that had been monodispersed in the glutaraldehyde-precipitated clot prior to freezing. Very large areas of protoplasmic fracture face (*PF*) and extracellular fracture face (*EF*) of plasma membrane can be seen, along with large areas of freeze-fractured nuclear membrane. Cross-fractured nucleoplasm (*N*), cytoplasm (*C*), inclusions (*I*), and hemolymph (*H*) appear as completely featureless areas by this method. **A** and **B** × 7,000; **C** and **D** × 10,000. (**A** and **B** from Baerwald, 1974; **C** and **D** from Teigler and Baerwald, 1972)

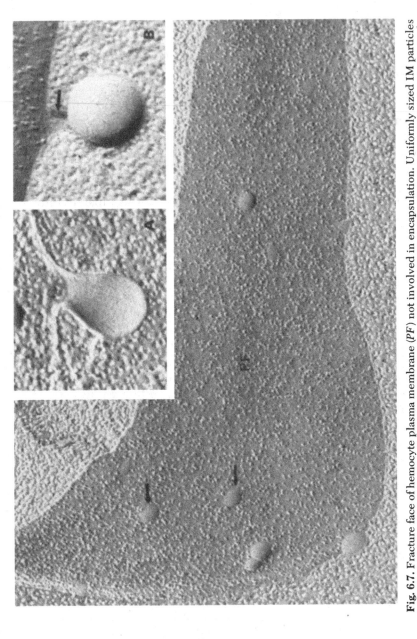

Fig. 6.7. Fracture face of hemocyte plasma membrane *(PF)* not involved in encapsulation. Uniformly sized IM particles can be clearly seen scattered randomly over fracture face, except where blebbing (possibly exocytosis) has occurred (arrows). At least 10 separate blebs can be seen in various stages of development. **Inset A** shows a convex profile of one of these vesicles budding from and almost completely free of P face, although still attached by a narrow "neck" (arrow in **B**). **Inset B** shows opposite fracture face of such a vesicle. Note absence of IM particles in both instances. **Main figure** ×61,500. **A** and **B** ×120,540. (**A** and **B** from Baerwald, 1974)

6.4. Internal hemocyte membranes

6.4.1. Nuclear envelope

The nuclear envelope of a hemocyte viewed in cross section is similar to that of other (Fig. 6.8A) cell types, displaying inner and outer unit membranes interrupted by, and continuous with, each other at the sites of nuclear pores. However, when viewed by grazing "en face" sections through the nuclear envelope complex (Fig. 6.8B) or by freeze fracture (Fig. 6.8C), clustering of nuclear pores was observed in the hemocytes of *P. americana* by thin sectioning (Baerwald and Boush, 1970) and by freeze fracture (Teigler and Baerwald, 1972). Nuclear pore clustering (functional significance unknown) was often observed in complete and incomplete hexagonal arrays (Fig. 6.8C), indicative of tight packing. However, the pores are restricted from touching each other by intervening membranes of the nuclear envelope. Clustering of nuclear pores in this fashion is not typical for cells in general and presumably not typical for hemocytes either, although the proof of this statement with regard to hemocytes will have to await further studies.

The distribution and size of the IM particles embedded in the membranes of the cockroach hemocyte nuclear envelope were typical of the asymmetric distributions of mammalian cell IM particles described by Branton (1969) and Branton and Deamer (1972). The P face bears $\simeq 1,500$ particles/μm^2 (80–100 Å in diameter) (Baerwald, 1974); the E face bears $\simeq 1,300$ particles/μm^2 (80–120 Å in diameter).

6.4.2. Membranes associated with cytoplasmic inclusions

Hemocytes are well known for their numerous and often elaborately structured cytoplasmic inclusions. Closer examination by standard thin-section techniques (Fig. 6.4) reveals that these inclusions are bounded by a typical unit membrane (Baerwald and Boush, 1970; Scharrer, 1972). However, when examined by the freeze-fracture method, these membranes, at least in *P. americana* hemocytes, are unusual in several respects. First, the typical asymmetric distribution of particles is not seen (Fig. 6.9A,B), with both E and P faces bearing the same ($\simeq 800$/μm^2) concentration of particles (Baerwald, 1974). Symmetric distributions of IM particles representing integral proteins on P and E fracture faces are seen less frequently than asymmetric distributions. Second, more recent and closer observations have shown that these membranes often reveal pits as well as particles on the fracture face (Fig. 6.9B) (see Fig. 6.5 for further explanation of fracture planes). A recent review by Rothman and Lenard (1977) deals with the asymmetry in membranes normally seen and the topological problems as-

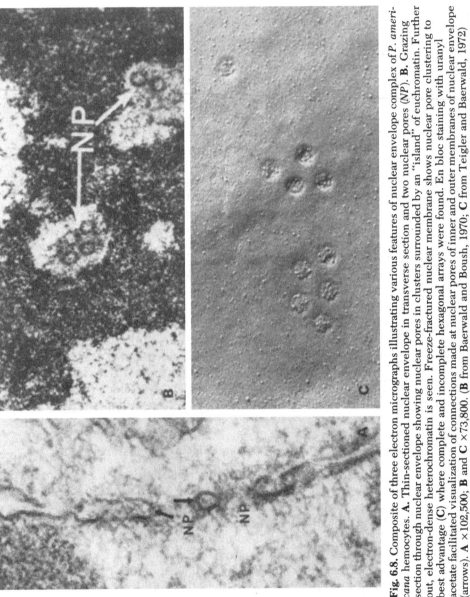

Fig. 6.8. Composite of three electron micrographs illustrating various features of nuclear envelope complex of *P. americana* hemocytes. **A.** Thin-sectioned nuclear envelope in transverse section and two nuclear pores (*NP*). **B.** Grazing section through nuclear envelope showing nuclear pores in clusters surrounded by an "island" of euchromatin. Further out, electron-dense heterochromatin is seen. Freeze-fractured nuclear membrane shows nuclear pore clustering to best advantage (C) where complete and incomplete hexagonal arrays were found. En bloc staining with uranyl acetate facilitated visualization of connections made at nuclear pores of inner and outer membranes of nuclear envelope (arrows). **A** ×102,500; **B** and **C** ×73,800. (**B** from Baerwald and Boush, 1970; **C** from Teigler and Baerwald, 1972)

sociated with synthesis and insertion of integral "ectoproteins" and "endoproteins" in the biological membrane.

One might reasonably expect pits to show up as frequently as particles from the fracturing process, but with few exceptions pits are not seen. At least two reasons have been suggested to explain the usual lack of pits when membrane fracture faces are examined (McNutt and Weinstein, 1973). First, the shadowing process will tend to increase the size of raised areas (particles) and diminish or obliterate pits formed after fracturing. Second, it is thought that, at least in some membranes, a partial fluidity or plasticity may still be present during the fracturing process, so that pits momentarily formed from the intercalated protein components being pulled from the phospholipid bilayer would be lost during the short delay between fracturing and shadowing.

Additional freeze-fracture studies of the membranes associated with cytoplasmic inclusions from other species should be done to determine whether this "unusual" symmetric membrane is common or not. Hemocytes lyse quickly upon exposure to air (Jones, 1962), which should facilitate inclusion isolation by relatively simple centrifuge techniques for biochemical analysis of this membrane. Many hemocytes are unique with respect to the high concentrations of inclusions they contain. This symmetric membrane may, therefore, represent a major component for hemocytes such as granulocytes and coagulocytes (= cystocytes).

6.5. Hemocyte capsules

6.5.1. General organization

The encapsulation reaction in insects is a cellular defense reaction to a living or nonliving foreign substance introduced into the hemocoel. This reaction involves only hemocytes, not other types of cells, and as such is unique to arthropods. The piling on in successive layers and flattening out of hemocytes onto metazoan parasites or plastic implants effectively sequester foreign substances away from the hemolymph (Fig. 6.10).

Although considerable early literature is available on the structure and development of insect capsules at the light microscope level (see Salt, 1963, for early review), not until 1967 (Grimstone et al., 1967; Mercer and Nicholas, 1967; Poinar et al., 1968; Reik, 1968, cited by Smith, 1968) were ultrastructural studies first performed on the insect capsule. More recent reviews have been done considering structure, development, and function of insect capsules at the light and/or electron microscope level (Robinson and Strickland, 1969; Salt, 1970;

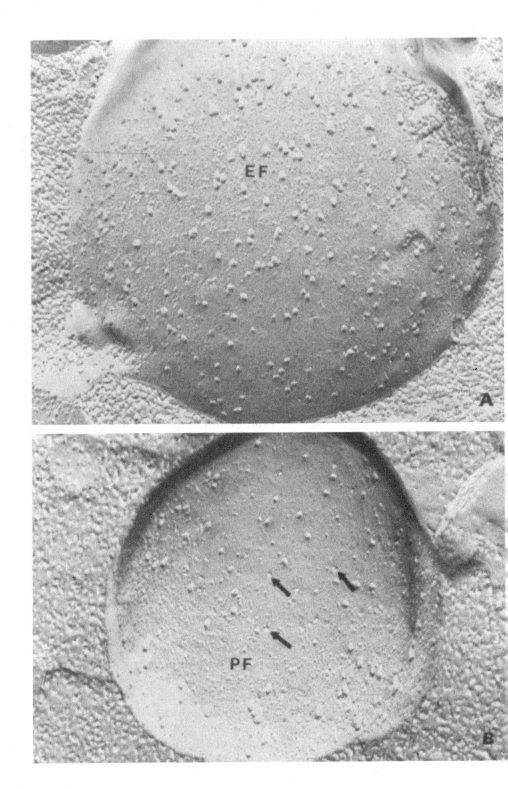

Sutherland, 1971; Vey, 1971; Vinson, 1971; Rotheram and Crompton, 1972).

The early work of Grimstone et al. (1967) deserves special mention because of the general high quality of the micrographs and the remarks concerning hemocyte structure and function. One of the objectives of this work was to determine whether syncytia were formed at any place in the capsule. Although the hemocytes were closely appressed to each other, especially near the core of the capsule, the plasma membranes remained intact. However, the inner cells showed evidence of degradation and melanization. All ultrastructural studies done to date indicate that syncytia are never formed and that the innermost cells are typically very closely appressed with little or no extracellular material existing between these cells. Complete or incomplete desmosomes, gap junctions, and possibly a septate junction were also described in the Grimstone et al. (1967) paper.

Since 1967, many additional papers on capsule ultrastructure have appeared in the literature. More recent capsule ultrastructural studies have been done by Poinar and Leutenegger (1971), Misko (1972), Baerwald (1975), Brehélin et al. (1975), François (1975), Ratcliffe and Gagen (1977), and others. However, most works have been largely descriptive, and more experimental and high-resolution substructure studies are needed before a reasonable understanding of the encapsulation reaction can be had.

An understanding of the substructure and function of the plasma membrane of the hemocyte, both free in circulation as well as involved in capsule formation, is central to a further elucidation of the process of encapsulation. Fortunately, in recent years, new techniques in electron microscopy (e.g., the freeze-fracture method) have contributed substantially to unit membrane structure analysis. Unfortunately, this new technique has been largely ignored in the study of hemocytes.

6.5.2. Growth and size of insect capsules

Encapsulation is a unique process for studying the factors that stimulate cell adhesion because the cells involved are normally monodispersed and free in the hemolymph and do not form cell junctions with each other unless a capsule is being formed. Hemocytes are added to a

Fig. 6.9. Convex (*EF*) (**A**) and concave (*PF*) (**B**) freeze-fractured profiles of unit membrane surrounding cytoplasmic inclusion bodies. This membrane was found to have an equal distribution of IM particles on two fracture faces – a rather unusual situation for glutaraldehyde-fixed and freeze-fractured membranes. Pits (**B**) as well as particles were found on P fracture face (arrows), also an unusual feature of *most* freeze-fractured membranes (see text). **A** and **B** × 130,000.

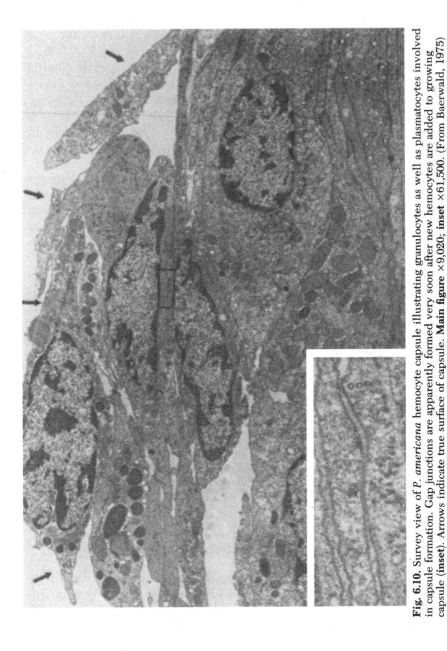

Fig. 6.10. Survey view of *P. americana* hemocyte capsule illustrating granulocytes as well as plasmatocytes involved in capsule formation. Gap junctions are apparently formed very soon after new hemocytes are added to growing capsule (**inset**). Arrows indicate true surface of capsule. **Main figure** ×9,020; **inset** ×61,500. (From Baerwald, 1975)

growing capsule from the outside, forming a multilayered capsule 20–50 cells thick, depending on the species and/or age of the capsule. As each layer of cells is attached, the surface properties (plasma membrane) of these cells must change so that they are recognized as "foreign" by newly attached hemocytes (Grimstone et al., 1967) in order to allow the capsule to continue growing. It is also clear that this changing surface property loses its effectiveness at some point, as capsules are limited in size after several days, depending on the species, and usually do not continue to grow longer than 3 days or so.

The number of cell layers making up 24- to 48-hr capsules in different species has been fairly similar: ≈20 in *Melolontha* (Brehélin et al., 1975), ≈30 in *Periplaneta* (Baerwald, 1975), and ≈20–25 in *Ephestia* (Grimstone et al., 1967). Usually, little or no increase is seen in cell layers after 48 hr (Baerwald, 1975; Brehélin et al., 1975), although up to ≈75 cell layers have been counted in 72-hr *Ephestia* capsules (Grimstone et al., 1967).

6.5.3. Hemocyte types making up capsules

Capsule ultrastructural studies published to date suggest that considerable variation exists among species with respect to the kinds of hemocytes that contribute to capsule formation. It has been shown that granulocytes (GRs) are the prime source of cells making up capsules in the flour moth (Smith, 1968), in *Melolontha* and *Locusta* (Brehélin et al., 1975), in *P. americana* (Baerwald, 1975), and in *Galleria mellonella* (Ratcliffe and Gagen, 1977). Plasmatocytes (PLs) are apparently the predominant cell type found in stick insect capsules (Grimstone et al., 1967). PLs (hemocyte types with few or no dense cytoplasmic granules) also have been observed in *Periplaneta* capsules (Fig. 6.10). A somewhat unusual capsule has been observed in *Calliphora* (Zachary et al., 1975), where the hemocytes, called "thrombocytoids," making up the capsule are apparently fragmented. It was determined by light microscopic studies of insect capsules that "lamellocytes" made up *Drosophila* capsules (Walker, 1959), and oenocytoids (OEs) were thought to participate in capsules studied by Nappi and Stoffolano (1971) (see also Chapter 13).

Interspecies variation in capsule cellular makeup is to be expected to the extent that free circulating hemocytes vary in type from species to species. In addition, it appears that not all types of hemocytes found in free circulation participate in capsule formation in any particular species. The significance of this will have to await a clearer understanding of the developmental relationship of one hemocyte type to another. Studies of hemocyte development in vitro and perhaps capsule formation in vitro may help shed light on these problems.

6.6. Intercellular junction types formed during encapsulation

Although the literature is still incomplete, it appears that a variety of cell junction types may occur in hemocyte capsules, depending on the insect species. It is anticipated that the list will grow as more species are studied and as we come to recognize more and more cell junctions as discrete and separate structures with respect to structure and function (see Gilula and Satir, 1971, and Staehelin, 1974, for reviews on intercellular junction types). Table 6.1 summarizes the distribution of intercellular junction types in hemocyte capsules studied so far.

6.6.1. Desmosomes

Desmosomes or desmosome-like junctions have been clearly demonstrated by standard electron microscope techniques in capsules of *P. americana* (Baerwald, 1975), *Melolontha melolontha* (Brehélin et al., 1975), *Locusta migratoria* (Brehélin et al., 1976), and *Galleria mellonella* (Ratcliffe and Gagen, 1977), but not *Calliphora erythrocephala* capsules (Zachary et al., 1975). Only partially formed desmosomes were described in *Ephestia kühniella* (Grimstone et al., 1967). Desmosomes or desmosome-like cell junctions formed between hemocytes look similar to desmosomes formed between other insect cell types (Hagopian, 1970). Desmosomes have been described in other insect tissues from many other insect species (see Satir and Gilula, 1973, for review) and are thought to function as local "weld spots" to hold neighboring cells together, but are not thought to contain channels for intercellular ionic communication (Satir and Gilula, 1973). Freeze-fracture studies of desmosomes have revealed that membrane changes often do not occur (Gilula and Satir, 1971). Insect hemocyte

Table 6.1. Interspecies distribution of hemocyte capsule intercellular junctions

Species	Junction type[a]			Authors
	Gap	Desmosome	Septate	
Ephestia kühniella	+	±	±	Grimstone et al. (1967)
Carausius morosus	±	+	−	Smith (1968)
Calliphora erythrocephala	−	−	−	Zachary et al. (1975)
Locusta migratoria	−	+	−	Brehélin et al. (1975)
Melolontha melolontha	−	+	−	Brehélin et al. (1975)
Periplaneta americana	+	+	±	Baerwald (1975)
Galleria mellonella	−	+	−	Ratcliffe and Gagen (1977)

[a] ± indicates that this junction may be present but confirmation is needed.

desmosomes were not recognized by the freeze-fracture method in the single report available (Baerwald, 1975), nor were IM particles seen associated with this type of cell junction. However, thin-section studies reveal that this same capsule contains many desmosomes (Fig. 6.11). More freeze-fracture studies are needed to determine if the hemocyte desmosomes observed by thin-section techniques can be observed by the freeze-fracture method. Although a paucity of literature now exists even with respect to freeze-fracture visualization of vertebrate desmosomes, those reports that are available (see Kelly and Shienvold, 1976, for review) indicate that clustered intramembranous particles are sometimes seen associated with these structures, similar to freeze-fractured gap junctions. More work needs to be done to determine whether invertebrate desmosomes are less likely than vertebrate desmosomes to display IM particles when viewed by the freeze-fracture method. This would constitute one basis for a structural comparison between vertebrate and invertebrate desmosomes.

6.6.2. Hemocyte gap junctions

The fine structural study of intercellular junctions has come of age with several excellent and recent reviews now available (McNutt and Weinstein, 1973; Satir and Gilula, 1973; Staehelin, 1974), with Satir and Gilula confining their remarks to invertebrate junctions. Insect cell junctions have also been reviewed by Lane et al. (1977) and insect gap junctions by Flower (1977). Gap junctions have been shown in general to be electrically low-resistance, cell-to-cell contacts that allow molecules of low molecular weight (300–500 MW) to pass freely, resulting in a practical cell-to-cell communication. This type of junction has received much attention because a junction that allows exchange of materials between living cells is important to the overall coordination of cells with a common purpose (e.g., heart tissue) and to the coordinated development of cells into tissues during growth of the organism.

Gap junctions have been identified in some insect capsules (Grimstone et al., 1967; Smith, 1968; Baerwald, 1975), but not all (Brehélin et al., 1975, 1976). Gap junctions formed between insect blood cells and viewed by thin-section techniques appear essentially indistinguishable from vertebrate gap junctions. Cross-section profiles in both types reveal a seven-layered membrane complex, with a central 20- to 30-Å electron-lucent gap, typically observed in high-magnification micrographs (Fig. 6.12A,B).

However, when viewed by the freeze-fracture method, a polarity shift is seen with respect to the intramembranous particle arrangement making up mammalian and insect gap junctions. Clustered IM particles associated with the hemocyte gap junction remain adhered

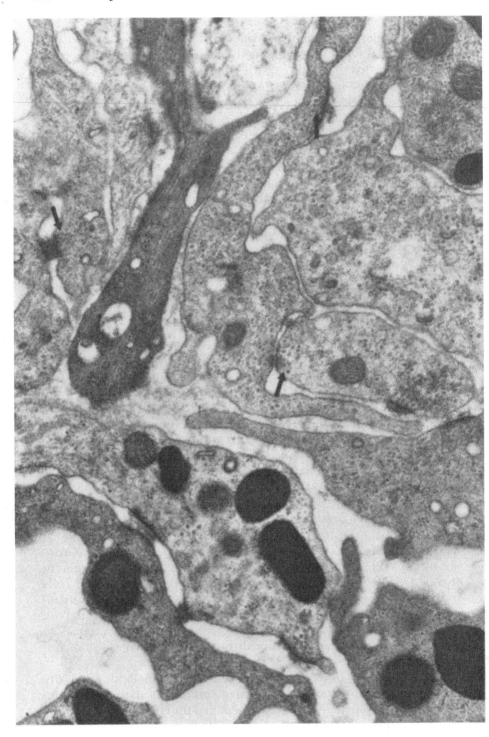

to the half membrane closest to the extracellular (Fig. 6.13A) space, with pits associated with the half membrane closest to the cytoplasm (Fig. 6.13B); the distribution of particles in mammalian gap junctions is reversed. Another interesting difference observed is that the particles making up the hemocyte gap junction are significantly larger– 130 Å (Baerwald, 1975) – than mammalian cell gap junction particles, which are typically 60–80 Å (McNutt and Weinstein, 1973). Insect gap junctions have been called "B"-type gap junctions by Satir and Gilula (1973) or "inverted" gap junctions by Flower (1972), with the gap junction type found in the mammal described as an "A" type junction.

In keeping with a recent renaming of the two fracture faces by Branton et al. (1975), it is proposed that mammalian gap junctions be referred to as "P-type gap junctions," as the IM particles making up these junctions usually remain adhered to the P fracture face (i.e. the half membrane closest to the protoplasm). Insect gap junctions, including those formed between hemocytes during encapsulation, should now be referred to as "E-type gap junctions" because the IM particles making up these complexes usually adhere to the extracellular (E) fracture face during the process of freeze fracturing (Fig. 6.13A). The reader is referred to a model of an E-type gap junction redrawn by the author to include the new nomenclature (Fig. 6.14).

E-type gap junctions in other invertebrates

E-type gap junctions have been observed in several arthropod species, including mantid and cockroach epithelial cells (Flower, 1971) and midgut of termites (Satir and Gilula, 1973). One report exists of E-type gap junctions formed between hemocytes during capsule formation (Baerwald, 1975). Although the number of arthropod species examined for E-type junctions so far is small, P-type junctions have not been found in this group of animals. However, P-type gap junctions have been found in other invertebrates (mollusca) (Flower, 1971; Gilula and Satir, 1971). It has been suggested that E-type gap junctions may be characteristic of Arthropoda (Staehelin, 1974; Baerwald, 1975). However, a recent study (Wood, 1977) has revealed E-type gap junctions in *Hydra*.

E-type gap junctions have also been described in the midgut of *E. kühniella* (Flower and Filshie, 1975), where it was shown that unfixed freeze-cleaved junctions often show particles adhering to both P and E faces. It would be of interest to determine whether hemocyte gap-junction particles also undergo cleavage face changes influenced by

Fig. 6.11. Forty-eight-hour capsule near outer boundaries of capsule showing many desmosome-like junctions between cells. Note microtubules apparently from displaced microtubule marginal bundles associated with some of desmosomes (arrows). × 36,000. (From Baerwald, 1975)

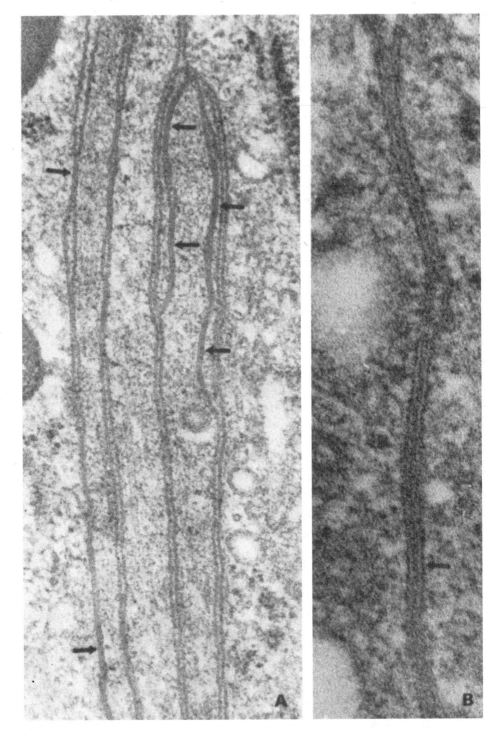

glutaraldehyde fixation or the lack of it. There is precedent for variability in P- and E-face particle distributions influenced by these factors in other membranes and intercellular junctions (desmosomes: Kelly and Shienvold, 1976; endothelial cell plasma membrane: Dempsey et al., 1973; electric tissue plasma membranes: Allen et al., 1977).

At least in one example (Wood, 1977), E-type junctions from fixed *Hydra* epidermis were thought to be "strongly" adhesive because fracture deviations were often shown associated with the E-type junctions. The structural integrity of E-type junctions may be strongly influenced by aldehyde fixation and other factors as compared with P-type junctions (Flower, 1977). Although much more work is needed, it now appears that the cleavage properties of the E-type gap junction may vary greatly depending on the cell or species in which the junction is found, as well as on the physical state of the junction before cleavage. The experimentally induced hemocyte capsule may prove an excellent model for studying these changes under controlled conditions.

6.7. Role of microtubules-microfilaments in capsule formation

Most cockroach hemocytes have an overall discoid shape in free circulation influenced by a marginal bundle of microtubules (Fig. 6.3) (Baerwald and Boush, 1970, 1971b). However, electron microscope studies from other species indicate that microtubule marginal bundles are not always present (Gupta, 1978) (see also Chapter 4). Encapsulation will occur in the presence of 10^{-4} M colchicine or vinblastine (pers. observ.), conditions where microtubules are completely disrupted (Baerwald and Boush, 1971b). Also, in the presence of 10^{-3} or -10^{-4} M vinblastine, in vitro hemocyte migrations occur in a manner indistinguishable from that seen in control hemocytes (Baerwald and Boush, 1971a).

The importance of microtubules in the normal functioning of the hemocyte in general and the plasma membrane in particular remains to be elucidated. However, as we have seen, microtubules, presumably from the marginal bundle, are often associated with desmosomes in insect hemocyte capsules. Nicolson (1974) has suggested that cytoplasmic contractile microtubule-microfilament systems may, at least in some cells, be in direct contact with membrane proteins, influencing membrane protein movements within the phospholipid fluid do-

Fig. 6.12. **A.** *P. americana* capsule illustrating numerous gap junctions seen in cross section (arrows). × 74,000. **B.** High-magnification view of a seven-layered hemocyte gap junction. A periodicity is seen in especially favorable areas (arrow) that may represent channel complexes traversing 20-Å gap. × 280,000. (From Baerwald, 1975)

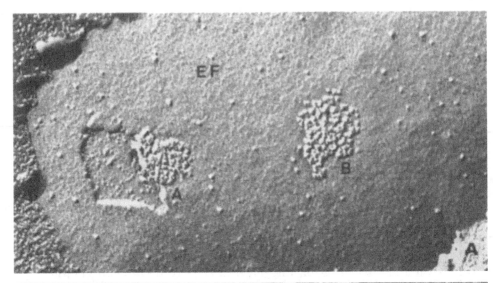

main of the membrane. Exhaustive clear-cut experimental evidence to support this suggestion remains to be obtained in both vertebrate and invertebrate cell systems.

Experiments dealing with the effect of cytochalasins on hemocyte activities such as fibroblast-like activities and capsule development should be done, because microfilaments rather than microtubules are more closely associated with the mechanisms of cell membrane movements (Spooner et al., 1971). Of particular note in this regard is a recent ultrastructural study by Bréhélin et al. (1975), showing an apparent "stimulation" of microfilaments in hemocytes engaged in encapsulation in *Melolontha*.

Studies similar in design should also be done to study the effect of plant lectins (conconavalin A) on capsule development in vivo, as

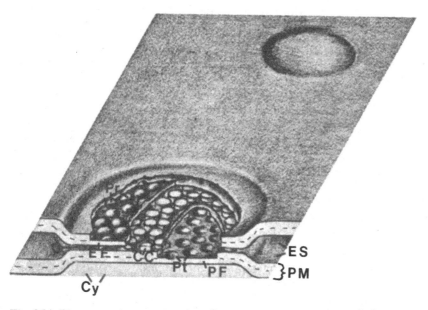

Fig. 6.14. Diagrammatic representation of an insect gap junction typical of type formed between hemocytes during encapsulation. Note that IM particles (*Pr*) adhere to extracellular (*EF*) half of freeze-fractured plasma membrane (*PM*), whereas pits (*Pt*) are associated with protoplasmic fracture face (*PF*), which is in contact with cytoplasm (*Cy*). Channel complexes (*CC*) probably bridge 20-Å gap (*G*), which is continuous with extracellular space (*ES*). Dotted line represents hydrophobic interior of plasma membrane where freeze fracturing is thought to occur. Drawing is not to scale. (Redrawn from Satir and Gilula, 1972)

(facing page)
Fig. 6.13. A. Two E-type gap junctions showing fused (*A*) and unfused IM particles (*B*) making up the macular area. Fused particles are most often seen; the hexagonal particle packing typically of mammalian cell P-type junctions is not found in insect gap junctions. × 80,000. **B.** Several concave areas that contain pits resulting from junction subunits being pulled from the face (*PF*) during fracture process. × 87,000.

these polypeptides have been shown to influence cell agglutination and in general are of use as specific molecular probes for studying membrane, cell, and tissue structure and organization (Nicolson, 1974). The mechanisms of capsule formation are surely complex, but may share some of the properties of mammalian cell agglutination, which have been extensively studied (Nicolson, 1974).

6.8. Hemocyte capsule as a model for isolating and studying gap junctions

Hemocyte capsule formation is a reaction restricted to Arthropoda and is unusual in that the cells (hemocytes) involved in gap junction formation are normally monodispersed and free in circulation. Hemocytes that fortuitously come in contact with one another have not been observed to form junctions of any type (pers. observ.). Although the factors responsible for inducing gap junction formation are unknown, it is felt that the cockroach hemocyte capsule offers unique potential for studying gap junction formation, both for isolation studies and for in vivo studies. The American cockroach adult is an amazingly hardy organism capable of surviving a wide variety of laboratory treatments. For example, the LD_{50} for vinblastine is 10^{-3} M (10 μl/adult) (Baerwald and Boush, 1971a,b). In terms of hardiness, cost, fecundity, and disposability, this organism is an excellent laboratory experimental animal. Other species of Blattidae (*Blaberus giganteus*) also show special promise as experimental animals for these studies because of their large size and ease of rearing in the laboratory.

Induced capsule formation in vivo lends itself to laboratory investigation in many convenient ways. For example, encapsulation can be induced with plastic implants instead of metazoan parasites. This should simplify laboratory procedures for isolation studies and for studying the factors inducing gap junction formation. Uniform capsule sizes and time zero determinations can be easily established by use of plastic implants of known surface areas. Varying the shape of the implant may demonstrate new features of capsule development, and certain shapes (flat plastic plate implants) should facilitate future freeze-fracture experiments regardless of whether a microtome device or a hinge device is used for fracturing.

As a time saver for a wide variety of laboratory investigations, it is convenient if a fully developed capsule can be secured in 48–72 hr. Quick formation of capsules should make studies using injected drugs and hormones more meaningful, as insects usually are very efficient at excreting dissolved foreign substances introduced into the hemocoel. Largely because of the potential, and now to some extent use, of insect hormones as pest control agents (Staal, 1975), juvenile hormone, ecdysone, and many analogs of these hormones are now readily available. Ecdysone was shown to influence gap junction formation in *Limulus* (Johnson et al., 1974). In spite of the extensive literature available on the action of juvenile hormones and ecdysone on insects, their mechanism of action at the cellular and subcellular levels remains unclear (Staal, 1975). Because of the nature of capsule formation and the ease of manipulating capsule development in the laboratory, hemocyte capsules may show real potential as a model for studying hormone effects at the cellular and subcellular levels, especially the effects of introduced insect hormones on intercellular junctions formed between hemocytes during capsule formation.

It is clear that, at least in cockroach capsules, gap junctions are probably formed very soon after new hemocytes have been added to the outside of the capsule (Fig. 6.10). Gap junctions have been seen, using thin-section techniques, between hemocytes from the outermost layers of cells, although the junctions are more numerous deeper within the capsule (Baerwald, 1975). On the other hand, hemocytes involved in nodule formation, which is apparently closely related to capsule formation, do not form desmosomes initially (5 min postinjection) (Ratcliffe and Gagen, 1976). However, desmosome-like junctions are formed in 24-hr-old nodules (Ratcliffe and Gagen, 1977) produced from killed *Bacillus cereus* injected into the hemocoel of *G. mellonella*.

Both thin-section studies (Fig. 6.12A) and freeze-fracture studies (Fig. 6.15) show high concentrations of gap junctions deep within the capsule. These two figures, taken together, suggest that the hemocyte capsule of the American cockroach offers excellent potential as a rich source of E-type gap junctions for isolation studies. P-type gap junctions have been isolated from mouse liver (Goodenough and Stoeckenius, 1972; Goodenough, 1974) and the principal protein characterized. The protein profiles described were not complex, with the principal protein called connexin (MW 20,000). To date E-type gap junctions have not been isolated. Studies similar in design to the P-type isolation studies should be done for a comparison of the proteins of E-type with P-type. Some interesting differences might be expected, as the fracturing process is obviously different between P-type and E-type gap junctions, as is the size of the intramembrane particles making up the two types of junctions.

Isolation studies would be greatly simplified if a capsule could be found where E-type junctions alone could be found. Alternatively, there is hope that by varying concentrations or types of detergents and/or centrifugation methods used for isolation studies, a procedure could be developed for enrichment of the gap junction fraction over a "contaminating" desmosome fraction or other junction type.

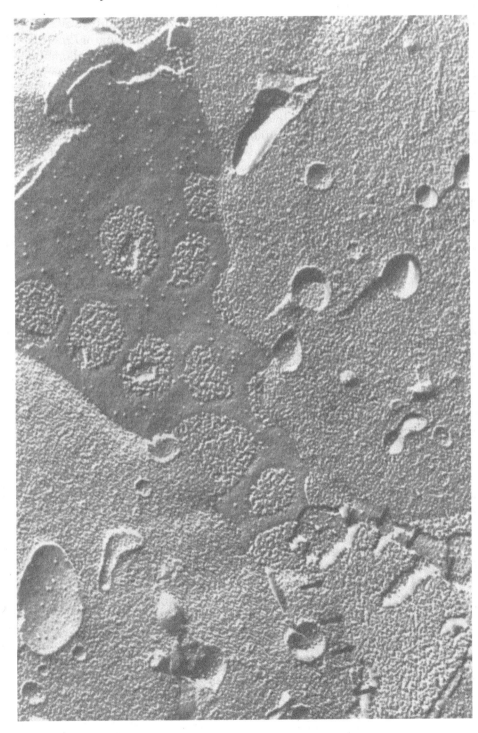

6.9. Summary

Hemocyte membrane activities and substructure have been reviewed at the light and electron microscope levels. Factors influencing hemocyte migrations in vivo and in vitro have been considered, including the effects of mitotic poisons. Membrane substructure in free circulating hemocytes and hemocytes involved in capsule formation has been considered in terms of the fluid mosaic model as revealed by thin sectioning and the freeze-fracture method. Elaborations of the plasma membrane such as exocytosis and internal membrane substructure have been reviewed, as have the nuclear envelope and membranes surrounding cytoplasmic inclusions.

Ultrastructural studies of hemocyte capsules have been reviewed, including general organization, growth, size, and hemocyte types making up capsules. The distribution, formation, and substructure of desmosomes and gap junctions in hemocyte capsules have been considered, along with the possible role of microtubules and microfilaments in capsule formation. Consideration has also been given to the substructure and subunit assemblies of arthropod E-type gap junctions compared with P-type gap junctions found in vertebrate cells. Finally, the hemocyte capsule has been considered as a model for future studies on gap junction formation and as source material for the isolation of arthropod gap junctions.

References

Abercrombie, M., J. E. M. Heaysman, and S. M. Pegrum. 1970. The locomotion of fibroblasts in culture. I. Movements of the leading edge. *Exp. Cell Res.* 59:393–8.

Allen, T., R. Baerwald, and L. T. Potter. 1977. Postsynaptic membranes in the electric tissue of *Narcine*. II. A freeze-fracture study of nicotinic receptor molecules. *Tissue Cell.* 9(4):595–608.

Arnold, J. W. 1959. Observations on living haemocytes in wing veins of the cockroach *Blaberus giganteus* (L.) (Orthoptera; Blattidae). *Ann. Entomol. Soc. Amer.* 52:229–36.

Arnold, J. W. 1972. A comparative study of the haemocytes (blood cells) of cockroaches (Insecta: Dictyoptera: Blattaria) with a view of their significance in taxonomy. *Can. Entomol.* 104:309–48.

Baerwald, R. J. 1974. Freeze-fracture studies on the cockroach hemocyte membrane complex: Symmetric and asymmetric membrane particle distributions. *Cell Tissue Res.* 151:383–94.

Baerwald, R. J. 1975. Inverted gap and other cell junctions in cockroach hemocyte capsules: A thin section and freeze fracture study. *Tissue Cell* 7(3):575–85.

Fig. 6.15. Freeze-fractured 48-hr *P. americana* hemocyte capsule. At least 13 discrete macular E-type gap junctions are evident. In addition, a much larger junctional region can be seen near bottom of picture outlined by arrows. × 50,000. (From Baerwald, 1975)

186   R. J. Baerwald

Baerwald, R. J., and G. M. Boush. 1970. Fine structure of the hemocytes of *Periplaneta americana* (Orthoptera: Blattidae) with particular reference to marginal bundles. *J. Ultrastruct. Res. 31:*151–61.

Baerwald, R. J., and G. M. Boush. 1971a. Time-lapse photographic studies of cockroach hemocyte migrations *in vitro. Exp. Cell Res. 63:*208–13.

Baerwald, R. J., and G. M. Boush. 1971b. Vinblastine-induced disruption of microtubules in cockroach hemocytes. *Tissue Cell 3:*251–60.

Borisy, G. G., and E. Taylor. 1967a. The mechanism of action of colchicine. I. Binding of colchicine-³H to cellular protein. *J. Cell Biol. 34:*525–33.

Borisy, G. G., and E. Taylor. 1967b. The mechanism of action of colchicine. II. Colchicine binding to sea urchin eggs and the mitotic apparatus. *J. Cell Biol. 34:*535–48.

Branton, D. 1969. Membrane structure. *Annu. Rev. Plant Physiol. 20:*209–23.

Branton, D., S. Bullivant, N. B. Gilula, M. J. Karnovsky, H. Moor, K. Muhlethaler, D. H. Northcote, L. Packer, B. Satir, P. Satir, V. Speth, L. A. Staehelin, R. L. Steere, and R. S. Weinstein. 1975. Freeze-etching nomenclature. *Science (Wash., D.C.) 190:*54–6.

Branton, D., and D. Deamer. 1972. *Membrane structure: Protoplasmatologia.* Springer-Verlag, Berlin.

Brehélin, M., J. A. Hoffmann, G. Matz, and A. Porte. 1975. Encapsulation of implanted foreign bodies by hemocytes in *Locusta migratoria* and *Melolontha melolontha. Cell Tissue Res. 160:*283–9.

Brehélin, M., D. Zachary, and J. A. Hoffmann. 1976. Functions of typical granulocytes in healing of the orthopteran *Locusta migratoria* L. *J. Microsc. Biol. Cell 25:*133–6.

Dempsey, G. P., S. Bullivant, and W. Watkins. 1973. Endothelial cell membranes: Polarity of particles as seen by freeze-fracturing. *Science (Wash., D.C.) 179:*190–2.

Flower, N. E. 1971. Septate and gap junctions between the epithelial cells of an invertebrate, the mollusk *Cominella maculosa. J. Ultrastruct. Res. 37:*259–68.

Flower, N. E. 1972. A new junctional structure in the epithelia of insects of the order Dictyoptera. *J. Cell Sci. 10:*683–91.

Flower, N. E. 1977. Invertebrate gap junctions. *J. Cell Sci. 25:*163–71.

Flower, N. E., and B. Filshie. 1975. Junctional structures in the midgut cells of Lepidopteran caterpillars. *J. Cell Sci. 17:*221–39.

François, J. 1975. L'encapsulation hémocytaine expérimentale chez le lepisme *Thermobia domestica. J. Insect Physiol. 21:*1535–46.

Gilula, N. P., and P. Satir. 1971. Septate and gap junctions in molluscan gill epithelium. *J. Cell Biol. 51:*869–72.

Goodenough, D. A. 1974. Bulk isolation of mouse hepatocyte gap junctions: Characterization of principal protein, connexin. *J. Cell Biol. 61:*557–63.

Goodenough, D. A., and W. Stoeckenius. 1972. The isolation of mouse hepatocyte gap junctions: Preliminary chemical characterization and X-ray diffraction. *J. Cell Biol. 54:*646–56.

Grégoire, Ch. 1974. Hemolymph coagulation, pp. 309–60. *In* M. Rockstein (ed.). *The Physiology of Insecta,* Vol. 5, 2nd ed. Academic Press, New York.

Grimstone, A. V., S. Rotheram, and G. Salt. 1967. An electron-microscope study of capsule formation by insect blood cells. *J. Cell Sci. 2:*281–92.

Gupta, A. P. 1979. Arthropod hemocytes and phylogeny, pp. 669–735. *In* A. P. Gupta (ed.). *Arthropod Phylogeny.* Van Nostrand Reinhold, New York.

Hagopian, M. 1970. Intercellular attachment of cockroach nymph epidermal cells. *J. Ultrastruct. Res. 33:*233–44.

Ingram, V. M. 1969. A side view of moving fibroblasts. *Nature (Lond.) 22:*641–4.

Johnson, G., D. Quick, R. Johnson, and W. Herman. 1974. Influence of hormones on gap junctions in horseshoe crabs. *J. Cell Biol. 63:*157a.

Jones, J. C. 1962. Current concepts concerning insect hemocytes. *Amer. Zool. 2:*209–46.

Kelly, D. E., and F. Shienvold. 1976. The desmosome: Fine structural studies with freeze-fracture replication and tannic acid staining of sectioned epidermis. *Cell Tissue Res. 172:*309–23.

Lane, N. J., H. Skaerand, and L. S. Swales. 1977. Intercellular junctions in the central nervous system of insects. *J. Cell Sci. 26:*175–99.

McNutt, N. S., and R. S. Weinstein. 1973. Membrane ultrastructure at mammalian intercellular junctions. *Prog. Biophys. Mol. Biol. 26:*45–101.

Mercer, E. H., and W. L. Nicholas. 1967. The ultrastructure of the capsule of the larval stages of *Moniliformis dubius* (Acanthocephala) in the cockroach *Periplaneta americana. Parasitology 57:*169–74.

Misko, I. S. 1972. The cellular defense mechanism of *Periplaneta americana* (L.). Ph.D. dissertation, Australian National University, Canberra.

Moor, H., and K. Muhlethaler. 1963. Fine structure in frozen-etched yeast cells. *J. Cell Biol. 17:*609–28.

Nappi, A. J., and J. G. Stoffolano. 1971. *Heterotylenchus autumnalis:* Hemocytic reactions and capsule formation in the host, *Musca domestica. Exp. Parasitol. 29:*116–25.

Nicolson, G. L. 1974. The interactions of lectins with animal cell surfaces. *Int. Rev. Cytol. 39:*90–190.

Poinar, G. O., and R. Leutenegger. 1971. Ultrastructural investigation of the melanization process in *Culex pipiens* (Culicidae) in response to a nematode. *J. Ultrastruct. Res. 36:*149–58.

Poinar, G. O., R. Leutenegger, and P. Gotz. 1968. Ultrastructure of the formation of a melanotic capsule in *Diabrotica* (Coloptera) in response to a parasitic nematode (Mermithidae). *J. Ultrastruct. Res. 25:*293–306.

Ratcliffe, N. A., and S. J. Gagen. 1977. Cellular defense reactions of insect hemocytes *in vivo:* An ultrastructural analysis of nodule formation in *Galleria mellonella. Tissue Cell 9:*73–85.

Robertson, J. D. 1960. The molecular structure and contact relationships of cell membranes, pp. 343–418. *In* J. A. Butler and B. Katz (eds.). *Progress in Biophysics.* Pergamon Press, New York.

Robinson, E. S., and B. C. Strickland. 1969. Cellular responses of *Periplaneta americana* to acanthocephalan larvae. *Exp. Parasitol. 26:*384–92.

Rotheram, S., and D. T. Crompton. 1972. Observations on the early relationship between *Moniliformis dubius* (Acanthocephala) and the hemocytes of the intermediate host, *Periplaneta americana. Parasitology 64:*15–21.

Rothman, J. E., and J. Lenard. 1977. Membrane asymmetry. *Science (Wash., D.C.) 195:*743–53.

Salt, G. 1963. The defense reactions of insects to parasites. *Parasitology 53:*527–642.

Salt, G. 1970. *The Cellular Defense Reactions of Insects.* Cambridge Monographs in Experimental Biology, No. 16. Cambridge University Press, Cambridge.

Satir, P., and N. B. Gilula. 1973. The fine structure of membranes and intercellular communication in insects. *Annu. Rev. Entomol. 18:*143–66.

Scharrer, B. 1972. Cytophysiological features of hemocytes in cockroaches. *Z. Zellforsch. Mikrosk. Anat. 12:*301–19.

Singer, S. J., and G. L. Nicolson. 1972. The fluid mosaic model of the structure of cell membranes. *Science (Wash., D.C.) 175:*720–31.

Smith, D. S. 1968. *Insect Cells: Their Structure and Function.* Oliver & Boyd, Edinburgh.

Spooner, B. S., K. M. Yamada, and N. K. Wessells. 1971. Microfilaments and cell locomotion. *J. Cell Biol. 49:*595–613.

Staal, G. B. 1975. Insect growth regulators with juvenile hormone activity. *Annu. Rev. Entomol. 20:*417–60.

Staehelin, L. A. 1974. Structure and function of intercellular junctions. *Int. Rev. Cytol. 39:*191–283.

Sutherland, D. J. 1971. A hemocytic disorder of the American cockroach *P. americana. J. Invertebr. Pathol.* 17:369–74.

Teigler, D. J., and R. J. Baerwald. 1972. A freeze-etch study of clustered nuclear pores. *Tissue Cell* 4(3):447–56.

Vey, A. 1971. Etudes des reactions cellulaires anticryptogamiques chez *Galleria mellonella:* Structure et ultrastructure des granulomes à *Aspergillus niger. Ann. Zool. Col. Anim.* 3:17–30.

Ville, C. A., and V. G. Dethier. 1976. *Biological Principles and Processes,* 2nd ed., Saunders, Philadelphia.

Vinson, S. B. 1971. Defense reaction and hemocyte changes in *Heliothis virescens* in response to its habitual parasitoid *Cardiochiles nigriceps. J. Invertebr. Pathol.* 18:94–100.

Walker, I. 1959. Die Abwehreaktion des Wintes *Drosophila melanogaster* gegen die zoophage Cynipide *Pseudocoila bochei* Weld. *Rev. Suisse Zool.* 66:529–32.

Wood, R. L. 1977. The cell junctions of *Hydra* as viewed by freeze-fracture replication. *J. Ultrastruct. Res.* 58:299–315.

Zachary, D., M. Brehélin, and J. A. Hoffmann. 1975. Role of the "Thrombocytoids" in capsule formation in the dipteran *Calliphora erythrocephala. Cell Tissue Res.* 162:343–8.

7   Controversies about the coagulocyte

CH. GRÉGOIRE AND G. GOFFINET

*Laboratory of General and Comparative Biochemistry, University of
Liège, 17 Place Delcour, B-4020 Liège, Belgium*

Contents

7.1. Introduction

The literature on the coagulocyte (CO) has been reviewed in several
papers (Grégoire, 1951, 1964, 1970, 1971, 1974; Hinton, 1954; Wigglesworth, 1959; Jones, 1962, 1964; Arnold, 1974; Ratcliffe and Price,
1974). The most recent review is by Crossley (1975).

The CO is a newcomer in the considerable and highly controversial
literature on insect hemocytes. Mentioned without details by Yeager
and Knight (1933) in the hemolymph of *Belostoma fluminea*, this labile hemocyte has been missed in works in which old, classic methods
of fixation and staining were used. The phase-contrast microscope
(PCM) made it possible to observe in thin films of hemolymph the

specificity of certain hemocytes in initiating plasma coagulation in the form of circular islets around themselves (Grégoire and Florkin, 1950). We named these cells coagulocytes; they were called cystocytes by Jones (1954, 1962). Further PCM studies in vitro on the hemolymph of about 1,600 species of Palearctic, African, and neotropical arthropods (Grégoire, 1951, 1955a,b, 1957, 1959a,b, and unpub. observ.) led to the recognition of four patterns of CO reactions in different taxonomic groups. Hemocyte studies by transmission electron microscopy (TEM) (especially those of Marschall, 1966; Hoffmann and Stoeckel, 1968; Stang-Voss, 1970; Moran, 1971; Scharrer, 1972; Ratcliffe and Price, 1974; Goffinet and Grégoire, 1975) furnished information on the ultrastructure of the cell components, but also opened new controversies. As pointed out by Ratcliffe and Price (1974), without integrated light and electron microscopic studies, correlation of results obtained by the two methods (PCM and TEM) is difficult or impossible.

Most data about the identity of the CO, its relation to other hemocytes, and its functions are still controversial. These controversies, although partly discussed in the reviews cited above, will be summarized in this chapter.

7.2. Techniques for in vitro study of coagulocytes

The extreme rapidity with which the CO reacts as soon as it comes in contact with a foreign surface is the most important factor in selecting a procedure for preparation. Procedures that permit observations before coagulation develops – such as at the stump of a sectioned appendage (Tait, 1910, 1911), lying on a slide (Loeb, 1903), hanging from a slide (Gupta and Sutherland, 1966), through artificial hemolymph blisters (Marschall, 1966), through the wing veins intact (Arnold, 1959, 1961, 1970) or wounded (Marschall, 1966) – simulate conditions found in nature. However, only the superficial layers of the veins or of the clots are suitable for photographic recording. Thin films of hemolymph from a freshly severed or cut appendage also permit observation under optimal conditions and with minimal manipulations. In *Tenebrio,* this method has given results identical to those obtained by Marschall (1966) with the blister method. The pitfalls of the film method have been previously examined in detail (Grégoire, 1951, 1955a).

The most reliable results are obtained when the alterations in the COs start after the expansion of the blood film has stopped. Some difficulties arise when the clotting material (e.g., islets of coagulation in formation) is stretched by the spreading hemolymph into straight filament-like structures. These structures may be confused with true cytoplasmic expansions. Accidental traction caused by the coverslip can make the interpretation of reactions unreliable. Except in insects whose blood coagulates slowly, collecting hemolymph in syringes and pipettes, transferring it onto glass slides, and agitating or spreading the hemolymph with needles (Yeager and Knight, 1933, and some recent authors) destroy the COs, interfere with the surrounding plasma reactions before any observation can be made, or frequently produce an artificial agglutination of the other more resistant hemocyte types. As will be shown below, use of such techniques has been responsible for the misinterpretations of the CO reactions. In insects whose blood coagulates instantly (e.g., Tenthredinidae, Hymenoptera), there is little possibility of observing the first changes in the COs unless mild anticoagulants are used (Grégoire, 1955a, Fig. 28).

7.3. In vitro hemolymph coagulation patterns in insects

The differences in the reactions and structural changes of the COs in different insect groups observed in thin films of hemolymph by PCM have been classified into four patterns:

Pattern I. Selective structural alterations of the CO, described in detail previously (Grégoire and Florkin, 1950; Grégoire, 1951), cause exudation or explosive discharge of small amounts of cytoplasmic substance, without cell destruction, into the surrounding plasma. These discharges are followed by development of circular islets owing to plasma coagulation around these cells. In addition, the fluid plasma in the channels that separate the islands clots into a granular substance (Figs. 7.2–7.4, 7.7) that is progressively organized into networks of granular fibrils (Fig. 7.4).

Pattern II. On contacting glass, COs extrude long, straight, threadlike cytoplasmic expansions, carrying with them granules. The expansions are highly adhesive to solid particles, to other categories of hemocytes, and to physical interfaces (air bubbles). These cytoplasmic expansions form meshworks with similar expansions produced by other COs, and in these meshworks the other categories of hemocytes are passively agglutinated. Confined to these systems of cytoplasmic meshworks, the reactions of the plasma produce transparent, elastic veils (Grégoire, 1951, 1955a).

Pattern III. COs form islets owing to plasma coagulation, as in pattern I, and send out expansions, as in pattern II. The two reactions may be superimposed or separated in the same blood sample (Figs. 7.8, 7.9). Starved or diseased specimens show only pattern II reaction.

Pattern IV. Hyaline hemocytes resembling the COs of the other three patterns do not alter on contact with foreign surfaces or become hyaline after ejection of cell substance. Under PCM, visible changes in the plasma cannot be detected in the vicinity of these CO-like hemocytes.

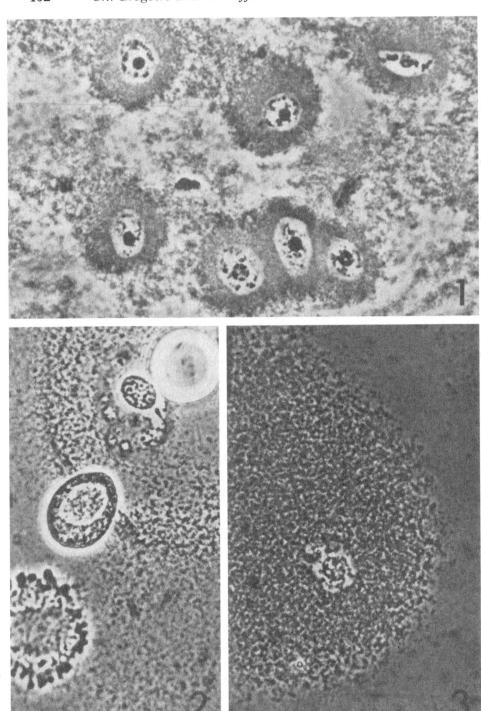

In all four patterns, the other categories of hemocytes do not take part in initiating coagulation of the plasma. They remain unaltered for long periods of time or undergo slow modifications (e.g., emission of spiky pseudopods) without cytolysis. They are passively entrapped at random in the plasma clots (Fig. 7.3) or in the veils or are gathered along the highly adhesive expansions of the CO.

7.4. Phase-contrast (PCM) and transmission electron microscopic (TEM) studies of the coagulocyte

7.4.1. PCM studies

The CO, before it starts changing in vitro, instantaneously or a few seconds after hemolymph withdrawal cannot generally be distinguished from other hemocytes. Some of the main alterations in the CO in various species are formation and retraction of vacuoles, radial expansions of cytoplasmic blebs, swelling of the nucleus (congealed appearance) in certain groups of insects, and exudation or explosive discharge of cytoplasmic and nuclear material into the surrounding plasma without cell disintegration.

Under PCM (Grégoire and Florkin, 1950; Grégoire, 1951, 1955a, 1957; Jones, 1954, 1962; Lea and Gilbert, 1966; Marschall, 1966; Hoffmann, 1967; Stang-Voss, 1970; Price and Ratcliffe, 1974; Rowley, 1977), the altered COs appear in the form of spherical or oval cells, some considerably swollen owing to imbibition. The eccentric, sharply outlined nucleus is comparatively smaller than the nuclei of other categories of hemocytes (except in Tettigoniidae and Acrididae). The chromatin is disposed in a cartwheel pattern. The pale, hyaline cytoplasm, in contrast to the dark appearance of the cytoplasm of the

Fig. 7.1. Coagulation pattern in *Meloe proscarabaeus*, 16 min after withdrawal of hemolymph. Seven islets of plasma coagulation centered by COs are shown. Clotting of plasma is extending to channels between islets in form of granular substance. In wide areas of hemolymph films, COs are only hemocytes seen. × 1,080. (From Grégoire, 1955a)

Fig. 7.2. Coagulation pattern in *Carausius morosus*, 5 min after hemolymph withdrawal. From the top right to bottom left, the following can be seen. At top is a bubble to which an irreversibly altered CO is adhering. Dark cytoplasmic granules are located in one of club-shaped evaginations of cytoplasm. Immediately around CO, plasma clot is more condensed than in surrounding area. Clotting process of plasma appears as a circular wave of coarse precipitate. This precipitate progressively embeds an inert GR surrounded by its characteristic bright diffraction halo. Below, another flow of clotting plasma, an extension of circular wave, is also progressively embedding an inert PL. × 800. (From Goffinet and Grégoire, 1975)

Fig. 7.3. Coagulation pattern in African Gryllotalpidae. Abundant circular islet of plasma coagulation is seen surrounding a CO. Reaction, already stabilized after 2–3 min, was recorded photographically 18 min after hemolymph withdrawal. Clotting of plasma in channels surrounding islet starts slowly. No other hemocyte except this CO could be cause of plasma clotting. × 800.

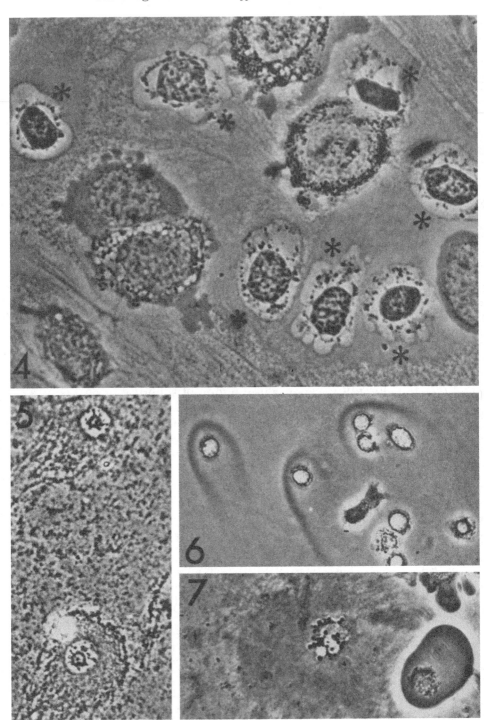

other categories of hemocytes, contains scattered granules undergoing intense brownian motion.

7.4.2. TEM studies

Imbibition of the hyaloplasm, nuclear swelling (Acrididae), considerable enlargement of the perinuclear cisternae, and ejection of nuclear and cytoplasmic substances into the plasma without destruction of the cell are the chief modifications observed in the ultrastructure of the COs in pattern I (Hoffmann and Stoeckel, 1968, in *Locusta migratoria;* Moran, 1971, in *Blaberus discoidalis;* Goffinet and Grégoire, 1975, in *Carausius morosus* and in *Gryllotalpa*, in preparation) and in pattern III (Marschall, 1966, and Stang-Voss, 1970, in *Tenebrio molitor*). Different modes of expulsion of structured or unstructured coagulins by the COs have been described. Hoffmann and Stoeckel (1968), in *L. migratoria,* did not mention rupture of the CO plasma membrane, but rather diffusion of the active substances. Stang-Voss (1970) in *T. molitor* and Scharrer (1972) in several cockroaches reported rupture of this membrane. Goffinet and Grégoire (1975) in *C. morosus* and in *Gryllotalpa* (unpub. observ., see Figs. 7.14–7.16) observed at high magnification a great number of microruptures immediately upon shedding of the hemolymph. These observations furnish a structural basis for explaining the experimental results of Belden and Cowden (1971), which will be examined below, under "Hemolymph Coagulation." Stang-Voss (1970) observed nucleocytoplasmic and nucleoplasmic exchanges through vesicles budding off from the inner nuclear membranes. Similar emissions of nuclear substance by alter-

Fig. 7.4. Coagulation pattern in *Periplaneta americana,* 10 min after hemolymph withdrawal. Dense plasma clot is visible around seven COs (asterisks). Coagulum extends in form of granular precipitates organized into fibrils. One PR, a transitional form, and three PLs are passively embedded in clot. (See Grégoire, 1953, Fig. 1, for another aspect of hemolymph coagulation in *Periplaneta*). × 1,080.

Fig. 7.5. Coagulation in *Vespula saxonica*, male. Contrary to statements in literature, COs are present in hymenopteran hemolymph. Here, two COs are surrounded by islets of plasma coagulation, with considerable extension of plasma clot, organized into granular fibrils. Coagulation pattern I is representative or highly predominant in several families of Hymenoptera. × 800.

Fig. 7.6. Coagulation in *Eriocheir sinensis* (Crustacea), 10 min after blood withdrawal. Islets of plasma coagulation developed within a few seconds exclusively around category of hemocytes functionally corresponding to Hardy's explosive corpuscles. One PL, surrounded by a diffraction halo, is not involved in initiation of plasma coagulation. Identical islets of plasma coagulation (see Grégoire, 1955b, Fig. 24) develop rapidly around similar corpuscles in *Homarus americanus* L. × 800. (From Grégoire, 1955b).

Fig. 7.7. Coagulation in *Nudaurelia richelmanni* larva. Islet of coagulation can be seen that developed a few seconds after collection of hemolymph and is centered by a CO. In vicinity of this islet an inert OE occurs, not yet hyaline, surrounded by a bright diffraction halo. × 800. (From Grégoire, 1955a).

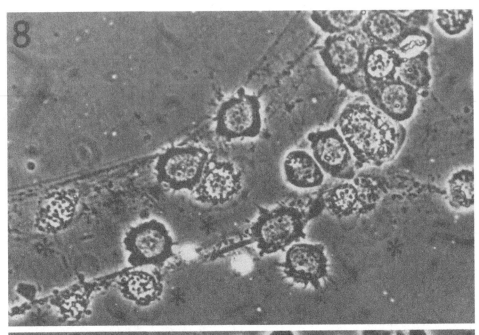

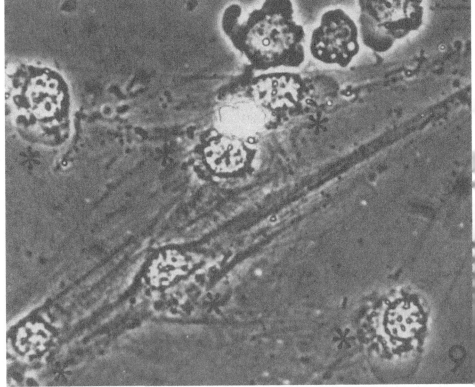

ing COs had been seen under the PCM (Grégoire, 1951). The granules contained in the COs and other hemocytes have received considerable attention in TEM studies. The role of these granules in plasma coagulation will be examined below, under "Functions of the Coagulocyte in Insects."

In contrast to the COs, the prohemocytes (PRs), plasmatocytes (PLs), and granulocytes (GRs) have perinuclear cisternae in the form of thin channels, and their plasma membranes remain intact (Figs. 7.18, 7.19) (see also Chapter 4).

7.5. Origin of the coagulocyte

COs are formed by differentiation in the isogenic islets of reticular cells of the dorsal, unorganized hemopoietic cell clusters (*L. migratoria*) or in true organs (*Gryllus bimaculatus*), which are the "organes phagocytaires" of Cuénot (Hoffmann, 1970a). Production and differentiation of the COs are decreased by X-irradiation localized to the hemopoietic organs (Hoffmann, 1972), blocked in inanition (Hoffmann, 1970a), and stimulated by an 80% bleeding of the fifth-instar larva of *L. migratoria* (Hoffmann, 1969) (see Chapter 2 and below, under "Terminology, Identity, and in Vitro Reaction Patterns of the Coagulocyte").

Jones (1970) and Crossley (1964, 1975) discussed the findings of Hoffmann in relation to the criteria for proof of a normal hemopoietic tissue in insects.

7.6. Terminology, identity, and in vitro reaction patterns of the coagulocyte

7.6.1. Terminology

The term CO was selected because of the spectacular role played by this cell in initiating coagulation, observed under the PCM. Jones (1962) and Ravindranath (1974b) pointed out that a morphological

---

Figs. 7.8, 7.9. Coagulation in *Aphrophora alni*. Pattern III (reactions of patterns I and II occurring in same sample, either superimposed or separated) characterizes hemolymph coagulation in several families of Homoptera. **7.8.** COs (asterisks) have projected, within 60 sec, adhesive cytoplasmic expansions along which other inert hemocytes surrounded by diffraction halos are scattered. Several COs, swollen by imbibition, expand spherical blisters. Plasma reaction takes place in form of fibrillar veils confined in spaces delimited by cytoplasmic expansions of Cos. × 800. **7.9.** View at higher magnification shows six COs (asterisks) and progressive condensation of plasma clot leading to formation of islets around two of these COs in center of picture. Note three inert hemocytes surrounded by diffraction halos at upper part of figure. × 1,200.

(and noncommittal) rather than a physiological nomenclature of the different categories of hemocytes is preferable. A physiological term, such as CO, tends to be too specific for cells with multiple functions. However, as noted by Crossley (1975), it is hard to avoid physiological criteria in the description of cell types. The use (Grégoire, 1951, 1955a, but not 1974) of the term "fragile, unstable hyaline hemocyte" for CO is confusing because the oenocytoids (OEs) (Poyarkoff, 1910), which are probably not involved in coagulation (see below), also appear hyaline after discharge of cytoplasmic material (see Grégoire, 1964, text Fig. 1, p. 172). The cystocytes, described by Jones (1954) and involved in coagulation, are obviously COs, though the equivalence of these two kinds of hemocytes has been questioned (e.g., Arnold, 1974) (see also Chapter 4).

7.6.2. Identity

The main difficulty in identifying the CO is that this hemocyte is recognizable only after it has undergone its alterations in vitro. This is well illustrated in the photographs of Arnold (1959, 1961, 1970), using his method of time-lapse cinephotomicrography of the movements of the living hemocytes in wing veins. Under these in vivo conditions, intact COs could not be recognized among the GRs of cockroaches, in which their role in initiating coagulation is especially observed in vitro. It is difficult to interpret the fiberlike ghosts found in the veins of moribund insects by Arnold (1961, Fig. 49, p. 761).

Gupta and Sutherland (1966) reported that when a drop of hemolymph (*Periplaneta*) is made to coagulate, several PLs instantaneously transform into cystocytes (COs). In our material of *Periplaneta* (Grégoire, 1953a, Fig. 2), PLs with large nuclei and slowly developing spiky cytoplasmic expansions could always be distinguished, without transitional forms, from the COs on the basis of the plasma reactions that developed earlier only around the COs (Figs. 7.2, 7.4). Comparison of Figs. 7 and 8 of Gupta and Sutherland (1966) (*Periplaneta:* hemolymph examined in physiological saline-versene) with Fig. 12 of Grégoire (1953a) (*Periplaneta:* hemolymph examined in physiological saline-sequestrene) does not show in the latter material any transition between PLs and cystocytes (COs), the latter appearing completely different in structure. Reversibility of the CO to the GR stage, as suggested by Gupta and Sutherland (1966) in *Periplaneta,* is not supported by observations that once the CO has undergone its alterations, it appears irreversibly inert (congealed aspect) (Grégoire and Florkin, 1950; Grégoire, 1951). After treatment with anticoagulants in *Gryllotalpa, Periplaneta,* and *Nepa* (Grégoire 1953a), or with dilute saline in *Periplaneta* (Price and Ratcliffe, 1974) (substances that preserve the cytological features and delay, reduce, or inhibit plasma coagulation), the COs appeared either loosely fixed to the glass or air bubbles, or as free-floating spherical, oval, cucumber-like bodies (Grégoire, 1953a; Price and Ratcliffe, 1974).

Marschall (1966) described in *T. molitor* two types of COs (Gerinnungszellen). His type II corresponds to the COs of pattern I of Grégoire's classification and induced in Marschall's material formation of islands of plasma coagulation. Marschall's type III corresponds to the COs of pattern II and shows the cytoplasmic expansions. These simultaneous changes characterize pattern III of coagulation, which is predominant in several Tenebrionidae and in the former supergroup of Heteromera (see below). As pointed out elsewhere (Grégoire, 1974, p. 339), it is still uncertain whether the COs involved in the formation of islands of coagulation and the emission of cytoplasmic expansions belong to the same or to different categories of hemocytes. However (Grégoire, 1955a), in Tenebrionidae islands of plasma coagulation were seen to develop around the reduced bodies of COs that had produced cytoplasmic expansions.

Scharrer (1972) did not recognize the selective alterations of the living COs, especially visible in the Blattodea under the PCM (Grégoire, 1955a, 1964). She considers the varying cytological features in insect hemocytes as manifestations of the functional versatility of a single cell type.

The identification of the CO with the GR, as suggested by Devauchelle (1971) on the basis of TEM studies in *Melolontha melolontha* (that the CO is a slightly differentiated GR), has been discussed by Neuwirth (1973), Lai-Fook (1973), and Costin (1975) (see also Chapter 4). According to Lai-Fook (1973), the GR is the cell type that has caused the greatest confusion in the literature owing to the tendency of authors to subdivide GRs on the basis of shape and presumed functions. Neuwirth (1973) synonymized the GR of *Galleria mellonella* with the clotting cell of *Tenebrio* (Stang-Voss, 1970) and with the CO of Hoffmann and Stoeckel (1968).

According to Neuwirth (1973), no evidence has been presented that clearly defines the function of the GRs, and therefore classifying them according to a possible function is misleading. However, Costin (1975) recognized the CO as a cell different from the GR on the basis of size, number, and composition of the granules. COs and GRs differ strikingly in their cytological features and in their reactions in the plasma (see Fig. 7.2). The labile COs in the islets of plasma clot contrast with the nonlabile GRs, which appear surrounded by intense halos and are obviously inert in the clotting process (see Grégoire, 1951, 1955a, 1957, 1964, 1970, 1971, 1974; Hoffmann 1970b; Goffinet and Grégoire, 1975). In the stick insect, *Clitumnus extradentatus*,

Rowley (1977) reports that the cystocytes (COs) and GRs are two morphologically distinct cell types, both involved in hemolymph coagulation (see below).

The problem of identification of the OE of Poyarkoff (1910) and its relation to the CO is highly complex and controversial. The related literature has been reviewed in previous papers (Grégoire, 1955a; Jones, 1962; Lea and Gilbert, 1966; Arnold, 1974; Crossley, 1975) (see also Chapter 4). In several groups of insects, OEs exhibit the sequences of alterations observed in the COs, including expulsion of cytoplasmic substance and sudden hyaline appearance (Grégoire, 1955a, 1964).

After becoming hyaline, OEs closely resemble COs, but are much larger. A veillike gelation of the plasma was reported to follow the changes in the OEs in some lepidopteran larvae, but not in others (e.g., Lea and Gilbert, 1966, in *Hyalophora cecropia*, *Samia cynthia*, and *Antheraea polyphemus*). Figure 47 from Grégoire (1955a) clearly shows in the hemolymph of the African saturniid, *Nudaurelia richelmanni* Weymer (larva), a typical islet of coagulation around a CO and, in the close vicinity, an inert OE surrounded by an intense halo (Fig. 7.7). In Pentatomidae and adult Diptera (Grégoire, 1955a), in *Prodenia* larvae (Lepidoptera) (Jones, 1959), in *Rhodnius* (Hemiptera) (Jones, 1962), in *H. cecropia* (Lepidoptera) (Lea and Gilbert, 1966), and in Diptera (Crossley, 1975) release of cytoplasmic material from the OEs did not modify the plasma consistency. A possible synonymy of OEs and COs has been tentatively suggested with the important restriction, however, that these categories of hemocytes differ by several characteristics (see Grégoire, 1955a, p. 131). If COs cannot obviously be identified morphologically from OEs, one cannot definitively exclude the possibility that, in some insects, OEs functionally collaborate with true COs in causing plasma gelation.

7.6.3. In vitro reaction patterns

COs have been recognized under the PCM in more than 1,000 species of insects from 20 orders (Grégoire, 1951–1974; see Table 1 in 1974), and recently in 28 species from 15 orders (not in Thysanura, Diptera, Lepidoptera, Hymenoptera) (Price and Ratcliffe, 1974). However, the review of the hemocyte complex in the different orders of insects (Arnold, 1974) shows that the COs are not mentioned in the hemograms of most insects. This absence illustrates well the uncertainties about the existence of this hemocyte.

Referring to the three coagulation patterns proposed by Yeager and Knight (1933) (coagulation as a feature of the plasma alone, coagulation by cellular agglutination, coagultion by plasma and cellular agglutination together), Arnold (1974) noted that since then there has

Table 7.1. Presence and reaction patterns of coagulocytes in various insect orders as reported in the literature

---

Thysanura
COs detected in Lepismatidae (Grégoire, 1955a; Gupta, 1969; François, 1975), but not in Machilidae (*Petrobius maritimus*) (Price and Ratcliffe, 1974).
Ephemeroptera
Reactions in plasma around COs did not appear in *Ephemerella danica* (larvae and adults) or in *Ephemera lineata* (unpub. observ.).
Odonata
COs reported in seven species, including larva of *Aeschna grandis* (Grégoire and Florkin, 1950; Grégoire, 1955a; Price and Ratcliffe, 1974).

Orthopteroid complex
Pattern I characterizes coagulation in the different orders of this complex. More than 200 species have been investigated by Grégoire (1951–1974).

Blattodea
Periplaneta americana has been studied by Grégoire (1953a, 1955a), Wheeler (1963), Gupta and Sutherland (1966), Gupta (1969), Ryan and Nicholas (1972), Price and Ratcliffe (1974). See also Fig. 7.4. In *P. americana* COs were not recognized as a special type by Wharton and Lola (1969), Baerwald and Boush (1970), Scharrer (1972), or in SEM studies by Olson and Carlson (1974). Belden and Cowden (1971, see below), in cockroaches, furnished clear experimental confirmation of the instantaneous reactions of COs, in contrast to the inertness of other hemocytes.
Mantodea
Grégoire (1951) and Price and Ratcliffe (1974) have studied this order.
Isoptera
Winged termites, workers, soldiers, and queens have been studied by Grégoire (1953b, 1954, 1955a, 1970, 1971, 1974) and by Price and Ratcliffe (1974).
Phasmida
Carausius (see Fig. 7.2) has been investigated by Grégoire and Florkin (1950), Czihak (1956), Price and Ratcliffe (1974), Goffinet and Grégoire (1975), and Rowley (1977). Several African and neotropical species have been studied by Grégoire (1957, 1959) and Grégoire and Jolivet (1957).
Orthoptera
General studies have been done by Grégoire (1955a, 1957, 1970, 1974) and Grégoire and Jolivet (1957). For *L. migratoria*, see Grégoire (1955a), Hoffmann (1967), Hoffmann and Stoeckel (1968), Brehélin and Hoffmann (1971), Hoffmann and Brehélin (1971), Price and Ratcliffe (1974).
Dermaptera
Grégoire and Florkin (1950), Grégoire (1957), and Price and Ratcliffe (1974) are the chief investigators.
Plecoptera
COs and pattern I were recorded in *Perla abdominalis* (Grégoire, 1955a).
Hemiptera-Heteroptera
More than 160 species have been examined by Grégoire (1951–1974). The COs do not produce plasma reactions visible under the PCM (pattern IV) in most specimens of the following families of Heteroptera: Reduviidae, including *Rhodnius* (Wigglesworth, 1955, 1959, 1965; Grégoire, 1957; Price and Ratcliffe, 1974); Nabidae;Gerridae; Coreidae (Grégoire, 1955a); Lygaeidae; Pentatomidae; Miridae; Naucoridae; Scutellaridae; Corixidae; Notonectidae; and Pyrrhocoridae. Three families (Belostomatidae, Nepidae, Ranatridae) strikingly contrast with the families cited above by showing a substantial pattern I reaction around their COs (Grégoire, 1953a, 1970, 1974; Grégoire and Jolivet, 1957).

Table 7.1. (Cont.)

Hemiptera-Homoptera

Pattern I characterizes coagulation in Fulgoridae and Cicadidae (Grégoire, 1957). Pattern III (I and II combined or dissociated in some samples) was the most representative reaction in more than 100 species of Cicadellidae, Dictyopharidae, Cercopidae (Figs. 7.8, 7.9) and Jassidae (Grégoire, 1955a, 1974). The reactions of the plasma, abundant in all these families, were considerable in Fulgoridae. Granados and Meehan (1973) did not recognize COs in hemograms of *Agallia constricta* (Cicadellidae). Presence of COs, exceeding GRs in number, in *Macrosiphum euphorbiae* (Aphididae) (which exhibit patterns I and III: islets, veils, and cytoplasmic filaments) is in agreement with the above findings (Boiteau and Perron, 1976).

Coleoptera

More than 400 species have been studied by Grégoire (1951–1974).

Adephaga

Reactions of the COs vary greatly in Carabidae, Hygrobiidae, and Dytiscidae. In several species, the COs did not react (pattern IV); in others, they produced long cytoplasmic expansions (pattern II or pattern III incomplete). Other species (e.g., *Carabus, Agra*) exhibited substantial reactions of complete or incomplete pattern III. In Dytiscidae, pattern I was the characteristic reaction of *Cybister* (Grégoire, 1951). In *Dytiscus*, the reactions were heterogeneous. In these groups, except possibly for *Cybister*, feeding conditions, starvation, and length of captivity might be some of the causes of the versatility in the reactions of COs. COs have been reported in *Dytiscus marginalis* by Price and Ratcliffe (1974).

Polyphaga

Reactions of COs have been recorded in more than 400 species from 35 families (Grégoire, 1951–1974). Homogeneous reactions or predominance of a pattern characterize Hydrophilidae (Grégoire, 1955a, 1974); Staphylinidae (pattern IV: Grégoire, 1955a, 1970, 1971, 1974); several Curculionidae (pattern IV); the former group of Heteromera (Tenebrionidae, Lagriidae, Monommidae), which showed predominance of pattern III (Grégoire, 1955a, 1957, 1970, 1974), with dissociation into its components in some samples; several African Passalidae (pattern I); Meloidae (see below); several subfamilies of Scarabaeidae, e.g., Rutelinae, Melolonthinae, Dynastinae, Trichiinae, Cetoninae (pattern II); subfamily Coprinae (pattern I); Cerambycidae (pattern I, abundant in Lamiinae and Prioninae: Grégoire, 1957; less abundant in Cerambycinae); Brenthidae (20 species: pattern IV). In other families (e.g., Canthariidae, Chrysomelidae), large variations in the reactions did not permit establishment of the predominant pattern. However, at the genus level, some reactions were highly characteristic: In Silphidae, *Necrophorus* (Grégoire, 1955a) had a very abundant clot; in Chrysomelidae, five species of *Timarcha* showed pattern I, and seven species of *Chrysolina* showed patterns I and III. Gupta and Sutherland (1966) recognized cystocytes (= COs) in Tenebrionidae. In *T. molitor*, Marschall (1966), using observations of wings of pupae and PCM microcinematography, confirmed the characteristics of pattern III in this insect (Grégoire, 1951). The problem of the unique or double nature of these COs is examined in the text. In Meloidae, characterized by an abundant plasma clot (Fig. 7.1 and Grégoire, 1955a; Grégoire and Jolivet, 1957), pattern I was representative in the genus *Meloe* (three species) and predominant in *Epicauta grammica;* patterns I and III were predominant in *Zonabris* (four species). Findings of pattern II in *Epicauta cinerea* by Gupta (1969) do not really contradict the preceding observations. As previously reported, pattern III can be incomplete and islets of coagulation can be absent (e.g., in poorly fed or diseased specimens). Inconsistent coagulation of the plasma with the characteristics of pattern II occurred in 2 out of 33 species of Curculionidae (*Rhynchophorus phoenicis, Brachycerus kumbanenenis*). As in Diptera (see below), plastic, pleomorphic PLs

Table 7.1. (Cont.)

developed cytoplasmic expansions in the shape of arborizations and of filaments of great length, some of them resembling dendritic expansions of nerve cells in vertebrates. These expansions form loose networks over wide areas of the hemolymph films (Grégoire, 1959a; McLaughlin and Allen, 1965).

Panorpoid complex

Megaloptera
Pattern I was observed in *Sialis flavilatera* (Grégoire, 1955a) and in *Sialis lutaria* (Selman, 1962), in which Price and Ratcliffe (1974) reported COs.

Neuroptera
Pattern I was recorded in *Chrysopa vulgaris* (Grégoire, 1955a) and in a larva of Myrmeleontidae (Grégoire, 1957). This pattern appeared instantaneously, with a very abundant clot, in *Corydalus* (Grégoire, 1959b). No reaction (pattern IV) appeared in Mantispidae (Grégoire, 1957, 1959b).

Mecoptera
Pattern I developed, though irregularly, in three species of *Panorpa* (Grégoire, 1955a).

Trichoptera
Pattern I developed in Limnophilidae, in *Anabolia nervosa* (Grégoire, 1955a), and in *Leptonema* (Grégoire, 1959b). Price and Ratcliffe (1974) reported COs in *Anabolia*.

Lepidoptera
Pattern II was predominant in 92 species (larvae) from 20 families (Grégoire, 1955a). Pattern I has been recorded in Palearctic and African Saturniidae (*Bunea caffraria, Lobobunea chrystyi, Cirina forda, Imbrasia macrothyris, Nudaurelia rubra, Nudaurelia (Gonimbrasia) richelmanni;* in Sphingidae (*Lophostethus dumolini*); and in Cossidae (*Cossus cossus*) (Fig. 7.7; Grégoire, 1955a, 1964). Gupta and Sutherland (1966) using *G. mellonella* and Breugnon and Le Berre (1976) using *Pieris brassicae,* confirmed presence of COs in hemolymph of Lepidoptera. In *Pieris,* explosion of COs, relatively scarce and frequently of large size (20 μm), produced clotting of surrounding hemolymph (Breugnon and Le Berre, 1976). In many papers, COs were not recognized in lepidopteran hemograms as a special hemocyte category, but were included among GRs or PLs. Examples are *Bombyx mori* (Nittono, 1960; Akai and Sato, 1973; Price and Ratcliffe, 1974); *Pseudaletia unipuncta* (Wittig, 1965); saturniid silkworms *A. polyphemus, S. cynthia, H. cecropia* (Walters, 1970); *Pieris rapae* (Takada and Kitano, 1971); *Calpodes ethlius* (Lai-Fook, 1973); *G. mellonella* (Neuwirth, 1973; Price and Ratcliffe, 1974; Rowley and Ratcliffe, 1976); *Pectinophora gossypiella* (Raina and Bell, 1974); *Malacosoma disstria* (Arnold and Sohi, 1974; see also Grégoire, 1955a concerning *M. neustria,* pattern II); *A. pernyi* (Beaulaton and Monpeyssin, 1976); the noctuid *Euxoa declarata* (Arnold and Hinks, 1976).

Diptera
Abundant gelation in the form of veils characterizes reactions in vitro of hemolymph of larval Tipulidae (Grégoire, 1955a). Jones (1956) also observed coagulation in larval *Sarcophaga.* The hemolymph of several adult Diptera (Tipulidae, *Bibio, Erystalomyia, Sarcophaga, Lucilia, Calliphora, Musca*) did not clot (pattern IV) (Grégoire, 1955a; 1959), nor did that of *Sarcophaga bullata* (Jones, 1956). In larvae of the following species, hemolymph did not clot: mature larvae of *Rhynchosciara americana* (Terra et al., 1974), *Gastrophilus equi* and *Hypoderma bovis,* Oestridae (Grégoire, 1955a), and *Psilopa petrolei* (Grégoire, 1975). Plate V, Fig. 1 of Grégoire (1959a) and Figs. 2K and 3C of Whitten (1964) show inert COs in hemolymph of adult *S. bullata.* The COs differ from hyaline (after discharge) OEs, which also appear in dipteran blood, by their smaller size. In *Gastrophilus equi* and *Hypoderma bovis,*

some of the clarified hemocytes resemble the lysing ADs free from fat globules described by Murray (1971). These corpuscles could correspond to the type B cell of *Calliphora* (Crossley, 1964). Price and Ratcliffe (1974) did not find COs in *Calliphora erythrocephala*. The hemolymph picture of several Diptera is complicated by development of extremely long, wriggling, filamentous expansions by the pleomorphic PLs (Grégoire, 1955a, p. 121; 1959a; Jones, 1956; Whitten, 1964; Zachary et al. 1972, 1975, who called these cells thrombocytoids; Seligman and Doy, 1973; Crossley, 1975). In larval Tipulidae, fragments of these expansions are seen embedded in great numbers in the gel (Grégoire, 1955a). These filamentous expansions form extensive loose meshworks by fusion and twisting of the cytoplasmic arborizations of several hemocytes. They also occur in wing hypodermis and are hormonally controlled (Seligman and Doy, 1973; Crossley, 1975). It is important to note that these filaments and their fragments have nothing in common with the pseudopodial threads produced by COs in pattern II.

Hymenoptera
COs have been found in 70 species from 9 families of this order (Grégoire, 1955a, 1957, 1959b). Patterns I (Fig. 7.5) and/or III characterize the reactions of the hemolymph in all the studied specimens from Tenthredinidae, Ichneumonidae, Formicidae, Vespidae, Pompilidae, and Sphecidae. In larval Tenthredinidae, the COs represent more than 50% of the hemocytes and in *Athalia lineolata*, 80%. Plasma coagulation is instantaneous. On the other hand, Price and Ratcliffe (1974) did not find COs in *Anoplonyx destructor* or in *Apis mellifera*. Hemolymph of Apidae does not clot (pattern IV: Grégoire, 1955a; Gilliam and Shimanuki, 1967, 1970). COs were not seen in *Apis* by Gilliam and Shimanuki (1967), Vecchi et al. (1972), or by Wille and Vecchi (1974) or Joshua et al. (1973) in *Vespa orientalis*, although they are present in Vespidae (Fig. 7.5). Joshua et al. (1973) observed in *V. orientalis* spontaneous gelation of the hemolymph in vitro, disintegration of the hemocytes during that gelation, and appearance of granular and fibrillar material around the disintegrated cells. Gelation was not suppressed by rapid elimination of the hemocytes. This description might correspond to the islets of coagulation that characterize the beginning of coagulation in pattern I. Owing to the great rapidity of the CO reactions in many Hymenoptera, these hemocytes might have already liberated their coagulins into the plasma when the hemocytes were removed. Stroganova and Guliy (1973) did not mention COs among the six identified types of hemocytes of *Diprion pini*, a species of Tenthredinidae characterized by an abundant reaction of pattern III (Grégoire, 1955a, p. 122).

been no cause to change their view. Somewhat similar coagulation patterns were suggested by Grégoire (1964). However, examination of both classifications shows that the two are quite different.   .

The first (coagulation of the plasma alone) of the three coagulation patterns of Yeager and Knight (1933) has been observed by Åkesson (1953) and Joshua et al. (1973) after elimination of the hemocytes by centrifugation. These results have been questioned by Hinton (1954) on the ground that a coagulation-inducing factor might have been liberated into the hemolymph during centrifugation.

The second type (cellular agglutination) of coagulation of Yeager and Knight (1933) can be directly produced by the technique used (see above, under "Techniques for in Vitro Study of Coagulocytes").

The random, mechanically induced packing of the hemocytes, leading to so-called cell coagulation, makes it impossible to discriminate between active and inactive hemocytes. The use of the slide and cover-slip technique, without any other procedure than spontaneous spreading out of the hemolymph into a thin film, shows that agglutination is a secondary process occurring along the adhesive expansions of the COs in pattern II. In pattern I, the hemocytes other than the COs were scattered or agglutinated at random in discrete clusters of a few cells and passively embedded in the clots.

The presence and reaction patterns of COs in various orders as reported in the literature are summarized in Table 7.1.

7.7. Functions of the coagulocyte in insects

Some aspects of the role of the COs have been reviewed by Grégoire (1974), Crossley (1975), and Goffinet and Grégoire (1975). The high percentage of COs in the hemograms of insects strongly suggests that these hemocytes must play a considerable role in the physiology of insects, which probably goes beyond the function of coagulation alone. In the hemograms of *T. molitor* (Jones, 1954), the cystocytes (= COs) represent 43.5% of the hemocytes. In the hemograms of 16 orthopteroids, Coleoptera, and Hymenoptera, the following percentages have been recorded (Grégoire, 1955a, 1974): *Heliocopris haroldi*, 100% in some samples; *Meloe proscarabaeus*, 95% in some samples; *Necrophorus vespilloides*, 84% in some samples; *Geotrupes* sp., 65–82%; *Lagochile sparsa*, 64%; *Pterostichus vulgaris*, 60%; *Lema melanopa*, 60%; *Tettigonia viridissima*, 57%; *Cetonia aurata, Locusta* sp., 54%; *Carabus monilis*, 52%; *Aphrophora alni*, about 50%; *Melolontha* larvae, 42%; *Gryllus domesticus*, 17–47%; *Gryllotalpa gryllotalpa*, 8–36%; *C. morosus*, 9–15%. Wheeler (1963) counted 37% in *P. americana* at ecdysis, and Hoffmann (1967) about 50% in *L. migratoria*, in rather close agreement with the amount cited above for *Locusta*.

As shown by Hoffmann (1970b) in *L. migratoria*, these hemograms are under endocrine control. Maturation of the ovary does not directly influence the hemograms, including the proportion of COs. The increased activity of the corpora allata during the first phase of ovary maturation is the only cause of the variations in hemograms during this process. Thus, the corpora allata stimulate the production and differentiation (COs, GRs) of hemocytes, whereas the pars intercerebralis exerts a moderating influence on the stimulation produced by the corpora allata. The prothoracic glands stimulate the production and the differentiation of the hemocytes (including the COs) during the last larval instar. Implantation of corpora cardiaca in normal animals does not change the proportion of COs (see also Chapter 3).

7.7.1. Hemolymph coagulation

Grégoire and Florkin (1950) pointed out the functional homology of the CO and the explosive hyaline amoebocytes of Hardy (1892), Tait (1910, 1911), and Tait and Gunn (1918) in *Astacus fluviatilis.* Hardy, Tait, and Gunn had suggested that disruption of the explosive corpuscles in Crustacea and dispersion of their granules would release a substance (fibrin ferment) affecting coagulable components of the plasma. As noted by Needham (1970) in a survey of the evolution of the hemostatic mechanism in invertebrates, the explosive cells are specialized for maximal production of coagulases and sensitiveness to triggering agents.

The present state of knowledge about the role of the COs in coagulation is summarized below in the form of 3 queries: (1) Is the CO the only hemocyte actually initiating coagulation of the plasma in insects? (2) Does the CO initiate coagulation of the plasma by releasing soluble substances or structured inclusions either qualitatively identical but in greater amounts or different from those of the other hemocytes? (3) If these differences do not appear between COs and the other hemocytes as regards active substances in coagulation, how can one explain the differences in reactions between the active CO and the other inert hemocytes? Let us now examine the evidences.

In regard to the first question, if we agree that coagulation of the plasma is induced by thromboplastic substances of the hemocytes, the various differences in opinion boil down to the identity of the hemocytes involved in initiating the process: either highly specialized hemocytes or all the hemocytes.

According to Gupta and Sutherland (1966), instantaneous transformation of PLs into cystocytes (= COs) in *Periplaneta* indicates that cystocytes are perhaps the effect rather than the cause of coagulation. This interesting suggestion is, however, not confirmed by examination of microcinematographic sequences (Grégoire and Florkin, 1950; Grégoire, 1951, 1974; Marschall, 1966). These sequences clearly showed that initiation of plasma coagulation takes place selectively around a category of hemocytes only after visible expulsion or explosive discharge by these hemocytes of cytoplasmic substance into a transparent plasma, which rapidly or instantaneously changes its consistency; the other categories of hemocytes remain inert during this process. In addition, as reported elsewhere (Grégoire, 1974, p. 341), parallel variations in the proportions of COs in the hemograms and in plasma coagulability furnish indirect evidence of the role of the CO. Such a parallelism has also been recorded at various stages of development in several insects (Wheeler, 1963; Brehélin, 1971). Total irradiation by X-rays (*P. americana, C. morosus, L. migratoria:* Grégoire, 1955c) or local irradiation of the hemopoietic tissues (*L. mi-*

gratoria: Hoffmann, 1972) produced simultaneously considerable decrease in the number of COs and in plasma coagulability. This direct correlation was ·also recorded in insects injected with bakers' yeast (*Gryllotalpa:* Grégoire, 1951) or with high amounts of iron saccharate (*Locusta:* Brehélin and Hoffmann, 1971; Hoffmann and Brehélin, 1971). In hemolymph characterized by an extremely high percentage of COs (see above: e.g., *Heliocopris, Meloe, Necrophorus, Athalia,* Tenthredinidae), plasma coagulation takes place rapidly or instantaneously. By contrast, in *Carausius,* which has a low percentage of COs, the clot is relatively scarce and slow to develop. Evidence of a connection between COs and plasma coagulation is less convincing in *Gryllotalpa,* in which the amount of clotted plasma is considerable and the proportion of COs relatively low. In summary, the selective role played by the CO in initiating hemolymph coagulation is demonstrated by microcinematographic records. Parallelism between plasma coagulability and proportion of COs in the hemograms in normal and experimental conditions supports indirectly this conclusion.

In the stick insect, *Clitumnus extradentatus,* Rowley (1977) reports that cystocytes (= COs) and GRs are both involved in hemolymph coagulation. Release of granules, which causes coagulation of the hemolymph, is extremely rapid in the COs and much slower in the GRs. He suggests that the COs produce the bulk of the coagulation and that the GRs are more concerned with the melanization of wounds and production of antibacterial factors (see below, under "Other Functions").

Let us now consider question 2. As previously reported (Grégoire, 1974), a biochemical demonstration of the presence of coagulation-inducing substances (coagulins) within the CO has been hampered until now by considerable technical difficulties, such as isolation from rapidly or instantaneously clotting hemolymph of a plasma-free hemocyte fraction, and from this fraction of another homogeneous CO fraction. Owing to the lability of the CO, the feasibility of the latter operation without considerable loss of specific, active coagulation-inducing substances or contamination of the other hemocytes by these substances (e.g., during centrifugation) could be questioned. In (unpublished) attempts to separate COs from the other hemocytes it was found that the sediments after centrifugation consisted of clusters of hemocytes in which the COs were no longer recognizable even when mild anticoagulants were used.

Before we examine the nature of the active thromboplastic substances released by the COs, ambiguous statements in literature should be clarified. Guthrie and Tindall (1958, p. 341), reporting Grégoire's results, wrote that CO breakdown releases clouds of granules into the plasma and that these clouds were perhaps the same as the masses of mitochondria that Wheeler (1964) believed were released by the cystocytes (= COs). Similarly, Arnold (1974, p. 143) wrote that COs act by bursting and releasing material, probably from mitochondria, which *forms* islets of coagulation. As previously pointed out by Grégoire and Florkin (1950, p. 129), it is obvious that, with the exception of the extremely small amounts of cytoplasmic substances discharged by the cell or exuded through the membrane, the granular precipitate ("granular clouds") forming the islets is much too abundant (see, for examples, Fig. 7.3) to have been contained within the cell; it originates from the plasma. This is probably also the opinion of Price and Ratcliffe (1974, p. 546), who wrote that the dense granular material that radiates outward from the cell forms the "isles of coagulation" and results from a phenoloxidase discharged from the cell that causes the hemolymph to precipitate.

Early PCM observations (Grégoire, 1951) suggested exudation of soluble cytoplasmic and nuclear materials through the CO membrane (in the form of sudden, multiple, amorphous puffs) or expulsion of a few granules without destruction of the CO.

The substrate of the coagulation-inducing material liberated into the plasma from the CO has been differently identified.

Recent TEM studies have considerably extended the information about the ultrastructure of the cytoplasmic organelles contained in the hemocytes, especially the granules. These granules are of several types: mucopolysaccharide granules in *T. molitor* (Marschall, 1966); two types of granules in *L. migratoria* – dense ovoid elements located at the periphery of the cell and bodies consisting of variously oriented bundles of fibrils (Hoffmann and Stoeckel, 1968); spherical bodies studded with ribosomes on their outer surfaces and granules with a highly organized, inner tubular structure in *T. molitor* (Stang-Voss, 1970); bodies filled with rows of 340-Å tubules (TCBs) in *B. discoidalis* (Moran, 1971); three types of inclusion bodies, one of them identical to the TCB of Stang-Voss (1970) and Moran (1971), and possibly derived from it in cockroaches (Scharrer, 1972). Among other kinds of intracytoplasmic structures, one can mention inclusions containing closely packed, single-, double-, or triple-walled cylinders in *P. americana* (Baerwald and Boush, 1970); inclusions displaying "regularly packed subunits," probably coresponding to the "multitubular granules" of other authors (Hagopian, 1971); "plastid-like structures" composed of parallel, elongated elements resembling the cylinders of Baerwald and Boush (Hagopian, 1971); and multivesicular bodies (Scharrer, 1972, in cockroaches). These various kinds of granules have been generally described as inclusions of GRs, not COs, by various authors (see also Chapter 4).

Marschall (1966), Stang-Voss (1970), Moran (1971), and Scharrer (1972) agree that structured bodies, including TCBs and vesicles, contain the active material presumed to participate after discharge in the formation of the clot. As these structured bodies are also contained in most of the other categories of hemocytes, and as no specific organelle has yet been found in COs, one must search for other factors to explain the specificity of the CO in initiating clotting of the plasma. In TEM studies, Hoffmann and Stoeckel (1968), Hoffmann et al. (1970) in *Locusta*, Goffinet and Grégoire (1975) in *Carausius* and in *Gryllotalpa* (unpub. observ.) noted the scarcity of disintegrated structured granules around the altered COs. A correlation between plasma coagulation and an early secretion by COs exclusively of the structured cytoplasmic material is not evident. Active coagulation-inducing substances could be soluble and may reach the plasma directly by diffusion through the plasma membrane ("exudation": Grégoire, 1951), or they may be enclosed in the small vesicles ejected through the multiple microruptures of this membrane (Goffinet and Grégoire, 1975).

The multiple microruptures in the membrane of the reacting CO of *Carausius* and *Gryllotalpa* (Goffinet and Grégoire, 1975 and in preparation), as shown in Figs. 7.14–7.16, are not found in the cell membranes of other inert hemocytes (Figs. 7.17, 7.18); this is another evidence of the specificity of the selective role of the CO in the initiation of coagulation.

The observations by Stang-Voss (1970) and by Goffinet and Grégoire (1975) of chains of vesicles budding off from the nuclear membrane and connecting with the plasma membrane of the CO suggest, in agreement with early PCM observations (Grégoire, 1951), that perhaps some nuclear substance ejected with the cytoplasmic material partly causes the induction of coagulation.

The evidence regarding question 3 is as follows. It has long been known (Schmidt, 1892, 1895; Nolf, 1908, 1909; Heilbrunn, 1961) that all cells contain thromboplastic substances. The release of cytoplasmic and nuclear contents by any insect hemocyte should induce coagulation of the plasma. Some data suggest that the problem of membrane permeability ("differential sensitiveness": Grégoire, 1951) is the basis for the behavioral differences in vitro between COs and the other categories of hemocytes, the former being active and the latter passive in the initiation of clotting. A considerable swelling of the cytoplasm and (in Tettigoniidae and Acrididae) of the nucleus (Grégoire, 1951, 1955a; Hoffmann and Stoeckel, 1968) has been observed in vitro under the PCM and illustrated in several papers (Grégoire, 1951, 1955a; imbibition of the hyaloplasm: Hoffmann and Stoeckel, 1968). This swelling does not occur in the other categories of hemocytes located in the same field. Differential permeability is also the basis of the results of Belden and Cowden (1971): Addition of fluorescent substances (8-alinino-1-naphthalene sulfonic acid) to the clotting hemolymph of *Blatta orientalis* produced instantaneous fluorescence of the CO cytoplasm and a slower reaction in the other categories of hemocytes. Older experiments (cited in Grégoire, 1971) tried to modify the permeability of the inactive COs of pattern IV (in *Hydrophilus piceus*) by injecting substances such as sodium oleate and urethan. Changes in the reactions of these COs were recorded in some samples in the form of emissions by these COs of threadlike expansions enclosing plasma veils, resembling the reaction of pattern II.

7.7.2. Other functions

The literature on phenol metabolism by insect hemocytes has been reviewed by Crossley (1975). The CO could participate through its phenol oxidase activity in melanin production and possibly in cuticle tanning (Hoffmann et al., 1970). In *L. migratoria*, the phenol oxidase activity is exclusively localized in the intact CO. This activity is bound to the presence of special dark granules characterized by fibrillar structure; these granules are rapidly exuded at the beginning of coagulation. The phenol oxidase activity is then transferred from the CO to the surrounding protidic clot. Because of this phenol oxidase activity the CO might be involved in the process of encapsulation of parasites (Stang-Voss, 1970), in which melanin-secreting cells are active (Salt, 1956). According to Brehélin et al. (1975), this participation of the COs in encapsulation of cellophane fragments in *L. migratoria* and *M. melolontha* is limited to a plasma coagulation immediately around the implant and involves only the COs. These cells do not participate in the later stages of encapsulation of the implanted material, which is accomplished by the GRs, forming multilayered capsules around these foreign bodies. Crossley (1975) also reported accumulation of disrupted labile hemocytes (COs?) at wound sites. These cells locally release their cytoplasmic contents, which are rich in phenoloxidizing enzymes.

After injection of suspensions of bakers' yeast into *Gryllotalpa*, the COs were observed to engulf yeast globules (Grégoire, 1951), which seems to indicate a phagocytic activity of the CO. However, later studies reported absence of phagocytosis in the CO (Jones, 1962; Lea and Gilbert, 1966; Hoffmann, 1967).

7.8. Reaction patterns of hemocytes in other arthropods

Except for the explosive corpuscles described by Hardy (1892) in Crustacea, information about hemocytes playing the role of insect COs in other groups of invertebrates is still scarce. The data available in this field and the old literature have been reviewed in previous reports (Grégoire, 1952a,b, 1955b, 1970, 1971; Grégoire and Tagnon, 1962).

7.8.1. Onychophora

Peripatus can bleed at a sectioned appendage for a long time (24 hr) without the formation of a clot at the wound site. In thin films of blood of 14 specimens observed under the PCM, most hemocytes remained unaltered or occasionally underwent slow swelling, spreading of the hyaloplasm, extrusion of short pseudopods, formation of small networks and clusters. In contrast to these hemocytes, other hemocytes with rounded or shuttle-shaped granules in their cytoplasm underwent rapid alterations (disintegration, intracytoplasmic dissolution, explosive discharge of their granules, and incidental extrusion of long pseudopodial processes). No plasma reaction, accompanying or following these alterations, appeared in any of the specimens studied (Grégoire, 1955b). Tuzet and Manier (1958) and Lavallard and Campiglia (1975) reported similar explosions of certain hemocytes, contrasting with lack of reaction in other cells. Lavallard and Campiglia (1975) noted that ruptured hemocytes discharged their granular contents, especially in specimens used on the day of ecdysis or near the molting period. Absence of COs in *Eoperipatus* has been reported by Sundara Rajulu et al. (1970).

7.8.2. Myriapoda

In the hemolymph of 12 Palearctic and African species and other unidentified chilopods and diplopods, studied in vitro in thin films under the PCM (Grégoire, 1955b, 1970; Grégoire and Jolivet, 1957), differences in adhesiveness ("differential sensitivity") to glass were found among hemocytes. Some hemocytes with large nuclei were shrunken, were surrounded by bright halos, or had developed lappet-like and spiky expansions of their hyaloplasm. These hemocytes were scattered or clustered into small meshworks. Other spherical, hyaline hemocytes, characterized by relatively small, cartwheel-shaped, eccentric nuclei and relatively scarce granules, were instantaneously or rapidly altered on contacting glass. After successive intracytoplasmic, explosive dissolution of their granules, followed by ejection of cytoplasmic substance into the surrounding plasma and hyalinization, these hemocytes appeared in the form of inert spherical cells (Figs. 7.12, 7.13), some of them considerably swollen by imbibition (Grégoire, 1970) and resembling COs of insects (pattern IV) (Grégoire, 1955b; Grégoire and Jolivet, 1957). These CO-like cells might correspond to hyaline leucocytes of type II described by Tuzet and Manier (1954). Other hemocytes of this category were destroyed, leaving naked nuclei with cytoplasmic fringes, a reaction not found in insect COs, except if destroyed by mechanical procedures. The discharge of the cytoplasmic substances by these hemocytes did not produce a distinct modification in the plasma that could be interpreted as a clot. Glassy, inelastic pellicles or veils, possibly caused by desiccation, appeared in the blood of the African diplopod, *Scaphiostreptus acuticonis* (Grégoire and Jolivet, 1957). In agreement with these findings, Ravindranath (1973, 1974c) observed in the millipede, *Thyropygus poseidon*, loss of opacity in cystocytes (= COs) after bursting and absence of visible plasma reaction. On the other hand (Gupta, 1968), the blood of *Scutigerella immaculata* exhibits alterations of pattern II: cytoplasmic strands sent by COs, more numerous than GRs, ADs, or SPs, and glassy plasma veils. Sundara Rajulu (1970), Vostal et al. (1968), and Vostal (1970) reported in centipede blood total lack of cystocytes (= COs), which they identified with GRs.

7.8.3. Xiphosura

Aggregation of GRs into networks and the disappearance and replacement of the granules by vacuoles have been consistently observed in *Limulus polyphemus* blood in vitro (Howell, 1885; Loeb, 1904, 1910; Alsberg and Clark, 1908; Grégoire, 1952a, 1955b). From extensive observations and detailed analyses, Levin and Bang (1964), Dumont et al. (1966), Solum (1970), and Mürer et al. (1975) concluded that the granules of the single category of hemocytes in *Limulus* contain all the factors required for coagulation of the blood, including the clottable protein. These conclusions support the previous suggestion (Schmidt, 1892, 1895; Nolf, 1908, 1909; Heilbrunn, 1961) that most kinds of cells provide coagulation-inducing substances (coagulins). On the other hand, upon blood shedding, instantaneous alterations (discharges of cytoplasmic substances) took place in certain hemocytes, scarce in numbers and characterized by coarser granulations than those of the predominant category of GRs. Simultaneously, gelation was seen in the plasma in the form of purple-blue, elastic, and retractile veils. Areas of greater density developed in the plasma surrounding these altered hemocytes, which resembled insect COs (Grégoire, 1952b, 1955b). These changes, found only in a few samples of blood (thin films), were followed by the well-known alterations in the predominant category of hemocytes. As Dumont et al. (1966) pointed out about these observations, apparently different cell types could be a single cell type in different physiological states. However, alterations such as in the hemocytes resembling COs (shown in Fig. 14, Grégoire, 1955b) could not be found among the serial changes of the GRs illustrated in the nine figures of Plate I of Dumont et al. (1966). In addition, the gelation appeared obviously before the degranulation of the predominant hemocytes, which seemed to be slower than that observed by Dumont et al. (1966). From these observations one cannot definitely rule out the possibility that in *Limulus* a

CO-like hemocyte might be active in initiating the clot before the cell aggregation and degranulation recognized by all the authors. Differences in the methods of blood collection (drawing by cardiac puncture, direct flowing between slide and coverglass) could possibly explain in part the differences in the reactions summarized above. In this respect, the suggestion of two types of hemocytes is supported by recent observations of Shishikura et al. (1977) on clotting of blood (immediately fixed) of the horseshoe crab *Tachypleus tridentatus* on the basis of a fluorescent antibody technique.

7.8.4. Arachnida

The hemolymph reactions observed in vitro in thin films under the PCM in 140 specimens (50 species from 18 families of spiders) (Grégoire, 1952b, 1955b, 1957) may be summarized as follows:

1. A category of labile, elliptical hemocytes with a pale, hyaline cytoplasm containing a few large, shuttle- or egg-shaped granules, underwent instantaneous alterations on contacting the glass. These reactions greatly varied and suggested gradations in the sensitiveness of these blood corpuscles in the different species and specimens: loose emission of large, club-shaped expansions (Fig. 7.10, *1* and *2*); intracellular dissolution and disappearance of the granules; successive explosive discharges of cell material without disintegration of the cell body; explosive cytolysis (Grégoire, 1955b); scattering of the organelles in the plasma; reduction of the cell to the naked nucleus (Figs. 7.10, *3;* 7.19; Grégoire 1955b); drawing out, in the blood streaming between glass and coverslip, of the cytoplasm into straight processes (Fig. 7.11; Grégoire, 1955b). In contrast to these reactions, those of the insect COs do not lead to destruction of the cell. On the other hand, emission of long cytoplasmic processes, along which other hemocytes are agglutinated, may be compared to the CO changes in pattern II of insects. After the discharge of their contents, the labile hemocytes that had not disintegrated resembled insect COs (Grégoire, 1970). It has not been possible to identify these labile hemocytes with the cells observed in the TEM by Sherman (1973). Hyaline hemocytes, which play a primary part in the aggregation of other hemocytes into clumps of cells in the blood of the Haitian tarantula (Deevey, 1941), might possibly correspond to the present labile hemocytes, which exhibit similar reactions (see Fig. 17 in Grégoire, 1970).

2. Other categories of hemocytes (PRs, GRs, PLs) remained well preserved for long periods of time, unless submitted to mechanical agents. These hemocytes, agglutinated along the adhesive cytoplasmic filaments of the altered labile hemocytes, also formed independent loose networks or strands in which they appeared frequently stretched, with elongated pseudopods connecting them with other he-

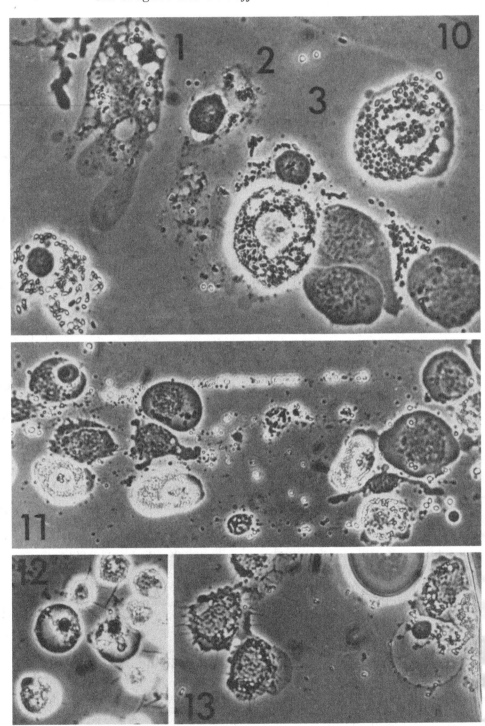

mocytes of the same category. This reaction of spider blood differs from that of insect blood in which clusters of hemocytes always occur along the adhesive expansions of the COs (pattern II). As shown in Figs. 7.21 and 7.22, the cell membrane of the unidentified fragile hemocytes presents – as does that of the insect COs – microruptures, which do not appear in the other categories of hemocytes.

3. The presence of OEs makes it difficult to understand the mechanism of blood coagulation in spiders. OEs are numerous in *Eurypelma, Argiope, Nephila, Lycosa* (Grégoire, 1955b), and *Eurypelma marxi* (Sherman, 1973). These large cells undergo, as in dipteran and lepidopteran larvae, a sudden loss of opacity after the discharge of cytoplasmic substance and transform into spherical hyaline hemocytes.

4. The plasma reactions were highly inconsistent in spiders and greatly differed in specimens of the same species and even in different samples from the eight legs of the same specimen. In many samples it seemed to remain fluid. In other samples (e.g., in *Aranea diademata*) instantaneous gelation occurred when the categories of stable hemocytes (PRs, PLs, GRs) were still intact. In films of blood spread out between the slide and coverglass, this reaction consisted of the development of transparent, elastic, and contractile veils, embedding all the cellular structures (Grégoire, 1955b, 1970). The reactions in spider blood cells, consisting of emission of pseudopods by the fragile cells and development of veils circumscribed by these pseudopods, resembled those that characterize pattern II in insects, especially in Scarabaeidae and in lepidopteran larvae (see Grégoire, 1955a). In a few Scorpionidae (*Opisthacantus elatus, Pandinus* sp.), Pedipalpi (*Tarantula fascimana*), and Ixodidae (*Ornithodorus rudis*) (Grégoire, 1955b), and in some unidentified specimens from these three groups,

Fig. 7.10. Three steps in alteration in fragile hemocyte (described in text) from *Brachypelma* sp. (Araneae, Theraphosidae), 5–10 sec after blood withdrawal. One of these cells (*1*) extends four club-shaped cytoplasmic expansions; a cell of same category (*2*) begins dislocation, while another (*3*), nearly reduced to nucleus, disintegrates with spreading out of its granules. Among other cells shown in this field are two GRs, each surrounded by a characteristic diffraction halo, and three PRs, one of them a transition form, also surrounded by halos. × 800.

Fig. 7.11. Photograph recorded a few seconds after blood withdrawal from adult *Araneus diadematus* (Araneae, Argiopidae). Two fragile hemocytes, reduced to their nuclei, have extended cytoplasmic filaments along which other categories of hemocytes, still intact, are scattered – PLs with short pseudopodia, GRs surrounded by characteristic bright diffraction halos. Identical reactions characterize hemolymph of lepidopteran larvae (pattern II). × 800.

Figs. 7.12, 7.13. Differential sensitiveness in blood cells of *Lithobius forficatus* (Myriapoda, Chilopoda). Difference in reaction to contact with glass and bubbles between spherical hyaline hemocytes, resembling COs of insects, and other hemocytes is apparent. Hyaline hemocytes are swollen by imbibition. In Fig. 7.13, other hemocytes with larger nuclei, have developed lappet-like peripheral expansions of their hyaloplasm. No plasma reaction was detected 90 min after blood withdrawal. **7.12** and **7.13** × 800.

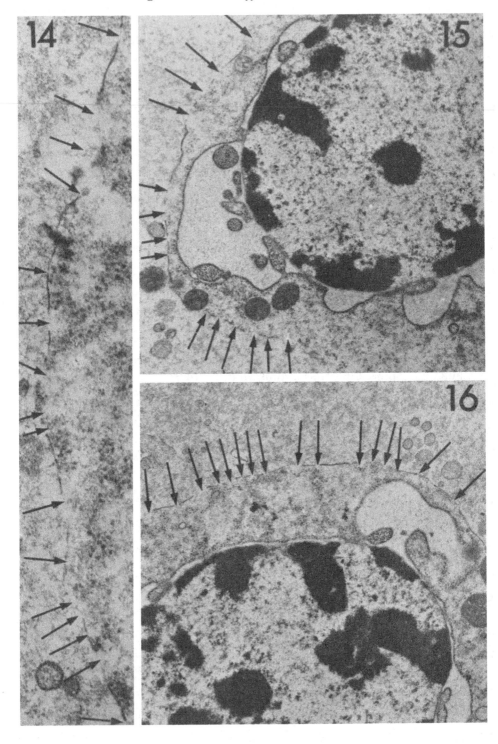

no plasma alteration occurred (pattern IV), except some scanty veils. Ravindranath (1974a) also reported that cystocytes (= COs) did not bring about any visible coagulation of the blood of the black scorpion, *Palamnaeus swammerdami.*

7.8.5. Crustacea

PCM findings in *Homarus americanus, Eriocheir sinensis,* and a specimen of Oniscoidea concerning initiation of plasma coagulation in the form of islets around a category of hemocytes (Fig. 7.6 and Grégoire, 1955b) extend to these species the classic observations of Hardy (1892) on *Astacus fluviatilis* (Tait, 1910, 1911, his type C coagulation; Tait and Gunn, 1918). In *Homarus,* Tait (1910) had reported only agglutination of the blood cells with subsequent gelation of the plasma (type B). Recent papers concerning the ultrastructure of hemocytes in Crustacea and the histochemistry of their granules (glyco- or mucoproteins) (Wood and Visentin, 1967; Hearing, 1969; Hearing and Vernick, 1969; Wood et al., 1971; Stang-Voss, 1971; Bauchau and De Brouwer, 1972, 1974) reported that certain blood cells are the first to react with great rapidity and to show ultrastructural alterations. Initiation of plasma coagulation in the form of circular islets around these cells (Fig. 7.6) does not seem to have been observed by the authors. These cells have been differently labeled. In the crayfish, *Orconectes virilis* (Wood and Visentin, 1967), hyaline cells with few or no granule appear to correspond to the explosive corpuscles of Hardy (1892) and to the explosive corpuscles and thigmocytes of Tait and Gunn (1918). These cells release coagulation-inducing substances composed of glyco- or mucoproteins related to the "fibrin ferment" described by Hardy (1892), Tait and Gunn (1918), and Wood et al. (1971). In *Homarus americanus* L. (Hearing and Vernick, 1967; Hearing, 1969),

Fig. 7.14. Plasma membrane of a CO from African Gryllotalpidae. A drop of hemolymph was immersed in fixative immediately upon shedding. Picture illustrates extreme rapidity of alterations undergone by COs in this and many other species. In this rather limited portion of plasma membrane, many microruptures are visible (arrows). Compare with plasma membranes of a GR and a PL (Figs. 7.17 and 7.18), still intact 60 sec after hemolymph withdrawal. TEM × 45,000.

Fig. 7.15. CO from African Gryllotalpidae. Hemolymph was allowed to clot on a slide during 60 sec; then fixed. No centrifugation. Alterations observed in this CO closely correspond to those seen under PCM (Fig. 7.3). Note considerable expansions of perinuclear cisternae and numerous microruptures (arrows) of plasma membrane. A small granule and a few vesicles are visible in surrounding plasma. Compare with intact plasma membranes of GR and PL shown in Figs, 7.17 and 7.18. TEM × 15,000.

Fig. 7.16. CO from African Gryllotalpidae (same preparation as in Fig. 7.15). Characteristic expansions of perinuclear cisternae of this CO are shown. Arrows indicate microruptures of plasma membrane. Note ejection from cytoplasm into plasma of small cytoplasmic vesicles. TEM × 15,000.

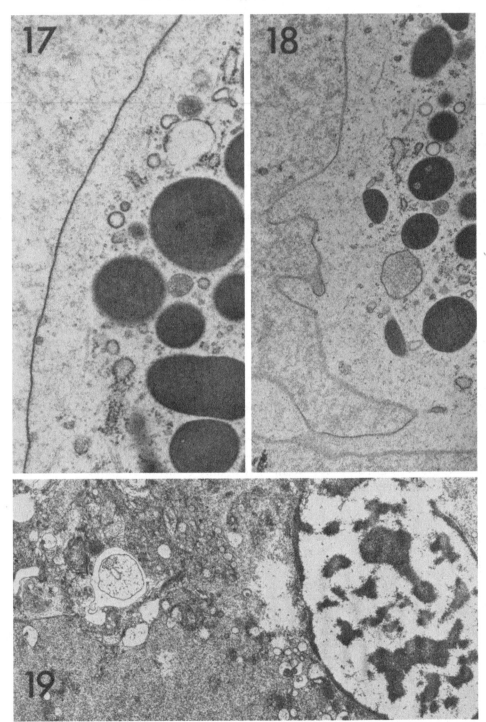

spindular basophil cells appeared to be involved primarily in coagulation. The acid phosphatase activity demonstrated in their granules is also localized in granules of the two other types of hemocytes.

In *Gecarcinus lateralis*, previously studied by Morrison and Morrison (1952), Stutman and Dolliver (1968) state that agglutination of cells (two main types), followed by degranulation and loss of cell membrane, are fundamental initiating steps in coagulation of crustacean hemolymph and that the astonishing characteristic of crab hemolymph is its ability to form a clot rapidly in the absence of large numbers of cells. Crab hemolymph clots 2–20 times faster than human lymph. Cells in the hemolymph of crabs are many times more active that are human cells. These statements of Stutman and Dolliver suggest that rapidly altering hemocytes, functionally analogous to COs and illustrated in Fig. 7.6, might be the actual elements responsible for the rapid coagulation, before the occurrence of agglutination, degranulation, and cytolysis of the other cells.

The "clotting cells" (Gerinnungszellen: Stang-Voss, 1971, TEM) of *Astacus astacus* are young forms from the three functional, age-dependent types of a single cell category present in this animal. These clotting cells contain a large number of granules with a tubular internal structure. Like the explosive cells of Hardy (1892), the clotting cells are very fragile and undergo cytolysis (in contrast to the COs in insects). Initiation of plasma coagulation is caused by release of small vesicles from the cytoplasm, not by the granules, which are freed during later cytolysis and could contribute only later to increase in the amount of coagulum.

In *E. sinensis* and *Carcinus maenas* (Bauchau and De Brouwer, 1972, 1974) all the categories of hemocytes are involved after agglutination in the formation of the clot. However, hyaline and semigranular cells react more rapidly and are then the most active. The reactions of the agglutinated GRs are not synchronous in all the cells of this category and are slower and last longer.

In the mole crab *Emerita asiatica*, Ravindranath (1975) observed in

Fig. 7.17. GR from African Gryllotalpidae (same preparation as in Fig. 7.15). Plasma membrane is intact and contrasts with that of CO shown in Figs. 7.14–7.16, which is perforated by multiple microruptures or holes. TEM × 28,500.

Fig. 7.18. PL from African Gryllotalpidae (same preparation as in Fig. 7.15). Plasma membrane is intact and contrast with that of CO shown in Figs, 7.14–7.16. Granules of various sizes are less numerous in this PL than in the GR shown in Fig. 7.17. Polyribosomes are scattered or aligned along wall of small vesicles. A few microtubules are also scattered in cytoplasm. TEM × 15,000.

Fig. 7.19. Hemocyte from *Tegenaria* sp. male subadult (Araneae, Agelenidae). Hanging drop of blood fixed 14 min after withdrawal. Cell's naked nucleus surrounded by cytoplasmic organelles, including many vesicles, corresponds to third stage of disintegration of fragile hemocytes, not yet clearly identified, observed in PCM (Fig. 7.11, cell 3). TEM × 10,500.

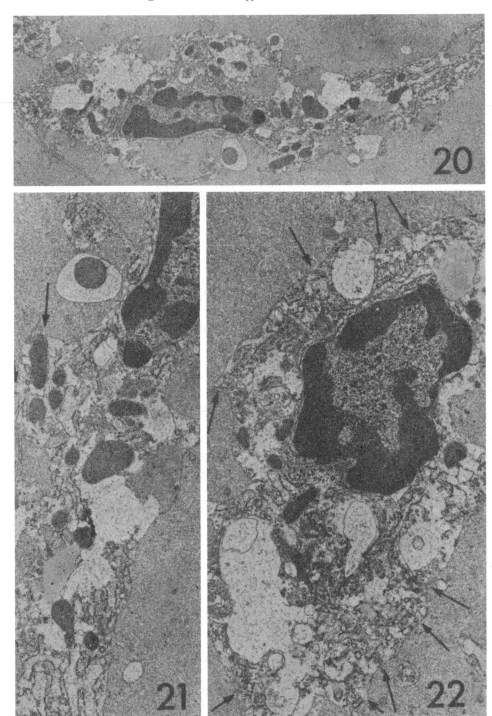

vitro quickly lysing cystocytes (= COs) and reported evidence of their role in plasma gelation. Ravindranath could not find correlation between alteration of GRs and clotting time at different temperatures: At lower temperature, gelation of the plasma occurs when the GRs remain unaltered. He recognized, in agreement with findings in insects (Grégoire, 1951, 1955a), that the process of cellular agglutination (reported wrongly in his text, p. 300, as "unlike" this process in insects) did not play any part in the phenomenon of plasma gelation in *Emerita*. Ravindranath suggested that in this material the COs may perform two functions: (1) promotion of plasma gelation and (2) initiation of the stepwise transformation of the GRs.

Ravindranath (1974b) did not observe plasma coagulation around cystocytes (= COs) of isopod, *Ligia exotica*. In another isopod (*Helleria brevicornis*), Hoarau (1976) did not find hemocytes corresponding to COs of insect.

7.9. Conclusions

It has long been known that all cells contain thromboplastic substances. Under these conditions, the release of cytoplasmic contents by any hemocyte should induce coagulation of the plasma. In insects, plasma coagulation is initiated by early alterations in a category of hemocytes, the COs; the other categories of hemocytes remain inert or release their contents later, after plasma coagulation is already established. This functional selectivity of the COs in the process of coagulation can be explained in several ways. The CO could contain a stock of coagulins qualitatively different from, and more efficient than, those of other hemocytes. An identical stock could be present in greater amounts in the COs. Testing this suggestion would meet considerable technical difficulties, especially because of the fragility of the CO. An answer may be found by establishing a biochemical inventory of all the active substances in the COs and in the other hemocytes, separated, for instance, by centrifugation. On the other hand, biophysical factors, related to differences in the permeability of the cell membrane, could explain the functional selectivity of the COs in

Figs. 7.20–7.22. First stage of disintegration of two fragile hemocytes from *Tegenaria atrica* (Araneae, Agelenidae) male. Hanging drop of blood fixed 8–12 min after withdrawal. No centrifugation. Cells are identical to cells 1 and 2 in Fig. 7.11. Cytoplasm of these cells contains a few mitochondria and is especially characterized by multiple vacuoles, some surrounded by aligned ribosomes. Vesicles, empty or containing rounded granules, are budding off at several places at periphery of hemocytes, detached from cell body, and scattered in surrounding plasma. As in COs of insects, several microruptures of plasma membrane are visible in hemocyte shown in Figs. 7.21 and 7.22 (arrows). TEM. **7.20** × 6,000; **7.21** (part of Fig. 7.20 at higher magnification) and **7.22** × 12,000.

initiating plasma coagulation. Previous attempts to modify the permeability of inert COs of pattern IV suggest that experimental work along this line might produce useful information. Existence in Crustacea of hemocytes functionally similar to insect COs (explosive corpuscles) is supported by the classic work of Hardy, Tait, and Gunn. In other groups of arthropods, there are reports of certain, not yet conclusively identified blood cells (GRs?), more sensitive to foreign surfaces than other blood corpuscles. These hemocytes could release, earlier than the other hemocytes, thromboplastic substances by altering rapidly or instantaneously without destruction or by disintegrating on contact with these foreign surfaces (*Peripatus,* Myriapoda and Araneae). These alterations are not necessarily followed by plasma coagulation. According to general agreement, *Limulus* blood clots exclusively by a process of cell agglutination and degranulation (types A and B of Tait, 1910). However, one could not definitely exclude an early gelation of the plasma produced by alterations in a category of fragile blood cells, much scarcer in number than the cells of the unique category reported in the literature.

7.10. Summary

A review of reliable methods, permitting undisturbed observation of the reactions of coagulocytes, and data obtained so far by PCM and TEM studies on the structure (PCM) and ultrastructure (TEM) of the CO and its different reactions in the patterns of hemolymph coagulation are presented. We have attempted to explain the discrepancies in the literature relating to the failure to find COs in the hemograms of species in which this cell plays a spectacular role in hemolymph coagulation. Use of the PCM is essential for understanding the kinetics and morphology of blood coagulation in insects on the basis of CO activity and for the identification of this cell in the TEM. The present state of research on the reactions of the CO in the different orders of insects and on the functions of this cell has also been reviewed. Tentative explanations of the functional selectivity of the CO in the process of plasma coagulation have been given. The most attractive of these explanations, based on morphological (multiple instantaneous microruptures of the plasma membrane and release of coagulins before the other categories of hemocytes) and experimental data (e.g., earlier diffusion of fluorescent substances than in the other hemocytes) is that biophysical factors, related to differences in the permeability of the plasma membrane, could be an important cause of this selectivity.

The presence in other groups of arthropods (Onychophora, Xiphosura, Myriapoda, and Arachnida) of blood cells functionally playing the same role as the explosive cells of Hardy in Crustacea and the COs in insects is still a controversial problem. Thus, observations report the presence of hemocytes more fragile than the other categories of blood cells. Like the COs of insects, these fragile hemocytes show instantaneous microruptures of their plasma membrane before other cells do. Their cytolysis or alteration coincides in some groups or species, but not in others, with rapid or instantaneous clotting of the plasma.

References

Akai, H., and S. Sato. 1973. Ultrastructure of the larval hemocytes of the silkworm, *Bombyx mori* L. (Lepidoptera, Bombycidae). *Int. J. Insect Morphol. Embryol.* 2(3):207–31.

Åkesson, B. 1953. Observations on the haemocytes during the metamorphosis of *Calliphora erythrocephala* (Meig.). *Ark. Zool.* 6(2):203–11.

Alsberg, C. L., and E. D. Clark. 1908–9. The blood clot of *Limulus polyphemus. J. Biol. Chem.* 5:323–9.

Arnold, J. W. 1959. Observations on amoeboid motion of living haemocytes in the wing veins of *Blaberus giganteus*. (L.) (Orthoptera: Blattidae). *Can. J. Zool.* 37:371–5.

Arnold, J. W. 1961. Further observations on amoeboid haemocytes in *Blaberus giganteus* (L.) (Orthoptera: Blattidae). *Can. J. Zool.* 39:755–65.

Arnold, J. W. 1970. Haemocytes of the Pacific beetle cockroach, *Diploptera punctata. Can. Entomol.* 102(7):830–5.

Arnold, J. W. 1974. The hemocytes of insects, pp. 201–54. *In* M. Rockstein (ed.). *The Physiology of Insecta*, Vol. 5, 2nd ed. Academic Press, New York.

Arnold, J. W., and C. F. Hinks. 1976. Haemopoiesis in Lepidoptera. I. The multiplication of circulating haemocytes. *Can. J. Zool.* 54:1003–12.

Arnold, J. W., and S. S. Sohi. 1974. Hemocytes of *Malacosoma disstria* Hübner (Lepidoptera, Lasiocampidae): Morphology of the cells in fresh blood and after cultivation in vitro. *Can. J. Zool.* 52(4):481–5.

Baerwald, R. J., and G. M. Boush. 1970. Fine structure of the hemocytes of *Periplaneta americana* (Orthoptera: Blattidae) with particular reference to marginal bundles. *J. Ultrastruct. Res.* 31:151–61.

Bauchau, A. G., and M.-B. De Brouwer. 1972. Ultrastructure des hémocytes d'*Eriocheir sinensis*, Crustacé Décapode Brachyoure. *J. Microsc. (Paris)* 15(2):171–80.

Bauchau, A. G., and M.-B. De Brouwer. 1974. Etude ultrastructurale de la coagulation de l'hémolymphe chez les crustacés. *J. Mirosc. (Paris)* 19(1):37–46.

Beaulaton, J., and M. Monpeyssin. 1976. Ultrastructure et cytochimie des hémocytes d'*Antheraea pernyi* Guér. (Lepidoptera, Attacidae) au cours du cinquième âge larvaire. I. Prohémocytes, plasmatocytes et granulocytes. *J. Ultrastruct. Res.* 55:143–56.

Belden, D. A., Jr., and R. R. Cowden. 1971. Detection of early changes in cockroach hemocytes during coagulation with 8-alinino-1-naphthalene sulfonic acid. *Experientia* 27:448–9.

Boiteau, G., and J.-M. Perron. 1976. Etude des hémocytes de *Macrosiphum euphorbiae* (Thomas) (Homoptera: Aphididae). *Can. J. Zool.* 54:228–34.

Brehélin, M. 1971. Coagulation of the hemolymph in larvae of *Locusta migratoria. C. R. Acad. Sci. Paris D* 273:1598–1601.

Brehélin, M., and J. A. Hoffmann. 1971. Effets de l'injection d'une poudre inerte (saccharate de fer) sur l'hémogramme et la coagulation de l'hémolymphe chez un insecte orthoptère: *Locusta migratoria* L. *C. R. Acad. Sci. Paris D* 272:1409–12.

224   Ch. Grégoire and G. Goffinet

Brehélin, M., J. A. Hoffmann, G. Matz, and A. Porte. 1975. Encapsulation of implanted foreign bodies by hemocytes in *Locusta migratoria* and *Melolontha melolontha*. *Cell Tissue Res.* 160:283–9.

Breugnon, M.-M., and J. R. Le Berre. 1976. Variation de la formule hémocytaire et du volume d'hémolymphe chez la chenille de *Pieris brassicae* L. *Ann. Zool. Ecol. Anim.* 8(1):1–12.

Costin, N. M. 1975. Histochemical observations of the haemocytes of *Locusta migratoria*. *Histochem. J.* 7:21–43.

Crossley, A. C. S. 1964. An experimental analysis of the origins and physiology of the hemocytes in the blue blow-fly *Calliphora erythrocephala*. *J. Exp. Zool.* 157:375–98.

Crossley, A. C. S. 1975. The cytophysiology of insect blood. *Adv. Insect Physiol.* 11:117–221.

Czihak, G. 1956. Die Lymphocyten der Insekten. *Mikrokosmos* 45:169–71.

Deevey, G. B. 1941. The blood cells of the Haitian tarantula and their relation to the moulting cycle. *J. Morphol.* 68(3):457–91.

Devauchelle, G. 1971. Etude ultrastructurale des hémocytes du Coléoptère *Melolontha melolontha* (L.). *J. Ultrastruct. Res.* 34:492–516.

Dumont, J. N., E. Anderson, and G. Winner. 1966. Some cytologic characteristics of the hemocytes of *Limulus* during clotting. *J. Morphol.* 119:181–208.

François, J. 1975. Hémocytes et organe hémopoïétique de *Thermobia domestica* (Packard) (Thysanura: Lepismatidae). *Int. J. Insect Morphol. Embryol.* 4(6):477–94.

Gilliam, M., and H. Shimanuki. 1967 Progress report: Studies on honey bee blood. *Amer. Bee J.* 107(7):256.

Gilliam, M., and H. Shimanuki. 1970. Coagulation of hemolymph of the larval honey bee (*Apis mellifera* L.) *Experientia* 26:908–9.

Goffinet, G., and Ch. Grégoire. 1975. Coagulocyte alterations in clotting hemolymph of *Carausius morosus* L. *Arch. Int. Physiol. Biochim.* 83(4):707–22.

Granados, R. R., and D. J. Meehan. 1973. Morphology and differential counts of hemocytes of healthy and wound tumor-virus-infected *Agallia constricta*. *J. Invertebr. Pathol.* 22:60–9.

Grégoire, Ch. 1951. Blood coagulation in Arthropods. II. Phase contrast microscopic observations on hemolymph coagulation in sixty-one species of insects. *Blood* 6(11):1173–98.

Grégoire, Ch. 1952a. Sur la coagulation du sang de *Limulus polyphemus* (Arachnide). *Arch. Int. Physiol.* 60(1):97–9.

Grégoire, Ch. 1952b. Sur la coagulation du sang des araignées. *Arch. Int. Physiol* 60(1):100–2.

Grégoire, Ch. 1953a. Blood coagulation in Arthropods. III. Reactions of insect hemolymph to coagulation inhibitors of vertebrate blood. *Biol. Bull.* (Woods Hole) 104(3):372–93.

Grégoire, Ch. 1953b. Sur la coagulation de l'hémolymphe des Termites. *Arch. Int. Physiol.* 61:391–3.

Grégoire, Ch. 1954. Sur la coagulation de l'hémolymphe des Termites (deuxième note). *Arch. Int. Physiol.* 62:117–19.

Grégoire, Ch. 1955a. Blood coagulation in Arthropods. V. Studies on hemolymph coagulation in 420 species of insects. *Arch. Biol.* 66(1):103–48.

Grégoire, Ch. 1955b. Blood coagulation in Arthropods. VI. A study by phase contrast microscopy of blood reactions *in vitro* in Onychophora and in various groups of Arthropods. *Arch. Biol.* 66(3):489–508.

Grégoire, Ch. 1955c. Coagulation de l'hémolymphe chez les insectes irradiés par les rayons X. *Arch. Int. Physiol.* 63(2):246–8.

Grégoire, Ch. 1957. Studies by phase-contrast microscopy on distribution of patterns of hemolymph coagulation in insects. *Smithson. Misc. Collect.* 134(6):1–35.

Grégoire, Ch. 1959a. Hemolymph of Curculionidae and Diptera. Observations by phase-contrast and electron microscopy. *Explor. Parc Nat. Albert Sér.* 2(10):1–17.

Grégoire, Ch. 1959b. Further studies on distribution of patterns of hemolymph coagulation in neotropical insects. *Smithson. Misc. Collect.* 139(3):1–23.

Grégoire, Ch. 1964. Hemolymph coagulation, pp. 153–88. *In* M. Rockstein (ed.). *The Physiology of Insecta*, Vol. 3, 1st ed. Academic Press, New York.

Grégoire, Ch. 1970. Haemolymph coagulation in arthropods, pp. 45–74. *In* R. G. Mac-Farlane (ed.). *The Haemostatis Mechanism in Man and Other Animals.* Symposium, Zoological Society of London, No. 27. Academic Press, New York.

Grégoire, Ch. 1971. Hemolymph coagulation in Arthropods, pp. 145–189. *In* M. Florkin and B. T. Scheer (eds.). *Chemical Zoology*, Vol. 6. Academic Press, New York.

Grégoire, Ch. 1974. Hemolymph coagulation, pp. 309–60. *In* M. Rockstein (ed.). *The Physiology of Insecta*, Vol. 5, 2nd ed. Academic Press, New York.

Grégoire, Ch. 1975. Reactions *in vitro* of the hemolymph of *Psilopa petrolei* (Larva) (Diptera, Ephydridae). *Arch. Int. Physiol. Biochim.* 83(1):127–9.

Grégoire, Ch., and M. Florkin. 1950. Blood coagulation in Arthropods. I. The coagulation of insect blood, as studied with the phase contrast microscope. *Physiol. Comp. Oecol.* 2(2):126–39.

Grégoire, Ch., and P. Jolivet. 1957. Coagulation du sang chez les Arthropodes. VIII. Réactions du sang et de l'hémolymphe in vitro, étudiées au microscope à contraste de phase, chez 210 espèces d'Arthropodes africains. *Explor. Parc Nat. Albert Sér* 2(4):1–45.

Grégoire, Ch., and H. J. Tagnon. 1962. Blood coagulation, pp. 435–82. *In* M. Florkin and H. S. Mason (eds.). *Comparative Biochemistry*, Vol. 4. Academic Press, New York.

Gupta, A. P. 1968. Hemocytes of *Scutigerella immaculata* and the ancestry of Insecta. *Ann. Entomol. Soc. Amer.* 61(4):1028–9.

Gupta, A. P. 1969. Studies of the blood of Meloidae (Coleoptera). I. The haemocyte of *Epicauta cinerea* (Forster) and a synonymy of haemocyte terminologies. *Cytologia* 34(2):300–44.

Gupta, A. P., and D. J. Sutherland. 1966. *In vitro* transformations of the insect plasmatocyte in some insects. *J. Insect Physiol.* 12:1369–75.

Guthrie, D. M., and A. R. Tindall. 1958. *The Biology of the Cockroach.* Edward Arnold, London.

Hagopian, M. 1971. Unique structures in the insect granular hemocytes. *J. Ultrastruct. Res.* 36:646–58.

Hardy, W. B. 1892. The blood corpuscles of the crustacea, together with a suggestion as to the origin of the crustacean fibrinferment. *J. Physiol.* 13:165–90.

Hearing, V. J. 1969. Demonstration of acid phosphatase activity in the granules of the blood cells of the lobster, *Homarus americanus. Chesapeake Sci.* 10:24–8.

Hearing, V. J., and S. H. Vernick. 1967. Fine structure of the blood cells of the lobster, *Homarus americanus. Chesapeake Sci.* 8:170–86.

Heilbrunn, L. V. 1961. The evolution of the haemostatic mechanism, pp. 283–301. *In* R. G. MacFarlane and A. H. T. Robb-Smith (eds.). *Functions of the Blood.* Academic Press, New York.

Hinton, H. E. 1954. Insect blood., *Sci. Prog.* 42:684–96.

Hoarau, F. 1976. Ultrastructure des hémocytes de l'oniscoïde *Helleria brevicornis* Ebner (Crustacé Isopode). *J. Microsc. Biol. Cell* 27(1):47–52.

Hoffmann, J. A. 1967. Etude des hémocytes de *Locusta migratoria* L. (Orthoptère). *Arch. Zool. Exp. Gén.* 108:251–91.

Hoffmann, J. A. 1969. Etude de la récupération hémocytaire après hémorragies expérimentales chez l'Orthoptère, *Locusta migratoria. J. Insect Physiol.* 15:1375–84.

Hoffmann, J. A. 1970a. Les organes hématopoïétiques de deux insectes orthoptères: *Locusta migratoria* et *Gryllus bimaculatus. Z. Zellforsch. Mikrosk. Anat. Anat.* 106:451–72.

Hoffmann, J. A. 1970b. Régulations endocrines de la production et de la différentiation des hémocytes chez un insecte orthoptère: *Locusta migratoria migratorïoides*. *Gen. Comp. Endocrinol. 15*(2):198–219.

Hoffmann, J. A. 1972. Modifications of the haemogramme of larval and adult *Locusta migratoria* after selective X-irradiations of the haemocytopoietic tissue. *J. Insect Physiol. 18:*1639–52.

Hoffmann, J. A., and M. Brehélin. 1971. Modifications de la protéinémie totale provoquées par l'injection de saccharate de fer chez des larves de stade V de *Locusta migratoria* (Orthoptère). *C. R. Acad. Sci. Paris D 272:*2932–3.

Hoffmann, J. A., A. Porte, and P. Joly. 1970. Sur la localisation d'une activité phénoloxydasique dans les coagulocytes de *Locusta migratoria* L. (Orthoptère). *C. R. Acad. Sci. Paris D 270:*629–31.

Hoffmann, J. A., and M. E. Stoeckel. 1968. Sur les modifications ultrastructurales des coagulocytes au cours de la coagulation de l'hémolymphe chez un insecte Orthoptéroïde: *Locusta migratoria* L. *C. R. Soc. Biol. 162*(12):2257–59.

Howell, W. H. 1885. Observations upon the blood of *Limulus polyphemus, Callinectes hastatus* and a species of Holothurian. *Stud. Biol. Lab. Johns Hopkins Univ. 3:*267–87.

Jones, J. C. 1954. A study of mealworm hemocytes with phase contrast microscopy. *Ann. Entomol. Soc. Amer. 47*(2):308–15.

Jones, J. C. 1956. The hemocytes of *Sarcophaga bullata* Parker. *J. Morphol. 99:*233–57.

Jones, J. C. 1959. The phase contrast study of the blood cells in *Prodenia* larvae (O. Lepidoptera). *Q. J. Microsc. Sci. 100:*17–23.

Jones, J. C. 1962. Current concepts concerning insect hemocytes. *Amer. Zool.2*(2):209–46.

Jones, J. C. 1964. The circulatory system of insects, pp. 1–107. *In* M. Rockstein (ed.). *The Physiology of Insecta*, Vol. 3, 1st ed. Academic Press, New York.

Jones, J. C. 1970. Hemocytopoiesis in insects, pp. 7–65. *In* A. S. Gordon (ed.). *Regulation of Hematopoiesis*, Vol. 1. Appleton, New York.

Joshua, H., J. Fischl, E. Henig, J. Ishay, and S. Gitter. 1973. Cytological, biochemical and bacteriological data on the hemolymph of *Vespa orientalis*. *Comp. Biochem. Physiol. 45 B:*167–75.

Lai-Fook, J. 1973. The structure of the haemocytes of *Calpodes ethlius* (Lepidoptera). *J. Morphol. 139:*79–104.

Lavallard, R., and S. Campiglia. 1975. Contribution à l'hématologie de *Peripatus acacioi* Marcus et Marcus (Onychophore). I. Structure et ultrastructure des hémocytes. *Ann. Sci. Nat. Zool. Biol. Anim. (Sér. 12) 17*(1):67–92.

Lea, M. S., and L. I. Gilbert. 1966. The hemocytes of *Hyalophora cecropia* (Lepidoptera). *J. Morphol. 118*(2):197–216.

Levin, J., and F. B. Bang. 1964. A description of cellular coagulation in the *Limulus*. *Bull. Johns Hopkins Hosp. 115:*337–45.

Loeb, L. 1903. Ueber die Bedeutung der Blutkörperchen für die Blutgerinnung und die Entzündung einiger Arthropoden und über mechanische Einwirkungen auf das Protoplasma dieser Zellen. *Virchows Arch. Pathol. Anat. Physiol. 173:*35–112.

Loeb, L. 1904. Ueber die Koagulation des Blutes einiger Arthropoden. *Beitr. Chem. Physiol. Pathol. 5:*191–207.

Loeb, L. 1910. Ueber die Blutgerinnung bei Wirbellosen. *Biochem. Z. 24:*478–95.

Marschall, K. J. 1966. Bau und Funktionen der Blutzellen des Mehlkäfers *Tenebrio molitor* L. *Z. Morphol Oekol. Tiere 58*(6):182–246.

McLaughlin, R. E., and G. Allen. 1965. Description of hemocytes and the coagulation process in the boll weevil, *Anthonomus grandis* Boheman (Curculionidae). *Biol. Bull. (Woods Hole) 128:*112–24.

Moran, D. T. 1971. The fine structure of cockroach blood cells. *Tissue Cell 3*(3):413–22.

Morrison, P. R., and K. C. Morrison. 1952. Bleeding and coagulation in some Bermudan Crustacea. *Biol. Bull.* (*Woods Hole*) 103:395–406.

Mürer, E. H., J. Levin, and R. Holme. 1975. Isolation and studies of the granules of the amebocytes of *Limulus polyphemus*, the horseshoe crab. *J. Cell Physiol.* 86(3):533–42.

Murray, V. I. E. 1971. The haemocytes of *Hypoderma* (Diptera, Oestridae). *Proc. Entomol. Soc. Ont.* 102:46–63.

Needham, A. E. 1970. Haemostatic mechanism in the invertebrata, pp. 19–44. *In* R. G. MacFarlane (ed.). *The Haemostatic Mechanisms in Man and Other Animals.* Symposium, Zoological Society of London, No. 27. Academic Press, New York.

Neuwirth, M. 1973. The structure of the hemocytes of *Galleria mellonella* (Lepidoptera). *J. Morphol.* 139:105–24.

Nittono, Y. 1960. Studies on the blood cells in the silkworm, *Bombyx mori* L. *Bull. Seric. Exp. Stn.* 16(4):171–266.

Nolf, P. 1908. Contribution à l'étude de la coagulation du sang (3e mémoire): Les facteurs primordiaux: Leur origine. *Arch. Int. Physiol.* 6:1–72.

Nolf, P. 1909. Contribution à l'étude de la coagulation du sang (8e mémoire): La coagulation chez les crustacés. *Arch. Int. Physiol.* 7:411–61.

Olson, K., and S. D. Carlson. 1974. Surface fine-structure of hemocytes of *Periplaneta americana:* A scanning electron microscope study. *Ann. Entomol. Soc. Amer.* 67(1):61–5.

Poyarkoff, E. 1910. Recherches histologiques sur la métamorphose d'un Coléoptère (La Galéruque de l'Orme). *Arch. Anat. Microsc.* 12:333–474.

Price, C. D., and N. A. Ratcliffe. 1974. A reappraisal of insect haemocyte classification by the examination of blood from fifteen insect orders. *Z. Zellforsch. Mikrosk. Anat.* 147:537–49.

Raina, A. K., and R. A. Bell. 1974. Haemocytes of the pink bollworm *Pectinophora gossypiella*, during larval development and diapause. *J. Insect Physiol.* 20:2171–80.

Ratcliffe, N. A., and C. D. Price. 1974. Correlation of light and electron microcopic hemocyte structure in the Dictyoptera. *J. Morphol.* 144(4):485–98.

Ravindranath, M. H. 1973. The hemocytes of a millipede, *Thyropygus poseidon. J. Morphol.* 141:257–68.

Ravindranath, M. H. 1974a. The hemocytes of a scorpion, *Palamnaeus swammerdami. J. Morphol.* 144:1–10.

Ravindranath, M. H. 1974b. The hemocytes of an isopod, *Ligia exotica. J. Morphol.* 144:11–22.

Ravindranath, M. H. 1974c. Changes in the population of circulating hemocytes during molt cycle phases of the millipede, *Thyropygus poseidon. Physiol. Zool.* 47:252–60.

Ravindranath, M. H. 1975. Effects of temperature on the morphology of hemocytes and coagulation process in the mole-crab *Emerita* (*Hippa*) *asiatica. Biol. Bull.* (*Woods Hole*) 148:286–302.

Rowley, A. F. 1977. The role of the haemocytes of *Clitumnus extradentatus* in haemolymph coagulation. *Cell Tissue Res.* 182:513–24.

Rowley, A. F., and N. A. Ratcliffe. 1976. The granular cells of *Galleria mellonella* during clotting and phagocytic reactions *in vitro. Tissue Cell* 8(3):437–46.

Ryan, M., and W. L. Nicholas. 1972. The reaction of the coakroach *Periplaneta americana* to the injection of foreign particulate material. *J. Invertebr. Pathol.* 19:299–307.

Salt, G. 1956. Experimental studies in insect parasitism. IX. The reactions of a stick insect to an alien parasite. *Proc. R. Soc. Lond.* 146:93–108.

Scharrer, B. 1972. Cytophysiological features of hemocytes in cockroaches. *Z. Zellforsch. Mikrosk. Anat.* 129:301–19.

Schmidt, A. 1892. *Zur Blutlehre.* Vogel, Leipzig.

Schmidt, A. 1895. *Weitere Beiträge zur Blutlehre.* Bergmann, Wiesbaden. (Quoted by Nolf, 1908, 1909.)

Seligman, I. M., and F. A. Doy. 1973. Hormonal regulation of disaggregation of cellular fragments in the haemolymph of *Lucilia cuprina. J. Insect Physiol. 19:*125–35.

Selman, B. J. 1962. The fate of the blood cells during the life history of *Sialis lutaria* L. *J. Insect Physiol. 8:*209–14.

Sherman, R. G. 1973. Ultrastructurally different hemocytes in a spider. *Can. J. Zool. 51:*1155–9.

Shishikura, F., J. Chiba, and K. Sekiguchi. 1977. Two types of hemocytes in localization of clottable protein in Japanese horseshoe crab, *Tachypleus tridentatus. J. Exp. Zool. 201*(2):303–8.

Solum, N. O. 1970. Coagulation in *Limulus:* Some properties of the clottable protein in *Limulus polyphemus* blood cells, pp. 207–16. *In* R. G. MacFarlane (ed.). *The Haemostatic Mechanism in Man and Other Animals.* Symposium, Zoological Society of London, No. 27. Academic Press, New York.

Stang-Voss, Chr. 1970. Zur Ultrastruktur der Blutzellen wirbelloser Tiere. I. Ueber die Haemocyten der Larve des Mehlkäfers *Tenebrio molitor* L. *Z. Zellforsch. Mikrosk. Anat. 103:*589–605.

Stang-Voss, Chr. 1971. Zur Ultrastruktur der Blutzellen wirbelloser Tiere. V. Ueber die Hämocyten von *Astacus astacus* (L.) (Crustacea). *Z. Zellforsch. Mikrosk. Anat. 122:*68–75.

Stroganova, V. K., and V. V. Guliy. 1973. The morphology of the haemocytes in some conifer-feeding Tenthredinoidae (Hym.). *Entomol. Obozr. 52*(2):256–9. (*Entomol. Abs.* 4 E 8087; in Russian.)

Stutman, L. J., and M. Doliver. 1968. Mechanism of coagulation in *Gecarcinus lateralis. Amer. Zool. 8:*481–9.

Sundara Rajulu, G. 1970. A study of haemocytes of a centipede *Ethmostigmus spinosus* (Chilopoda: Myriapoda). *Curr. Sci. 20:*324–5.

Sundara Rajulu, G., N. Krishnan, and M. Singh. 1970. The haemocytes of *Eoperipatus weldoni* (Onych.). *Zool. Anz. 184:*220–5.

Tait, J. 1910. Crustacean blood caogulation as studied in the Arthrostraca. *Q. J. Exp. Physiol. 3:*1–20.

Tait, J. 1911. Types of crustacean blood coagulation. *J. Mar. Biol. Assoc. U.K. 9:*191–8.

Tait, J., and J. D. Gunn. 1918. The blood of *Astacus fluviatilis:* A study in crustacean blood, with special reference to coagulation and phagocytosis. *Q. J. Exp. Physiol. 12:*35–80.

Takada, M., and H, Kitano. 1971. Studies on the larval haemocytes of *Piereis rapae crucivora* Boisduval, with special reference to haemocyte classification, phagocytic activity and encapsulation capacity (Lep. Pieridae). *Kontyû 39:*385–94.

Terra, W. R., A. G. de Bianchi, and F. J. S. Lara. 1974. Physical properties and chemical composition of the hemolymph of *Rhynchosciara americana* (Diptera) larvae. *Comp. Biochem. Physiol. 47 B:*117–29.

Tuzet, O., and J. F. Manier. 1954. Les organes hématopoïétiques et le sang des Myriapodes Diplopodes (étude par le microscope à contraste de phase). *Bull. Biol. Fr. Belg. 88*(1):90–8.

Tuzet, O., and J. F. Manier. 1958. Recherches sur *Peripatopsis moseleyi* Wood-Mason, péripate du Natal. *Bull. Biol. Fr. Belg. 92*(1):7–23.

Vecchi, M. A., M. M. Bragaglia, and H. Wille. 1972. Etudes sur l'hemolymphe de l'abeille adulte (*Apis mellifica* L.). 3. Observations ultrastructurales sur deux éléments cellulaires. *Mitt. Schweiz. Entomol. Ges. 45:*291–8.

Vostal, Z. 1970. Contribution to the classification of the hemocytes of Tracheata. *Biológia (Bratislava) 25:*811–18. (In Czech.)

Vostal, Z., Z. Denek, and E. Pircŏva. 1968. Zur Kenntnis der Hämocyten der Vielfüsser (Diplopoda). *Biológia (Bratislava)*, 23:161–5. (in Czech.)

Walters, D. R. 1970. Hemocytes of saturniid silkworms: Their behavior *in vivo* and *in vitro* in response to diapause, development and injury. *J. Exp. Zool.* 174:441–50.

Wharton, D. R. A., and J. E. Lola. 1969. Lysozyme action on the cockroach, *Periplaneta americana*, and its intracellular symbionts. *J. Insect Physiol.* 15:1647–58.

Wheeler, R. E. 1963. Studies on the total haemocyte count and haemolymph volume in *Periplaneta americana* (L.), with special reference to the last moulting cycle. *J. Insect Physiol.* 9(2):223–35.

Whitten, J. M. 1964 Haemocytes and the metamorphosing tissues in *Sarcophaga bullata, Drosophila melanogaster* and other cyclorraphous Diptera. *J. Insect Physiol.* 10:447–69.

Wigglesworth, V. B. 1955. The role of the haemocytes in the growth and moulting of an insect, *Rhodnius prolixus* (Hemiptera). *J. Exp. Biol.* 32:649–63.

Wigglesworth, V. B. 1959. Insect blood cells. *Annu. Rev. Entomol.* 4:1–16.

Wigglesworth, V. B. 1965. *The Principles of Insect Physiology*, 6th ed. Methuen, London.

Wille, H., and M. A. Vecchi. 1974. Untersuchungen über die Hämolymphe der Honigbiene (*Apis mellifica* L.) *Mitt. Schweiz. Entomol. Ges.* 47(3,4):133–49.

Wittig, G. 1965. Phagocytosis by blood cells in healthy and diseased caterpillars. I. Phagocytosis of *Bacillus thuringiensis* Berliner in *Pseudaletia unipunctata* (Haworth). *J. Invertebr. Pathol.* 7:474–88.

Wood, P. J., J. Podlewski, and Th.E. Schenk. 1971. Cytochemical observations of hemolymph cells during coagulation in the crayfish, *Orconectes virilis*. *J. Morphol.* 134:479–88.

Wood, P. J., and L. P. Visentin. 1967. Histological and histochemical observations of the hemolymph cells in the crayfish *Orconectes virilis*. *J. Morphol.* 123(12):559–68.

Yeager, J. F., and H. H. Knight. 1933. Microscopic observations on blood coagulation in several different species of insects. *Ann. Entomol. Soc. Amer.* 26:591–602.

Zachary, D., M. Brehélin, and J. A. Hoffmann. 1975. Role of the "Thrombocytoids" in capsule formation in the dipteran *Calliphora erythrocephala*. *Cell Tissue Res.* 162:343–8.

Zachary, D., J. A. Hoffmann, and A. Porte. 1972. Sur un nouveau type de cellule sanguine (le thrombocytöide) chez *Calliphora erythrocephala*. *C. R. Acad. Sci. Paris D* 275:393–5.

8 Controversies about hemocyte types in insects

J. W. ARNOLD

Biosystematics Research Institute, Research Branch, Agriculture Canada, Ottawa, Ontario K1A 0C6, Canada

Contents

8.1. Introduction

It seems strange that after almost a century of research in insect hemocytology, there is still no consensus on one of its most basic questions: Do the different forms of cells we find in the blood represent transient morphological variants of one kind of cell or constitute a number of distinctive and immutable types? If our indecision on this point seems strange, perhaps it can be forgiven if we consider that the answer requires at least some knowledge of the origins and biology of the hemocytes in a wide range of insect species. Such knowledge cannot yet be claimed, for in only a few species among about 100 genera have even superficial studies been made, and in only about 20 others has the research continued in any depth. Surprisingly, almost the entire body of research on insect hemocytes is concentrated among only 6 orders (Coleoptera, Dictyoptera, Diptera, Hemiptera, Lepidoptera, and Orthoptera) of the 25 or more in the class, and there are widely varied opinions on the status of hemocyte types in any one of them. Little wonder, then, that the subject is a contentious one.

Although this question cannot be answered to everyone's satisfaction even now, it can be examined fairly in the light of recent observations and with an attempt to eliminate ambiguities and misinterpretations that are perpetuated in the literature. This will be the approach taken here.

The status of the hemocyte types is inseparable from hemocyte classification and is bound to studies of the nature of the cells. Unfortunately, hemocyte classification is constantly in disrepute. Critics reiterate that the hemocytes are classified in different ways by different authors, by different terminologies, based on different criteria, and for different purposes to the point where new workers are forced to create new systems that further aggravate the problem. To a large extent these criticisms are valid. Most classification schemes are based on some aspect of cell morphology, often with an attempt also to relate cell structure to cell processes or functions. These relationships are often suggested only by the cell shape, by cytochemistry of inclusions, or by the reaction of the cell to particular conditions. Some authors view such a classification, based on cell form and function, as the ultimate scheme. No doubt it has considerable merit for some purposes, but it leads to confusion where a hemocyte type is multifunctional or where a particular function is served by different cells in different insects. Such inconsistencies defeat the purpose of hemocyte classification.

Given the variations in form and functions of hemocytes in different insects, and the difficulty of creating a consistent classification on these bases, it seems logical to look to the source of the cells as an alternative criterion. Surely the site or the kind of origin of a line of cells is basic to its recognition as a class and is likely to be constant for many insects. Perhaps our question on the status of the types can be answered on this basis, if we can recognize the one or the several sites of their origin and relate this to our knowledge of cell form and functions. A classification from this basis should serve widely in Insecta and provide a degree of constancy not found in other schemes. Let us examine the possibility.

8.2. Hemocyte types

There is no need to recount here the history of hemocyte classification or to dwell on disputes about hemocyte nomenclature and synonymies. There are many such accounts in the literature: Jones (1962, 1964), Gupta (1969), Arnold (1974). Instead, it seems more useful to examine the most commonly recognized types and try to evaluate their status from the standpoint of their apparent origins and basic characteristics.

Jones (1962, 1964) noted that nine seemingly valid types of hemocytes had been described from all insects whose blood had been examined. He still holds this view (Jones, 1977), with an addition of the granulocytophagous cell, which he finds only in *Rhodnius*. The nine types he proposes are prohemocyte, plasmatocyte, granular hemocyte, cystocyte, spherule cell, oenocytoid, adipohemocyte, podocyte, and vermiform cell (nematocyte). I propose to deal here with only five of these, the ones that are most commonly recognized by authors and show clear morphological distinctions with a variety of techniques: prohemocytes (PRs), plasmatocytes (PLs), granulocytes (GRs), spherulocytes (SPs), and oenocytoids (OEs).

Two of Jones's nine types, the podocytes (POs) and the nematocytes (NEs) (VEs) (Figs. 8.7, 8.8) are recognized as peculiar to only one or two species and are disregarded here. They are considered to represent species-specific morphological variants that develop in midlarval life from the PL line of cells. They seem similar to, but more exaggerated than, some forms found in *Euxoa* species (Arnold and Hinks, 1975; Arnold, 1976). Another of the types, the GR, is considered here in the broad sense, to include a variety of forms of cells with the one common characteristic of relatively small but obvious cytoplasmic inclusions of various kinds (for details, see Chapter 4). It therefore includes the adipohemocytes (ADs) of Jones's classification and is synonymous with the spheroidocytes of Yeager's work (1945), where much of the controversy originated. It has been noted (Arnold, 1952a,b; Raina, 1976) that GRs in some insects accumulate lipid droplets progressively to the point where they can be referred to as ADs, but with little reason to categorize them separately. Perhaps a more worthwhile distinction could be made between the GRs of holometabolous and hemimetabolous insects, where the inclusions and the form of the cells are quite different (Fig. 8.9, 8.17, 8,18). Finally, the cystocytes (coagulocytes or COs of Grégoire, 1974) are disregarded here because their separate identity seems questionable (for a different view, see Chapter 7). Under light microscopy they are evident only in wet films, where they are recognized with assurance only when they erupt and cause gelation of hemolymph in the immediate vicinity. Although some authors claim to identify cystocytes by ultramicroscopy (e.g., Hoffmann et al., 1968), others find little distinction at this level between these and GRs (Devauchelle, 1971). We are left, then, with only five of the nine types for discussion: PRs, PLs, GRs, SPs, and OEs (Figs. 8.1–8.19). They are found commonly in Insecta, but not all are common to every species.

In many insects these five types of hemocytes are quite distinctive in vivo, in vitro, in fixed and stained blood films, and ultramicroscopically. One would expect their separate status to be quite clear. Yet apparently there are no cytological characters that distinguish each with certainty in all insects – not, at least, to the satisfaction of different authors. Nor are there activities that are exclusive to any one type in all insects, as far as we know. One might well wonder whether the types

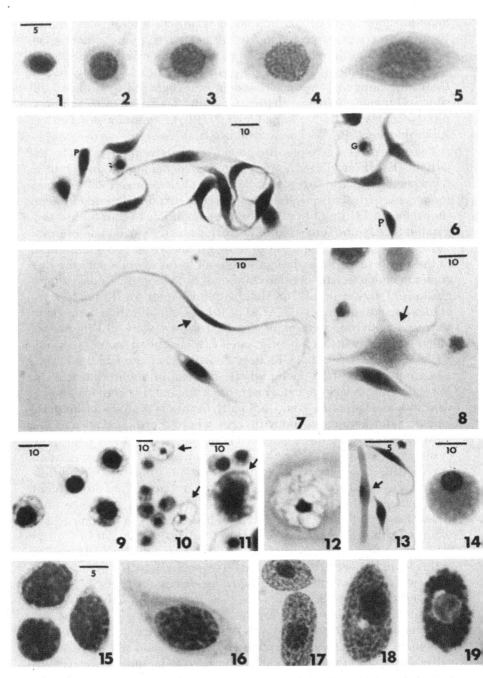

Figs. 8.1–8.14. Hemocytes of Lepidoptera (heat-fixed blood, Giemsa-stained). Magnification in microns, as indicated or in sequence. **8.1–8.5.** Simulated progression from prohemocyte (**8.1**) through immature stages (**8.2, 8.3**) to mature plasmatocytes (**8.4, 8.5**) in *Euxoa campestris* (Noctuidae). **8.6.** Group of PLs, along with PRs (*P*) and granulocytes (*G*) in *E. declarata*, illustrating diversity of form in this type of cell in

are mere illusions or else are not homologous in different groups of insects. The latter speculation is also part of the controversy.

8.3. Controversies

The controversies about hemocyte types stem from many sources, most importantly perhaps from the inherent differences in the hemocyte complex in different taxa, the inadequacy of our definition of a hemocyte type, and undoubtedly also from the proclivity of hemocytes to react drastically to any change in their environment and thus to the different techniques used to prepare them for observation. These and other causes of discrepancy have been discussed frequently in the literature and need no elaboration here. More pertinent to this discussion is the way these discrepancies affect the viewpoints of different authors in this field. Some seem unaffected and devote themselves to the description and naming of types in particular species, as though this taxonomy of hemocytes was an end in itself. Their contribution is in new descriptions, with no question of validity of types. At the other extreme are several mostly recent authors, who bring new techniques to the field and suggest that hemocyte classification be abandoned entirely, for as one of them says, "transitional features of hemocyte morphology seem to favour a concept of functional flexibility by one type of cell" (Scharrer, 1972). Between these extremes are authors who seem satisfied that the types are real, and they proceed to assess the roles of each type in the life of the insect.

The two extremes of view symbolize the two main concepts or theories on the status of hemocyte types, which may be summarized as follows: (1) the hemocyte types are morphological representations of phases in the life of a blood cell, each phase with a different function; (2) there are several distinct types of hemocytes, usually about five, which are immutable and serve different roles in insect physiol-

some species. **8.7.** Nematocyte (arrow) beside a typical PL in *Prodenia eridania* (Noctuidae), seemingly a very attenuated form of PL peculiar to this species at certain stages. **8.8.** Podocyte (arrow) in *P. eridania,* seemingly a large, multiramous form of PL peculiar to this species at certain stages. **8.9.** Typical GRs in *E. declarata.* Note compact, often eccentric nucleus and vesiculate cytoplasm, which features mucoid or lipoid inclusions. **8.10–8.12.** Spherulocytes (arrows) with GRs in *Euxoa* sp., showing cells with spherule contents absent, present, and dispersed around cell. **8.13, 8.14.** Oenocytoids of different sizes and forms in *Agrotis vetusta* (**8.13,** arrow) and in *E. declarata* (**8.14**).
Figs. 8.15–8.19. Hemocytes of various hemimetabolous insects (heat- or glutaraldehyde-fixed and Giemsa-stained). Magnification in microns, as indicated or in sequence. **8.15.** PRs of *Diaphemora femorata* (Phasmidae). **8.16.** PL of *D. femorata.* Note heavy chromatin in nuclei of both types. **8.17, 8.18.** GRs of *Acroneuria arenosa* (Perlidae) (**8.17**) and of *Blaberus discoidalis* (Blaberidae) (**8.18**). Note distinctions in cell size, form, and cytoplasmic contents between these and lepidopteran GRs. **8.19.** SP in *B. discoidalis,* distinct in details also from lepidopteran cells of this type.

ogy. For easy reference, these may be termed the "single-cell theory" and the "multiple-cell theory," respectively. There are variations on both these themes, but they relate to one or other of these principles.

8.3.1. Single-cell theory

This is a tidy concept, supported by many general observations and by a number of ultramicroscopic and cultural studies with hemocytes of a range of species. It has been accepted in principle by many authors, including some recent ones: Moran (1971), Scharrer (1972), Olson and Carlson (1974), Ratcliffe and Price (1974), Crossley (1975), Landureau and Grellet (1975), and Beaulaton and Monpeyssin (1977). Fundamentally, the theory takes into account the seemingly transitional forms of hemocytes that share characteristics of more than one type. It accounts also for the general shift in the relative numbers of the different types during larval life, from an early complex that favors PRs and PLs to a complex near the end of larval development that is composed largely of GRs and SPs. Beyond this, some of the recent ultramicroscopic research seems to indicate a considerable overlap in the fine structure of the different types (Moran, 1971; Olson and Carlson, 1974), and some cultural studies show only PRs and PLs in subcultures in vitro (Mitsuhashi, 1966; Sohi, 1971). In other words, these ultramicroscopic studies suggest that the electron images demonstrate different morphological expressions that a hemocyte can assume during its life, whereas the cultural studies indicate that germinal PRs transform to PLs in vitro and would likely proceed to other types were it not for the unfavorable conditions in culture.

One might be persuaded by these observations alone that this is a reasonable theory and might prove correct in the final analysis. To make it seem even more acceptable, there is also an autoradiographic study (Shrivastava and Richards, 1965) that demonstrates a direct line of development from PR to PL to AD (GR) in the blood of a lepidopteran, *Galleria mellonella*. This evidence would seem to clinch the theory for this insect, and it has been quoted widely in this context for other species.

In examining the validity of this theory, it is interesting to note that the concept of one cell is held most commonly by authors whose research has been limited to hemocytes of pauro- or hemimetabolous insects. Several have worked exclusively with cockroaches, and some have concentrated on species from the primitive blattoid group alone. This was the case for Smith (1938), who expressed the view early for hemocytes of one of the best known representatives, the American cockroach, *Periplaneta americana*, from his studies of the cells in fixed and stained blood films. The same conclusion was reached with the same species in much later studies of the fine structure of the cells (Olson and Carlson, 1974) and in studies of the cells in culture (Landureau and Grellet, 1975). Again this conclusion was reached for a number of other species of cockroaches, including some from the modern blaberoid group, mostly from research in ultramicroscopy (Moran, 1971; Ratcliffe and Price, 1974). Other authors reached similar conclusions working with species of Orthoptera. Ogel (1955) seemed to have a single-cell concept for hemocytes of 22 species, and Hoffmann and his associates developed a modified view of this theory from work with *Locusta migratoria.* They suggested first (Hoffmann, 1967) that the PRs were the stem cells from which PLs were derived, whereas the GRs and perhaps the OEs developed from undifferentiated PLs. Later (Hoffmann et al., 1968), they considered that there were three fundamental types, PLs, GRs, and COs, with the OE as a specialized type of GR. It is interesting that among these groups of insects the simple, morphological distinctions among the five types of hemocytes are comparatively minor and that in some species not all the types occur.

Nevertheless, not all proponents of the single-cell theory have restricted their research to insects in the lower orders. Aside from the previously mentioned work of Shrivastava and Richards (1965) with a lepidopteran, there are a number of other examples, including Devauchelle's (1971) work with the beetle *Melolontha melolontha,* the research of Lai-Fook (1973) and Beaulaton and Monpeyssin (1976, 1977) with other Lepidoptera, and Selman's investigations (1962) with hemocytes of the neuropteran, *Sialis lutaria.* It should be noted perhaps that the first three authors depended almost entirely on ultramicroscopy for their conclusions and that the last one employed a blood smear technique that causes hemocyte distortion. This comment requires some explanation.

Inadequate techniques of fixation and staining of blood films have received much comment in the literature. The references often refer to attempts to apply vertebrate blood smear techniques directly to insect blood. These seldom function well without modification, presumably because of the high concentrations of protein in insect hemolymph and of special features of the hemocyte membrane. Inevitably, they cause artifacts and encourage misinterpretations. Some of the seemingly transitional forms of hemocyte types are likely products of fixation or the lack of it, and so also are certain named types (e.g., the lamellocytes of some authors and some of the podocytes of others).

Ultramicroscopy is used increasingly in insect hemocytology, and some authors feel it is the ultimate means of distinguishing types (Crossley, 1975). Nevertheless, a review of the main papers that include this discipline or use it exclusively reveals a wide range of interpretation by the different authors, sometimes with the same or similar species. I have grouped them here according to their conclusions on the number of hemocyte types in the complex.

1. *Blood with a single type of hemocyte.* Authors of four papers, all dealing with cockroach blood, found only one basic type of hemocyte, as mentioned previously in another context. Two of these papers dealt only with cells in *P. americana:* One employed scanning electron microscopy (SEM)(Olson and Carlson, 1974), and the other used cultural methods as well as transmission electron microscopy (TEM)(Landureau and Grellet, 1975). As for the other two papers, one dealt with one blaberoid species, *Blaberus discoidalis* (Moran, 1971), and the other with several species from both the blattoid and blaberoid families (Ratcliffe and Price, 1974).

2. *Blood with three types of hemocytes.* Authors of four papers found three types of hemocytes in different insects, but the types were not the same in each case. Hoffmann et al. (1968) identified PLs, GRs, and COs in *L. migratoria* by TEM, and considered OEs to be a form of GR. François (1975) found the same three types in the thysanuran, *Thermobia domestica,* but gave no indication of OEs. Akai (1969) found two types of GRs and a third type that seemed to be a PL in the lepidopteran, *Philosamia cynthia ricini.* Zachary and Hoffmann (1973) distinguished PLs, thrombocytoids, and OEs in the dipteran, *Calliphora erythrocephala.*

3. *Blood with four types of hemocytes.* Authors of three different papers agreed quite closely on the occurrence of four types of hemocytes in various insects. Harpaz et al. (1969) found plasmatocytoids, GRs, ADs, and OEs in a lepidopteran, *Spodoptera littoralis.* Neuwirth (1973) identified plasmatocytoids, GRs, SPs, and OEs in another lepidopteran, *G. mellonella.* François (1974) described PLs, two types of GRs, and COs in *T. domestica.* Note that PLs and GRs are the only cells recognized by each of the authors and that in each case, at least one other type is different.

4. *Blood with five types of hemocytes.* Authors of four papers concluded that there are five types of hemocytes in insects, the five standard ones considered in the present paper: PRs, PLs, GRs, SPs, and OEs. Each paper dealt only with Lepidoptera. This was true for the silkworm, *B. mori* (Akai and Sato, 1973) and for *C. ethlius* (Lai-Fook, 1973), although in the latter paper the author considered that only the GRs, SPs, and OEs could be identified positively with both light and electron microscopy. The same five cells were recognized in *Pectinophora gossypiella* (Raina, 1976) and in *Antheraea pernyi* (Beaulaton and Monpeyssin, 1976, 1977), although in the last case the authors conceived a modified view of the single-cell theory. Nevertheless, this is the only grouping in which all the authors agree on the types in the hemocyte complex, and the conclusion is supported by the recent

work by Akai and Sato (1976) where SEM was used exclusively (see also Chapter 5).

5. *Blood with six types of hemocytes.* Finally, authors of three papers found evidence of six types of hemocytes by ultramicroscopy, but applied a variety of names to them. In an early paper, Crossley (1964) described six types in *C. erythrocephala*, but referred to them only by number. Wittig (1968) described microplasmatocytes, macroplasmatocytes, POs, SPs, PRs, and GRs in *Pseudaletia unipuncta*. Devauchelle (1971) found the standard five types along with ADs in *M. melolontha*.

A number of observations can be made from this survey. It is clear that the technique of ultramicroscopy is not yet standardized to the point where interpretations are reliable. For example, some types such as the COs and ADs are recognized by some authors and ignored by others. Similarly, there is wide variation in the interpretation of the number of types in the hemocyte complex, even for the same or closely related species; examples are the variation from three to six types by different authors for *C. erythrocephala* and for different species of Lepidoptera. Some authors recognize only one type of GR; others see two types. Some identify PRs; others ignore them. And throughout this group of papers there is an aura of confusion created by different terminologies.

These observations and comments are not meant to denigrate the value of ultramicroscopy in insect hemocytology. Nor are they designed simply to play the devil's advocate. Rather, they are intended to call attention to a lack of standardization in this discipline not less than that in light microscopy. This belies the contention of Crossley (1975) that a cytophysiological identification of types by ultramicroscopy can overcome the inconsistencies introduced into hemocyte classification by other techniques. Nevertheless, there is no doubt that ultramicroscopy can help to clarify the status and functions of the hemocyte types, but at this stage and in different hands it cannot serve alone.

The autoradiographic study mentioned previously also leaves the question of the status of hemocyte types unanswered, despite the innovative nature of the research. The authors, Shrivastava and Richards (1965), recognized the standard five types of hemocytes in the blood of *G. mellonella*, but their evidence showed transitions through only three of them: PRs to PL to AD (GR). The SPs and OEs were not labeled by the tritiated thymidine (Shrivastava and Richards, 1964) and were unidentifiable in the preparations. Furthermore, there is some question about the identity of the GRs, which the authors describe as PLs that contained fat droplets. This is not a suitable description of GRs in a species where they are much smaller than the PLs and have a small, distinctive nucleus, even though allowance must be made for the technical difficulty of preserving the character of cells exposed to the process of development of autoradiographs.

In continuing to question the single-cell theory, we must examine also its feasibility. A number of questions might be asked: (1) Is the life of a hemocyte long enough to allow the prescribed changes in form and function? (2) What course of change would the primary cell follow? (3) What is the ultimate cell in the series of transformations? (4) How are the transformations regulated in the blood?

Probably the best estimate we have of the length of hemocyte life comes from the work of Shrivastava and Richards (1965). They found that the total cycle from the uptake of tritiated thymidine in the cells to its presence in the hemolymph after cell degeneration occupied less than 6 days. This agrees fairly well with a number of more general observations and is within the range of life span of some leucocytes in vertebrates. We must ask, then, whether 6 days is a long enough period for a cell to grow from germinal size and change its form and functions as many as four times. No doubt the answer depends on the extent of the changes and on what mechanisms are involved. If we consider only the comparatively simple and limited steps from PR to PL to GR in a hemimetabolous insect, where these types are fairly similar in size and form, the time scale seems not unrealistic. However, if we consider the complete series of changes through the standard five types, especially in a holometabolous insect, where the types are very distinctive in size, form, and structure, the time seems much too short. In either case, the system of changes would have to be staggered in some way to account for the shift in the composition of the hemocyte complex during the longer-than-6-day period from early to late larval life.

The course of hemocyte transformation is usually given as PR to PL, and perhaps then to GR. Authors seldom speculate further, even though the next steps are critical in a test of the feasibility of the single-cell theory. It is interesting to speculate on the possible course of events. Do the germinal cells change directly and inevitably from one type to the next through the entire series, or is there some system of alternate paths of development? For example, do some of the germinal cells stop differentiating at the PL stage, while others bypass this step and proceed directly to one of the other stages? Or do the paths of transformation bifurcate at certain points, e.g., does the sequence stop for some cells at the PL stage, while others continue either toward the GR stage or the OE stage? Or, indeed, are we correct in assuming that the series starts with the PR, when some authors suggest otherwise (Gupta and Sutherland, 1966) or else ignore the existence of the PR (Hoffmann et al., 1968; Akai, 1969; Zachary and Hoffmann, 1973) (see also Chapter 4)?

Although it is easy to speculate on these various possibilities, it is difficult to imagine what initiates any one of them and how they might be regulated in the blood. Are the changes generated by a regular need for a particular mixture of hemocyte types in the blood at certain stages of development, and can this need be changed, as happens under certain conditions? Can we assume that the germinal cells are programmed for change from the start and that each cell responds on schedule in its own way, perhaps with the decline of activity of one enzyme and the activation of another? Or can we simply consider that there are a number of types of PRs, which we do not recognize until they begin to differentiate in separate directions? Or again, can we imagine that a single type of cell at the start is affected in different ways by changing titers of hormones in the hemolymph and differentiates accordingly? There are obvious objections to any one of these suggestions, but let us look now at the alternative.

8.3.2. Multiple-cell theory

The multiple-cell theory appears in several forms, but most of its proponents see the PRs, or their equivalent germinal cells in hemopoietic tissue, as the source of all the other types. It differs from the single-cell theory mainly in considering that each of the differentiated types is a separate cell line and that the cells grow, mature, and degenerate without losing their identity.

The theory derives from four common observations: (1) the PR or germinal cell has undisputed germinal characteristics (e.g., it divides fairly often by mitosis, it is small, it has a high nucleus-to-cytoplasm ratio, with a nucleus rich in chromatin, a cytoplasm rich in ribosomes but poor in mitochondria, and Golgi bodies little developed); (2) there are always some cells that appear transitional between the germinal and the differentiated types; (3) there are usually marked cytological distinctions between mature cells of each type, at both the gross and the fine structural levels; (4) there are usually examples of young, mature, and old cells of each type in the blood at all times, with the exception of the PRs.

Such observations seem to have convinced the majority of authors, perhaps partly because most have dealt with holometabolous insects, whose hemocyte types are especially distinctive in appearance. However, the theory is not restricted to this group. For example, Hoffmann (1970) described four distinct cell types in *L. migratoria;* Arnold (1966) found the same number in the stonefly, *Acroneuria arenosa;* and several authors described a variety of types in species from Hemiptera (Khanna, 1964; Jones, 1965; Granados and Meehan, 1973). Even in Collembola, one author has indicated that there is more than one type of hemocyte (Barra, 1969).

Beyond these general observations, the theory is supported by a number of ultrastructural, histochemical, and cultural studies, some of which were referred to previously. The majority of ultramicroscopic studies has demonstrated significant differences in the fine structure of the main types. They include works by Wittig (1968), Harpaz et al. (1969), Akai and Sato (1973), Lai-Fook (1973), Neuwirth (1973), Zachary and Hoffmann (1973), François (1974), Raina (1976), and Beaulaton and Monpeyssin (1976, 1977). The differences include the size and form of the cells, the extent of cellular organization, and special cytoplasmic features that suggest special functions for particular types. In other areas, some authors have found distinctions in the cytochemistry of cytoplasmic inclusions in different types (Arnold and Salkeld, 1967; Lai-Fook, 1973; Beaulaton and Monpeyssin, 1976, 1977), and one cultural study (Arnold and Sohi, 1974) indicated two major cell lines in vitro in the forest tent capterpillar, *Malacosoma disstria*. Some support for the theory comes also from observations on the size characteristics of the different types (Fig. 8.39) in the noctuid, *Euxoa declarata* (Arnold and Hinks, 1976), where there is no evidence of progression in size from one type to another. With these observations in mind, it seems possible that morphological parameters for hemocyte classification and for the determination of cell maturity might be generated for insect hemocytes at the species level, as they have been for leucocytes in human blood (Miller et al., 1972).

One might conclude from these observations that this theory too is a reasonable one, but here too there are some puzzling aspects. Perhaps the most puzzling aspect again is the question of a regulatory mechanism for cell differentiation from a common source. It is difficult to understand how a germinal cell can be triggered to differentiate along different lines while it is circulating with the other types in a relatively homogeneous solution. We might speculate again that hormonal fluctuations in the hemolymph influence the germinal cells in different ways at different times or that a demand process exists to stimulate differentiation along different lines when one type or other has declined in numbers below its normal complement. We might also suggest that all the types originate separately in hemopoietic tissue at one or more sites in the body, as some authors have indicated in some hemimetabolous species. Or we might ask whether our long-held views on events in the blood should be reexamined.

8.4. Possible solution to the controversy

It is obvious from the preceding discussion that the various criteria for distinguishing types of hemocytes or their functions do not resolve the question of the status of the types. It should be clear also that this question can be answered only from a knowledge of the origin of the types in the major groups of Insecta. In other words, the validity of types will continue to be questioned until the origin and course of development of the one or the several hemocyte lines are clearly demonstrated.

It seems likely that hemocyte origins will differ within the Insecta – that different groups of insects will have developed different ways of generating blood cells and of adapting the cells to their particular needs. Such is the case with many other tissues and structures. We chose to examine the process in Lepidoptera, where the variety of hemocyte form and structure seems to culminate. Our findings may be true, at least in principle, for some other orders, but not likely for all of them.

8.4.1. Postembryonic hemopoiesis in Lepidoptera

Our study began with the observation by my colleague, Dr. C. F. Hinks, that the long-known "hemopoietic organs" associated with the wing disks in the meso- and metathorax of lepidopteran larvae (Schaeffer, 1889) seemed to be composed almost entirely of PRs and PLs. This was not an original observation, for Nittono (1964) had reported much the same for these tissues in *B. mori*, and he added (Jones, 1970) that OEs might also occur there. Nevertheless, the observation was intriguing because other types of hemocytes were absent. This led to speculation on the origin of the types and to a search for the source of the missing ones. We were faced, as in the earlier discussion here, with the question of whether the PRs and PLs in the organs entered the blood and differentiated there into the other types, or whether the other types had a separate source.

We set out to examine hemopoiesis from two viewpoints on the basis of common knowledge of mitosis of some hemocytes in the blood and on the apparent source in the organs: (1) hemocyte production by mitosis of different types in the blood, (2) hemocyte production by the organs as evidenced by histology and by the effect of organ elimination on the numbers of the different hemocyte types in the blood. The research was done with one species of noctuid cutworm, *Euxoa declarata*, and checked in principle only in species from several other families of Lepidoptera. Separate papers on both aspects are published elsewhere (Arnold and Hinks, 1976; Hinks and Arnold, 1977). The report here will summarize our findings.

8.4.2. Mitotic activity among circulating hemocytes

We found that three of the five types of hemocytes in this species – the PRs, the GRs, and the SPs – divide mitotically in the blood (Figs. 8.20 –8.29). Binucleate plasmatocytes and oenocytoids are seen (Figs. 8.30 –8.33), but not in cytokinesis. These observations were novel to some

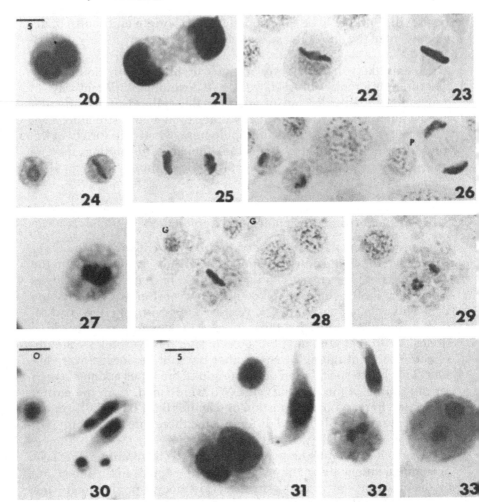

Figs. 8.20–8.29. Mitotic hemocytes in various insects (Giemsa or acetocarmine stain). Magnification in microns, as indicated or in sequence. **8.20–8.23.** Mitotic PRs: **8.20**, early telophase in *B. discoidalis*; **8.21**, late telophase in *D. femorata*; **8.22**, metaphase in *E. declarata*; **8.23**, mitotic spindle dislodged from cell in *E. declarata*. **8.24–8.26.** Mitotic GRs in *E. declarata*: **8.24**, metaphase; **8.25**, late telophase; **8.26**, daughter cells, along with late anaphase in a PR (*P*). Note clear distinction in cell and nuclear bundle size in PRs and GRs. **8.27–8.29.** Mitotic SPs in *E. declarata*: **8.27**, binucleate cell; **8.28**, metaphase; **8.29**, telophase. Note distinctions in size and cytoplasmic structure between SPs and GRs (*G*) with both staining methods.

Figs. 8.30–8.33. Binucleate hemocytes in *E. declarata* Giemsa-stained). **8.30, 8.31.** PLs. Note abnormal size of cell in **8.31**. **8.32, 8.33.** OEs, showing different degrees of nuclear separation. Neither of these cell types was seen in cytokinesis.

extent, for although there have been some reports of mitosis in GRs (e.g., Arnold, 1952a; Lea and Gilbert, 1966; Shapiro, 1968), and one report for the SPs (Nittono, 1960), most workers seem to consider that mitoses are exclusive to PRs (e.g., Cheng, 1964; Jones and Liu, 1968). In fact, various phases of mitosis can be seen in GRs after normal staining procedures with Giemsa and in SPs especially after suppressing staining of the cytoplasmic inclusions so that the nucleus is clearly visible.

In order to evaluate hemocyte production from this source, we estimated the total number of hemocytes in circulation at specified times during larval development, along with the numbers of dividing cells of each of the three types. From these data we calculated the number of hemocytes that could be produced daily by this method and in total during larval development. Our calculations utilized the following determination for each instar: blood volume (V), total numbers of hemocytes per cubic milliliter (THC), relative numbers of each type of hemocyte (DHC), and mitotic index of each type (MI = cells in division per 1,000). The total number of hemocytes in circulation (HIC) for any instar is derived from the formula THC × V, and the number of each type from HIC × DHC. Estimates of the productivity of each type by mitosis required certain assumptions on the duration and frequency of mitosis. These were based on the findings of Clark and Harvey (1965) and Lea and Gilbert (1966), which agreed that the period from metaphase through cytokinesis required 20–30 min. We assumed that a complete mitosis would require approximately 1 hr and that daily productivity at its peak would equal (MI × HIC × 24)/1,000. From our knowledge of the number of days in each instar, we could then estimate the maximum hemocyte production or the maximum number of cells of any one type produced for any instar. In theory, we could also judge how many cells should be in circulation if mitosis of circulating cells was the only source of hemocytes.

Unfortunately, the comparison of potential and actual numbers of hemocytes in circulation is more than a simple accounting procedure. There are several sources of error: (1) the estimates of productivity are maximum projections and not likely to be achieved regularly, (2) the counts of mitotic cells are much less than normal because we rarely see phases of mitosis prior to metaphase, (3) we know little of the length of the life of hemocytes or of the effect of various natural factors on the inception and rate of mitosis. Clearly, these are factors that can seriously influence calculations of hemocyte numbers, and they are areas that cry for research.

Our estimates of hemocyte numbers and production by mitosis in the blood are summarized in Tables 8.1 and 8.2. Table 8.1 shows the results of differential counts of hemocytes and of the dividing hemocyte types. It shows two significant relationships: (1) a decline in the

Table 8.1. Differential counts and mitotic indexes of circulating hemocytes

Instar	No. of larvae	Proportion of hemocytes of each type (%)				
		PR	PL	GR	SP	OE
1	4	8.0	33.9	49.3	8.1	0.7
2	4	6.0	42.4	30.8	18.8	2.0
3	7	10.2 ± 4	34.6 ± 5	34.6 ± 7	19.2 ± 11	1.4 ± 1
4	10	4.4 ± 3*	27.4 ± 11*	45.7 ± 9*	20.8 ± 8	1.7 ± 1*
5	10	2.8 ± 1*	26.6 ± 7	56.0 ± 11*	12.3 ± 6*	2.3 ± 1
6	35	3.2 ± 2	29.8 ± 7*	50.8 ± 9*	12.8 ± 6	3.3 ± 2*

Differential counts show average figures from counts of indicated numbers of larvae in each instar. Mitotic indexes for first and second instars are adjusted to mitoses per 1,000 from actual counts of 3,788 cells from 4 larvae and 2,482 cells from 3 larvae, respectively. Binucleate cells not included.
* Significantly different from figures in preceding instar (t-test).

MI of PRs after the fourth instar that coincides with a decline in the relative numbers of both PRs and PLs, and (2) continued mitotic activity of the GRs and SPs that seems appropriate to the maintenance of their numbers during larval life. We consider that discrepancies in the data for SPs reflect difficulty in recognizing mitotic figures in this type and the different timing of mitotic activity in these two types, which came to our attention later. Table 8.2 gives estimates of the numbers of each type in circulation at the end of each instar, along with projections of daily productivity by mitosis of circulating cells. Despite the aforementioned sources of error, a number of conclusions can be drawn: (1) there are more PRs in the blood after the third instar than can be accounted for by their mitoses alone; (2) in the absence of mitosis, there is no accounting for the continued increase in numbers of PLs without considering other means; (3) there is reason to believe that the numbers of GRs and SPs in the blood are maintained by self-multiplication, at least during their mitotically active periods; (4) in the absence of mitosis, there is no accounting for the continued increase in numbers of OEs without considering other means.

In other words, these data suggest that GRs and SPs are self-sustaining by division in the blood, whereas some of the PRs and all the PLs and OEs must derive from another source. The next aspect of our program suggested some modification of this statement.

8.4.3. Activity of the hemopoietic organs

Evidence of hemocyte production in the "hemopoietic organs" was sought first through histology – from whole mounts in vitro, from organ squashes stained in Giemsa and in acetocarmine, and from fixed

No. of larvae	Number of mitotic cells per 1,000					
	PR	PL	GR	SP	OE	Total
4	2.6	—	4.0	—	—	6.6
3	2.1	—	2.0	—	—	4.1
7	4.0 ± 3	—	3.4 ± 3	1.3 ± 1	—	8.7
8	2.6 ± 3	—	2.5 ± 2	1.0 ± 1	—	6.1
13	1.4 ± 1*	—	3.2 ± 3	2.3 ± 2*	—	7.1
8	0.5 ± 0.4*	—	1.0 ± 0.7*	2.5 ± 2	—	4.0

and stained sections. The squashes and sections were used to identify the cells in the organs and to check on mitotic activity. We also measured the size of the organs at intervals during larval life. Then we proceeded to examine total and differential hemocyte populations in portions of the larval body that were separated by ligatures to include or exclude one or both pairs of organs, with or without association of the organs with the head (Fig. 8.40). Each aspect contributed to our understanding of the role of the organs in hemopoiesis and merits extension in this and other species.

The hemopoietic organs are simple, lobulated structures associated closely with the germinal wing disks. Each lobe contains numerous clusters of cells, mainly germinal, in association with variable numbers of free cells that lie in the lumen (Figs. 8.34–8.38). The differentiated cells there are mostly PRs and PLs, but among them are occasional clusters of "micronucleocytes," which we identified tentatively as OEs (Fig. 8.38). Details are given in a recent paper (Hinks and Arnold, 1977).

The thin membrane that encloses the organs tends to rupture easily, and we assume this action is involved in the intermittent release of the free cells from the lumen into the blood. This is not supported by the work of Akai and Sato (1971), which suggests that hemocytes are released from the organs prior to each ecdysis through openings in the acellular sheath. Whichever method functions, there is little doubt that releases of cells from the organs account for the regular changes in the size of the organs and that major releases in the fifth and sixth instars account for the large changes in organ size at those times. They likely also account in part for the normal variability in the total and differential counts of hemocytes in the blood during larval develop-

Table 8.2. Estimates of numbers of circulating hemocytes and of daily hemocyte production by mitosis in each instar of larval life

Type of hemocytes	Total hemocyte counts (cells/mm³) in instars					Total number of hemocytes in circulation near end of instar					Maximum daily production of hemocytes in instars				
	2	3	4	5	6	2	3	4	5	6	2	3	4	5	6
All types	6,460	4,140	8,587	11,613	20,000	4,457	16,560	145,979	940,653	2,540,000	792	3,458	21,371	158,030	243,840
PR	388	422	378	325	660	268	1,688	6,423	26,325	83,820	14	162	401	885	1,005
PL	2,739	1,432	2,353	3,089	5,960	1,890	5,728	39,998	250,209	756,920	—	—	—	—	—
GR	1,990	1,432	3,924	6,503	10,160	1,373	5,728	66,712	526,766	1,290,520	66	467	4,003	40,456	30,968
SP	1,214	795	1,786	1,428	2,560	838	3,180	30,363	115,700	325,120	—	99	729	6,387	18,507
OE	129	58	146	267	660	89	232	2,482	21,635	83,820	—	—	—	—	—

Estimates were derived from measurements of blood volume (see methods and data below), total hemocyte counts, differential counts (Table 8.1), and mitotic indexes (MI) (Table 8.1.) in the following ways:

total number of hemocytes in circulation (HIC) = total hemocyte counts × blood volume:

maximum daily production = (MI × HIC × 24)/1,000.

Blood volume, second to sixth instars = 0.69, 4.0, 17, 81, and 127 mm³, respectively.

Total hemocyte counts for second instar larvae were made by direct count of all cells in a blood film made from a measured volume of blood. The counts for others were made by standard dilution method with a hemocytometer.

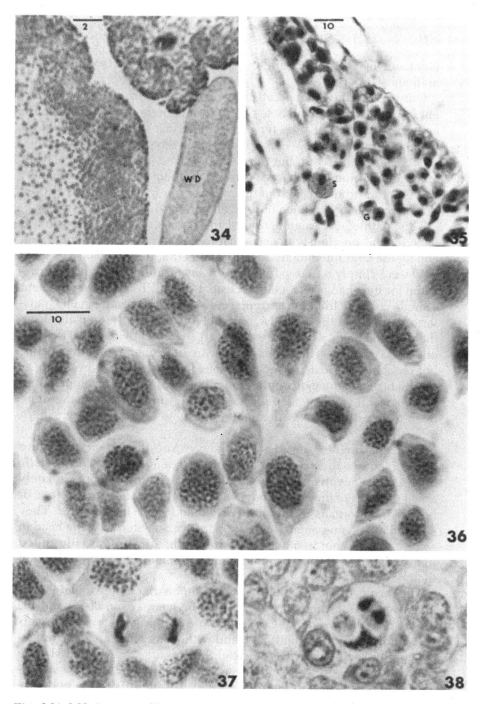

Figs. 8.34–8.38. Structure of hemopoietic organs (acetocarmine stain). Magnification in microns, as indicated or in sequence. **8.34.** Portion of two lobes, showing germinal tissue and free cells in lumen (*wd* = wing disk). **8.35.** Ruptured area of free cells, showing invasion of GRs (*G*) and SPs (*S*) among mainly PRs and PLs. **8.36.** Cell mass in region of differentiation. **8.37.** Mitotic PR in region of differentiation. **8.38.** Germinal "micronucleocyte" or OE in region of differentiation.

ment and for the apparent influx of large PLs in the last two instars (Arnold and Hinks, 1976).

Our view of the hemopoietic organs in *E. declarata* is quite different from that of Akai and Sato (1971) for the organs of *B. mori*. These authors identified all five types in the organs and assumed that all were generated there. We also noted all five types in organ squashes on occasion (Fig. 8.39), but the GRs and SPs appeared in such small and irregular numbers that we considered them to be "contaminants" that entered ruptured areas of the organs prior to or during dissection. We saw no clear evidence of the generation of all five types there.

This histological examination of the organs seemed to account at once for the previously unaccountable numbers of PRs in the blood and for the continued increase of PLs and OEs. It seemed obvious that they were simply released from the organs in sufficient numbers to account for the changes. It seemed clear also that PRs and PLs represent a single cell line from the germinal cells (i.e., that germinal cells transform to PRs, and these to PLs, and that the OEs likely stem also from the germinal cells as a separate line). It seemed obvious also that the other two cell types, the GRs and SPs, are not generated in the organs, but probably originate as separate lines in the embryo. Nevertheless, this picture of hemocyte origins and relationships was developed from histological evidence alone, and the interpretations are therefore open to question. It was incumbent on us to test the theory experimentally.

Our experiments were designed to compare the hemocyte picture in the blood in the presence and in the absence of the hemopoietic organs. They involved comparable hemocyte counts in approximately equal portions of the larval body, which were ligated to include one or both pairs of organs in the one portion and none in the other. They allowed the organ-containing portion to be associated with the head in some cases and not in others. The sites of the ligatures are illustrated in association with the resultant hemocyte picture in each portion of the body (Fig. 8.40).

The results of these experiments are summarized here and detailed elsewhere (Hinks and Arnold, 1977). The comparison of PR and PL numbers in the blood in the presence and in the absence of the organs was highly significant, both with and without association of the organ-containing segments with the head. We found no difference in the numbers of OEs in these circumstances, despite the histological evidence of their origin in the organs. We attribute this to the normally small numbers of OEs in the blood, in conjunction with our less than ideal method of comparison. We found no significant change in the numbers of GRs and SPs that could indicate an involvement of the organs in their production.

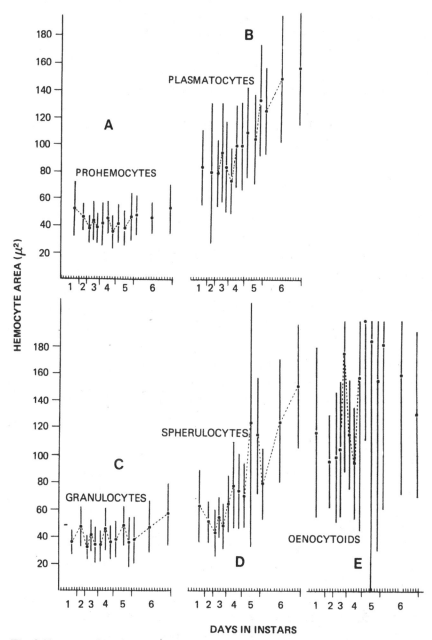

Fig. 8.39. Area values for each hemocyte type in *E. declarata*, at stages during larval life, showing size distinctions between cells of each. (From Hinks and Arnold, 1977).

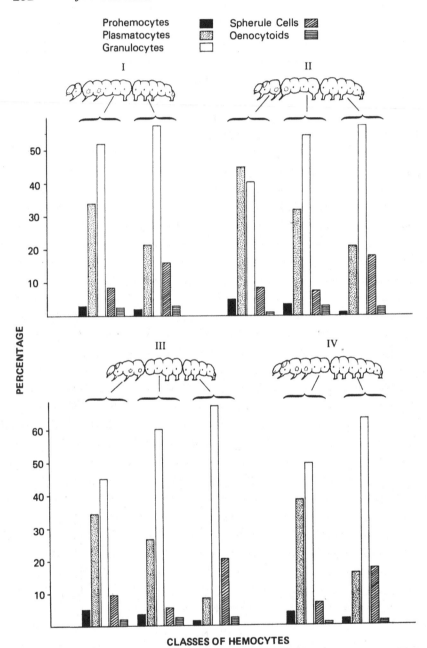

Fig. 8.40. Percentage of hemocytes of each type in sections of larvae separated by ligatures to include or exclude hemopoietic organs, 4 days postligature. (From Hinks and Arnold, 1977)

The combined evidence from histology and experimentation provides a strong basis for recognizing four immutable types of cell lines among the hemocyte population of *E. declarata:* PR-PL (or simply PL), GR, SP, and OE. The presence of similar hemopoietic organs and of the same kinds of mitotic hemocytes in the blood of a number of other species of Lepidoptera suggests that this applies generally to insects in this order. The same checking procedure may well reveal a similar system in other orders of insects. We conclude that the four types originate and multiply in the manner illustrated (Fig. 8.41). The PLs derive from germinal cells in the hemopoietic organs, where their immediate progenitor is a partly differentiated germinal cell commonly called the PR. Both PRs and PLs are free in the lumina of the organs and are released into the blood. Here too the PRs can divide and differentiate to form PLs alone, and these in turn mature and degenerate without further transformations. GRs and SPs seem to be separate clones that presumably originate in the embryo and continue in the blood as self-generating lines through mitotic division. They too pass through the cycle of growth, maturation, and degeneration,

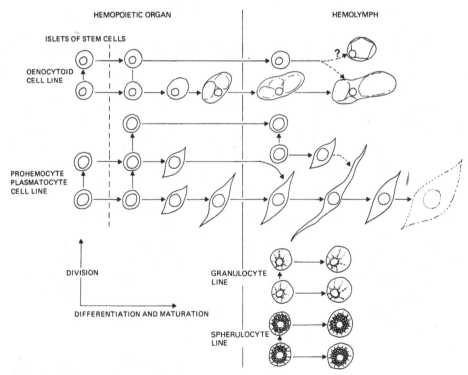

Fig. 8.41. Suggested relationships between types of hemocytes during postembryonic development in Lepidoptera. (From Hinks and Arnold, 1977)

without losing their identity. Although both are involved in metabolizing materials from the hemolymph, and their separate identity in cockroaches has been questioned (Arnold and Salkeld, 1967), there is no doubt of their separate status in Lepidoptera. Aside from distinctions in their stainability with different dyes, including fluorescent ones, and in morphological features, the two seem also to have a different sequence for mitosis (unpub. observ.). Finally, the OEs seem to be generated in isogenic islets in the hemopoietic organs, but we have only histological evidence for this (see also Chapter 2).

8.5. Discussion

Although most of the work presented here indicates a particular pattern of hemocyte development to support the concept of immutable types, there is evidence also that a somewhat different system functions among some other insects, particularly among Hemimetabola. This system, demonstrated by Nittono et al. (1964) and especially by Hoffmann and his colleagues, indicates that in Orthoptera (and in one dipteran) each of the hemocyte types is generated separately in isogenic islets (see Chapter 2) or regions of germinal tissue situated dorsally in the abdomen. In other words, they suggest an entirely different scheme for hemocyte origin in these insects and an entirely different hemopoietic tissue than we find in Lepidoptera, while holding to a concept of immutable hemocyte types. Their findings are supported to some extent by the apparent absence of mitosis among GRs and SPs in the blood of such other hemimetabolous insects as *Rhodnius prolixus* (Jones, 1967) and species of cockroaches (Arnold, 1972). They may be supported also in a different way by differences in comparable types of hemocytes in hemi- and holometabolous species. Especially notable in this respect are the GRs in the two groups, large and with prominent granules in the lower orders, and small and with variable inclusions in higher orders. In this case there seems some justification for distinguishing them by name, as was done by some authors (e.g., distinguishing GRs from spheroidocytes).

Perhaps the one factor most responsible for delays in the development of insect hemocytology has been the absence of standard criteria for classifying the types of cells. This lack continues to dilute efforts to understand the significance of the various cell forms and the functions of the types. A return to the concept of the clonal nature of hemocyte types should help to obviate this difficulty, but there is an obvious need to expand this line of research in depth and with other insects.

Although histological and experimental methods, as employed here, can be rewarding in future investigations, other procedures can define the types more precisely. Histochemistry and ultramicroscopy have much to offer in this respect, along with fluorescence microscopy. Serology may yet provide a clear statement on distinctions among hemocyte types, as it has among cells of other tissues (Krywienczyk and Sohi, 1976). Perhaps eventually it will be possible to separate the types from the hemolymph and study their separate biology in vitro.

For the present, it should be acknowledged that the fundamental basis for hemocyte classification and the acceptance of a particular form of cell as a type lies in the cell's origin. True hemocyte types are isogenic clones. Modifications in the forms of the types in different species can be recognized without altering this basic concept. The view that there is only one cell line among hemocytes is not supported by the present evidence.

8.6. Summary

Two views are held generally on the status of hemocyte types. One considers the types to be distinctive phases with separate functions in the life of one line of cells. The other considers that they represent separate, immutable cell lines, each with a separate role or roles in the life of the insect. The feasibility of both concepts is examined.

Usually the hemocyte types are distinguished by cytological features, often allied to cell processes or functions. The origin of the types has been largely overlooked as a basic criterion of cell lineage.

Hemocyte origins were examined in one lepidopteran species, *Euxoa declarata* (Noctuidae), and checked in principle for species in other families in the same order. It was found that two cell lines, prohemocyte-plasmatocyte and oenocytoid, originate separately in hemopoietic organs associated with germinal disks in the meso- and metathorax and are released from there into the blood. Two other lines of cells, the granulocytes and spherulocytes, seem to sustain their numbers postembryonically by mitosis in the blood. These conclusions were reached from an evaluation of selective hemocyte production by mitosis in the blood, from histological studies of the organs, and from analyses of hemocyte populations in the blood in the presence and in the absence of the organs. There was no evidence to support the concept of a single line of hemocytes.

Although the concept of separate cell lines in the hemocyte complex seems clear for Lepidoptera, and may well be true for some other orders of insects, there is reason to believe that other systems may have developed in other groups. At least, there are distinction in the site and structure of hemopoietic tissue reported for the two main groups, Hemimetabola and Holometabola, and distinctions also between comparable hemocyte types in these groups. It would be most surprising if the organization of the hemocyte complex was the same for all insects.

Acknowledgments

Special acknowledgment is extended to Dr. C. F. Hinks, co-worker and co-author in the original research on hemopoiesis in Lepidoptera.

References

Akai, H. 1969. Ultrastructure of haemocytes observed on the fat-body cells in *Philosamia* during metamorphosis. *Jpn. J. Appl. Entomol. Zool. 13*:17–21.

Akai, H., and S. Sato. 1971. An ultrastructural study of the haemopoietic organs of the silkworm, *Bombyx mori. J. Insect Physiol. 17*:1665–76.

Akai, H., and S. Sato. 1973. Ultrastructure of the larval hemocytes of the silkworm, *Bombyx mori* L. (Lepidoptera: Bombycidae). *Int. J. Insect Morphol. Embryol. 2*:207–31.

Akai, H., and S. Sato. 1976. Surface ultrastructure of the larval hemocytes of the silkworm, *Bombyx mori* L. (Lepidoptera: Bombycidae). *Int. J. Insect Morphol. Embryol. 5*:17–21.

Arnold, J. W. 1952a. The haemocytes of the Mediterranean flour moth, *Ephestia kühniella* Zell. (Lepidoptera: Pyralididae). *Can. J. Zool. 30*:352–64.

Arnold, J. W. 1952b. Effects of certain fumigants on haemocytes of the Mediterranean flour moth, *Ephestia kühniella* Zell. (Lepidoptera: Pyralididae). *Can. J. Zool. 30*:365–74.

Arnold, J. W. 1966. An interpretation of the haemocyte complex in a stone-fly, *Acroneuria arenosa* Pictet (Plecoptera: Perlidae). *Can. Entomol. 98*:394–411.

Arnold, J. W. 1972. A comparative study of the haemocytes (blood cells) of cockroaches (Insecta: Dictyoptera: Blattaria), with a view of their significance in taxonomy. *Can. Entomol. 104*:309–48.

Arnold, J. W. 1974. The hemocytes of insects, pp. 201–55. *In* M. Rockstein (ed.). *The Physiology of Insecta*, Vol 5, 2nd ed. Academic Press, New York.

Arnold, J. W. 1976. Biosystematics of the genus *Euxoa* (Lepidoptera: Noctuidae). VIII. The hemocytological position of *E. rockburnei* in the "declarata group." *Can. Entomol. 108*:1387–90.

Arnold, J. W., and C. F. Hinks. 1975. Biosystematics of the genus *Euxoa* (Lepidoptera: Noctuidae). III. Hemocytological distinctions between two closely related species, *E. campestris* and *E. declarata. Can. Entomol. 107*:1095–1100.

Arnold, J. W., and C. F. Hinks. 1976. Haemopoiesis in Lepidoptera. I. The multiplication of circulating haemocytes. *Can. J. Zool. 54*:1003–12.

Arnold, J. W., and E. H. Salkeld. 1967. Morphology of the haemocytes of the giant-cockroach, *Blaberus giganteus*, with histochemical tests. *Can. Entomol. 99*:1138–45.

Arnold, J. W., and S. S. Sohi. 1974. Hemocytes of *Malacosoma disstria* Hübner (Lepidoptera: Lasiocampidae): Morphology of the cells in fresh blood and after cultivation *in vitro. Can. J. Zool. 52*:481–5.

Barra, J. A. 1969. Tégument des Collemboles: Présence d'hémocytes à granules dans le liquide exuvial au cours de la mue. *C. R. Acad. Sci. Paris 269*:902–3.

Beaulaton, J., and M. Monpeyssin. 1976. Ultrastructure et cytochimie des hémocytes d'*Antheraea pernyi* Guér. (Lepidoptera: Attacidae) au cours du cinquième âge larvaire. 1. Prohémocytes, plasmatocytes et granulocytes. *J. Ultrastruct. Res. 55*:143–56.

Beaulaton, J., and M. Monpeyssin. 1977. Ultrastructure et cytochimie des hémocytes d'*Antheraea pernyi* Guér. (Lepidoptera: Attacidae) au cours du cinquième âge larvaire. 2. Cellules au sphérules et oenocytoïds. *Biol. Cell. 28*:13–18.

Cheng, C. 1964. A phase contrast study of blood cells in *Pseudaletia separata* larvae. *Acta Entomol. Sinica 13*:536–41.

Clark, R. M., and W. R. Harvey. 1965. Cellular membrane formation by plasmatocytes of diapausing cecropia pupae. *J. Insect Physiol.* 11:161–75.

Crossley, A. C. S. 1964. An experimental analysis of the origins and physiology of haemocytes in the blue blow-fly, *Calliphora erythrocephala* (Meig.). *J. Exp. Zool.* 157:375–97.

Crossley, A. C. S. 1975. The cytophysiology of insect blood. *Adv. Insect Physiol.* 11:117–221.

Devauchelle, G. 1971. Etude ultrastructurale de hémocytes du Coléoptère, *Melolontha melolontha* (L.). *J. Ultrastruct. Res.* 34:492–516.

François, J. 1974. Etude ultrastructurale des hémocytes du Thysanoure *Thermobia domestica* (Insecte, Apterigote). *Pedobiologia* 14:157–62.

François, J. 1975. Hémocytes et organe hématopoïétique de *Thermobia domestica* (Packard) (Thysanura: Lepismatidae). *Int. J. Insect Morphol. Embryol.* 4:477–94.

Granados, R. R., and P. J. Meehan. 1973. Morphology and differential counts of hemocytes of healthy and wound tumor virus-infected *Agallia constricta. J. Invertebr. Pathol.* 22:60–9.

Grégoire, Ch. 1974. Hemolymph coagulation, pp. 309–60. *In* M. Rockstein (ed.). *The Physiology of Insecta*, Vol 5, 2nd ed. Academic Press, New York.

Gupta, A. P. 1969. Studies of the blood of Meloidea (Coleoptera). 1. The haemocytes of *Epicauta cinerea* (Forster), and a synonymy of haemocyte terminologies. *Cytologia* 34:300–44.

Gupta, A. P., and D. J. Sutherland. 1966. *In vitro* transformations of the insect plasmatocytes in some insects. *J. Insect Physiol.* 12:1369–75.

Harpaz, I., N. Kislev, and A. Zelcer. 1969. Electron-microscopic studies on hemocytes of the Egyptian cottonworm, *Spodoptera littoralis* (Boisduval) infected with a nuclear-polyhedrosis virus, as compared to noninfected hemocytes. I. Noninfected hemocytes. *J. Invertebr. Pathol.* 14:175–85.

Hinks, C. F., and J. W. Arnold. 1977. Haemopoiesis in Lepidoptera. II. The role of the haemopoietic organs. *Can. J. Zool.* 55:1740–55.

Hoffmann, J. A. 1967. Etude des hémocytes de *Locusta migratoria* L. (Orthoptère). *Arch. Zool. Exp. Gén.* 108:251–91.

Hoffmann, J. A. 1970. Régulations endocrines de la production et de la différentiation des hémocytes chez un insecte orthoptère: *Locusta migratoria migratoroides. Gen. Comp. Endocrinol.* 15:198–219.

Hoffmann, J. A., M.-E. Stoekel, A. Porte, and P. Joly. 1968. Ultrastructure des hémocytes de *Locusta migratoria* (Orthoptère). *C. R. Acad. Sci. Paris* 266:503–5.

Jones, J. C. 1962. Current concepts concerning insect hemocytes. *Amer. Zool.* 2:209–46.

Jones, J. C. 1964. The circulatory system of insects, pp. 1–107. *In* M. Rockstein (ed.). *The Physiology of Insecta*, Vol 3, 1st ed. Academic Press, New York.

Jones, J. C. 1965. The hemocytes of *Rhodnius prolixus* Stal. *Biol. Bull. (Woods Hole)* 129:282–94.

Jones, J. C. 1967. Normal differential counts of haemocytes in relation to ecdysis and feeding in *Rhodnius. J. Insect Physiol.* 13:1133–41.

Jones, J. C. 1970. Hemocytopoiesis in insects, pp. 7–65. *In* A. S. Gordon (ed.). *Regulation of Hematopoiesis*, Vol. 1, Appleton, New York.

Jones, J. C. 1977. *The Circulatory System of Insects.* Thomas, Springfield, Illinois.

Jones, J. C., and D. P. Liu. 1968. A quantitative study of mitotic divisions of haemocytes of *Galleria mellonella* larvae. *J. Insect Physiol.* 14:1055–61.

Khanna, S. 1964. The circulatory system of *Dysdercus koenigii* (Fabr.) (Hemiptera: Pyrrhocoridae). *Indian J. Entomol.* 26:404–10.

Krywienczyk, J., and S. S. Sohi. 1976. Serologic characterization and identification of four lepidopteran cell lines. *Can. J. Zool.* 54:1559–64.

Lai-Fook, J. 1973. The structure of the haemocytes of *Calpodes ethlius* (Lepidoptera). *J. Morphol.* 139:79–104.

Landureau, J. C., and P. Grellet. 1975. Obtention de lignées permanentes d'hémocytes de blatte: Caractéristiques physiologiques et ultrastructurales. *J. Insect Physiol.* 21:137–51.

Lea, M. S., and L. I. Gilbert. 1966. The hemocytes of *Hyalophora cecropia* (Lepidoptera). *J. Morphol. 118:*197–216.

Miller, M. N., M. Partin, and E. L. Weinreb. 1972. Computerized classification of nucleated blood cells by morphological criteria. *Proc. F.A.S.E.B. 31*(a):926.

Mitsuhashi, J. 1966. Tissue culture of the rice stem borer, *Chilo suppressalis* Walker (Lepidoptera: Pyralidae). II. Morphology and *in vitro* cultivation of hemocytes. *Appl. Entomol. Zool. 1:*5–20.

Moran, D. T. 1971. The fine structure of cockroach blood cells. *Tissue Cell 3:*413–22.

Neuwirth, M. 1973. The structure of the hemocytes of *Galleria mellonella* (Lepidoptera). *J. Morphol. 139:*105–24.

Nittono, Y. 1960. Studies on the blood cells in the silkworm, *Bombyx mori* L. *Bull. Seric. Exp. Stn. (Tokyo)* 16:171–266.

Nittono, Y. 1964. Formation of hemocytes near the imaginal wing disc in the silkworm, *Bombyx mori* L. *J. Seric. Sci. Jap. 33:*43–5.

Nittono, Y., S. Tomabechi, N. Onodera, and S. Hayasaka. 1964. Total haemocyte counts from the larvae and moths cauterized on their imaginal wing discs and the neighboring parts in their larval stage. *18th Seric. Congr., Japan Seric. Soc., Tohoku Branch,* p. 20. (In Japanese.)

Ogel, S. 1955. A contribution to the study of blood cells in Orthoptera. *Comm. Fac. Sci. Univ. Ankara 4:*15–88.

Olson, K., and S. D. Carlson. 1974. Surface fine-structure of hemocytes of *Periplaneta americana:* A scanning electron microscope study. *Ann. Entomol. Soc. Amer.* 67:61–5.

Raina, A. K. 1976. Ultrastructure of the larval hemocytes of the pink bollworm, *Pectinophora gossypiella* (Saunders) (Lepidoptera: Gelechiidae). *Int. J. Insect Morphol. Embryol. 5:*187–95.

Ratcliffe, N. A., and C. D. Price. 1974. Correlation of light and electron microscopic hemocyte structure in the Dictyoptera. *J. Morphol. 144:*485–98.

Schaeffer, C. 1889. Beiträge zur Histologie der Insekten. *Zool. Jahrb. Anat. Ontog. 3:*611–52.

Scharrer, B. 1972. Cytophysiological features of hemocytes in cockroaches. *Z. Zellforsch. Mikrosk. Anat. 129:*301–19.

Selman, B. J. 1962. The fate of the blood cells during the life history of *Sialis lutaria* L. *J. Insect Physiol. 8:*209–14.

Shapiro, M. 1968. Changes in the haemocyte population of the wax moth, *Galleria mellonella,* during wound healing. *J. Insect Physiol. 14:*1725–33.

Shrivastava, S. C., and A. G. Richards. 1964. The differentiation of blood cells in the wax moth, *Galleria mellonella. Amer. Zool. 4:*182.

Shrivastava, S. C., and A. G. Richards. 1965. An autoradiographic study of the relation between hemocytes and connective tissue in the wax moth, *Galleria mellonella. Biol. Bull. (Woods Hole) 128:*337–45.

Smith, H. W. 1938. The blood of the cockroach, *Periplaneta americana* L.: Cell structure and degeneration, and cell counts. *New Hampshire Agric. Exp. Stn. Tech. Bull. 71:*1–23.

Sohi, S. S. 1971. *In vitro* cultivation of hemocytes of *Malacosoma disstria* Hübner (Lepidoptera: Lasiocampidae). *Can. J. Zool. 49:*1355–8.

Wittig, G. 1968. Electron microscopic characterization of insect hemocytes. *26th Annu. Meeting E.M.S.A. 1968:*68–9.

Yeager, J. F. 1945. The blood picture of the southern armyworm (*Prodenia eridania*). *J. Agric. Res. 71:*1–40.

Zachary, D., and J. A. Hoffmann. 1973. The haemocytes of *Calliphora erythrocephala* (Meig.) (Diptera). *Z. Zellforsch. Mikrosk. Anat. 141:*55–74.

9 Hemocyte cultures and insect hemocytology

S. S. SOHI

Insect Pathology Research Institute, Department of Fisheries and Forestry, Sault Sainte Marie, Ontario, P6A 5M7, Canada

Contents

9.1. Introduction

The importance of insect tissue and cell cultures has long been recognized by entomologists because cells, tissues, and organs grown in vitro in controlled physical environment and nutrient media provide ideal systems for studying insect pathology, developmental biology, hemocytology, physiology, and cell biology.

Insect hemocyte cultures are of much value in the study of the origin and differentiation of hemocyte types. They can help elucidate the origin and function of the hemopoietic organs in insects and the interrelationships among various hemocytes, and their use could settle the controversies about insect hemocyte types.

Unfortunately, however, in vitro cultivation of insect tissues and cells has been very difficult (Vaughn, 1968; Brooks and Kurtti, 1971; Feir and Pantle, 1971). Although the first known attempt to grow insect tissue cultures was made in 1915 (Goldschmidt, 1915), it was not until almost 50 years later that the first continuous insect cell line was obtained by Grace (1962). Culturing insect hemocytes in vitro has been even more difficult than culturing other insect tissues (Arnold, 1974; Mazzone, 1976). But considerable progress has been made in recent years in growing insect tissue cultures. Over 100 continuous lines of insect cells, including six of hemocytes, have been developed since 1962 (Hink, 1976).

Although six continuous cell lines of insect hemocytes have been obtained, it is not always possible to start a successful primary culture of hemocytes at will, let alone develop a new continuous cell line. In recent years insect tissue culture has been reviewed by Brooks and Kurtti (1971), Hink (1972), Stanley (1972), and Granados (1976). Other useful publications on invertebrate tissue culture are Vago (1963, 1971, 1972), Barigozzi (1968), Weiss (1971), Rehacek et al. (1973), Maramorosch (1976), and Kurstak and Maramorosch (1976).

9.2. Current status of hemocyte culture

9.2.1. Culture vessels

Most of the earlier work on hemocyte cultures was done with hanging-drop cultures (Glaser, 1917; Taylor, 1935; Arvy and Gabe, 1946; Mosolov et al., 1967), although roller tubes were also used (Horikawa and Kuroda, 1959). Later on, small chambers assembled with coverglasses and microslide rings were used by Mitsuhashi (1966) and Sohi (1971), and small commercial containers such as Cooper vessels and Sykes-Moore tissue culture chambers were employed by Kurtti and Brooks (1970). In recent years, 30-ml disposable plastic flasks have been used mostly (Chao and Ball, 1971; Sohi, 1971, 1973a; Landureau and Grellet, 1975; Granados and Naughton, 1975, 1976). Landureau and Grellet (1975) started their cockroach hemocyte cultures in small Carrel flasks and subsequently transferred them to plastic flasks when they were multiplying rapidly. In my work on hemocyte cultures of the forest tent caterpillar (Sohi, 1971), I compared microslide ring chambers, 35-mm diameter tissue culture plastic dishes, and 30-ml plastic flasks, and obtained best growth in 30-ml flasks. At present these flasks are the most commonly used vessels for insect tissue cultures. They are available from several manufacturers (Corning Glass Works, 717 Fifth Avenue, New York, New York, 10022 U.S.A.; Dynatech Laboratories, Inc., 900 Slaters Lane, Alexandria, Virginia 22314, U.S.A.; Falcon Plastics, Division of Bioquest, 1950 Williams Drive, Oxnard, California 93030, U.S.A.; A. S. Nunc, 8 Algade, DK-4000 Roskilde, Denmark).

9.2.2. Culture media

In earlier work, hemocyte cultures were grown in the blood plasma of the host insect with or without the addition of a balanced salt solution

(Glaser, 1917; Taylor, 1935; Arvy and Gabe, 1946). However, in recent years various synthetic media have been used (Horikawa and Kuroda, 1959; Mitsuhashi, 1966, 1967; Landureau and Grellet, 1975). A medium designed by Wyatt (1956), and subsequently modified by Grace (1962), has been used by several workers to grow hemocyte cultures of Lepidoptera (Chao and Ball, 1971; Sohi, 1971, 1973a; Granados and Naughton, 1975, 1976). This medium, commonly known as Grace's medium, is commercially available from Grand Island Biological Company, Grand Island, New York 14052, U.S.A.

The basal media have to be supplemented with vertebrate sera. Fetal bovine serum (FBS) is the most commonly used, although chicken and turkey sera also have been used (Goodwin, 1975). The concentration of FBS varies from 5% to 20% of the medium, but 5–10% was sufficient for Grace's cell lines of *Antheraea eucalypti* and *Bombyx mori* (Sohi and Smith, 1970).

Previously, insect tissue cultures could not be grown without addition of insect hemolymph serum (IHS) to the medium, but it has recently been shown that IHS is not essential (Yunker et al., 1967; Sohi, 1969; Hink, 1976). Insect hemocyte cultures have also been successfully initiated and grown in IHS-free media (Mitsuhashi, 1966, 1967; Sohi, 1973a; Landureau and Grellet, 1975).

9.2.3. Preparation and maintenance of primary cultures

The procedure described here for explanting and cultivating hemocytes in vitro is based on the works of Mitsuhashi (1966, 1967), Chao and Ball (1971), Sohi (1971, 1973a), Granados and Naughton (1975), and Landureau and Grellet (1975). Their methods have produced the most successful hemocyte cultures so far.

In this procedure, the insects are surface-sterilized by immersion in ethyl alcohol (75%), quadramine (0.1–0.2%), or sodium hypochlorite (2%) for a few minutes. Then they are washed in several changes of sterile distilled water and allowed to dry in sterile dishes. A sterile insect is held carefully in hand, and an antenna (in an adult) or a proleg (in a larva) is cut with sterile scissors. In the case of a pupa, a puncture is made at the wingpad with a sterile needle. The oozed blood is allowed to drip into a culture vessel containing either a balanced salt solution or culture medium. The amount of blood explanted depends on the size of the culture vessel. If too much blood is used, there is an excessive amount of melanization, which kills all the cells because of toxicity. If, however, too little blood is explanted, the number of hemocytes is too low to start a culture. In my work on the forest tent caterpillar, *Malacosoma disstria*, blood from five larvae per 30-ml Falcon flask gave best results (Sohi, 1971). Ten larvae per flask were too many and one larva per flask was not enough. Melanization occurred when two to five drops of larval blood of *Estigmene acrea* were explanted per flask (Granados and Naughton, 1975). These workers added one drop of blood per flask every other day to get a total of three or four drops per culture.

After the insects have been bled, the culture vessels are left undisturbed for 10–15 min in the horizontal position. The hemocytes settle to the bottom of the vessels, and most of them attach to the vessels. The supernatant fluid (saline or medium) is removed, the attached cells are rinsed several times with saline or medium, and then the growth medium is added to the culture vessels. The cultures are maintained at 25–28 °C.

Discarding the medium or saline into which the blood is explanted and rinsing the cultures several times helps reduce melanization in cultures. This reduction is most likely owing to the partial removal of tyrosinase and extracellular substances responsible for melanin formation. In my work (Sohi, 1971, 1973a), primary cultures always have had melanization in varying degrees. Cultures with extensive melanization died out, but those with less melanization survived and started to grow. Granados and Naughton (1975) used 0.001 M cysteine in the medium to retard melanization.

The medium is replenished in the cultures once every 1–2 weeks. Too frequent and complete renewal of medium in the initial stages of the cultures can do more harm than good. Therefore, for several weeks only a part of the spent medium is replaced. In some cases the cells started to multiply in cultures within 1–2 weeks after explantation (Chao and Ball, 1971; Granados and Naughton, 1975, 1976), but in others multiplication began 3–6 months after initiation of the cultures (Mitsuhashi, 1966, 1967; Sohi, 1971, 1973a). The cultures are split and transferred to new containers as the cell number warrants.

9.2.4. Subculturing and development of continuous cell lines

Primary cultures of insect hemocytes have been grown by several workers (see Mitsuhashi, 1966). However, their cultures survived only a few months at the most. Although mitoses were observed in primary cultures by some of the workers, the hemocytes could not be successfully transferred from the culture vessels in which they were originally explanted.

It was Mitsuhashi (1966, 1967) who for the first time succeeded in subculturing insect hemocytes and developing a continuous cell line from them. Since then, Chao and Ball (1971), Sohi (1971, 1973a), Granados and Naughton (1975), and Landureau and Grellet (1975) also have developed continuous cell lines of hemocytes.

The hemocytes in cultures can be either freely suspended in the medium (Chao and Ball, 1971; Sohi, 1971) or attached to the bottom of the flask (Mitsuhashi, 1966; Sohi, 1973a; Granados and Naughton,

1975). Suspended cells can be subcultured by simply transferring a small volume of the cell suspension to a new culture vessel. Attached cells, however, have to be dislodged mechanically with repeated pipetting (Mitsuhashi, 1967) or with a rubber scraper (Sohi, 1973a). More strongly attached hemocytes in cultures are treated with trypsin solution (0.05–0.125%) prior to mechanical dislodging. This reduces mechanical damage to the cells as a result of scraping and increases the yield of cells from the cultures. Early researchers on insect tissue cultures felt that trypsin was injurious to insect cells (see Schneider, 1969), but in recent times many workers have successfully used trypsin (Singh, 1967; Schneider, 1969; Varma and Pudney, 1969; Sohi, 1973b; Goodwin, 1975; Granados and Naughton, 1975; Kurtti and Brooks, 1977). *Malacosoma disstria* hemocytes of the attached cell line IPRI-108 (Sohi, 1973a) from the early passages of cultivation in vitro are illustrated in Fig. 9.1.

9.2.5. Contamination

Because of the rich nutrient media, all cell cultures are very susceptible to contamination with bacteria, fungi, mycoplasmas, and other parasites (Fogh et al., 1971; Fogh, 1973). Also, cells of foreign types and species may be accidentally introduced into the tissue and cell cultures, resulting in incorrect identity of cell lines (Herrick et al., 1970; Greene and Charney, 1971; Greene et al., 1972; Krywienczyk and Sohi, 1973).

The introduction of airborne contaminants into cell cultures is prevented by good housekeeping in the laboratory and by the use of laminar flow hoods in the transfer rooms (Coriell et al., 1967; Coriell and McGarrity, 1970; McGarrity and Coriell, 1971; McGarrity, 1975). If microbial contamination occurs in cultures in spite of all precautions, it is best to discard the cultures. If, however, the cultures cannot be easily replaced, one or a combination of antibiotics can be added to the medium. It is wise to be selective and moderate in the use of antibiotics. Their continued and excessive use in cell cultures can cause many problems (Coriell, 1973; Merchant, 1973; Perlman, 1976). However, antibiotics are very useful in initiating primary cultures because the probability of contamination during explantation of tissues is fairly high. Penicillin and streptomycin have been frequently used in primary cultures to suppress bacterial contamination (see Coriell, 1973). Recently gentamicin has been found very effective against bacteria, but harmless to vertebrate and insect cell cultures (Rudin et al., 1970; Schafer et al., 1972; Sohi, unpub. data). Therefore, in our work we now use mostly gentamicin if we need an antibacterial drug. A recent discussion of cell culture facilities by McGarrity (1977) provides useful information on preventing microbial and chemical contamination of cultures and on setting up new tissue culture facilities.

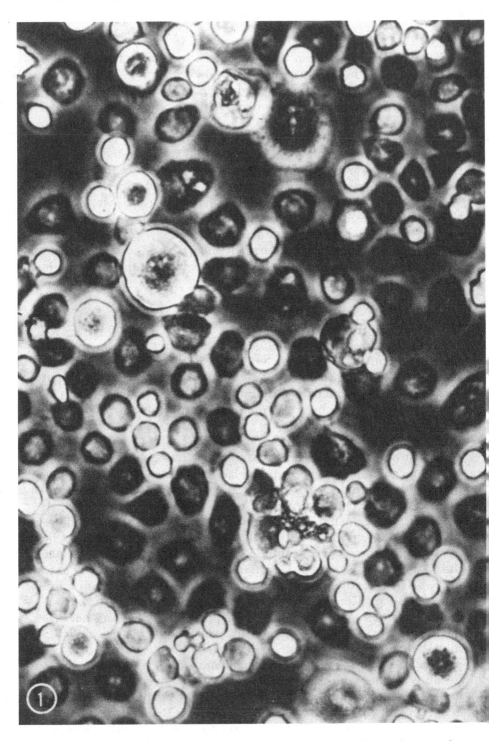

Fig. 9.1. *M. disstria* hemocytes (IPRI-108) 41 days after fifth passage in culture, total in vitro time 217 days. Bright phase contrast. × 710.

9.2.6. Deep-freeze storage

While working with cell cultures it is always desirable and sometimes necessary to freeze-preserve cells, for the following reasons: (1) cells grown in vitro for long periods tend to change in certain characters, (2) unnecessary labor and materials are spent in maintaining cultures during periods they are not used for experimentation, and (3) there is the constant risk of losing cultures owing to toxicity or contamination by microorganisms and cells of other strains and species.

To overcome these and similar problems, cells from early passages in vitro are freeze-preserved at a temperature of −80 °C or lower. The procedure consists of suspending actively growing cells in growth medium containing a cryoprotective agent such as glycerol or di-methylsulfoxide, cooling them at a slow rate (approximately 1 °/min) to about −40 °C or lower, and storing in liquid nitrogen. The frozen insect cells are revived by immersing the ampules in 30 °C water (Sohi et al., 1971).

The slow cooling of cells can be accomplished by using either so-phisticated and expensive programmed cooling equipment (Shannon and Macy, 1973) or one of the improvised versions (Sohi et al., 1971; Schroy and Todd, 1976; Waymouth and Varnum, 1976; Moklebust et al., 1977).

9.3. Applications in hemocytology

Most of the work on hemocyte cultures has been done by investigators interested primarily in using them for studying pathogens in vitro (Glaser, 1917; Martignoni and Scallion, 1961a,b; Kurtti and Brooks, 1970, 1971; Mazzone, 1971; Sohi, 1971, 1973a; Sohi and Cunningham, 1972; Sohi and Bird, 1976; Sohi and Wilson, 1976; Granados and Naughton, 1975, 1976; Wilson and Sohi, 1977). These researchers did not conduct detailed studies of hemocyte types and their interrela-tionships in vitro. Whenever possible, they have reported the hemo-cyte types observed in vitro, but their observations are based mainly on phase-contrast microscopic examination of live hemocytes in cul-tures. Consequently, their contribution to hemocytology is rather lim-ited. However, a few insect pathologists and some hemocytologists have made detailed studies of hemocytes in vitro; this has produced quite interesting information and has demonstrated the potential of in vitro investigations for solving the problems encountered in insect he-mocytology.

9.3.1. Hemocyte types

The confusion in hemocyte classification and the resulting controver-sies about the hemocyte types are well known (Wigglesworth, 1959;

Jones, 1962; Gupta, 1969; Arnold, 1974). Morphology and functions of the hemocytes, and their staining and histochemical reactions, are some of the criteria used in classifying them. Some of these criteria are subject to interpretation because of the variability of the parameters. Consequently, quite often the same hemocyte or its various forms in different species are referred to by different names, which results in confusing terminology (see Chapters 4, 8, 10).

Studies on tissue culture cells indicate that the shape and size of the cells, and the presence or absence of granulation and vacuolation in them, are not reliable parameters for classifying hemocytes. Suitor et al. (1966), working with a clone of *Aedes aegypti* cells, observed that a spindle-shaped cell rounded up in preparation for mitosis and returned to spindle shape after dividing a few times. Further, those working with cell cultures commonly observe that culture conditions (pH, osmolarity, nutrient deficiency, toxicity) have profound effects on the presence or absence of granulation and vacuolation. As cultures get crowded toward the end of the growth cycle, the cells become granular and smaller. Also, some cells become much larger because of polyploidy. Thus, we should be looking for some more reliable criteria. In the meantime, extreme care and caution should be used in classifying hemocytes on the basis of shape and general appearance.

Glaser (1917) observed ordinary amoebocytes (= plasmatocytes) and minute amoebocytes (other hemocytes) in hemocyte cultures of *Malacosoma americanum, Pseudaletia (Cirphis) unipuncta, Spodoptera (Laphygma) frugiperda,* and *Porthetria dispar,* but only the plasmatocytes multiplied in culture. Martignoni and Scallion (1961a) reported plasmatocytes (PLs), prohemocytes (PRs), and spheroidocytes (= adipohemocytes) in the cultures of *Peridroma saucia* blood cells. Chao and Ball (1971) observed only one type of cell, which they believed to be PRs. They also reported that these cells changed shape and growth characteristics in culture, but they did not elaborate further on this aspect.

In cultures of American cockroach blood, amoebocytes (= PLs) were most numerous and had a higher mitotic rate than chromophils (other hemocyte types) of all sizes (Taylor, 1935). In two continuous cell lines established from nymphal and adult cockroach hemolymph, the cells showed characteristics of PLs (Landureau and Grellet, 1975). The PLs transformed into small dividing cells (PRs) and large storing cells (e.g., granulocytes, GRs; adipohemocytes, ADs) under different culture conditions. On the basis of light and electron microscopic examinations, Granados and Naughton (1976) concluded that the cells in the continuous hemocyte lines of *Estigmene acrea* appeared to be PLs.

Kurtti and Brooks (1970) observed PRs, GRs, PLs, and spherulocytes (SPs) in primary cultures of *Trichoplusia ni, Choristoneura fu-*

miferana, and *Malacosoma disstria.* Only the PRs were actively dividing in the cultures. On the basis of cell morphology, as revealed by phase-contrast microscopic examination of live cells in culture, Sohi (1971, 1973a) recognized PRs, PLs, and SPs in two continuous cell lines (Md66, Md108) of *M. disstria* hemocytes, and he felt that only PRs multiplied and persisted in them.

Subsequently, Arnold and Sohi (1974) studied the hemocyte complex of the full-grown *M. disstria* larva and reexamined the hemocyte types in the above two cell lines developed by Sohi (1971, 1973a). The hemocyte cultures were initiated from the blood of full-grown *M. disstria* larvae. Cultures of different ages were examined. Some had been maintained in continuous culture by serial passage; the others had been preserved in liquid nitrogen for several months and then revived for examination. They had been maintained in vitro through 39, 52, 73, 77, and 148 passages when examined in this study. Identification of hemocytes in culture, as in fresh larval blood, was based on the appearance and stainability of the nucleus, cytoplasm, and inclusions, and on the cell/nucleus size ratio.

Arnold and Sohi (1974) found that the fresh blood of *M. disstria* contained five types of hemocytes: PRs (2%), PLs (10–15%), GRs (ca. 80%), SPs (10%), and oenocytoids (OEs) (2%). In the hemocyte cultures they observed three categories: PRs (20–71%), PLs (25–45%), and GRs (5–39%). The hemocytes in cultures were not morphologically identical to the same kinds of hemocytes in fresh larval blood, nor did they occur in the same relative numbers. But their characteristics were sufficiently distinctive to permit identification.

Mitsuhashi (1966) studied the hemocyte complex of diapausing larvae of *Chilo suppressalis* in fresh blood and in hemocyte cultures, using size, shape, staining properties, and behavior as distinguishing criteria. In fresh blood, he observed seven types: PRs, PLs, GRs, SPs, OEs, podocytes (POs), and vermicytes (VEs). Differential hemocyte counts were not made, but he felt that the SPs predominated. PRs, PLs, and GRs were the next in abundance and almost equal in number. There were few OEs. POs and VEs were rare. The ratio of various types seemed to vary in different individuals.

In hemocyte cultures of this insect, Mitsuhashi (1966) observed that only PLs and PRs survived and multiplied for long periods. The GRs did not survive for more than 2 days. The SPs, OEs, POs, and VEs also gradually degenerated. During cultivation, cells with various forms between PRs and PLs appeared, suggesting the transformation of PRs into PLs. With prolonged culturing in vitro, PLs gradually disappeared, and PRs became predominant (Mitsuhashi, 1967).

Ritter and Blissit (1969) observed 12 morphologically distinct hemocyte types in the blood of the cockroach, *Gromphadorhina portentosa,* in vivo and in the hemocyte cultures maintained for over a year.

It would appear from the in vitro studies of hemocytes that in general there are fewer hemocyte types in vitro than in vivo. This could be because the more specialized hemocytes do not survive in vitro or because the specialized hemocytes – GRs, SPs, ADs, etc. – are just the basic cells, probably PLs, that have picked up nutrient materials or metabolites in vivo. As long as the culture conditions are favorable and optimal, only one or two cell types are present. As soon as culture conditions are altered, different hemocyte types appear in the cultures. Landureau and Grellet's (1975) work supports such an interpretation. This suggests that the various hemocyte types do not originate separately as immutable, self-sustaining cell lines, as is most often proposed, but that they are merely different stages of a single cell type.

9.3.2. Interrelationships among hemocytes

Differential hemocyte counts of several species by various authors indicate that the relative number of the different hemocyte types within the same species varies with the age and physiological condition of the insect. Literature on this topic has been reviewed by Jones (1962), Gupta and Sutherland (1966), and Arnold (1974). The occurrence of this phenomenon suggests that specific physiological conditions trigger the transformation of one type of hemocyte into another. There seems to be agreement that such transformations perhaps do occur, but there is no consensus on their extent and direction.

Some experimental information is now available on such transformations from in vivo studies. Rizki (1962) observed the transformation of a PL into a lemellocyte in *Drosophila melanogaster* larvae. He reported that the transformation was influenced by experimental alterations in the hormone balance of the larvae. Using autoradiography, Shrivastava and Richards (1965) reported a hemocyte transformation sequence of PR to PL to AD in the larvae and pupae of *Galleria mellonella*. Indirect support for the occurrence of transformation was provided by Devauchelle (1971) from his electron microscopic investigations on the hemocytes of *Melolontha melolontha*. He observed many intermediate stages between PRs, PLs, GRs, and ADs, and concluded that these four types are derived from each other. Also, on the basis of indirect evidence, Arnold and Hinks (1976) concluded that in the noctuid *Euxoa declarata* PRs differentiate into PLs. According to them, this is the main source of supply of PLs in this insect up to the fourth instar.

More direct evidence of such transformations has been provided by in vitro studies. Gupta and Sutherland (1966) reported the in vitro transformation of PLs into GRs, SPs, cystocytes (= coagulocytes or COs), OEs, POs, and VEs. All these transformations seemed to be governed by specific physiological conditions. They concluded that the PL is the basic type of hemocyte in insects.

Glaser (1917) and Taylor (1935) reported that only PLs multiplied in their primary cultures of hemocytes. Landureau and Grellet (1975) and Granados and Naughton (1976) observed that the cells in their established hemocyte lines had the characteristics of PLs. Landureau and Grellet further reported that under different conditions of cultivation, these cells transformed into small dividing cells (PRs) and large storing cells (GRs, ADs, etc.).

Mitsuhashi (1966) reported that cells with various forms intermediate between PRs and PLs appeared in his *C. suppressalis* hemocyte cultures, suggesting the transformation of PRs into PLs.

These studies, particularly those of Rizki (1962), Gupta and Sutherland (1966), and Landureau and Grellet (1975), demonstrate that a change in physiological conditions could result in the transformation of one type of hemocyte into another. Although there are conflicting reports on the direction of the transformation, there seems to be a strong indication from the in vitro studies (Glaser, 1917; Taylor, 1935; Gupta and Sutherland, 1966; Landureau and Grellet, 1975; Granados and Naughton, 1976) that the PL is the basic hemocyte. As a corollary of this, the direction of transformation would have to be from plasmatocyte to the other types (see also Chapter 4).

9.3.3. Hemopoietic organs

It is generally accepted that hemocytes originate from the mesodermal band in the embryo, but there is some disagreement about the exact location on the band from which they arise (Arnold, 1974) (see also Chapter 1). With refined tissue culture techniques it should be possible to isolate and grow in vitro the various regions of the embryonic mesodermal band and to determine from which region the hemocytes originate. When embryonic tissues obtained from 3- to 5-day-old eggs of *Choristoneura fumiferana* were explanted in vitro, several types of cells started to grow in the cultures (Sohi, 1968). Representative cells are illustrated in Figs. 9.2 and 9.3. These cultures survived for 6–7 weeks. Although I did not attempt to identify the cells as hemocytes or other types, it appeared that the small, round and slightly spindle-shaped, freely scattered cells in these cultures were hemocytes. This study indicates that it is possible to grow embryonic hemocytes in vitro.

Postembryonically, hemocytes have been reported to be supplied by two sources: (1) multiplication of the circulating hemocytes, and (2) hemopoietic organs (Arnold and Hinks, 1976). The fact that the hemocytes can grow and multiply in vitro for an indefinite period in the absence of any hemopoietic organ (Mitsuhashi, 1966, 1967; Chao and

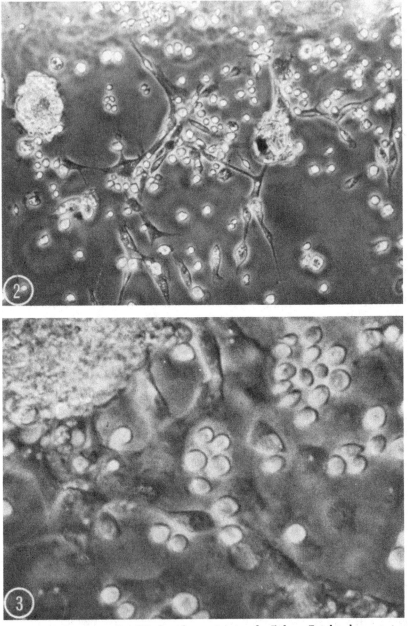

Fig. 9.2. Embryonic cells of *C. fumiferana* in vitro for 7 days. Bright phase contrast. × 180.

Fig. 9.3. Embryonic cells of *C. fumiferana* in vitro for 11 days. Bright phase contrast. × 500.

Ball, 1971; Sohi, 1971, 1973a; Landureau and Grellet, 1975; Granados and Naughton, 1975, 1976) indicates that they can maintain their number in vivo through division without the involvement of the hemopoietic organs.

Not much is known about the role of hemopoietic organs in the production of hemocytes in vitro. The only study on such cultures is that of Mitsuhashi (1972). He initiated cultures of the hemopoietic tissue of *Papilio xuthus*. A rapid migration of cells occurred within 24 hr of explantation, and in some cultures cell migration was so complete that no structure of the explant was left behind. Cinematographic observations revealed that, in most cultures, the migrating cells were spherical, but soon after migration most of them became spindle-shaped. In some cultures, especially toward the end of cell migration, spindle-shaped cells migrated directly from the explants. The migrated cells attached to the culture vessel and developed pseudopodia. Mitoses were rarely seen in the cultures, and the cells survived for approximately a month. Although Mitsuhashi felt that a few cells looked like GRs or POs, he did not report any other hemocyte types in the cultures. Further studies on hemopoietic organ cultures should provide more information on the function of the hemopoietic organs. They could elucidate whether these organs produce all types of hemocytes or just a certain type. Also, if the hemopoietic organs do produce hemocytes in vitro, it would be of interest to determine if changes in the composition of medium would affect the types of hemocytes produced (see also Chapters 2, 3, 8).

9.4. Suggestions for future research

Most of the work on hemocyte cultures has been done by insect pathologists, who were not primarily interested in hemocytology and have not, therefore, addressed themselves to the problems of hemocytology. Although considerable good work has been done in culturing hemocytes, and some very interesting and useful information has been obtained using in vitro cultivation techniques, the art and science of hemocyte culture need much improvement to realize the full potential of the method. The following types of work should be quite productive in answering some of the questions hemocytologists are trying to answer:

1. In the present methods of initiating primary cultures of hemocytes, the medium and the suspended cells are discarded within a few minutes after setting up the cultures, and the flasks are rinsed several times with fresh medium to reduce melanization. The cells that do not attach to the flasks are eliminated from the cultures, or at least their relative numbers are reduced as compared with the cells that attach to the flasks. This could alter the differential counts of the hemocytes in

culture as compared with fresh blood. Therefore, the method of starting primary cultures should be improved so that the suspended cells are not differentially removed from the cultures.

2. The currently available insect tissue culture media are far from perfect. These media favor the growth of certain hemocyte types, but are not so good for others. Improved media that will support the growth of all or most hemocyte types should be formulated. Such media would also facilitate the cloning of hemocyte cultures.

3. Clones should be isolated from existing and new lines of hemocyte cultures to determine whether a single cell will produce only cells of its kind and whether its progeny could differentiate into other hemocyte types under proper conditions of cultivation. Also, primary cultures (clones) should be started from known single hemocytes obtained directly from insects. Initiating clones like this will not be easy, because such single hemocytes would need a culture medium completely duplicating the in vivo conditions. But if such clones could be started, it would be easy to determine whether there is only one basic hemocyte type, giving rise to the others under specific physiological conditions, or whether all the types are derived and maintained as separate cell lines independent of the others.

4. Hemocytes from different stages of the insect, including embryos, should be grown in culture to determine whether the cultures maintain the same relative numbers of the various types as are found in the donor developmental stage, or whether all the cultures develop a similar hemocyte complex regardless of the age of the donor insect.

5. Organ cultures of whole intact hemopoietic organs and cell cultures from minced hemopoietic organs should be grown to see if they produce hemocytes in vitro. The production of hemocytes in such cultures under controlled conditions would prove the organs' hemopoietic function beyond any doubt.

6. With improved culture techniques, it may some day be possible to grow in vitro the groups of cells isolated from the different regions of the mesodermal band of the embryo. This could answer questions about the exact site of origin of hemocytes on the mesodermal band.

7. As mentioned earlier in this chapter, the size, shape, and appearance of cells can and do change in tissue cultures. The same cell may be round at one time and spindle-shaped at another. It may be free of granules and vacuoles at one time, but loaded with them at another, depending on the density of cell population and nutrient conditions in the culture. Similarly, the shape, size, and appearance of the hemocytes in vivo can vary from time to time, depending on the physiological and nutritional condition of the insect. Therefore, we should not rely too heavily on shape, size, and appearance of the hemocytes in classifying them into various categories. Some better and more constant parameters should be used for their classification.

This kind of work with hemocyte cultures should provide useful and reliable information on the embryonic origin and postembryonic multiplication of hemocytes. Also, it would perhaps simplify the confused hemocyte classification.

In our laboratory, we have cells of two continuous hemocyte lines of *M. disstria* (Sohi, 1971, 1973a) and would be pleased to send a starter culture to anyone interested in using these cells.

9.5. Summary

This review of in vitro cultivation of insect hemocytes has been confined mainly to the methods used for initiating, developing, and maintaining hemocyte cultures, and to their applications in hemocytology. In addition, suggestions for future research have been made.

It has been extremely difficult to grow insect hemocytes in vitro. Although short-term primary cultures of hemocytes were obtained by several workers in earlier attempts, they could not be successfully subcultured. But considerable success has been achieved in recent years, and continuous cell lines have been developed from hemocytes of the moths, *Chilo suppressalis*, *Samia cynthia*, *Malacosoma disstria*, and *Estigmene acrea*, and of a cockroach. Also, short-term primary cultures of the hemopoietic tissue of *Papilio xuthus* have been grown.

Most of the work on hemocyte cultures has been done by those primarily interested in using them for studying pathogens in vitro. Consequently, hemocytes per se have not been studied by these researchers. However, a few hemocytologists and others have used hemocyte cultures in investigations on hemocyte transformation, terminology, and classification. Their works have produced quite useful and interesting information and have demonstrated the potential of hemocyte cultures for solving many hemocytological problems, such as embryonic origin of hemocytes, their postembryonic multiplication, hemopoiesis, and hemocyte types. For example, the development of continuous hemocyte lines amply demonstrates that hemocytes do multiply by division without the involvement of hemopoietic organs. Also, extensive and rapid in vitro migration of cells from the hemopoietic tissue has been demonstrated, although the multiplication and differentiation of migrated cells were not observed. With improvements in technique and culture media, it should be possible to obtain multiplication of these cells and to determine if they differentiate into the various types of hemocytes.

The major hemocyte type in hemocyte cultures has been reported as plasmatocyte by several workers and prohemocyte by others. Plasmatocytes have been reported to transform into other hemocyte types in vitro and in vivo. Although the evidence is not yet conclusive, it indicates that the plasmatocyte is the basic hemocyte type and that under proper physiological conditions this cell can transform into the other hemocyte types. ·

References

Arnold, J. W. 1974. The hemocytes of insects, pp. 201–54. In M. Rockstein (ed.). The Physiology of Insecta, Vol. 5, 2nd ed. Academic Press, New York.

Arnold, J. W., and C. F. Hinks. 1976. Haemopoiesis in Lepidoptera. 1. The multiplication of circulating haemocytes. Can. J. Zool. 54(6):1003–12.

Arnold, J. W., and S. S. Sohi. 1974. Hemocytes of Malacosoma disstria Hübner (Lepidoptera: Lasiocampidae): Morphology of cells in fresh blood and after cultivation in vitro. Can. J. Zool. 52(4):481–5.

Arvy, L., and M. Gabe. 1946. Sur la multiplication in vitro des cellules sanguines de Forficula auricularia L. C. R. Soc. Biol. 140(19–20):787–9.

Barigozzi, C. (ed.). 1968. Proceedings, Second International Colloquium on Invertebrate Tissue Culture. Instituto Lombardo di Scienze e Lettere, Milan.

Brooks, M. A., and T. J. Kurtti. 1971. Insect cell and tissue culture. Annu. Rev. Entomol. 16:27–52.

Chao, J., and G. H. Ball. 1971. A cell line isolated from hemocytes of Samia cynthia pupae. Curr. Topics Microbiol. Immunol. 55:28–32.

Coriell, L. L. 1973. Methods of prevention of bacterial, fungal, and other contaminations, pp. 29–49. In J. Fogh (ed.). Contamination in Tissue Culture. Academic Press, New York.

Coriell, L. L., and G. J. McGarrity. 1970. Evaluation of the Edgegard laminar flow hood. Appl. Microbiol. 20(3):474–9.

Coriell, L. L., G. J. McGarrity, and J. Horneff. 1967. Medical applications of dust-free rooms. I. Elimination of airborne bacteria in a research laboratory. Amer. J. Pub. Health 57(10):1824–36.

Devauchelle, G. 1971. Etude ultrastructurale des hémocytes du Coléoptère Melolontha melolontha (L.). J. Ultrastruct. Res. 34(5–6):492–516.

Feir, D., and C. R. Pantle. 1971. In vitro studies of insect hemocytes. J. Insect Physiol. 17(4):733–7.

Fogh, J. (ed.). 1973. Contamination in Tissue Culture. Academic Press, New York.

Fogh, J., N. B. Holmgren, and P. O. Ludovici. 1971. A review of cell culture contaminations. In Vitro 7(1):26–41.

Glaser, R. W. 1917. The growth of insect blood cells in vitro. Psyche 24(1):1–7.

Goldschmidt, R. 1915. Some experiments on spermatogenesis in vitro. Proc. Natl. Acad. Sci. Wash. 1(4):220–2.

Goodwin, R. H. 1975. Insect cell culture: Improved media and methods for initiating attached cell lines from the Lepidoptera. In Vitro 11(6):369–78.

Grace, T. D. C. 1962. Establishment of four strains of cells from insect tissues grown in vitro. Nature (Lond.) 195(4843):788–9.

Granados, R. R. 1976. Infection and replication of insect pathogenic viruses in tissue culture. Adv. Virus Res. 20:189–236.

Granados, R. R., and M. Naughton. 1975. Development of Amsacta moorei entomopoxvirus in ovarian and hemocyte cultures from Estigmene acrea larvae. Intervirology 5(1–2):62–8.

Granados, R. R., and M. Naughton. 1976. Replication of Amsacta moorei entomopoxvirus and Autographa californica nuclear polyhedrosis virus in hemocyte cell lines from Estigmene acrea, pp. 379–89. In E. Kurstak and K. Maramorosch (eds.). Invertebrate Tissue Culture: Applications in Medicine, Biology, and Agriculture. Academic Press, New York.

Greene, A. E., and J. Charney. 1971. Characterization and identification of insect cell cultures. *Curr. Topics Microbiol. Immunol. 55*:51–61.

Greene, A. E., J. Charney, W. W. Nicols, and L. L. Coriell. 1972. Species identity of insect cell lines. *In Vitro 7*(5):313–22.

Gupta, A. P. 1969. Studies of the blood of Meloidae (Coleoptera). 1. The haemocytes of *Epicauta cinerea* (Forster), and a synonymy of haemocyte terminologies. *Cytologia 34*(2):300–44.

Gupta, A. P., and D. J. Sutherland. 1966. *In vitro* transformations of the insect plasmatocyte in some insects. *J. Insect Physiol. 12*(9):1369–75.

Herrick, P. R., G. W. Baumann, D. J. Merchant, M. C. Shearer, C. Shipman, Jr., and R. G. Brackett. 1970. Serologic and karyologic evidence of incorrect identity of an animal cell line (guinea pig spleen). *In Vitro 6*(2):143–47.

Hink, W. F. 1972. Insect tissue culture. *Adv. Appl. Microbiol. 15*:157–214.

Hink, W. F. 1976. A compilation of invertebrate cell lines and culture media, pp. 319–69. *In* K. Maramorosch (ed.). *Invertebrate Tissue Culture: Research Applications.* Academic Press, New York.

Horikawa, M., and Y. Kuroda. 1959. *In vitro* cultivation of blood cells of *Drosophila melanogaster* in a synthetic medium. *Nature (Lond.) 184*(4704):2017–18.

Jones, J. C. 1962. Current concepts concerning insect hemocytes. *Amer. Zool. 2*(2):209–46.

Krywienczyk, J., and S. S. Sohi. 1973. Serologic characterization of a *Malacosoma disstria* Hübner (Lepidoptera: Lasiocampidae) cell line. *In Vitro 8*(6):459–65.

Kurstak, E., and K. Maramorosch. (eds.). 1976. *Invertebrate Tissue Culture: Applications in Medicine, Biology and Agriculture.* Academic Press, New York.

Kurtti, T. J., and M. A. Brooks. 1970. Growth of lepidopteran epithelial cells and hemocytes in primary cultures. *J. Invertebr. Pathol. 15*(3):341–50.

Kurtti, T. J., and M. A. Brooks. 1971. Growth of a microsporidian parasite in cultured cells of tent caterpillars (*Malacosoma*). *Curr. Topics Microbiol. Immunol. 55*:204–8.

Kurtti, T. J., and M. A. Brooks. 1977. Isolation of cell lines from embryos of the cockroach, *Blattella germanica. In Vitro 13*(1):11–17.

Landureau, J. C., and P. Grellet. 1975. Obtention de lingnées permanentes d'hémocytes de blatte: Caractéristiques physiologiques et ultrastructurales. *J. Insect Physiol. 21*(1):137–51.

Maramorosch, K. (ed.). 1976. *Invertebrate Tissue Culture: Research Applications.* Academic Press, New York.

Martignoni, M. E., and R. J. Scallion. 1961a. Preparation and uses of insect hemocyte monolayers *in vitro. Biol. Bull. (Woods Hole) 121*(3):507–20.

Martignoni, M. E., and R. J. Scallion. 1961b. Multiplication *in vitro* of a nuclear polyhedrosis virus in insect amoebocytes. *Nature (Lond.) 190*(4781):1133–4.

Mazzone, H. M. 1971. Cultivation of gypsy moth hemocytes. *Curr. Topics Microbiol. Immunol. 55*:196–200.

Mazzone, H. M. 1976. Influence of polyphenol oxidase on hemocyte cultures of the gypsy moth, pp. 275–8. *In* E. Kurstak and K. Maramorosch (eds.). *Invertebrate Tissue Culture: Applications in Medicine, Biology and Agriculture.* Academic Press, New York.

McGarrity, G. J. 1975. Control of microbiological contamination. *Tissue Culture Assoc. Manual 1*(4):181–4.

McGarrity, G. J. 1977. Cell culture facilities. *Tissue Culture Assoc. Manual 3*(4):679–83.

McGarrity, G. J., and L. L. Coriell. 1971. Procedures to reduce contamination of cell cultures. *In Vitro 6*(4):257–65.

Merchant, D. J. 1973. Summary, pp. 257–69. *In* J. Fogh (ed.). *Contamination in Tissue Culture.* Academic Press, New York.

Mitsuhashi, J. 1966. Tissue culture of the rice stem borer, *Chilo suppressalis* Walker (Lepidoptera: Pyralidae). II. Morphology and *in vitro* cultivation of hemocytes. *Appl. Entomol. Zool. 1*(1):5–20.

Mitsuhashi, J. 1967. Establishment of an insect cell strain persistently infected with an insect virus. *Nature (Lond.) 215*(5103):863–4.

Mitsuhashi, J. 1972. Primary culture of the haemocytopoietic tissue of *Papilio xuthus* Linne. *Appl. Entomol. Zool. 7*(1):39–41.

Moklebust, R., N. Diaz, and I. E. Goetz. 1977. An inexpensive method of freezing human skin fibroblasts at a controlled cooling rate. *Tissue Culture Assoc. Manual 3*(3):671–3.

Mosolov, A. N., V. V. Gulii, T. A. Batalina, and L. S. Berdichevskaya. 1967. Kul'tura kletok iz gemol imfy pil il'shika. (A culture of cells from the hemolymph of the sawfly.) *Tsitologiia 9*(2):241–4.

Perlman, D. 1976. Some guidelines for selection of antibiotics for elimination of unwanted microbial contaminants in tissue cultures. *Tissue Culture Assoc. Manual 2*(3):383–6.

Rehacek, J., D. Blaskovic, and W. F. Hink (eds.). 1973. *Proceedings, Third International Colloquium on Invertebrate Tissue Culture.* Slovak Academy of Sciences, Bratislava.

Ritter, H., Jr., and L. Blissit. 1969. Cell transformation *in vitro:* Crescent cell origin in cockroach blood. *J. Cell Biol. 43* (2 Pt. 2):117a.

Rizki, T. M. 1962. Experimental analysis of hemocyte morphology in insects. *Amer. Zool. 2*(2):247–56.

Rudin, A., A. Healey, C. A. Philips, D. W. Gump, and B. R. Forsyth. 1970. Antibacterial activity of gentamicin sulfate in tissue culture. *Appl. Microbiol. 20*(6):989–90.

Schafer, T. W., A. Pascale, G. Shimonaski, and P. E. Came. 1972. Evaluation of gentamicin for use in virology and tissue culture. *Appl. Microbiol. 23*(3):565–70.

Schneider, I. 1969. Establishment of three diploid cell lines of *Anopheles stephensi* (Diptera: Culicidae). *J. Cell Biol. 42*(2):603–6.

Schroy, C. B., and P. Todd. 1976. A simple method for freezing and thawing cultured cells. *Tissue Culture Assoc. Manual 2*(1):309–10.

Shannon, J. E., and M. L. Macy. 1973. Freezing, storage, and recovery of cell stocks, pp. 712–18. *In* P. F. Kruse, Jr., and M. K. Patterson, Jr. (eds.). *Tissue Culture: Methods and Applications.* Academic Press, New York.

Shrivastava, S. C., and A. G. Richards. 1965. An autoradiographic study of the relation between hemocytes and connective tissue in the wax moth *Galleria mellonella* L. *Biol. Bull. (Woods Hole) 128*(2):337–45.

Singh, K. R. P. 1967. Cell cultures derived from larvae of *Aedes albopictus* (Skuse) and *Aedes aegypti* (L.). *Curr. Sci. (India) 36*(19):506–8.

Sohi, S. S. 1968. *In vitro* cultivation of *Choristoneura fumiferana* (Clemens) (Lepidoptera: Tortricidae) tissues. *Can. J. Zool. 46*(1):11–13.

Sohi, S. S. 1969. Adaptation of an *Aedes aegypti* cell line to hemolymph-free culture medium. *Can. J. Microbiol. 15*(10):1197–1200.

Sohi, S. S. 1971. *In vitro* cultivation of hemocytes of *Malacosoma disstria* Hübner (Lepidoptera: Lasiocampidae). *Can. J. Zool. 49*(10):1355–8.

Sohi, S. S. 1973a. Establishment of cultures of *Malacosoma disstria* Hübner (Lepidoptera: Lasiocampidae) hemocytes in a hemolymph-free medium, pp. 27–39. *In* J. Rehacek, D. Blaskovic, and W. F. Hink (eds.). *Proceedings, Third International Colloquium on Invertebrate Tissue Culture.* Slovak Academy of Sciences, Bratislava.

Sohi, S. S. 1973b. *In vitro* cultivation of larval tissues of *Choristoneura fumiferana* (Clemens) (Lepidoptera: Tortricidae), pp. 75–92. *In* J. Rehacek, D. Blaskovic, and W. F. Hink (eds.). *Proceedings, Third International Colloquium on Invertebrate Tissue Culture.* Slovak Academy of Sciences, Bratislava.

Sohi, S. S., and F. T. Bird. 1976. Replication of a nuclear polyhedrosis virus of *Choristoneura fumiferana* (Lepidoptera: Tortricidae) in *Malacosoma disstria* (Lepidoptera: Lasiocampidae) hemocyte cultures, pp. 361–367. *In* E. Kurstak and K. Maramorosch (eds.). *Invertebrate Tissue Culture: Applications in Medicine, Biology, and Agriculture.* Academic Press, New York.

Sohi, S. S., and J. C. Cunningham. 1972. Replication of a nuclear polyhedrosis virus in serially transferred insect hemocyte cultures. *J. Invertebr. Pathol. 19*(1):51–61.

Sohi, S. S., and C. Smith. 1970. Effect of fetal bovine serum on the growth and survival of insect cell cultures. *Can. J. Zool. 48*(3):427–32.

Sohi, S. S., C. R. Sullivan, and C. L. Bodley. 1971. Simple controlled-rate freezing device. *Lab. Pract. 20*(2):127–8.

Sohi, S. S., and G. G. Wilson. 1976. Persistent infection of *Malacosoma disstria* (Lepidoptera: Lasiocampidae) cell cultures with *Nosema (Glugea) disstriae* (Microsporida: Nosematidae). *Can. J. Zool. 54*(3):336–42.

Stanley, M. S. M. 1972. Cultivation of arthropod cells, pp. 327–370. *In* G. H. Rothblat and V. J. Cristofalo (eds.). *Growth, Nutrition and Metabolism of Cells in Culture,* Vol. 2. Academic Press, New York.

Suitor, E. C., Jr., L. L. Chang, and H. H. Liu. 1966. Establishment and characterization of a clone from Grace's in vitro cultured mosquito (*Aedes aegypti* L.) cells. *Exp. Cell Res. 44*(2–3):572–8.

Taylor, A. 1935. Experimentally induced changes in the cell complex of the blood of *Periplaneta americana* (Blattidae: Orthoptera). *Ann. Entomol. Soc. Amer. 28*(1): 135–45.

Vago, C. (ed.). 1963. Proceedings, First International Colloquium on Invertebrate Tissue Culture. *Ann. Epiphyties 14*(3):1–222.

Vago, C. (ed.). 1971. *Invertebrate Tissue Culture,* Vol. 1. Academic Press, New York.

Vago, C. (ed.). 1972. *Invertebrate Tissue Culture,* Vol. 2. Academic Press, New York.

Varma, M. G. R., and M. Pudney. 1969. The growth and serial passage of cell lines from *Aedes aegypti* (L.) larvae in different media. *J. Med. Entomol. 6*(4):432–9.

Vaughn, J. L. 1968. A review of the use of insect tissue culture for the study of insect-associated viruses. *Curr. Topics Microbiol. Immunol. 42*:108–28.

Waymouth, C., and D. S. Varnum. 1976. Simple freezing procedure for storage in serum-free media of cultured and tumor cells of mouse. *Tissue Culture Assoc. Manual 2*(1):311–13.

Weiss, E. (ed.). 1971. Arthropod cell cultures and their application to the study of viruses. *Curr. Topics Microbiol. Immunol. 55*:1–288.

Wigglesworth, V. B. 1959. Insect blood cells. *Annu. Rev. Entomol. 4*:1–16.

Wilson, G. G., and S. S. Sohi. 1977. Effect of temperature on healthy and microsporida-infected continuous cultures of *Malacosoma disstria* hemocytes. *Can. J. Zool. 55*(4):713–17.

Wyatt, S. S. 1956. Culture *in vitro* of tissue from the silkworm, *Bombyx mori* L. *J. Gen. Physiol. 39*(6):841–52.

Yunker, C. E., J. L. Vaughn, and J. Cory. 1967. Adaptation of an insect cell line (Grace's *Antheraea* cells) to medium free of insect hemolymph. *Science (Wash., D.C.) 155*(3769):1565–6.

10   Pathways and pitfalls in the classification and study of insect hemocytes

J. C. JONES

Department of Entomology, University of Maryland, College Park, Maryland 20742, U.S.A.

Contents

10.1. Introduction

Like other cells, hemocytes are generally differentiated anatomically by differences in size and shape of the soma and nucleus; by type, number, size, and staining affinities of their inclusions; and by the general appearance and staining of the cytoplasm. They can also be distinguished by inherent behavioral differences in their ability to divide, in speed of degranulation (or vacuolization) of their inclusions, in their general cytoplasmic fragility, by the development (and types) of pseudopodia (or surface extensions), and by their tendencies to stick to each other or to different surfaces. They may also be classed according to their relative abundance at specific periods during the life span of a given species. Finally, hemocytes may be identified with some definite activity, such as phagocytosis, coagulation of the hemoplasma, or trephocytosis (Jones, 1962, 1964, 1977).

10.2. Hemocytopoiesis

The pitfalls in this area of insect hematology are particularly numerous. They begin with the acceptance of an imprecise vertebrate terminology. Because the words "hemopoiesis," "hematopoiesis," and "hematogenesis" all refer to the production or formation of blood (Dorland, 1974), they can refer to the formation of both the liquid fraction (plasma and lymph) and the formed elements (erythrocytes and leucocytes). Because insects have a combination of blood and lymph (hemolymph) and because they do not have red blood cells, it is better to use the word "hemocytopoiesis" to refer specifically to the formation of hemocytes (Jones, 1970) and to coin some other word, like "hemoplasmopoiesis," to refer only to the formation of the liquid fraction of the hemolymph – hemoplasma. Hemolymph includes the hemoplasma and its indigenous cells, the hemocytes.

It is sometimes assumed that we have hard evidence that some specific sessile tissue or organ in insects differentiates into specific types of hemocytes from a single stem cell and that these then mature in situ and are subsequently released into the hemolymph to circulate, either to replace dying cells or to keep a constant number of cells circulating in a hemolymph that is steadily increasing in quantity as the insect goes through much of its development. In a few insects there is preliminary, but still incomplete and mostly histological, evidence that such tissues supply some, but not all, of the circulating hemocytes (see Jones, 1977). In most insects there is no evidence that they have any such tissue (Jones, 1970). In them, it has been suggested that sessile hemocytes or those already in circulation divide sufficiently to account for the observed changes in the number of hemocytes (Jones, 1967a; Jones and Liu, 1968). Although the phagocytic organs of the various orthopterans are said to be hemocytopoietic (Hoffmann, 1972), the tissue seems far too extensive and the numbers of circulating hemocytes far too few for this to be their primary function. If the phagocytic organs of *Locusta* normally supply hemocytes during adult life, does this mean they are renewing those hemocytes that have a short life span? It seems probable that the tissue's primary function is phagocytosis and that it plays only a minor role in supplying hemocytes (Jones, 1977).

The tissue around the developing wings of the Lepidoptera (Akai and Sato, 1971; Arnold and Hinks, 1976; Hinks and Arnold, 1977) and at the posterior ends of some of the larvae of the higher Diptera (Arvy, 1954) is also said to be hemocytopoietic. Although it is indeed possible that cells from these tissues are required to supply many circulating hemocytes during much of larval life, during the last stage of larval development in both the Diptera and Lepidoptera, the organ steadily enlarges (Arvy, 1954; Akai and Sato, 1971; Hinks and Arnold, 1977), and all the cells in these tissues are then suddenly released en masse into the hemolymph at the moment of pupation (Jones, 1970, 1977).

It is entirely possible that the tissue surrounding the wing disks in the Lepidoptera supplies many circulating hemocytes (rather than replacing worn-out cells) and yet still grows in numbers of sessile cells, which are then released at pupation.

We still do not know the life span of even one type of hemocyte. It seems probable that plasmatocytes (PLs) and granulocytes (GRs) have long life spans in some insects. It seems likely that some other types, for example, vermicytes (VEs), live for only a short time.

We need to investigate the relationship between the rate of formation of hemocytes in some compact tissue and of hemocytes in circulation in relation to the rate of elimination of hemocytes from circulation under normal (not under abnormal) conditions. We need to study those tissues that may be forming hemocytes with regard to the problem of whether the cells in them divide cyclically or slowly but continuously, and we need specific data relative to numbers of cells in the tissue and in the circulating hemolymph. We especially need to study hemocytes in relation to changes in the amount of hemolymph in the whole insect.

Future students of hemocytopoiesis in insects should study the volume edited by Wolstenholme and O'Connor (1960) and Gordon's (1970) book on regulation of hematopoiesis.

10.3. Hematological methods

Even though it is important to examine hemocytes with different techniques before arriving at conclusions about them (Jones, 1962), most workers still examine them with only one technique. A few researchers have studied them with two methods, but no one has yet systematically and simultaneously examined the hemocytes of even a single stage of a single species (1) in unfixed wet whole mounts with phase-contrast microscopy, (2) in air-dried and stained smears with ordinary light microscopy, and (3) in fixed material with electron microscopy. Many workers collect hemolymph from untreated insects, allowing it to drip from a severed appendage into a fixative, despite the fact that certain hemocytes may very rapidly change in appearance as the hemolymph is being collected. It would be valuable to inject an insect with a fixative or some anticoagulant (like versene) prior to withdrawing hemolymph or at least to collect the hemolymph directly *under* a fixative (Jones, 1962).

Use of the electron microscope does not resolve many of the problems of identifying hemocytes, and in some cases the problems are made more difficult. The primary reason most workers use only phase-contrast microscopy and unfixed hemolymph is that the method demonstrates the hemocytes quickly and allows one to distinguish different types easily and rapidly. There is one major problem: The cells may quickly change their shape (withdraw their spindle ends or spread out), may vacuolate or degranulate, or disintegrate in vitro. Further, the cells may continue to degenerate with time so that differential counts are radically altered.

If total counts are made on unfixed (untreated) insects, they should be contrasted with counts from heat-fixed animals because the latter method is reported to preserve the number of cells in actual circulation (Jones, 1962).

Hemolymph volume determinations (using either amaranth red or ^{14}C-inulin) allow one to calculate the numbers of hemocytes circulating in an entire insect's hemocoele. Such determinations are critical to estimating how great a role hemocytopoietic organs play in releasing hemocytes.

10.4. Classification of hemocytes

Using multiple criteria, the hemocytes of different insects may be classified into from two to nine distinctive types (Yeager, 1945; Jones, 1962). In some insects the differences among the cells are obscured or minor, and this has led those insect hematologists who have worked with only one species to conclude that only one or possibly two types of hemocytes exist in all orders of insects. In some other species, many differences appear and are very striking, and this has led other hematologists to conclude that there are many different types present among insects. Some workers have examined hemocytes in only one or two stages during an insect's entire life span and concluded that only the cells they saw existed in that insect, whereas if they had examined, for example, a newly hatched larva or a pupa or an adult, they might well have encountered additional types.

At one time or another specific hemocytes of insects have been called lymphocytes (Ries, 1932) and have even been compared to neutrophils, eosinophils, and megakaryocytes of vertebrates. Although they may be tempting, such comparisons are fraught with peril because so much more is known about the cytology, behavior, and functions of vertebrate blood cells and because use of vertebrate terms implies there are no great differences between an insect and a vertebrate. The cytoplasmic free filaments that break away from the podocytes of larvae of the higher Diptera (Jones, 1956) may or may not be concerned with gelation of the plasma, whereas the platelets that are shed from the very large sessile megakaryocytes of vertebrates are specifically associated with coagulation of the blood.

Using many criteria, the hemocytes that can be identified as occurring among some, but not all, insects include prohemocytes (PRs); plasmatocytes (PLs); (Figs. 10.1A, 10.2, 10.3, 10.5, 10.7, 10.15); podocytes (POs) (Figs. 10.3, 10.8); vermiform cells, here termed vermicytes (VEs) (Figs. 10.5, 10.16); granular hemocytes, here termed granulocytes (GRs) (Figs, 10.3, 10.5, 10.9); cystocytes (= coagulocytes, COs) (Figs. 10.2, 10.5, 10.10, 10.15); spherule cells, here termed spherulocytes (SPs) (Figs, 10.1D, 10.12); adipohemocytes (ADs) (Figs.

10.1*B,C*, 10.4, 10.11, 10.14); and oenocytoids (OEs) (Figs, 10.1*E*, 10.3, 10.13) (Jones, 1962). Although it was suggested earlier that the crescent cells (Figs. 10.5, 10.6) of *Gromphadorhina* might be a specialized kind of SP (Jones, 1975), recent electron microscopic studies indicate that these cells are too distinctive to place in that category. I have observed crescent cells in only a few species of cockroaches and in no other order of insects. It is suggested that crescent cells be placed in a separate category (see also Chapter 8).

It may be useful to subdivide certain categories where the information is consistent and may lead to useful findings. The ADs of *Galleria* larvae were subdivided into immature and mature varieties (Jones, 1967a), and the GRs of *Rhodnius* were counted as intact and lysing forms (Jones, 1967b).

In addition to tabulating data on numbers of the basic cell types, it is also useful to count mitotically dividing cells, cells that degranulate or vacuolate, already degenerating cells (including naked nuclei), unidentifiable cells, and nonhemocytic cells (e.g., fat body cells). It may

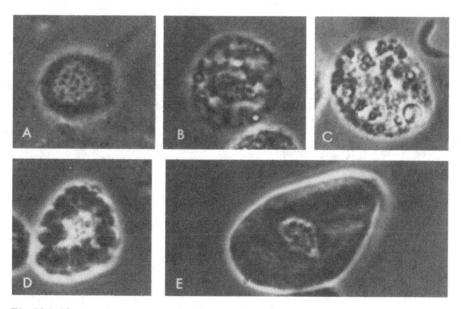

Fig. 10.1. Phase-contrast appearance of main types of hemocytes of *Galleria mellonella*, as seen under oil immersion in unfixed hemolymph from last-stage larva. **A.** Cell that could be identified as either a large prohemocyte or a small plasmatocyte. **B.** Immature adipohemocyte with four sharply outlined lipid droplets, larger and more vaguely outlined watery vacuoles that can form in these cells in vitro, and some smaller phase-dark granules. **C.** Mature adipohemocyte with many lipid droplets and other vaguely outlined granular inclusions that obscure nucleus. Some workers have identified these cells (**B** and **C**) as granulocytes. **D.** Spherulocyte with large phase-dark spherular inclusions surrounding and obscuring nucleus. **E.** Oenocytoid with thick, completely structured cytoplasm and slightly eccentric nucleus. × 1,000.

also be valuable to count fusiform cells and hemocytes with and without pseudopodia.

The hemocyte classification of Jones (1962) is considerably less complex than that proposed by Yeager (1945). In *Spodoptera,* however, the hemocyte picture is extremely complex, especially when viewed in stained films (Fig. 10.16).

It is often useful or convenient to pool different cells together, especially when some types vary enormously in quantitative counts. Jones (1956) counted both PLs and POs in *Sarcophaga* larvae, but found the POs enormously variable and difficult to identify accurately because of their tendency to fragment into multiple pieces. He pooled the two cells together as "plasmatocytes." Pooling PRs and PLs in the category "plasmatocytoids" was used by Jones (1967a) because the two types could not be easily distinguished. Recently ADs and GRs were

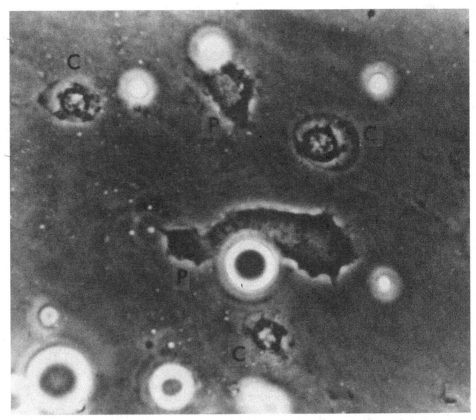

Fig. 10.2. Phase-contrast appearance of two main types of hemocytes of *Tenebrio molitor,* as seen under oil immersion in unfixed hemolymph from last-stage larva. *P* = plasmatocytes. Note three cystocytes (*C*), each with characteristic hyaline cytoplasm, a few large coarse granules, and a sharply outlined nucleus. × 980.

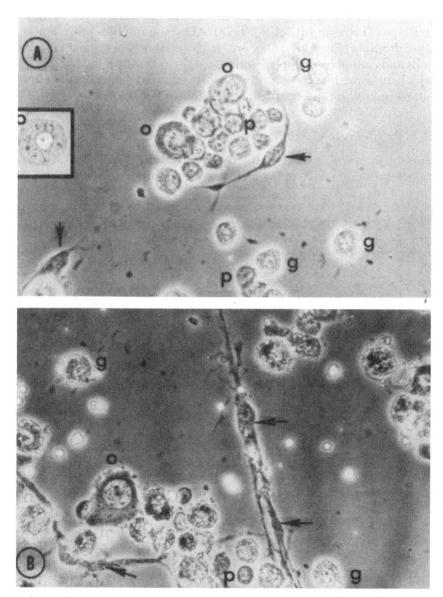

Fig. 10.3. Phase-contrast appearance of main types of hemocytes of *Calliphora vicina* (= *C. erythrocephala*), as seen with high dry objective in unfixed hemolymph from last-stage larva. Small cells (*p*) could be identified as prohemocytes or small plasmatocytes. Large plasmatocyte-like cells are oenocytoids (*o*); they have an eosinophilic cast, are not ameboid, and frequently undergo lysis, as shown in **inset.** Granulocytes (*g*) contain many small, spherical inclusions. Podocytes are indicated by arrows. In **3A,** podocytes typically show a thin, phase-dark strap of cytoplasm extending over elongated nucleus. In **B,** podocyte at right shows long, irregular, and jagged filamentous extensions that may break away as separate fragments or plastids × 400.

pooled, with the implication that there were no differences between two types (Price and Ratcliffe, 1974). ADs by definition must contain lipid droplets (Fig. 10.4); they are much smaller than typical fat body cells and cannot be confused with those very large, true fat body cells, which, in some insects (e.g., *Pyrrhocoris apterus*), may actually circulate normally in the hemolymph (see Jones, 1977).

10.5. Problems in classifying hemocytes

Unpublished studies comparing the phase-contrast appearance of hemocytes in unfixed wet whole mounts, in histological sections of

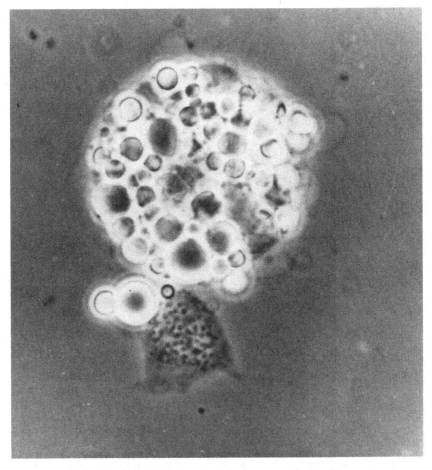

Fig. 10.4. Phase-contrast view of two types of hemocytes of *Pseudaletia unipunctata*, as observed under oil immersion in unfixed hemolymph from pupa. Larger cell is a mature adipohemocyte filled with large lipid inclusions that obscure small nucleus. Small cell is a plasmatocyte with a large nucleus filled with sharply outlined punctate chromatin granules. × 980. (Courtesy of Dr. Gertraude Wittig)

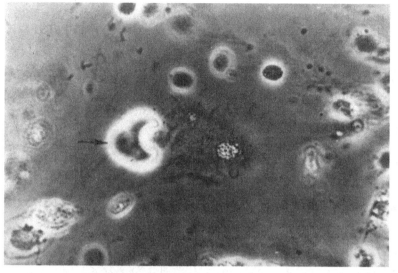

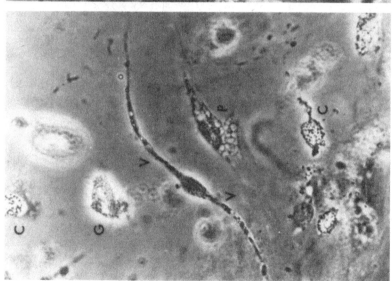

Fig. 10.5. Four types of hemocytes of adult hissing cockroach, *Cromphadorhina portentosa*, as seen in unfixed hemolymph under oil immersion with phase-contrast microscope. *V* = hemocyte that could be identified either as elongated plasmatocyte or as small vermicyte. *P* = highly vacuolated ameboid plasmatocyte; note very fine threadlike filaments extending from cell. *G* = ameboid granulocyte densely filled with many small granules that obscure nucleus. Cystocytes (*C*) at left have lost almost all their cytoplasm. Cystocyte (*C*) at right is surrounded by finely granular precipitate in hemoplasma. A crescent cell is indicated by arrow. × 260.

287

whole animals, in well-stained smears (air-dried films from heat-fixed insects, treated with methanol and stained with Giemsa or Wright stain), indicate that it is often relatively easy to identify PRs, PLs, SPs, and GRs in wet mounts and in stained smears, but very difficult to identify ADs and COs accurately in stained films. Although POs are conspicuous in larvae of the higher Diptera when they are studied in fresh whole mounts with phase-contrast microscopy, I have not been able to recognize them at all in high-quality histological sections of intact larvae of *Musca, Dacus,* and *Sarcophaga.* The sections clearly showed PLs, GRs, and either SPs or OEs. As shown in Fig. 10.3, the small hemocytes without many inclusions could be identified as ei-

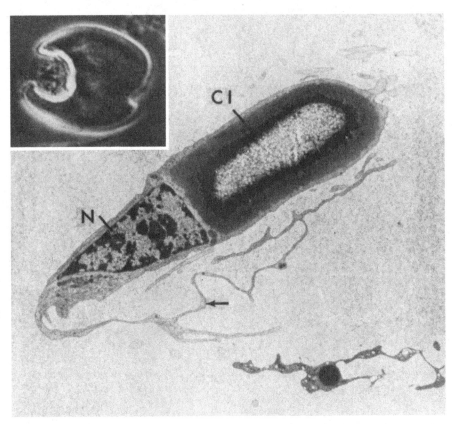

Fig. 10.6. Crescent cell of *G. portentosa*. **Inset** shows cell as it appears under oil immersion. Note enormous C-shaped inclusion and extremely eccentric nucleus at far left. × 650. **Main figure,** electron micrograph of cross section through a crescent cell, shows well-formed eccentric nucleus (*N*) and large, complex cytoplasmic inclusion (*CI*). Crescent cells often send out long filamentous extensions (arrow). Electron microscope shows that crescent inclusion has a densely and finely vacuolated, lightly stained inner core, which is surrounded by an intensely electron-dense layer that is enclosed within a smooth, finely granular substance. ×5,075. (Courtesy of Dr. T. M. Tadkowsky)

ther PRs or small PLs and the large OEs could be mistaken for true PLs unless one knew they were eosinophilic cells and frequently undergo lysis. In wet whole mounts studied with the phase-contrast microscope, I have observed SPs and no OEs in *Sarcophaga* larvae, but in two other species of Diptera (*Musca* and *Calliphora*), I have seen OEs and no SPs.

Using only a single technique and the criteria given by Jones (1962), it is often very difficult to distinguish between PLs and GRs and between GRs and ADs (Fig. 10.14) in many species of insects. Two workers looking at exactly the same material may still classify the

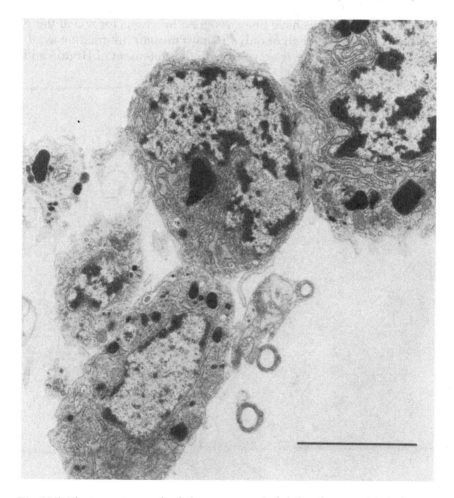

Fig. 10.7. Electron micrograph of plasmatocytes of adult female mosquito, *Aedes aegypti*. Cells contain small microtubules, well-formed Golgi bodies, a few mitochondria, some crystals, and lysosomes. Cells are densely filled with long, tortuous canaliculi, which are formed from rough endoplasmic reticulum. Canaliculi are densely filled with finely granular secretory material. Bar represents 5 μm.

same cell as belonging to two or even three different categories (Fig. 10.14). The two cells shown in the upper portion of Fig. 10.15 are identified by Rowley (1978, pers. comm.) as PLs and the darker cells in the lower part of the figure as COs. Without additional information, it would have been possible to identify these cells in just the opposite manner. Rowley (pers. comm.) observed cells similar to the PLs in tissue culture and found them to be phagocytic. The COs were not found in tissue cultures.

10.6. Hemocyte functions

Many times hemocytes have been assigned functions for which there is either no evidence at all or only the most meager information available. For example, there is no validity to the statement of Hinton and

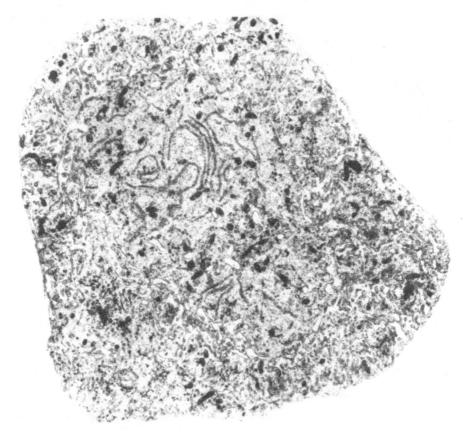

Fig. 10.8. Low-power electron micrograph of podocyte of *Calliphora vicina* (× *C. erythrocephala*) larva, showing numerous clear, tortuous extracellular channels, which elaborately and deeply incise cell. × 7,000. (Courtesy of Dr. J. A. Hoffmann)

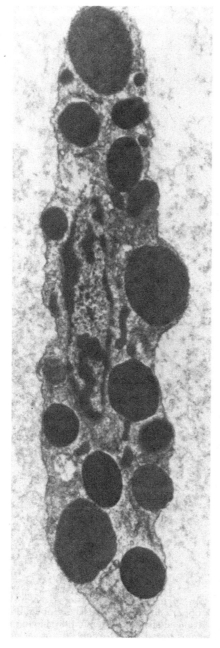

Fig. 10.9. Electron micrograph of granulocyte of cockroach, *Periplaneta americana.* × 17,000. (Courtesy of Dr. R. J. Baerwald)

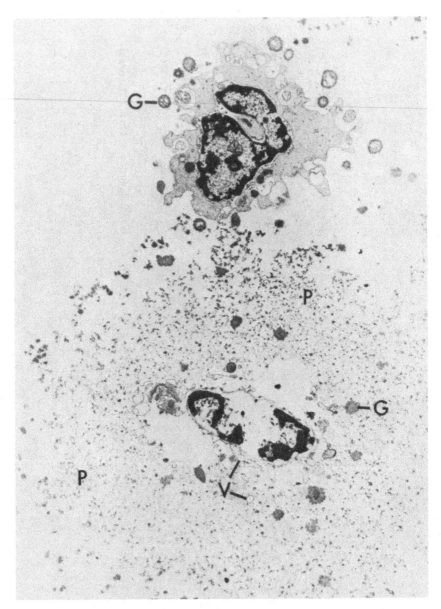

Fig. 10.10. Low-power electron micrograph of two cystocytes of a stick insect, *Clitumnus extradentatus*. Dark upper cell is rapidly discharging granules (G) into hemoplasma; lower cell is in advanced stage of degeneration, having released a number of granules (G) and numerous round, clear vesicles (V) of various sizes. Lower cell is surrounded by precipitated hemoplasm (P). × 6,750. (Courtesy of Dr. A. F. Rowley)

Mackerras (1970) that young hemocytes are totipotent mesenchymatous cells that normally transform into connective tissue, basement membrane, fat body cells, and skeletal muscles (see Jones, 1962, 1977). One of the main assigned functions of hemocytes is said to be trephocytosis; yet, only morphological data are available to support this idea, and physiological studies are seriously needed.

The primary pitfall in studies on function is to study one type of cell in only one stage with only one test condition. Each of the proposed functions of hemocytes needs much further investigation, especially with regard to possible changes occurring at specific stages of an insect's development. Several cell types may be normally few or ab-

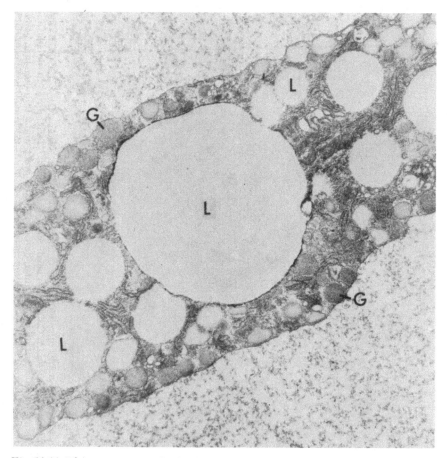

Fig. 10.11. Electron micrograph of portion of mature adipohemocyte of *Galleria mellonella* larva, showing structured granular inclusions (G) transforming into lipid droplets (L) of various sizes. Note that cell is densely filled with many ribosomes and has long tracks of rough endoplasmic reticulum. × 17,160. (Courtesy of Dr. Maria Neuwirth)

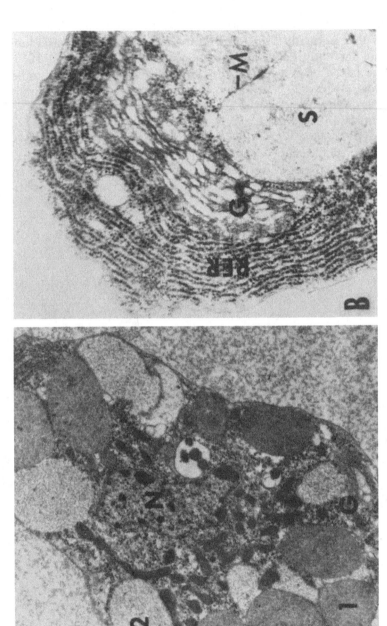

Fig. 10.12. Electron micrographs of spherulocyte of larva of pink boll-worm, *Pectinophora gossypiella*. **A.** Cell with a small nucleus (N) and two kinds of spherules: those with finely granular, densely packed material (1), and those with loosely packed flocculent material (2). Cell contains many long, well-formed mitochondria and possesses a Golgi body (G). × 12,960. **B.** Portion of spherulocyte containing stacks of well-formed rough endoplasmic reticulum (RER), elaborate Golgi body (G), and a thin membrane (M) that can sometimes be seen around spherules (S) in well-fixed material × 28,305. (Courtesy of Dr. A. K. Raina)

sent at a given stage, thus allowing a researcher to test the few types that are present.

Although several different types of hemocytes (mostly PLs, GRs, and ADs) may engage in phagocytosis, no one has critically compared possible differences in this ability using particles of uniform size. It is important to be aware that phagocytosis in vivo is very different from in vitro conditions where some types of cells stick to glass, other disin-

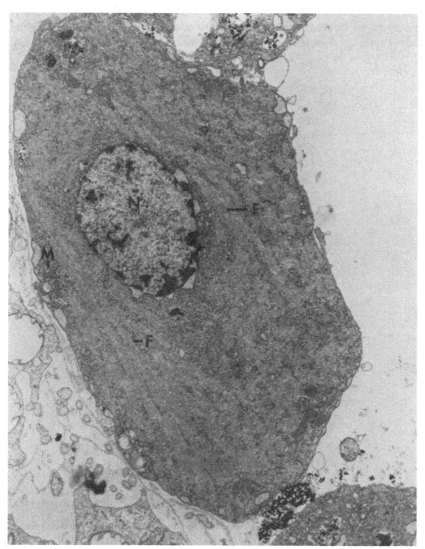

Fig. 10.13. Electron micrograph of oenocytoid of *Calpodes ethlius* larva, showing characteristically eccentric nucleus (*N*), some peripheral mitochondria (*M*), and long, complex, finely granular filamentous inclusions (*F*). Cell contains free ribosomes and some small microtubules. × 8,100. (Courtesy of Dr. Maria Neuwirth)

tegrate, and the washed cells that remain are probably in a state of shock.

We also need to investigate the ability of different kinds of hemocytes to encapsulate objects of uniform sizes and after they have been treated with specific substances known to alter surface properties.

There are at least three different kinds of hemostasis in insects. In one type, a specific cell (the cystocyte) releases material that causes precipitation of the plasma locally around each cell (Grégoire, 1974). This form of hemostasis is "coagulation of the hemolymph." In another types of hemostasis, one or more types of hemocytes may

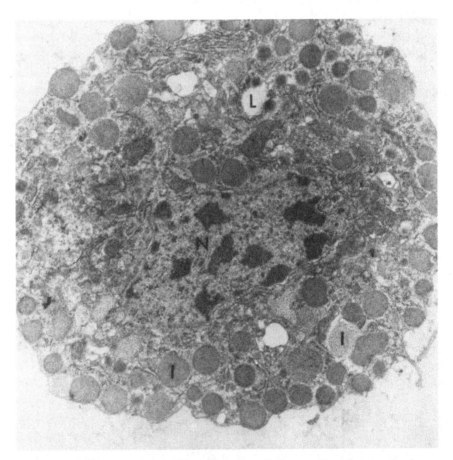

Fig. 10.14. Electron micrograph of hemocyte from *Galleria mellonella* larva that has received at least three different names, each describing a different aspect. It has been referred to as a micronucleocyte because of its relatively small nucleus (*N*), a granulocyte because of its many round granular inclusions (*I*) of various sizes, and an immature adipohemocyte because of the few developing lipid droplets (*L*) and because many of the granules will later develop into lipid droplets. × 16,055. (Courtesy of Dr. Maria Neuwirth)

quickly clump together to physically occlude a wound. This form of hemostasis is termed "cell clotting" (Yeager et al., 1932). It seems probable that in some insects the plasma alone can solidify without the participation of hemocytes. This form of hemostasis is referred to as "hemoplasmal gelation." It can be studied by using chilled insects and centrifuging ice-cold hemolymph to separate the hemoplasma from the hemocytes. Every aspect of hemostasis in insects needs careful restudy.

10.7. Primary pitfalls in studying hemocytes

The primary pitfalls in the study of insect hemocytes are (1) changing the nomenclature and classification system after studying the hemo-

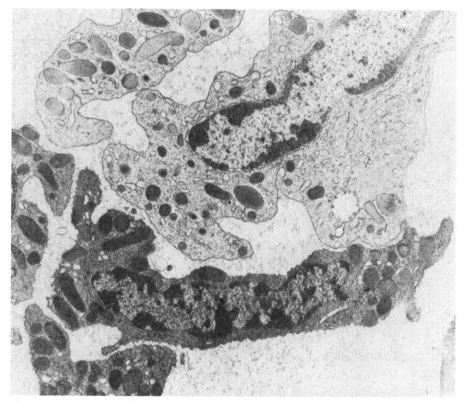

Fig. 10.15. Electron micrograph of hemocytes of *P. americana*. Rowley classifies light-staining upper cells as plasmatocytes and darker-staining cells as cystocytes. Both types have a large and well-formed nucleus, mitochondria, patches of rough endoplasmic reticulum, clear vacuoles, free ribosomes, and large dense granules. Dark cells have more numerous dense granules, many more free ribosomes, more extensive granular endoplasmic reticulum, and fewer vacuoles than light cells. Plasmalemma of dark cells is crenated; that of light cells is smooth. × 10,640. (Courtesy of Dr. A. F. Rowley)

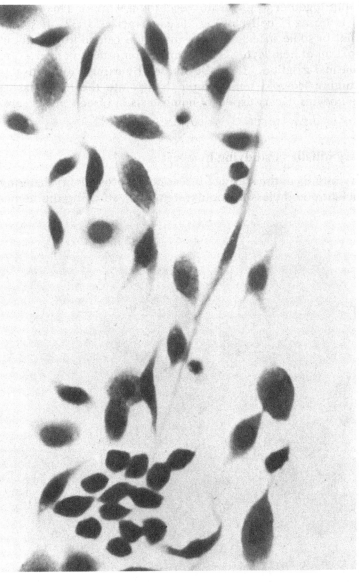

Fig. 10.16. Appearance of some of many types of hemocytes of *Spodoptera eridania* (= *Prodenia eridania*) in air-dried, Wright-stained hemolymph smear of heat-fixed last-stage larva. Smallest cells are prohemocytes; one of them is dividing. Next largest cells were classified by Yeager (1945) as liocytes; they are transitional cells that can be classified with plasmatocytes. Spindle-shaped cells are typical mature plasmatocytes. Large cells filled with vaguely outlined eosinophilic inclusions were classified by Yeager as cystocytes and by Jones (1959) as granulocytes. Most elongated hemocyte is a vermicyte. Hemocyte with three extensions is a podoyte. (Courtesy of Dr. John Arnold)

cytes in only one species or only one stage of many species, (2) studying the cells with only one technique, and (3) making only qualitative observations on cells from insects of unknown stage, age, and physiological status. Other pitfalls include (1) counting too few cells in smears and in hemocytometers, (2) counting or identifying cells at the upper or outer edges of preparations, (3) using inadequate or no fixation cells, (4) mistaking two discrete but superimposed cells for binucleate forms, (5) believing that a simple mass of agglutinated or disintegrating cells represents a single multinucleate cell, (6) identifying sessile clumps of hemocytes in histological sections as hemocytopoietic tissue, and (7) assuming that one type of cell has only one function throughout its own life and the life of the insect.

The hemocytic variability of individual insects is very great, even when the animals have been laboriously selected for uniformity in size, sex, and physiological status. This variability, the many seemingly transitional forms between clear-cut types of cells in the same insect, and the striking differences in cytology with different techniques combine to make studies of hemocytes peculiarly difficult. We are still in need of statistically valid data on hemocyte numbers in relation to their origin from some possible noncirculating source and with reference to specific functions.

10.8. Summary

Hemocytes are differentiated by differences in their anatomy, staining reactions, and activities in vitro and in vivo. They should be studied with more than one technique. The six types of hemocytes that are relatively common among insects include prohemocytes, plasmatocytes, granulocytes, adipohemocytes, spherulocytes, and oenocytoids. Podocytes, vermicytes, and crescent cells occur relatively rarely, some in only a very few species belonging to one or two orders. Hemocytopoiesis refers specifically to the production of circulating hemocytes and thus may refer to mitoses of free or temporarily sessile hemocytes or to the production and release of hemocytes from some compact tissue. Although phagocytic organs, which are present on either side of the heart in the abdomens of some, but not all, orthopteroid insects, may mature and release hemocytes even during adult life, the number of cells present in this tissue seems far too great for this to be the organs' main function, and it seems far more probable that phagocytosis remains their primary function. Although some of the larval Diptera and Lepidoptera may have compact organs that produce and release large numbers of hemocytes as the amount of hemolymph is steadily increasing during development, the size of these tissues steadily grows throughout larval life, and all the cells are suddenly released into the hemolymph shortly before or just after pupation. Ten pitfalls in studying hemocytes are briefly listed.

References

Akai, H., and S. Sato. 1971. An ultrastructural study of the haemopoietic organs of the silkworm, *Bombyx mori. J. Insect Physiol.* 17:1665–76.

Arnold, J. W., and C. F. Hinks. 1976. Haemopoiesis in Lepidoptera. I. The multiplication of circulating haemocytes. *Can. J. Zool.* 54:1003–12.

Arvy, L. 1954. Données sur la leucopoïèsis chez *Musca domestica* L. *Proc. R. Entmol. Soc. Lond.* A 29:39–41.

Dorland, A. W. N. 1974. *The American Illustrated Medical Dictionary.* Saunders, Philadelphia.

Gordon, A. S. (ed.). 1970. *Regulation of Hematopoiesis,* Vol. 2. Appleton, 1974. New York.

Grégoire, Ch. 1974. Hemolymph coagulation, pp. 309–60. *In* M. Rockstein (ed.). *The Physiology of Insecta,* Vol. 5, 2nd ed. Academic Press, New York.

Hinks, C. F., and J. W. Arnold. 1977. Haemopoiesis in Lepidoptera. II. The role of the haemopoietic organs. *Can. J. Zool.* 55:1740–55.

Hinton, H. E., and I. M. Mackerras. 1970. Reproduction and metamorphosis. *In* D. F. Waterhouse (ed.). *The Insects of Australia.* Melbourne University Press, Melbourne.

Hoffmann, J. A. 1972. Modifications of the haemogramme of larval and adult *Lousta migratoria* after selective x-irradiations of the haemocytopoietic tissue. *J. Insect Physiol.* 18:1639–52.

Jones, J. C. 1956. The hemocytes of *Sarcophaga bullata* Parker. *J. Morphol.* 99:233–57.

Jones, J. C. 1962. Current concepts concerning insect hemocytes. *Amer. Zool.* 2:209–46.

Jones, J. C. 1964. The circulatory system of insects, pp. 1–107. *In* M. Rockstein (ed.). *The Physiology of Insecta,* Vol. 3, 1st ed. Academic Press, New York.

Jones, J. C. 1967a. Changes in the hemocyte picture of *Galleria mellonella* (Linnaeus). *Biol. Bull.* (Woods Hole) 132:211–21.

Jones, J. C. 1967b. Normal differential counts of haemocytes in relation to ecdysis and feeding in *Rhodnius. J. Insect Physiol.* 13:1133–41.

Jones, J. C. 1970. Hemocytopoiesis in insects, pp. 7–65. *In* A. S. Gordon (ed.). *Regulation of Hematopoiesis,* Vol. I. Appleton, New York.

Jones, J. C. 1975. Forms and functions of insect hemocytes, pp. 119–28 *In* K. Maramorosch and R. E. Shope (eds.). *Invertebrate Immunity.* Academic Press, New York.

Jones, J. C. 1977. *The Circulatory System of Insects.* Thomas, Springfield, Illinois.

Jones, J. C., and D. P. Liu. 1968. A quantitative study of mitotic divisions of haemocytes of *Galleria mellonella* larvae. *J. Insect Physiol.* 14:1055–61.

Price, C. D., and N. A. Ratcliffe. 1974. A reappraisal of insect haemocyte classification by examination of blood from fifteen insect orders. *Z. Zellforsch. Mikrosk. Anat.* 147:537–49.

Ries, E. 1932. Experimentelle Symbiosestudien. II. Mycetomtransplantationen. *Z. Wiss. Biol.* A 25:184–234.

Wolstenholme, G. E. W., and M. O'Connor (eds.). 1960. *Haemopoiesis.* Little Brown, Boston.

Yeager, J. F. 1945. The blood picture of the southern armyworm (*Prodenia eridania*). *J. Agric. Res.* 71:1–40.

Yeager, J. F., W. E. Shull, and M. D. Farrar. 1932. On coagulation of blood from the cockroach, *Periplaneta orientalis* (Linn.) with special reference to blood smears. *Iowa State Coll. J. Sci.* 6:325–45.

III. Functions

11 Hemocytes and growth in insects

V. B. WIGGLESWORTH

Department of Zoology, University of Cambridge, Cambridge, CB2, 3EJ, U.K

Contents

11.1. Introduction

The purpose of this chapter is to review the contribution of the hemocytes to the processes, and particularly to the control, of insect growth and metamorphosis. But there are certain ancillary matters that must be briefly considered.

11.2. Hemocytes as a source of "embryonic cells"

Hemocytes arise as residual mesodermal cells at the conclusion of embryonic growth. Berlese (1900, 1901) regarded their differentiation to form muscles and other mesodermal tissues as being their main function during postembryonic growth in higher Diptera. Likewise, in *Galerucella* (Col.: Chrysomelidae), Poyarkoff (1910) included the imaginal myoblasts among the hemocytes. Nowadays, although embryonic cells (notably myoblasts, derived from imaginal histoblasts) may be moving freely in the hemocoel at certain stages in development, it is usual not to include them among the hemocytes (Crossley, 1964, 1975). However, Penzlin (1963), studying regeneration of limbs in *Periplaneta*, concluded that the new muscles are formed by blastema cells derived from the hemocytes, not from the epidermis. And there are many claims for fat body formation during postembryonic

303

development from "mesodermal leucocytes" (see Wigglesworth, 1959, and Chapter 1).

11.3. Hemocyte types

There is a vast descriptive literature on the varied types of hemocytes in different groups of insects (Wigglesworth, 1959; Jones, 1962; Gupta, 1969; Crossley, 1975). In recent years the tendency has been to simplify these descriptions and to put the emphasis on a limited number of more or less well-defined types. It is hoped that this procedure will lead to the recognition of true homologies between blood cells of different insects, with the allocation of definite functions to common types of cells.

After studying a range of insects along with other invertebrates, Liebmann (1946) recognized only two main groups of hemocytes originating from a common stem cell: (1) phagocytes (lymphoidocytes) and (2) nonphagocytic cells containing granular or spherical inclusions (trephocytes), the oenocytoids being regarded as perhaps trephocytes that have discharged their inclusions. I have much sympathy with this approach and propose to look at a number of recent descriptions from this point of view.

In *Rhodnius* (Wigglesworth, 1933, 1955b), I described (1) "prohemocytes" (PRs), the universally recognized stem cells; (2) phagocytic amoebocytes of highly varied form, which in common with most authors I would now call "plasmatocytes" (PLs); and (3) nonphagocytic "oenocytoids" (OEs). In addition, I recognized adipocytes, which are very few in number and which I should now regard as free fat body cells, and two very large forms, which I now regard as derivatives of PLs and OEs, both of which are subject to great changes in form during the molting cycle (see below, under "Functions of the Oenocytoids").

The "adipocytes" seen in *Rhodnius* are not to be confused with the adipocytes of Lepidoptera. In *Rhodnius*, the PLs rarely contain droplets of triglyceride, whereas in Lepidoptera such droplets are almost invariably present – and when they become unusually conspicuous the cells are called adipocytes (Shrivastava and Richards, 1965).

The OEs also present some difficulty in terminology. The name was first used by Poyarkoff (1910) and adopted by Hollande (1911) for nonphagocytic cells in Coleoptera and Hemiptera that, after staining with hematoxylin and eosin, have a uniform weakly acidophilic cytoplasm resembling that of small oenocytes similarly treated. In such preparations these cells contrast sharply with the phagocytic PLs, which contain various granular contents (lysosomes, mucopolysaccharide inclusions, vacuoles) as well as basophilic ergastoplasm (ribonucleoprotein) (Wigglesworth, 1973). In the same preparations the OEs show inconspicuous colorless "vacuoles" of varying size, usually in contact with the cell surface. On the other hand, when examined with the electron microscope after staining with uranyl and lead, the OEs have a dense homogeneous cytoplasm that contains very dense spherical inclusions (the colorless "vacuoles" of the light microscope) (Wigglesworth, 1973). They were therefore termed by Lai-Fook (1970) "granulocytophagous hemocytes." These inclusions are also very conspicuous in living cells under phase contrast and the cells are therefore termed "granular cells" (GRs) by Jones (1965), whereas Jones uses the term OEs for the scarce large aberrant forms that commonly lack "granules."

Granulation thus depends on the method of preparation and examination. The PLs of *Rhodnius* also contain "granules" that stain with lipid stains, remain PAS-positive after treatment with saliva, and are often made up of parallel rods or tubules as seen in the electron microscope (Lai-Fook, 1970; Wigglesworth, 1973).

If we assume that the "granular" nonphagocytic hemocytes of other insects correspond to the OEs of Hemiptera and Coleoptera, we shall have come some way toward unification. Thus, Walters (1970) recognizes two predominant classes of hemocytes in Saturniidae: pleomorphic PLs and completely passive "granulocytes" (GRs). Zachary and Hoffmann (1973) recognize in *Calliphora*, in the electron microscope, three cell lineages: PLs, OEs, and thrombocytoids (see below). Hoffmann (1966) describes two cell types in *Locusta;* one containing globules with tubular fascicles (cf. PLs of *Rhodnius*) and the other showing densely staining granules in the electron microscope (cf. OEs of *Rhodnius*). McLaughlin and Allen (1965) in *Anthonomus* recognize PLs (adipohemocytes = ADs) and spherule cells (SPs). Scharrer (1972) in *Periplaneta* goes further and favors the "concept of functional flexibility of one basic cell type rather than a strict classification into definitely separate cellular types."

Coagulocytes (COs) (thrombocytoids), which fragment readily in shed blood, though absent in *Rhodnius* and other insects in which clotting of hemolymph does not occur, form a well-defined type in *Calliphora* (Zachary and Hoffmann, 1973) and in many other insects (Grégoire, 1955; Crossley, 1975). Their relation to the two main types is uncertain, but in *Tenebrio* the granules of the COs have a tubular structure (Stang-Voss, 1970) that suggests a possible relation with the PLs.

As to the naming of the two major hemocytes types, I propose, for the purpose of this review, to employ the widely adopted term PLs for the phagocytic type and to retain the term OEs for the nonphagocytic type I shall be considering in *Rhodnius* because these cells are indubitably those so named by Poyarkoff (1910) and Hollande (1911) and this name has been so used in my earlier publications. This term is not entirely satisfactory because it has been taken by some authors to imply an ontological relation with the oenocytes, which are wholly unrelated cells of ectodermal origin. The name "granulocytes" is unsatisfactory for the reasons set out above. Perhaps Liebmann's noncommittal term "trephocytes" would be suitable – if the cells' general homology in insects can be accepted. For further discussion on hemocyte types, see Chapters 3, 4, 7, 8, 10.

11.4. Phagocytosis and insect growth

At the turn of the century, following upon Metchnikoff's (1892) theory of inflammation and phagocytosis, the phagocytic activity of the hemocytes was considered a prime factor in metamorphosis. This belief was fostered by the classic work of Pérez (1910) on the higher Diptera. However, at the present time it is generally agreed that the blood cells, even in *Calliphora* etc., do not initiate the process of cell death. They are engaged only in the phagocytosis of cytolyzing tissues (e.g., Crossley, 1964).

Even in *Rhodnius* there is a striking cycle of development in the intersegmental abdominal muscles in readiness for each ecdysis, followed by their rapid involution after the molt. But the hemocytes play no part at all in the process of breakdown; they do not even accumulate on the surface of the degenerating muscles (Wigglesworth, 1956b). In the reversible degeneration of flight muscle in adult *Leptinotarsa* (Stegwee et al., 1963), lysosomes in the muscle cells, not external hemocytes, are responsible; and lysosomes are likewise responsible for the degeneration of the abdominal muscles of adult Lepidoptera soon after eclosion (Beaulaton and Lockshin, 1976).

11.5. Hemocytes and growth hormones

It has long been established that the initiation of the molting cycle in insects is brought about by a two-stage process of hormone secretion: Neurosecretory cells in the pars intercerebralis release a prothoracotropic hormone by way of the corpora cardiaca, and this activates the prothoracic gland to secrete ecdysone. I pointed out the possibility that "yet further links in the chain of endocrine organs, located for example in the abdomen" might exist (Wigglesworth, 1955b). I had earlier excluded the dermal glands and the oenocytes (suggested by von Buddenbrock, 1931, and Koller, 1929, as sources of growth hormones) on the grounds that their secretory activity becomes apparent long after the growth changes in the epidermis have begun (Wigglesworth, 1933, 1948). I therefore focused attention on the pericardial cells and the hemocytes.

Because these tissues cannot be readily removed by surgical means,

the experiments consisted in observing the effects of injected substances that are specifically taken up by the cells in question – on the
assumption that cells laden with such substances may thereby be
"blocked" and precluded from performing their other functions.

The experiments were mostly done on the fourth-instar nymph of
Rhodnius. Blocking the pericardial cells with trypan blue at doses that
did not affect the hemocytes caused no delay in molting; heavier
doses of trypan blue, as well as iron saccharate and India ink (India
ink is not accumulated by the pericardial cells), which fill the PLs and
cause them to aggregate, invariably induced a complete arrest of molting when injected during the early days after feeding. This arrest, although complete, was only temporary, the initiation of molting being
delayed for 1–3 weeks (Wigglesworth, 1955b).

This inhibitory effect of agents blocking the hemocytes was 100%
effective at 1 day, 2 days, and 3 days after feeding. At 4 days after feeding it was 0% effective: All the nymphs molted at the normal interval
of 14 days (at 25 °C) after feeding. The PLs appear necessary for some
purpose during the early stages of the molting process. As the inclusions in the PLs showed the same staining reactions as the neurosecretory material of the pars intercerebralis with the chrome-hematoxylin method of Gomori or with the fuchsin-paraldahyde method of
Gabe, and as the PLs were plentiful around the corpus cardiacum and
the prothoracic glands, I at first suspected that the hemocytes might
be responsible for transport of neurosecretion to the prothoracic
glands. But in sections stained with Masson's trichrome, whereas the
neurosecretory substance stains red with the fuchsin, the inclusions,
in common with the connective tissues, stain green (Wigglesworth,
1956a). (It may be noted that on similar grounds Dogra (1970) has proposed that the hemocytes in *Gryllotalpa* transfer neurosecretory material to the target organs, notably during maturation of the ovaries in
the adult female.)

It may be said at once that the role of the PLs in the initiation of
molting is still obscure. Their probable role in the formation of connective tissue is reviewed later in this Chapter. The "critical period"
for arrest by blocking of the hemocytes is not only more sharply defined than the critical period for decapitation or for isolation of the abdomen, but it is a little later in time. In a preliminary note (Wigglesworth, 1955a) it was therefore suggested that under the action of
ecdysone the hemocytes could be the source of a hormone acting
directly on the epidermis. In the light of present knowledge, this hormone might be β-ecdysone (ecdysterone) and the PLs might be a
major site for the conversion of α-ecdysone to this active principle.
The fat body is claimed by Koolman (1976) as the chief site of this conversion – but the PLs are in fact closely associated to a varying extent
with most other tissues, including the fat body.

However, in the completed publication (Wigglesworth, 1955b) it was pointed out that implantation of a fully active prothoracic gland or injection of 1 μg of α-ecdysone per fourth-instar nymph induces immediate continuation of the molting process, which is not delayed by blocking the hemocytes. It was therefore concluded that some substance secreted by the hemocytes is necessary for full activation of the prothoracic glands.

In reviewing the problem a year or two later (Wigglesworth, 1959), I pointed out that "injury of any kind, sufficiently early in the moulting process, caused a delay in growth. In any such injury the hemocytes are involved. Perhaps it is a mechanism which operates through the hemocytes that ensures the consequent arrest of growth; by the blockage of the hemocytes we may be bringing this machinery into action." But this leaves the precise role of the hemocytes quite obscure. This question will be discussed further below, under "Hemocytes and Connective Tissue Formation."

The importance of the hemocytes in the early stages of molting has been confirmed in *Locusta*, in which the hemocytes become aggregated at the level of the dorsal diaphragm to form a rudimentary hemopoietic tissue (Hoffmann, 1968, 1970). X-irradiation of this tissue causes a rapid fall in blood cells (Hoffmann, 1972); and if this is done at an early stage in the last nymphal instar of *Locusta*, the treated nymphs survive for months without ever initiating a molt. The prothoracic glands show histological signs of continued activity and are able to induce molting on transplantation into nymphs deprived of their own glands (Joly et al., 1973; Hoffmann and Weins, 1974). These authors conclude that the radiosensitive cells of the hemopoietic tissue produce a substance which, while having no overt influence on the functioning of the prothoracic glands, is necessary for molting and probably acts in synergy with the hormone from the prothoracic glands. (Incidentally, these authors believe the product of the prothoracic gland is not the steroid ecdysone, but a protein.)

Weir (1970) observed that in the last-instar larva of *Calpodes* (Lepidoptera), whereas the prothoracic glands will normally induce pupation in the headless larva they fail to do so in an isolated thorax. This failure is attributed to the absence of an essential interaction with some component in the abdomen, and the oenocytes are proposed as the most likely source of the abdominal factor. In view of the role of the hemocytes as demonstrated in *Rhodnius* and in *Locusta*, it is possible that they are involved likewise in Lepidoptera. In *Bombyx*, at the time of molting, the hemopoietic organs release all types of hemocytes into the hemolymph (Akai and Sato, 1971). In *Sarcophaga* the failure of the larva to pupate after ligaturing off the posterior segments is attributed by Ohtaki (1972) to some agent derived from this region. This agent could well be the hemocytes liberated from the hemopoietic organs located there (cf. Zachary and Hoffmann, 1973)

11.6. Hemocytes and reproduction

The PLs in Lepidoptera are actively engaged in the uptake of hemolymph by pinocytosis (Grimstone et al., 1967). This activity is very evident in *Rhodnius* and leads to the formation of mucopolysaccharide globules and probably other inclusions (Wigglesworth, 1973). Indeed it has long been assumed that the hemocytes are an active site of protein synthesis. In the cockroach *Nauphoeta* they will take up labeled amino acids and utilize these in protein synthesis, including the production of vitellogenin (Bühlmann, 1974). In *Locusta*, X-irradiation of the hemopoietic tissue affects both the growth of oocytes and the synthesis of hemolymph proteins, effects resembling those of juvenile hormone deficiency (Goltzené and Hoffmann, 1974).

Schmidt and Williams (1953) demonstrated the importance of a heat-sensitive macromolecular (protein) factor in the maturation of sperm in *Hyalophora*, and Kambysellis and Williams (1971) found that this factor is necessary both for spermatogenesis and for the maintenance of insect cells in culture. There is evidence that it is synthesized by some or other of the hemocytes. Likewise, Landureau and Szöllösi (1974) have shown that hemocytes of *Periplaneta* in culture liberate a protein factor that will initiate sperm maturation and spermatogenesis in isolated cysts of saturniids in diapause. This factor is secreted by the hemocytes of both sexes at all stages of the life history.

11.7. Hemocytes in wound healing and encapsulation

Any mechanical injury to the integument of insects results in a great increase in hemocytes in the circulating blood. I even suggested that this increase in hemocyte numbers might be the source of the rise in cyanide-sensitive respiration, which Harvey and Williams (1961) had shown to follow injury in the diapausing pupa of *Hyalophora*. But the increased demand for phagocytosis hardly seems to explain these changes in cell numbers.

Wounds in *Rhodnius* that break through the cuticle and basement membrane cause a rapid aggregation of hemocytes around the injury (Wigglesworth, 1937). The same reaction occurs if a small piece of the integument is excised and the wound sealed with a glass plate or with wax. But in the actual process of repair the hemocytes in *Rhodnius* seem to play only a subsidiary role. The cells that migrate over the gap, always maintaining contact with one another by means of cytoplasmic processes, are epidermal cells; the supply of such cells is met by mitosis in the area round the wound where the cells are depleted and sparse as the result of centripetal migration toward the wound. This process of migration and mitosis continues until epidermal continuity and normal cell density have been restored. If an excised fragment of the integument is implanted in the body cavity of another insect, the cells spread over the outer surface of the cuticle to unite with cells migrating from the other end of the implant to form a closed capsule (Wigglesworth, 1936, 1937). Meanwhile, the PLs merely furnish a protective sheet below the migrating epidermal cells. Whether they play a part in the restoration of the basement membrane will be discussed later.

More recently Bohn (1975) has claimed that in fragments of the *Leucophaea* integument in culture, the hemocytes are engaged in drawing out the epidermal cells mechanically from the margin of the cuticle; he has suggested that the hemocytes serve also as a chemical guide in the closure of the wound.

The vigorous reaction of the hemocytes to a wound in *Rhodnius*, as described above, is seen when the entire integument, including the basement membrane, is broken through. If the cells are killed by local high temperature, without injury to the basement membrane, the surrounding epidermal cells multiply and spread inward to replace the dead cells, but there is no conspicuous response by the hemocytes. The normal intact basement membrane provides a surface that does not excite their activity.

Similarly, as has been brought out particularly by the work of Salt (1968), the hemocytes are highly sensitive to the surfaces of foreign bodies, necrotic tissues, abnormal parasites, etc., all of which evoke aggregation of blood cells. It was observations on the encapsulation of foreign bodies by hemocytes that led Lazarenko (1925) to suggest that the hemocytes may be responsible for the formation of the normal connective tissue sheaths around insect tissues.

11.8. Hemocytes and connective tissue formation

As Pipa and Cook (1958) emphasized, the entire body cavity of an insect is lined by an unbroken connective tissue sheath that invests all the organs and tissues within the body cavity. This sheath forms the basement membrane of the integument, which is reflected over the muscles, nerves, tracheae, etc., when they enter the epidermal compartment. The neural lamella that coats the central nervous system is likewise continuous with the connective tissue sheath.

It seems highly probable that the neural lamella, containing collagen fibrils, is largely a product of the perineural cells that are best regarded as glial cells (Ashhurst, 1968). But it is an attractive idea that the PLs contribute to the connective tissue sheaths throughout the body. That would provide an explanation of the fact that the PLs do not react with and aggregate upon these sheaths; they are not regarded as "foreign," whereas whenever any of the tissue cells themselves become freely exposed the PLs condense upon them.

In *Rhodnius*, the rounded, lozenge-shaped, or flattened inclusions in the PLs show the same staining reactions as the basement membranes: Both stain green with Masson's trichrome, dark blue with chrome-hematoxylin, purple with fuchsin-paraldehyde. Both are stained red after the periodic acid–Schiff reaction (PAS), and this reaction is not eliminated by saliva. Both show some degree of lipid staining with osmic acid and ethyl gallate, and with Sudan black B in 60% pyridine (Wigglesworth, 1956a, and unpub. work; Pipa and Cook, 1958).

These inclusions, stained by the PAS method, appeared in the light microscope to be in process of incorporation into the PAS-positive basement membrane (Wigglesworth, 1956a). In the later stages of the molting process, the PLs insinuate themselves between the fibers of the newly developed muscles, where they appear to be discharging their mucopolysaccharide inclusions to form or add to the sarcolemma (Wigglesworth, 1956a,b). In the electron microscope they can occasionally be seen in process of discharge and apparent fusion with the basement membrane (Wigglesworth, 1973), although more numerous and better electron micrographs of this process are desirable. It is possible, of course, that the hemocytes are not themselves adding to the basement membrane, but are supplying the lipomucopolysaccharide for incorporation by the epidermal cells. But improved sections for the electron microscope provide good evidence of direct membrane secretion by the PLs (Wigglesworth, *Tissue Cell* 11(1), in press).

My belief is that the PLs do contribute to the basement membranes, but they are not the sole source of these membranes. This view is supported by Whitten (1964, 1969) from observations on *Sarcophaga* in which the SPs lose their PAS-positive granules when the connective tissues of the adult are being produced (see Ashhurst, 1968); by Beaulaton (1968) from observations on the hemocytes and the tunica propria of the prothoracic gland in *Antheraea* and *Bombyx* in which the PLs (his "adipohemocytes") contain occasional fat droplets along with mucopolysaccharide granules; and by Scharrer (1972), who noted that in cockroaches the dominant feature of blood cells engaged in the deposition of connective tissue is greatly distended cisternae of the rough endoplasmic reticulum and accumulations of banded fibrils at the interface between cytoplasm and extracellular space.

It is readily understandable that blocking of the PLs with India ink or other foreign matter should interfere with the preparations for laying down the connective tissues. It is at 4 days after feeding (when, as we have seen earlier, molting is no longer delayed by blocking the hemocytes) that the PLs are becoming spread out on the surface of the basement membrane and are no longer caused to aggregate by injection of India ink (Wigglesworth, 1955b).

It is during the period from 4 days to 8–9 days after feeding that the basement membrane undergoes a threefold increase in thickness

(Wigglesworth, 1973). By 9–10 days many of the PLs are forsaking the surface of the basement membrane, and those that remain spread out upon it show a marked reduction in the typical mucopolysaccharide inclusions.

Apolysis and the beginning of cuticle deposition do not occur until the new basement membrane is fully established at about 9 days after feeding. The question is whether the arrest of this process by blocking the PLs with foreign matter is an adequate cause for the arrest of molting, or whether the PLs have some other secretory activity of even greater importance (see also Chapter 12).

11.9. Function of the oenocytoids

In considering the functions of the hemocytes so far in this review, we have referred solely to the PLs. Little is known about the functions of the OEs. I have recently carried out a new study of the changes in form and distribution of the hemocytes in *Rhodnius* during the molting cycle. This study is not yet complete with respect to the OEs; the following account is based on the preliminary results.

While the insect is immersed in ice-cold 2.5% buffered glutaraldehyde, the abdomen is cut along the line of the spiracles. The sternites and tergites are lifted away from the blood-filled gut, fixed in the ice-cold mixture of 1% osmium tetroxide in 5% glutaraldehyde for 1 hr (Hinde, 1971), and studied in whole mounts and in sections for the light and electron microscopes. Whole mounts of the sternites have proved particularly useful for general observations of the blood cells because the hemolymph that occupies the narrow space between the distended gut and the body wall lies undisturbed in small chambers bounded by the longitudinal intersegmental muscles, the transverse nerves, tracheae, and fat body lobes.

At the time of feeding in the fourth-stage larva, after this has fasted for some weeks, both PLs and OEs tend to be aggregated in mixed clumps, particularly along the lateral margin of the abdomen. Both are small, the OEs oval or round (10–12 μm in their longest diameter) without cytoplasmic processes. During the first day of two after feeding, the hemocytes become more evenly dispersed. Mitoses are frequent in the PRs, and forms intermediate between PRs and the two mature forms can be seen. In material fixed as described and stained with Sudan black B, the OEs show a uniform, deep blue-gray cytoplasm with small nucleus; the PLs have a larger nucleus and almost unstained cytoplasm, containing darkly staining rhomboidal or rounded inclusions. This difference in staining is evident in quite small cells; indeed it may be that two forms of PRs with different cytoplasmic staining already exist.

By 4 days after feeding (25 °C), when the PLs are applying themselves everywhere to the basement membrane, the OEs are mainly seen at rest along the tracheae, the nerves, and on the surface of the muscles. Small forms of both types recently derived from PRs are plentiful on those tissues and in the body fluid. Only occasional OEs occur alongside the PLs on the epidermal basement membrane.

By 8–9 days, when the oenocytes, lying below the epidermal cells (distal to the basement membrane of the abdomen) have become greatly enlarged and stain deeply with osmium and Sudan B, the OEs appear in large numbers on the epidermal basement membrane. Each is commonly attached to the membrane by a single slender cytoplasmic process; many of them are becoming enlarged (up to 24 μm). They are most prominent in the areas where the oenocytes are concentrated. Usually each group of oenocytes shows a number of swollen, pedunculated OEs attached to the basement membrane below them.

By 9–10 days the oenocytes begin to contract down. Their lipidstaining contents are being discharged. As this occurs, the trichogen and tormogen cells develop intense lipid staining, and lipid-staining rudiments of the new setae are extruded (Wigglesworth, 1975). Then the wave of lipid staining involves all the epidermal cells – just before the new epicuticle is secreted.

The process of involution of the oenocytes continues up to the time of ecdysis at 14–15 days after feeding. By that time the OEs are widely distributed throughout the tissues and at ecdysis are becoming reduced in size.

It is evident that the OEs are concerned with events taking place during the final stages of the molting process. There are two major events occurring at that time: the incorporation of lipid into the waterproof outer layers of the cuticle and the incorporation of polyphenols in preparation for sclerotization and pigmentation (Wigglesworth, 1975).

With regard to the incorporation of lipid, it has long been supposed that the oenocytes are concerned with the metabolism of lipoprotein or other lipid-containing polymers for the outer layers of the cuticle (Wigglesworth, 1933, 1947, 1970), and this belief has been reinforced by the discovery by Diehl (1975) that in *Schistocerca* synthesis of hydrocarbons for the cuticle takes place in the oenocytes attached to the fat body. This function of the oenocytes must serve the integument of the whole insect; but in *Rhodnius*, as in *Periplaneta*, the oenocytes are confined to the epidermal compartment distal to the basement membrane of the abdomen. The OEs, as we have seen, apply themselves to the basement membrane particularly in areas where oenocytes are concentrated and they increase greatly in size during the period when the oenocytes are dispersing their secretion and undergoing involution. It is possible therefore that the OEs are concerned in the transfer of lipoprotein from the oenocytes of the abdominal wall to supply the integument elsewhere throughout the body.

The incorporation of phenols, notably N-acetyldopamine, which are concerned in sclerotization of the cuticle, takes place for the most part immediately before and shortly after ecdysis, under the action of the neurosecretory hormone (bursicon) that initiates hardening and darkening of the cuticle (Cottrell, 1962; Fraenkel and Hsiao, 1965). There is increasing evidence that the hemocytes are actively involved in this process (Crossley, 1975): They take up tyrosine from the hemolymph and convert this via tyramine to dopamine (Whitehead, 1969), and this process is greatly accelerated in the presence of bursicon (Whitehead, 1970). It is likely that dopamine may be further converted to N-acetyldopamine by the hemocytes before transfer to the cuticle (Mills and Whitehead, 1970). It may well be the OEs (or their equivalent in other insects) that are concerned in this metabolism of phenols for sclerotin formation (see Monpeyssin and Beaulaton, 1977, on *Antheraea*). Indeed, many years ago Dennell (1947) observed that in the larva of *Sarcophaga* certain large oval hemocytes (34–40 μm in length) termed by him "oenocytoids," are the only ones that give a deep blue color with the Nadi reagent. These cells appear in the last larval stage as feeding ceases, and Dennell suggested that they secreted tyrosinase into the blood and then disappeared before pupation. These cells were described under the name "spherule cells" by Jones (1956).

It is not necessary to regard the metabolism of lipids (wax formation) and the metabolism of phenols (sclerotin formation) as mutually exclusive functions. The two processes are intimately associated during cuticle formation (Wigglesworth, 1975); the OEs may well be involved in both.

11.10. Conclusions

From observations on *Rhodnius*, supported by published work on other insects, it is claimed that there are two sorts of hemocytes: (1) phagocytic PLs containing lipomucopolysaccharide inclusions and (2) OEs with uniform lipid-rich cytoplasm (weakly acidophilic in hematoxylin and eosin preparations) with spherical inclusions densely opaque, after metallic staining, in the electron microscope. Equivalent cells in other insects may be variously named OEs, SPs, or GRs. Perhaps Liebmann's (1946) term "trephocytes" would be an acceptable name for this cell type. (The relation of the COs, or thrombocytoids, to these cell types is not discussed.)

The molting process is arrested, temporarily or permanently, if the PLs are put out of action early in the instar, as after overloading the cells with India ink, iron saccharate, etc. in *Rhodnius* or X-irradiation of the hemopoietic aggregations of hemocytes in *Locusta*.

It is suggested that the essential factor, which is eliminated by inactivation of the PLs in *Rhodnius*, may be their contribution to the basement membrane, for this shows a threefold increase in thickness during the critical stage of molting; it may be a necessary prerequisite for deposition of the new cuticle. But other possibilities, such as conversion of α- to β-ecdysone or provision of specific proteins essential for molting, remain open.

New observations on the oenocytoids of *Rhodnius* show that they reach the height of their secretory activity during the period when the oenocytes are liberating their contents and the new epicuticle is being formèd. It is suggested that they may be concerned in the transfer of the lipid secretions of the oenocytes to the integument in other parts of the body and/or that they may be concerned in phenol metabolism preliminary to sclerotization of the cuticle.

It is claimed that in *Rhodnius* and perhaps in most insects there are only two types of hemocytes: phagocytic plasmatocytes and non-phagocytic forms called oenocytoids in *Rhodnius* and "granulocytes" or "spherulocytes" in some other insects.

Inactivation of plasmatocytes early in the molting process (by blocking them with foreign matter or destroying them by X-irradiation of hemopoietic centers) arrests growth temporarily or permanently. Observations in *Rhodnius* suggest that growth changes in the epidermis and cuticle deposition are delayed if the plasmatocytes are not free to contribute to the basement membrane. Other possible functions are reviewed.

The oenocytoids in *Rhodnius* become enlarged and active in the later stages of the molt. It is suggested that they take up the lipid-containing secretion from the oenocytes and distribute it to the epidermal cells throughout the body to provide the lipid components of the new cuticle. They may also perhaps be concerned in the phenol metabolism necessary for sclerotization of the cuticle.

References

Akai, H., and S. Sato. 1971. An ultrastructural study of the haemopoietic organs of the silkworm, *Bombyx mori. J. Insect Physiol. 17*:1665–76.

Ashhurst, D. E. 1968. The connective tissues of insects. *Annu. Rev. Entomol. 13*:45–74.

Beaulaton, J. 1968. La tunica propria et ses relations avec les fibres conjonctives et les hémocytes. *J. Ultrastruct. Res. 23*:474–98.

Beaulaton, J., and R. A. Lockshin. 1976. La dégénérescence cellulaire controlée: Un processus adaptif. *Recherche 7*:172–5.

Berlese, A. 1900. Considerazioni sulla fagocitosi negli insetti metabolici. *Zool. Anz. 23*:441–9.

Berlese, A. 1901. Vorgänge, welche während der Nymphosis der metabolishen Insekten vorkommen. *Zool. Anz.* 24:515–21.

Bohn, H. 1975. Growth-promoting effect of haemocytes on insect epidermis *in vitro. J. Insect Physiol.* 21:1283–93.

von Buddenbrock, W. 1931. Untersuchungen über die Häutungshormon der Schmetterlingsraupen. *Z. Vgl. Physiol.* 14:415–28.

Bühlmann, G. 1974. Vitellogenin in adulten Weibchen der Schabe *Nauphoeta cinerea*: Immunologische Untersuchungen über Herkunft und Einbau. *Rev. Suisse Zool.* 81:642–7.

Cottrell, C. B. 1962. The imaginal ecdysis of blowflies: Detection of the blood-borne darkening factor and determination of some of its properties. *J. Exp. Biol.* 39:413–30.

Crossley, A. C. 1964. An experimental analysis of the origins and physiology of haemocytes in the blue blow-fly *Calliphora erythrocephala* (Meig.). *J. Exp. Zool.* 157:375–98.

Crossley, A. C. 1975. The cytophysiology of insect blood. *Adv. Insect Physiol.* 11:117–221.

Dennell, R. 1947. A study of an insect cuticle: Formation of the puparium of *Sarcophaga falculata* Pand. (Diptera). *Proc. R. Soc.* (B) 134:79–110.

Diehl, P. A. 1975. Synthesis and release of hydrocarbons by the oenocytes of the desert locust, *Schistocerca gregaria. J. Insect Physiol.* 21:1237–46.

Dogra, G. S. 1970. Functional significance of the neurosecretory material in the haemocytes of the adult female *Gryllotalpa africana* Beauvois (Orthoptera: Gryllotalpidae). *Anat. Anz.* 126:355–62.

Fraenkel, G., and C. Hsiao. 1965. Bursicon, a hormone which mediates tanning of the cuticle in the adult fly and other insects. *J. Insect Physiol.* 11:513–56.

Goltzené, F., and J. A. Hoffmann. 1974. Control of haemolymph protein synthesis and oocyte maturation by the corpora allata in female adults of *Locusta migratoria* (Orthoptera): Role of the blood-forming tissue. *Gen. Comp. Endocrinol.* 22:489–98.

Grégoire, Ch. 1955. Blood coagulation in Arthropods. V. Studies on haemolymph coagulation in 420 species of insects. *Arch. Biol.* 66:103–48.

Grimstone, A. V., S. Rotheram, and G. Salt. 1967. An electron-microscope study of capsule formation by insect blood cells. *J. Cell Sci.* 2:281–92.

Gupta, A. P. 1969. Studies of blood of Meloidae (Coleoptera). I. The haemocytes of *Epicauta cinerea* (Forster), and a synonymy of haemocyte terminologies. *Cytologia* 34:300–44.

Harvey, W. R., and C. M. Williams. 1961. The injury metabolism of the Cecropia silkworm. I. Biological amplification of the effects of localized injury. *J. Insect Physiol.* 7:81–99.

Hinde, R. 1971. The fine structure of the mycetome symbiotes of the aphids *Brevicoryne brassicae*, *Myzus persicae*, and *Macrosiphum rosae. J. Insect Physiol.* 17:2035–50.

Hoffmann, J. A. 1966. Etude ultrastructurale de deux hémocytes à granules de *Locusta migratoria* (Orthoptère). *C. R. Acad. Sci. Paris* 263:521–4.

Hoffmann, J. A. 1968. Présence d'un tissue hématopöiétique au niveau du diaphragme dorsal de *Locusta migratoria. C. R. Acad. Sci. Paris* 266:1882–3.

Hoffmann, J. A. 1970. Les organes hématopöiétique de deux insectes orthoptères: *Locusta migratoria* et *Gryllus bimaculatus. Z. Zellforsch. Mikrosk. Anat.* 106:451–72.

Hoffmann, J. A. 1972. Modifications of the haemogramme of larval and adult *Locusta migratoria* after selective X-irradiations of the haemocytopoietic tissue. *J. Insect Physiol.* 18:1639–52.

Hoffmann, J. A., and M. J. Weins. 1974. Activité protéosynthétique des glandes prothoraciques et titre d'ecdysone chez les larves permanentes de *Locusta migratoria* obtenues par irradiation sélective du tissu hématopoïeique. *Experientia* 30:821–2.

Hollande, A. C. 1911. Etude histologique comparée du sang des insectes à hémorrhée et des insectes sans hémorrhée. *Arch. Zool. Exp. Gén.* 6:283–323.

Joly, L., M. J. Weins, J. A. Hoffmann, and A. Porte. 1973. Evolution des glandes prothoracique de larves permanentes de *Locusta migratoria* obtenues par irradiation sélective du tissu hématopoïétique. *Z. Zellforsch. Mikrosk. Anat.* 137:387–97.

Jones, J. C. 1956. The hemocytes of *Sarcophaga bullata. J. Morphol.* 99:233–58.

Jones, J. C. 1962. Current concepts concerning insect hemocytes. *Amer. Zool.* 2:209–46.

Jones, J. C. 1965. The hemocytes of *Rhodnius prolixus. Biol. Bull. (Woods Hole)* 129:282–94.

Kambysellis, M., and C. M. Williams. 1971. *In vitro* development of insect issues. I. A macro-molecular factor prerequisite for silkworm spermatogenesis. *Biol. Bull. (Woods Hole)* 141:527–40.

Koller, G. 1929. Die innere Sekretion bei wirbellosen Tieren. *Biol. Rev. (Cambridge)* 4:269–306.

Koolman, J. 1976. Ecdysone oxidase, an enzyme of the ecdysone metabolism in insects, pp. 403–412. *In* M. Durchon (ed.). *Actualités sur les hormones d'invertébrés.* Colloq. Int. C.N.R.S. No. 251, Paris.

Lai-Fook, J. 1970. Haemocytes in the repair of wounds in an insect (*Rhodnius prolixus*). *J. Morphol.* 130:297–314.

Landureau, J. C., and A. Szöllösi. 1974. Démonstration, par la méthode de culture *in vitro*, du rôle des hémocytes dans la spermatogénèse d'un insecte. *C. R. Acad. Sci. Paris* 278:3359–62.

Lazarenko, T. 1925. Beiträge zur vergleichenden Histologie des Blutes und des Bindegewebes. 2. Die morphologische Bedeutung der Blut- und Bindegewebeelemente der Insekten. *Z. Mikrosk. Anat. Forsch.* 3:409–99.

Liebmann, E. 1946. On trephocytes and trephocytosis: A study of the role of leucocytes in nutrition and growth. *Growth* 10:291–330.

McLaughlin, R. E., and G. Allen. 1965. Description of hemocytes and the coagulation process in the boll weevil, *Anthonomus grandis* Boheman (Curculionidae). *Biol. Bull. (Woods Hole)* 128:112–24.

Metchnikoff, E. 1892. *Leçons sur la pathologie comparée de l'inflammation.* Masson, Paris.

Mills, R. R., and D. L. Whitehead. 1970. Hormonal control of tanning in the American cockroach: Changes in blood cell permeability during ecdysis. *J. Insect Physiol.* 16:331–40.

Monpeyssin, M., and J. Beaulaton. 1977. Données sur la localisation ultrastructurale d'une activité phénol-oxydasique dans les hémocytes circulants d'*Antheraea pernyi* au dernier âge larvaire. *J. Insect Physiol.* 23:939–43.

Ohtaki, T. 1972. A possible role of the posterior half of larval body on *Sarcophaga peregrina. Jpn. J. Med. Sci. Biol.* 25:33–41.

Penzlin, H. 1963. Ueber die Regeneration bei Schaben (Blattaria). I. Das Regenerationsvermögen und die Genese des Regenerats. *Wilhelm Roux' Arch. Entwickl. Mech. Org.* 154:434–65.

Pérez, C. 1910. Recherches histologiques sur la métamorphose des Muscides (*Calliphora erythrocephala* Mg.). *Arch. Zool. Exp. Gén.* (5 Sér.) 4:1–274.

Pipa, R. L., and E. F. Cook. 1958. The structure and histochemistry of the connective tissue of the sucking lice. *J. Morphol.* 103:353–85.

Poyarkoff, E. 1910. Recherches histologiques sur la métamorphose d'un coléoptère (la Galeruque de l'orme). *Arch. Anat. Microsc.* 12:333–474. ·

Salt, G. 1968. The resistance of insect parasitoids to the defence reactions of their hosts. *Biol. Rev. (Cambridge)* 43:200–32.

Scharrer, B. 1972. Cytophysiological features of haemocytes in cockroaches. *Z. Zellforsch. Mikrosk. Anat.* 129:301–19.

Schmidt, E. L., and C. M. Williams. 1953. Physiology of insect diapause. V. Assay of the growth and differentiation hormone of Lepidoptera by the method of tissue culture. *Biol. Bull. (Woods Hole)* 105:174-87.

Shrivastava, S. C., and A. G. Richards. 1965. An autoradiographic study of the relation between haemocytes and connective tissues in the wax moth, *Galleria mellonella* L. *Biol. Bull. (Woods Hole)* 128:337-45.

Stang-Voss, C. 1970. Zur Ultrastruktur der Blutzellen wirbelloser Tiere. I. Ueber die Haemocyten der Larve des Mehlkafers *Tenebrio molitor* L. *Z. Zellforsch. Mikrosk. Anat.* 103:589-605.

Stegwee, C., E. C. Kimmel, J. S. der Boer, and S. Henstra. 1963. Hormonal control of reversible degeneration of flight muscle in the Colorado beetle, *Leptinotarsa decemlineata* Say (Coleoptera). *J. Cell Biol.* 19:519-27.

Walters, D. R. 1970. Haemocytes of Saturniid silkworms: Their behaviour *in vivo* and *in vitro* in response to diapause, development, and injury. *J. Exp. Zool.* 174:441-50.

Weir, S. B. 1970. Control of moulting in an insect. *Nature (Lond.)* 228:580-1.

Whitehead, D. L. 1969. New evidence for the control mechanism of sclerotization in insects. *Nature (Lond.)* 224:721-3.

Whitehead, D. L. 1970. L-dopa decarboxylase in the haemocytes of Diptera. *FEBS Lett.* 7:263-6.

Whitten, J. M. 1964. Haemocytes and the metamorphosing tissues in *Sarcophaga bullata, Drosophila melanogaster*, and other cyclorrhaphous Diptera. *J. Insect Physiol.* 10:447-60.

Whitten, J. M. 1969. Haemocyte activity in relation to epidermal cell growth, cuticle secretion, and cell death in a metamorphosing cyclorrhaphan pupa. *J. Insect Physiol.* 15:763-78.

Wigglesworth, V. B. 1933. The physiology of the cuticle and of ecdysis in *Rhodnius prolixus* (Triatomidae, Hemiptera), with special reference to the function of the oenocytes and of the dermal glands. *Q. J. Microsc. Sci.* 76:269-318.

Wigglesworth, V. B. 1936. The function of the corpus allatum in the growth and reproduction of *Rhodnius prolixus* (Hemiptera). *Q. J. Microsc. Sci.* 79:91-121.

Wigglesworth, V. B. 1937. Wound healing in an insect, *Rhodnius prolixus* (Hemiptera). *J. Exp. Biol.* 14:364-81.

Wigglesworth, V. B. 1947. The epicuticle in an insect, *Rhodnius prolixus* (Hemiptera). *Proc. R. Soc. B.* 134:163-81.

Wigglesworth, V. B. 1948. The structure and deposition of the cuticle in the adult mealworm, *Tenebrio molitor* L. (Coleoptera). *Q. J. Microsc. Sci.* 89:197-217.

Wigglesworth, V. B. 1955a. The endocrine chain in an insect. *Nature (Lond.)* 175:338.

Wigglesworth, V. B. 1955b. The role of the haemocytes in the growth and moulting of an insect, *Rhodnius prolixus* (Hemiptera). *J. Exp. Biol.* 32:649-63.

Wigglesworth, V. B. 1956a. The haemocytes and connective tissue formation in an insect, *Rhodnius prolixus* (Hemiptera). *Q. J. Microsc. Sci.* 97:89-98.

Wigglesworth, V. B. 1956b. Formation and involution of striated muscle fibres during the growth and moulting cycles of *Rhodnius prolixus* (Hemiptera). *Q. J. Microsc. Sci.* 97:465-80.

Wigglesworth, V. B. 1959. Insect blood cells. *Annu. Rev. Entomol.* 4:1-16.

Wigglesworth, V. B. 1970. Structural lipids in the insect cuticle and the function of the oenocytes. *Tissue Cell* 2:155-79.

Wigglesworth, V. B. 1973. Haemocytes and basement membrane formation in *Rhodnius*. *J. Insect Physiol.* 19:831-44.

Wigglesworth, V. B. 1975. Incorporation of lipid into the epicuticle of *Rhodnius* (Hemiptera). *J. Cell Sci.* 19:459-85.

Zachary, D. and J. A. Hoffmann. 1973. The haemocytes of *Calliphora erythrocephala* (Meig.) (Diptera). *Z. Zellforsch. Mikrosk. Anat.* 141:55-73.

12 Hemocytes and connective tissue: a critical assessment

DOREEN E. ASHHURST

Department of Structural Biology, St George's Hospital Medical School, University of London, London, SW17 ORE, U.K.

Contents

12.1. Introduction   *page* 319
12.2. Connective-tissue-synthesizing cells in vertebrates   320
12.3. Connective-tissue-synthesizing cells in insects   321
    12.3.1. Cells that produce fibrous tissue
    12.3.2. Basement membrane formation
12.4. Concluding remarks   327
12.5. Summary   328
References

12.1. Introduction

There has been much speculation in the past about the possible role of hemocytes in the production of the connective tissues in insects. At the time hemocytes were first implicated (Lazarenko, 1925; Wermel, 1938), our knowledge of insect connective tissues was rudimentary. It is only since 1955 that information about the biochemistry and fine structure of these tissues has accumulated.

Before insect connective tissues and the hemocytes are discussed in more detail, it is pertinent to consider current concepts about connective tissues in other animals, especially those of mammals and birds. Any notion that these are irrelevant to this discussion must be immediately dispelled. Recent work on insect tissues has provided much evidence of their biochemical, structural, and functional similarities to the tissues of higher animals.

The term "connective tissue" is currently used to include all the material found between the cells, whether it is that filling the large spaces in matrixes such as bone or cartilage or the small narrow spaces between cells. The constituents of the matrix are collagenous proteins, other glycoproteins, and glycosaminoglycans (formally termed acid mucopolysaccharides). The detailed structure of collagen molecules is now almost fully known (Miller, 1976); at least four molecular species of collagen exist in vertebrates. Type I collagen is found in bone, skin, and tendon. Type II collagen occurs in cartilage. Type III collagen

319

forms the thin fibrils associated with the loose networks of connective tissue supporting organs such as the spleen, and in embryonic skin and Type IV is the nonfibrous basement membrane collagen. The other glycoproteins are not yet well characterized. The glycosaminoglycans, hyaluronic acid, chondroitin 4- and 6-sulfates, dermatan sulfate, and keratan sulfate are all found in the extracellular matrixes. The chondroitin, dermatan, and keratan sulfates, in varying proportions, are always found with fibrous collagen and appear to be necessary for fibril formation (Mathews, 1975). Basement membrane collagen is alone is not being associated with glycosaminoglycans (Kefalides, 1971).

Examination of insect tissues by electron microscopy has revealed that basement membranes and typical connective tissues containing collagen fibrils and, in some instances, a kind of elastic fiber as well (Locke and Huie, 1972), are of widespread occurrence in insects (Ashhurst, 1968). The number of biochemical studies are limited owing to the very small amounts of tissue available. It is becoming apparent, nevertheless, that the collagen molecules of locust tissues are similar in length and amino acid sequence to those of rat-tail-tendon collagen (Ashhurst and Bailey, unpub. observ.).

There is also histochemical evidence that glycosaminoglycans are present in the tissues containing collagen fibrils (Ashhurst and Costin, 1971a–c). It is relevant to the subsequent discussion to point out that the glycosaminoglycans can bind alcian blue under specific conditions, but do not give a positive reaction with the standard periodic acid–Schiff (PAS) test. The positive PAS reaction of connective tissue matrixes is attributable to the presence of various glycoproteins. The basement membranes do not bind alcian blue, but give a positive PAS reaction (see Chapter 22 for the rationale of these histochemical reactions).

12.2. Connective-tissue-synthesizing cells in vertebrates

The constituents of the matrixes are, for the most part, produced by the fibroblasts or their more specialized descendants in bone and cartilage, osteoblasts and chondroblasts (Ross, 1969). Fibroblasts are of mesodermal origin. When a cell is actively synthesizing collagen and glycosaminoglycans, it has a characteristic appearance (Fig. 12.1). The rough endoplasmic reticulum (RER) is very dilated and the cisternae contain an electron-dense, amorphous material. The Golgi complexes are irregular and consist of vacuoles, rather than organized lamellae. Secretion granules are not found because the constitutents of the matrix are passed into the intercellular space as soon as they are synthesized; there is no storage phase.

The basement membranes under epithelial cells or around muscle fibers are secreted, not by fibroblasts, but by the cells with which they are associated (Hay and Dodson, 1973). All epithelial cells can secrete basement membrane (type IV) collagen, as can muscle cells also.

The concept that the ability to synthesize collagen is a property shared by most cells is rapidly gaining ground. It is well established that smooth muscle cells secrete collagen and that they acquire the typical cytological features of fibroblasts while this synthesis occurs (Ross and Klebanoff, 1971; Gerrity et al., 1975). Evidence of collagen synthesis by other cell types, such as hepatocytes and cells of the notochord during development (Lauscher and Carlson, 1975; Seyer et al., 1977), is also accumulating. Thus collagen synthesis is a fundamental property of many cell types with differing embryonic origin; It is not confined to specialized mesodermal cells.

12.3. Connective-tissue-synthesizing cells in insects

The purpose of this chapter is to discuss the evidence from which it has been deduced that hemocytes might secrete the constituents of some of the connective tissue matrixes in insects. The fibrous tissues will be considered first, the basement membranes later.

In order to prove that a particular cell is secreting the collagen and glycosaminoglycans of the surrounding matrix, certain criteria must be met. In insects the matrixes are often produced over a very short period of time, and it must be shown that the cells are present at this time and that they display the typical morphology of a cell actively secreting collagen and glycosaminoglycans; that is a well-developed, dilated RER and active Golgi complexes, but no secretory granules.

12.3.1. Cells that produce fibrous tissue

Very few insect tissues have been examined with the primary aim of deducing which cells produce the connective tissue matrix. In all instances, such studies have demonstrated that cells other than hemocytes are involved. During pupation in the Lepidoptera, the larval connective tissues around the central nervous system are removed to allow tissue reorganization to occur; later, new connective tissues are laid down. The hemocytes, notably the adipohemocytes (ADs) in *Galleria mellonella* and *Manduca sexta* (Ashhurst and Richards, 1964a; Pipa and Woolever, 1965; McLaughlin, 1974), are involved in the breakdown of the larval tissues, but the connective tissue around the adult nervous system is produced by the glial cells of the perineurium. These cells form a layer around the central nervous system, and in the abdominal region they proliferate to form a mass on the dor-

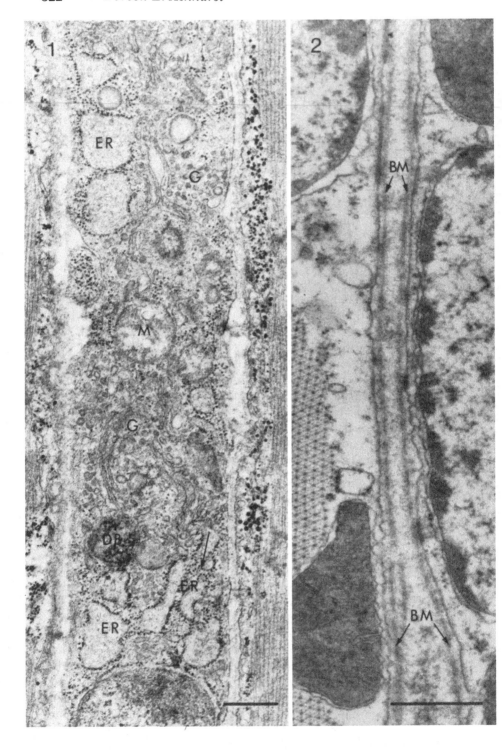

sal side of the connectives. These cells do not resemble any of the hemocytes of *Galleria* described by Ashhurst and Richards (1964b) or Neuwirth (1973), nor are they derived from hemocytes (Shrivastava and Richards, 1965). An occasional AD is seen trapped among the cells of the dorsal mass in *Galleria* or near the neural lamella in *Manduca* (Ashhurst, unpub. observ.; McLaughlin, 1974), but these do not display the morphological features of synthetic cells. At the time of active synthesis of the connective tissue matrix, the glial cells possess large aggregations of RER associated 'with very small Golgi complexes, but much of the cytoplasm is occupied by glycogen (Ashhurst and Costin, 1976). That these cells produce the collagen of the matrix is shown conclusively by the fact that they take up ³H-proline, which is later found incorporated into the collagen (Ashhurst and Costin, 1976). In a similar way, the cells of the perineurium of *Schistocerca gregaria* produce the neural lamella of the central nervous system; at the time of its development in the late embryo, no hemocytes are seen in contact with the central nervous system and the perineurial cells possess the dilated RER typical of a fibroblast (Ashhurst, 1965). The involvement of the perineurial cells in the production of the neural lamella was first suggested by Scharrer (1939) on circumstantial evidence. It is of interest that the perineurial cells are derived from the neural tissue during the initial differentiation of the neurons and hence are of ectodermal origin.

Another example is provided by the ejaculatory duct of the adult male locust, *Locusta migratoria*. The cells that produce the thick layer of connective tissue around the adult duct are first found lying around the duct in the second instar. They remain in this position, proliferate, and produce a small amount of connective tissue in the fourth and fifth instars, but most of the collagen and glycosaminoglycan is produced just after the adult molt (Ashhurst and Costin, 1971a, 1974). The cells in the young adult display all the features typical of a fibroblast (Fig. 12.3) and have been shown to incorporate ³H-proline into the collagen (Ashhurst and Costin, 1976). It is thought that these cells are of mesodermal origin. They are sessile and possess none of the typical structural or histochemical characteristics of locust hemocytes (Costin, 1975). Yet another example is provided by the colla-

Fig. 12.1. Part of a fibroblast between two muscle fibers of a young chicken. Fibroblast is elongated owing to its position. Endoplasmic reticulum (*ER*) is dilated and contains electron-dense material. In some places (arrow) ribosomes are arranged as spiral polyribosomes. Golgi complexes (*G*) are large, but unorganized. Mitochondria (*M*) and dense bodies (*DB*), which are probably lysosomes, are also present in cytoplasm. Bar equals 0.5 μm.

Fig. 12.2. Basement membranes (*BM*, arrows) between two fibrillar flight muscle fibers of *Lethocerus maximus*. Bar equals 0.5 μm.

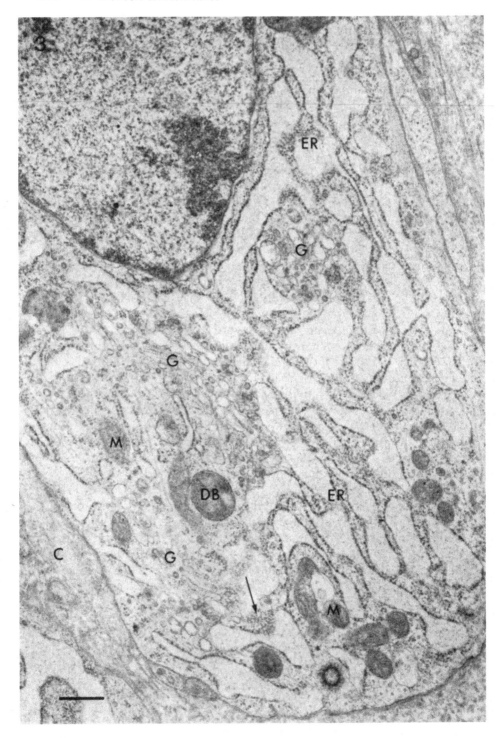

genous endoskeletal structures of the fire brat, *Thermobia domestica*, which are secreted by typical fibroblasts present within the connective tissue (François, 1977).

The cells of the prothoracic glands and corpora allata of cockroaches are separated by layers of fibrous connective tissue (stroma), which may be quite wide. Cells, described as sessile hemocytes by Scharrer (1972), may be found in this tissue. She suggests that this spatial relationship is "an expression of the active participation by these cells in the formation and reorganisation of the a cellular connective tissue framework which involve synthesis, maintenance, and breakdown of parts thereof." Further evidence for their involvement in the synthesis of the matrix is provided by a micrograph (Scharrer, 1972, Fig. 11) that shows a small part of one cell with dilated RER and a larger area of another cell with many mitochondria, little RER, and a discontinuous plasma membrane. The former cell appears to possess one of the features typical of a "fibroblast-like" cell, but the other does not. In the early 1960s, many fibroblasts with discontinuities in the plasma membrane were described, but with the advent of glutaraldehyde fixation, such discontinuities were recognized as artifacts of preparation. A second micrograph (Scharrer, 1972, Fig. 12) shows collagen fibrils apparently within the cytoplasm of a hemocyte. Unfortunately, the preservation of the membrane surrounding the fibrils is poor, and it is impossible to determine whether they are extracellular, i.e., in a depression in the cytoplasm such as occurs in irregularly shaped cells, or in a vacuole within the cell. It should be remembered that hemocytes are involved in connective tissue breakdown, and intracellular fibrils have been observed in such instances (Pipa and Woolever, 1965). A further observation is pertinent. Insects are peculiar in the relationship of the blood to the tissues. In vertebrates, many cells, such as macrophages, lymphocytes, and mast cells, move about the connective tissues, especially to sites of trauma. In all instances, they must move through the connective tissue matrix. Thus close proximity to the matrix does not imply an involvement in its production.

The tunica propria around the prothoracic gland of *Antheraea pernyi* was examined during the larval instars by Beaulaton (1968). The tunica consists of an amorphous material in which fibrils of 30 nm diameter are embedded; it is not a basement membrane (see below). At the time when the tunica propria increases in thickness in the fourth

Fig. 12.3. Part of a fibroblast from connective tissue around ejaculatory duct of a 6-day adult locust. Endoplasmic reticulum (*ER*) is dilated and contains electron-dense material; spiral polyribosomes are present (arrow). Golgi complexes (*G*) consist of an unorganized array of small vesicles. Mitochondria (*M*) and dense bodies (*DB*), which are probably lysosomes, are also present in cytoplasm. Collagen fibrils (*C*) are present in matrix. Bar equals 0.5 μm.

and fifth instars, ADs with inclusions containing an electron-dense, flocculent material are nearby and are seen to discharge this material near the tunica propria. The evidence that this material is incorporated into the tunica propria is based on the histochemical evidence that both are PAS-positive, but that whereas the tunica propria gives a negative reaction to the p-dimethylaminobenzaldehyde test for tryptophan, the material in the inclusions is strongly positive. Neither binds alcian blue, but the staining conditions are not given (see Chapter 22). Because the tunica propria contains collagen fibrils, more rigorous tests for glycosaminoglycans should be performed. Thus, the only common histochemical property is PAS positivity, which indicates merely the presence of *vic*-glycol groups, i.e., a carbohydrate-containing material, in both sites. In addition, the ADs do not display the morphological features of a connective-tissue-synthesizing cell. Further study is necessary to resolve the possible relationship of the tunica propria and the ADs.

12.3.2. Basement membrane formation

A basement membrane is a thin amorphous layer, between 50 and 80 nm thick, which is found under the basal surface of the epithelial cells and around muscle cells. Its density increases away from the cell membrane, giving the appearance of a gap between the cell and basement membrane. In vertebrates, it is secreted by the cells next to which it is found (Hay and Dodson, 1973). True basement membranes are not often encountered in insects; frequently they occur as a preliminary stage in the formation of a thicker fibrous sheath (Ashhurst, 1965). A typical basement membrane is, however, found around the muscle fibers of the belostomatid water bugs (Ashhurst, 1967) (Fig. 12.2). No further connective tissue and no hemocytes or fibroblast-like cells were observed between the muscle fibers of the water bugs. The precise classification of basement membranes is recent (Kefalides, 1971), and hence the thin layers of connective tissue containing banded fibrils that were included in this category by Ashhurst (1968) should no longer be regarded as basement membranes.

In light of the above, the interpretation of Wigglesworth's (1956, 1973) findings is difficult. The layers of connective tissue he describes under the epidermal cells of *Rhodnius prolixus* are between 150 and 500 nm in thickness. They appear to be fibrillar, though banded fibrils cannot be clearly observed. On both these criteria, their characterization as basement membranes is dubious; they might more properly be described simply as layers of fibrous connective tissue. He suggests that the plasmatocytes (PLs) discharge the contents of large granules onto the surface of the connective tissue and so contribute to its thickening. The similarity of the material in the granules and that of the connective tissue is based on their shared positive reactions to the PAS test and further tests for glycosaminoglycans were performed. The PLs, however, do not have the characteristic abundant RER of an actively synthesizing cell; indeed, it is suggested that the material in the inclusions is taken up in small amounts from the hemolymph by pinocytosis and that the large inclusions are formed by the repetitive fusion of small vesicles. This is not a process normally associated with the production of a connective tissue. Jones (1976) found no evidence for the participation of hemocytes in basement membrane formation in *Rhodnius* and several other insects (see also Chapter 11).

12.4. Concluding remarks

The main thesis I have tried to develop in this chapter is that insect connective tissues are essentially similar to those of all other animals, and therefore it is logical to assume that they are produced in a similar manner by cells with a similar cytological structure. That this is in fact so has been demonstrated by several examples, in particular the locust ejaculatory duct. Connective tissue materials are always synthesized de novo before they are secreted into the extracellular spaces; hence, the cells concerned must have the cytological features of protein-synthesizing cells, that is, they must possess much RER and active Golgi complexes. The products are not stored within the cell to await secretion on an appropriate stimulus, but are passed directly into the matrix; hence, large secretory granules would not be expected to contain connective tissue substances. The final posttranscriptional modifications of collagen molecules are, in fact, known to occur after the procollagen molecules have passed into the extracellular space. The only examples of storage are found in certain pathological conditions in which glycosaminoglycans are present in intracellular granules (Danes et al., 1970).

It is on morphological criteria such as those in the preceding paragraph that it is suggested that the involvement of hemocytes in connective tissue formation has not so far been proved. Much more information is required before the similarity of the material in inclusions and the nearby connective tissues is established. As has been stressed, PAS positivity means little except that the material possesses carbohydrate moieties.

It may be wondered why I have not discussed the work of Lazarenko (1925) and Wermel (1938), who considered that the hemocytes lay down connective tissue. At the time of their work, little was known about insect connective tissues, and the techniques available for study were restricted, so that their conclusions, though justified at that time, would not be borne out by similar studies done now.

Lazarenko investigated the reaction to foreign implants; it has since been shown by many workers that the capsules formed around foreign bodies are cellular (e.g., Grimstone et al., 1967; Brehélin et al., 1975; François, 1975). On aging, the hemocytes in the capsule may become necrotic or melanized, and it was the appearance of such capsules in light microscope preparations that led to the idea that the hemocytes might transform into connective tissue. It is now obvious that this is impossible. The process that occurs in the cells of aging capsules is probably akin to the keratinization of mammalian epidermal cells. Small amounts of extracellular material are present in the narrow spaces between the hemocytes in a capsule. This material has not been characterized, but it is quite consistent with the arguments advanced here that hemocytes on differentiation into the flattened cells of a capsule secrete their own basement membrane or other extracellular material, such as heparan sulfate, which is associated with many cell surfaces.

The conclusion reached is that there is no indisputable evidence for the production of the collagen, glycosaminoglycans, or glycoproteins of the connective tissues by circulating or temporarily sessile hemocytes.

12.5. Summary

It has been suggested that hemocytes move to areas of developing connective tissues and become involved in the production of the connective tissue matrix. These findings are reviewed in the light of our present knowledge of the similarities between insect and vertebrate connective tissues. It is concluded that there is not as yet any indisputable evidence that hemocytes contribute collagenous protein, glycosaminoglycans, or glycoproteins to the matrixes.

References

Ashhurst, D. E. 1965. The connective tissue sheath of the locust nervous system: its development in the embryo. *Q. J. Microsc. Sci.* 102:61–73.

Ashhurst, D. E. 1967. The fibrillar flight muscles of giant water-bugs: an electron microscope study. *J. Cell Sci.* 2:435–44.

Ashhurst, D. E. 1968. The connective tissues of insects. *Annu. Rev. Entomol.* 13:45–74.

Ashhurst, D. E., and N. M. Costin. 1971a. Insect mucosubstances. I. The mucosubstances of developing connective tissue in the locust, *Locusta migratoria*. *Histochem. J.* 3:279–95.

Ashhurst, D. E., and N. M. Costin. 1971b. Insect mucosubstances. II. The mucosubstances of the central nervous system. *Histochem. J.* 3:297–310.

Ashhurst, D. E., and N. M. Costin. 1971c. Insect mucosubstances. III. Some mucosubstances of the nervous systems of the wax-moth (*Galleria mellonella*) and the stick insect (*Carausius morosus*). *Histochem. J.* 3:379–87.

Ashhurst, D. E., and N. M. Costin. 1974. The development of a collagenous connective tissue in the locust, *Locusta migratoria*. *Tissue Cell* 6:279–300.

Ashhurst, D. E., and N. M. Costin. 1976. The secretion of collagen by insects: Uptake of ³H-proline by collagen-synthesizing cells in *Locusta migratoria* and *Galleria mellonella. J. Cell Sci. 20:*377–403.

Ashhurst, D. E., and A. G. Richards. 1964a. A study of the changes occurring in the connective tissue associated with the central nervous system during the pupal stage of the wax-moth, *Galleria mellonella* L. *J. Morphol. 114:*225–36.

Ashhurst, D. E., and A. G. Richards. 1964b. Some histochemical observations on the blood cells of the wax-moth, *Galleria mellonella* L. *J. Morphol. 114:*247–54.

Beaulaton, J. 1968. Etude ultrastructurale et cytochimique des glandes prothoraciques de vers à soie aux quatrième et cinquième âges larvaires. I. La tunica propria et ses relations avec les fibres conjonctives et les hémocytes. *J. Ultrastruct. Res. 23:*474–98.

Brehélin, M., J. A. Hoffmann, G. Matz, and A. Porte. 1975. Encapsulation of implanted foreign bodies by hemocytes in *Locusta migratoria* and *Melolontha melolontha. Cell Tissue Res. 160:*283–9.

Costin, N. M. 1975. Histochemical observations of the hemocytes of *Locusta migratoria. Histochem. J. 7:*21–43.

Danes, B. S., J. E. Scott, and A. G. Bearn. 1970. Further studies on metachromasia in cultured human fibroblasts. Staining of glycosaminoglycans (mucopolysaccharides) by alcian blue in salt solutions. *J. Exp. Med. 132:*765–74.

François, J. 1975. L'encapsulation hémocytaire expérimentale chez le lépisme *Thermobia domestica. J. Insect Physiol. 21:*1535–46.

François, J. 1977. Development of collagenous endoskeletal structures in the firebrat, *Thermobia domestica* (Packard) (Thysanura: Lepismatidae). *Int. J. Insect Morphol. Embryol. 6:*161–70.

Gerrity, R. G., E. P. Adams, and W. J. Cliff. 1975. The aortic tunica media of the developing rat. II. Incorporation by medial cells of ³H-proline into collagen and elastin: autoradiographic and chemical studies. *Lab. Invest. 32:*601–9.

Grimstone, A. V., S. Rotheram, and G. Salt. 1967. An electron microscope study of capsule formation by insect blood cells. *J. Cell Sci. 2:*281–92.

Hay, E. D., and J. W. Dodson. 1973. Secretion of collagen by corneal epithelium. I. Morphology of the collagenous products produced by isolated epithelia grown on frozen-killed lens. *J. Cell Biol. 57:*190–213.

Jones, J. C. 1976. Do insect hemocytes normally transform into basement membrane or fat-body? *Amer. Zool. 16:*220.

Kefalides, N. A. 1971. Chemical properties of basement membranes. *Int. Rev. Exp. Pathol. 10:*1–39.

Lauscher, C. K., and E. C. Carlson. 1975. The development of proline-containing extracellular connective tissue fibrils by chick notochordal epithelium *in vitro. Anat. Rec. 182:*151–68.

Lazarenko, T. 1925. Beiträge zur vergleichenden Histologie des Blutes und des Bindegewebes. *Z. Mikrosk. Anat. Forsch. 3:*409–99.

Locke, M., and P. Huie. 1972. The fiber components of insect connective tissue. *Tissue Cell 4:*601–12.

Mathews, M. B. 1975. *Connective Tissue: Macromolecular Structure and Evolution.* Springer-Verlag, Berlin.

McLaughlin, B. J. 1974. Fine structural changes in a lepidopteran nervous system during metamorphosis. *J. Cell Sci. 14:*369–87.

Miller, E. J. 1976. Biochemical characteristics and biological significance of the genetically distinct collagens. *Mol. Cell. Biochem. 13:*165–93.

Neuwirth, M. 1973. The structure of the hemocytes of *Galleria mellonella* (Lepidoptera). *J. Morphol. 139:*105–24.

Pipa, R. L., and P. S. Woolever. 1965. Insect neurometamorphosis. II. The fine structure of perineurial connective tissue, adipohemocytes, and the shortening ventral nerve cord of a moth, *Galleria mellonella* (L.). *Z. Zellforsch. Mikrosk. Anat.* 68:80–101.

Ross, R. 1969. The connective tissue fibre-forming cell, pp. 1–82. *In* B. Gould (ed.). *Treatise on Collagen.* Academic Press, New York.

Ross, R., and S. J. Klebanoff. 1971. The smooth muscle cell. I. *In vivo* synthesis of connective tissue proteins. *J. Cell Biol.* 50:159–71.

Scharrer, B. 1939. The differentiation between neuroglia and connective tissue sheath in the cockroach (*Periplaneta americana*). *J. Comp. Neurol.* 70:77–88.

Scharrer, B. 1972. Cytophysiological features of hemocytes in cockroaches. *Z. Zellforsch. Mikrosk. Anat.* 129:301–19.

Seyer, J. M., E. T. Hutcheson, and A. H. Kang. 1977. Collagen polymorphism in normal and cirrhotic human liver. *J. Clin. Invest.* 59:241–8.

Shrivastava, S. C., and A. G. Richards. 1965. An autoradiographic study of the relation between hemocytes and connective tissue in the wax-moth, *Galleria mellonella* L. *Biol. Bull.* (*Woods Hole*) 128:337–45.

Wermel, E. M. 1938. Die Explantation des Blutes von Seidenspinnerraupen. *Bull. Biol. Méd. Exp. USSR* 5:6–9.

Wigglesworth, V. B. 1956. The hemocytes and connective tissue formation in an insect, *Rhodnius prolixus* (Hemiptera). *Q. J. Microsc. Sci.* 97:89–98.

Wigglesworth, V. B. 1973. Hemocytes and basement membrane formation in *Rhodnius*. *J. Insect Physiol.* 19:831–44.

13 Role of hemocytes in defense against biological agents

N. A. RATCLIFFE AND A. F. ROWLEY

Department of Zoology, University College of Swansea, Singleton Park, Swansea, SA2 8PP, U.K.

Contents

13.1. Introduction

Over three-quarters of animal species known to man are insects, and it has been estimated that there may be as many as 3 million different species (Williams, 1960) living in almost every habitat on earth. During their evolution and colonization of the terrestrial niches, the insects became open to attack from a formidable array of predators, parasites, parasitoids, and microorganisms, and recent research indicates that they have effective defense mechanisms capable of resisting, containing, and even eliminating many would-be invaders (Whitcomb et al., 1974). These defenses include the physicochemical barriers formed by the tough outer cuticle and gut and the internal humoral and cellular defense mechanisms. The cellular defenses are mainly mediated by the hemocytes and include the processes of phagocytosis, nodule formation, encapsulation, and hemolymph coagulation (Salt, 1970), which together form the main subject of this review. However, because most of our knowledge of insect defense mechanisms has been derived from laboratory investigations and little is known about the effectiveness of these defenses in the field, brief consideration must also be given to the natural biological control agents themselves, their mode of entry into the host, and the type of host response most likely to be elicited. This approach, it is hoped, will give the results of laboratory experiments a new dimension and prompt future investigators to accord more consideration to the defense mechanisms of natural insect populations.

13.2. Diseases of insects

These are produced by the pathogenic viruses, bacteria, fungi, protozoa, rickettsiae, and nematodes. In addition, predatory and parasitic insects are of enormous significance in controlling vast numbers of pest and potentially injurious insect species (van den Bosch and Messenger, 1973) (see also Chapter 16).

13.2.1. Microbial pathogens

Viruses

The best known insect-attacking viruses are the nuclear polyhedroses, the cytoplasmic polyhedroses, the granuloses, and the entomopox viruses, which are often grouped together as the occluded viruses and which, in particular, are responsible for high mortality of insects in the field (Stairs, 1972). The principal mode of infection is probably by ingestion of contaminated food, as sometimes occurs during mating with an infected partner (Zelazny, 1976).

Little is known about the effectiveness of insect defense reactions against viruses. A recent study by Kalmakoff et al. (1977) failed to induce interferon-like substances in cultured *Aedes aegypti* cells or to detect a hemocytic response to Hd virus injected into *Bombyx mori* larvae. It has, however, been estimated that there is a 50–60% reduction in infectivity of virus in the hostile conditions of the insect midgut (Paschke and Summers, 1975), so the "gut barrier" is probably of prime importance in preventing invasion of the virus particles into the hemocoel. Once in the hemocoel, any defensive reactions elicited do not act immediately (Tinsley, 1975), and even if phagocytized, the virus may make use of the engulfing phagocyte as a site for replication (Whitcomb et al., 1974).

Rickettsiae
Much less is known about rickettsiae than about other microorganisms that cause disease in insects; they have many features characteristic of both viruses and bacteria. Rickettsiae appear to have a wide host range (Carter and Luff, 1977), and rickettsioses have been reported from natural populations of Coleoptera and Diptera larvae (Krieg, 1963). Infection probably takes place by ingestion of soil particles contaminated by dead larvae, by cannibalism (Krieg, 1963), or by vertical transmission through the egg (Krieg, 1971).

Little has been published about the defense mechanisms of insects against rickettsiae; no doubt the physicochemical barriers are of great importance. Krieg (1958), however, described a reduction in hemocyte numbers, especially of plasmatocytes, in the blood of *Melolontha* spp. infected with *Rickettsiella melolonthae* that was correlated with the formation of numerous melanized nodules attached to the fat body.

Bacteria
Bucher (1960) classified the bacteria pathogenic to insects as obligate pathogens, crystalliferous spore formers, facultative pathogens, and potential pathogens. These four groups in turn can be more simply divided into the spore formers and non-spore formers (Falcon, 1971). As is true of the pathogenic viruses, the principal route of entry into the insect is via the gut following feeding on contaminated food or on infected individuals.

The defense mechanisms of insects against bacteria have, in particular, been the subject of intensive research in recent years (see excellent reviews by Salt, 1970; Lafferty and Crichton, 1973; Whitcomb et al., 1974; Chadwick, 1975; Crossley, 1975). As with viral and other microbial infections, the gut barrier with its acid/alkaline reaction, enzymes, bactericidal substances, peritrophic membrane, and cuticular lining forms an effective first line of defense and prevents bacterial invasion of the hemocoel. There is, however, considerable potential risk to insects from gut rupture, as illustrated by Bucher's (1959) classic study of the bacterial flora in the intestines of grasshoppers. Following gut rupture and entry into the hemocoel, 65 different types of bacteria were isolated, of which 13 were potentially pathogenic and 52 nonpathogenic. This study also provides a rare illustration of the efficiency of the insect host defenses, which in this case were capable of dealing with over 50 different types of bacteria.

Once in the hemocoel, the invading bacteria are exposed to a variety of defense machanisms. These may include naturally occurring and inducible humoral factors, such as antibacterial substances (Boman et al., 1972, 1974), agglutinins (Gilliam and Jeter, 1970; Scott, 1971b), lysozyme (Mohrig and Messner, 1968), complement-like substances (Anderson et al., 1972; Aston et al., 1976), and melanin (Taylor, 1969). At least some of these factors probably originate from leakage and/or lysis of the hemocytes or other cells in the insect body (Whitcomb et al., 1974). The cellular defenses are also active in containing or eliminating intruding bacteria by means of phagocytosis, nodule formation, and hemolymph coagulation. Phagocytosis probably deals effectively with small bacterial insults; massive influxes, such as those following gut rupture, are contained by hemolymph coagulation and nodule formation (Ratcliffe et al., 1976a).

The question then arises as to the importance, if any, of the various defense mechanisms in overcoming infection in natural insect populations. Fortunately, evidence of a host response is sometimes retained within the body for some time. In Bucher's (1959) study of gut rupture in natural grasshopper populations, the subsequent hemocoel contamination resulted in the formation of nodulelike structures enclosing bacteria in as many as 10% of individuals. In a similar study (Bywater and Ratcliffe, unpub.), nodules with segregated bacteria have also been recorded in 8% of grasshoppers (*Chorthippus parallelus*) and 12% of earwigs (*Forficula* sp.) (Fig. 13.1). The results of experiments on phagocytic blockage and the subsequent development of septicemia (Bettini et al., 1951; Anderson, 1976b) further indicate that phagocytosis is important in preventing infection not only in the laboratory but also in the field.

Fungi

All four major classes of fungi (Phycomycetes, Ascomycetes, Basidiomycetes, and Fungi Imperfecti) contain species that are commonly associated with insects. Unlike the bacteria and viruses, the main route of entry of fungi into their hosts is through the integument (Roberts and Yendol, 1971); the gut barrier in insects appears to be highly effective in preventing fungal invasion.

Once in the hemocoel, however, the cellular defenses are soon mobilized, and phagocytosis of fungal elements by the hemocytes usually, but not always (Hurpin and Vago, 1958), occurs and is sometimes more intense against nonpathogenic than pathogenic fungi (Sussman, 1952). Frequently, however, phagocytosis is ineffectual, and the pathogen may break out and even parasitize the surrounding blood cell (Paillot, 1930). Nodules or capsules are also formed around intruding fungi (Fig. 13.2), and in some cases have been shown to be effective in at least retarding pathogen development (Vey and Fargues, 1977). In some dipteran larvae, an unusual type of encapsulation has been observed around fungal spores in which the humoral components of the blood appear to participate in the formation of melanized capsules (Fig. 13.3) (Götz and Vey, 1974; Vey and Götz, 1974).

Protozoa

The main pathogenic protozoan species are found among the flagellates, the amebas, the sporozoa, the microsporidans, and the ciliates (Brooks, 1974). The primary mode of infection is by ingestion of spores or cysts by the insect host and subsequent attachment to or invasion of the gut epithelium and/or malpighian tubules (McLaughlin, 1971) (Figs. 13.4–13.6).

The incidence of protozoan infections is sometimes quite high in the field (Merritt et al., 1975; Nordin, 1975), and probably at least some insects can influence the outcome of infection by means of their

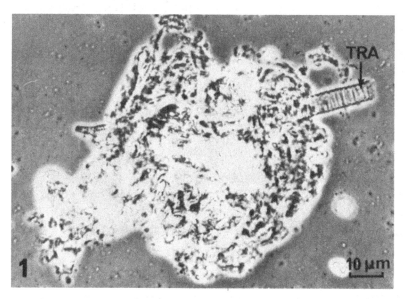

Fig. 13.1 Nodule excised from hemocoel of an earwig, *Forficula* sp. Note attached tracheae (*TRA*). Phase-contrast optics. (From Bywater and Ratcliffe, unpub.)

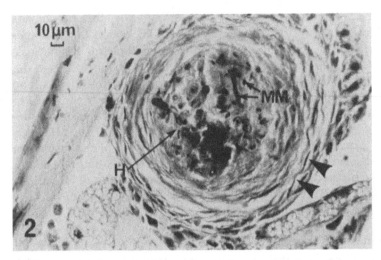

Fig. 13.2. Nodule (granuloma) formed in *Galleria mellonella* in response to the fungus, *Aspergillus niger*. Note central core containing melanized mycelia (*MM*) and broken-down hemocytes (*H*) surrounded by a distinct capsule (arrows) consisting of flattened hemocytes. (From Vey and Vago, 1971)

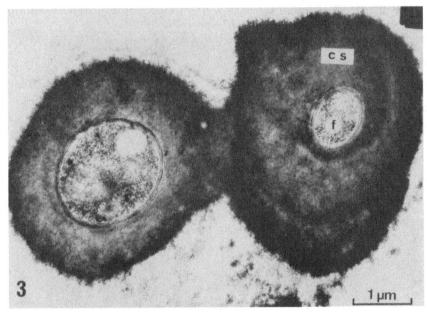

Fig. 13.3. Final stage of humoral encapsulation of *Beauveria bassiana* fungal spores (*f*) after their injection into *Chironomus luridus* larvae. Note thick deposit of melanized capsule substance (*CS*) around spores. (From Götz and Vey, 1974)

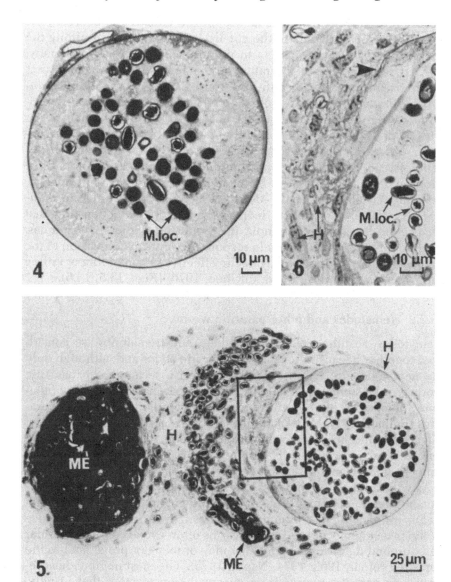

Figs. 13.4–13.6. Stages in infection of malpighian tubules of adult locust, *Schisto-cerca gregaria*, by *Malamoeba locustae*, and cellular defenses against these parasites. **13.4.** Malpighian tubule in section showing cysts and trophozoites (*M.loc.*) in lumen. Note that cells of tubule appear to have been partially eroded as result of multiplication of parasite. **13.5.** Encapsulation of *M. locustae* infective stages by thick sheath of hemocytes (*H*). Tubule is highly degenerate and breakdown in one area (boxed) has apparently initiated a hemocytic reaction. Some cysts and trophozoites that escaped into hemocoel are trapped within a nodule and heavily melanized (*ME*). **13.6.** Enlargement of boxed area in Fig. 13.5, showing site of breakage (arrow) of basement membrane surrounding tubule. Infiltrating hemocytes (*H*), cysts, and trophozoites (*M. loc.*) can be seen. (From Rowley and Ratcliffe, unpub.)

host defense mechanisms. As is true for the viruses and bacteria, the physicochemical barriers in the gut are important in determining the susceptibility of many insects to parasitic protozoans (Weathersby, 1975). Unfortunately, the literature on the defense reactions of insects to protozoan invaders is confusing. For example, *Trypanosoma rangeli* in *Rhodnius prolixus* (Tobie, 1968) and *Crithidia* spp. in *Drosophila virilis* (Schmittner and McGhee, 1970) use the hemocytes as sites of division and are apparently unaffected by ingestion. In contrast, however, phagocytized *Trypanosoma cruzi* in *R. prolixus* (Tobie, 1968) and *T. rangeli* in *Triatoma infestans* (Zeledón and De Monge, 1966) are disposed of by the hemocytes without additional development. Zeledón and De Monge (1966) also showed that in *T. infestans* phagocyte numbers increased following infection and that phagocytic blockade with India ink produced higher parasitemias. Nodule formation also occurs in response to protozoan invasion of the hemocoel, and this may or may not be effective in containing or killing the parasites (Schmittner and McGhee, 1970) (Figs. 13.5, 13.6).

13.2.2. Nematodes and other parasitic worms

Nematodes, acanthocephalans, cestodes, and trematodes are considered together because they have similar life styles and, although only the nematodes are important pathogens of insects, all the parasitic worms evoke similar host defense reactions. In natural insect populations, nematodes often occur much more frequently than other parasites (Poinar, 1971). Most of the parasitic worms have active modes of infection, and entry may occur through the gut, directly through the host cuticle, or occasionally through external openings such as the spiracles (Triggiani and Poinar, 1976).

As with most of the microbial pathogens, the physicochemical barriers formed by the insect integument and gut are probably important factors influencing host immunity to parasite invasion. However, once in the insect hemocoel, invading worms often evoke complex cellular and humoral defense reactions that may or may not prove fatal to the parasite (Poinar, 1969, 1974; Nappi, 1975). The most obvious host response to parasitic worms is an encapsulation reaction that usually, but not always, involves the blood cells in the formation of a multicellular sheath (Fig. 13.7).

Humoral defense reactions may also be important in insect immunity to parasitic worms. These include a melanization reaction (Poinar and Leutenegger, 1971; Poinar, 1974) similar to that observed around fungal spores by Götz and Vey (1974) (Fig. 13.3).

Finally, it is also apparent from studies of parasitic worms in their "normal" hosts that the worms may have the ability to circumvent or inhibit the normal host defense reactions (Brennan and Cheng, 1975).

13.2.3. Insect parasitoids

Tremendous numbers of insects parasitize other insects, and the term "parasitoid" has been coined to describe this special type of relationship. The insect parasitoids are all Endopterygota, and the bulk of the species occurs in the Hymenoptera and Diptera, although parasitic forms are also found in the Neuroptera, Lepidoptera, Coleoptera, and Strepsiptera. The parasitoids gain entry to their hosts by deposition of the eggs either directly into the host, via the ovipositor, or on or away from the host, in which case the larvae may actively seek out and bore their way into the host.

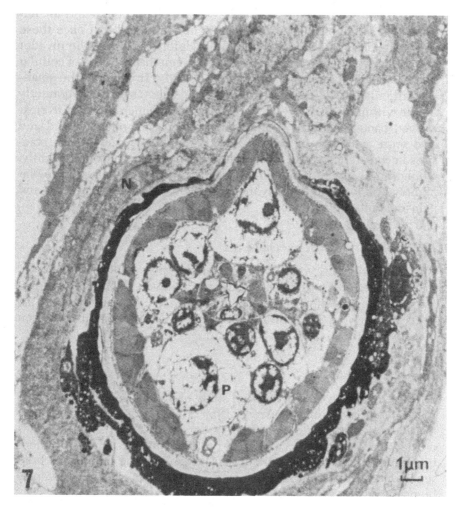

Fig. 13.7. Cross section of parasite, *Filipjevivermis leipsandra* (P), encapsulated by hemocytes of *Diabrotica larva*. Note flattened nuclei (N) and cytoplasm of hemocytes and underlying black pigment, melanin. (From Poinar et al., 1968)

The main defense mechanism of insects against metazoan endoparasites is encapsulation, and this reaction is the same against parasitoids as it is against most parasites, including the nematodes, described above. The capsule formed is also cellular and/or humoral in origin (Poinar, 1974) and appears to be more effective at killing parasitoids than parasitic worms (Salt, 1970). The habitual parasitoids, like the parasitic worms, have also evolved intricate methods of avoiding this host reaction (Salt, 1970) (see "Encapsulation," below). Capsule formation is thus only effective against alien parasitoids. Although no quantitative data exist, insect hosts are probably frequently attacked by alien parasitoids in nature (Salt, 1970).

In summary, it is apparent that the physicochemical barriers are all-important in defending insects against infection and that once these are breached both humoral and cellular defenses combat the invader to a greater or lesser extent. It can also be seen that the type of cellular defense reaction mounted varies considerably. In general, the smallest particulate invaders, the viruses, are phagocytized and frequently divide within the blood cells. Larger particulate microbes such as rickettsiae, bacteria, fungal spores, and protozoa are phagocytized and also segregated within nodules; fungal hyphae and metazoan parasites become enclosed within multicellular capsules. Concomitantly, wounding results in hemolymph coagulation, which helps to prevent hemolymph loss and to delay the penetration of infectious agents and parasites into the hemocoel.

13.3. Cellular defense mechanisms

The main brief of this review is to consider the role of the hemocytes in the insect host's defense reactions. However, it must be emphasized that any division of these defenses into cellular and humoral aspects may be purely arbitrary: In some cases interactions between these two components probably occur.

13.3.1. Hemocytes

Before embarking on any detailed study of the insect cellular defense reactions, it is absolutely essential for the research worker to clarify the cell types present in the insect under study. Failure to do this has created many misnomers and great problems for the readers of scientific papers. Price and Ratcliffe (1974) examined the hemocytes from representatives of 15 insect orders and recognized six cell types: prohemocytes (PRs) (Figs. 13.8, 13.9), plasmatocytes (PLs) (Figs. 13.10–13.13), granulocytes (granular cells, GRs) (Figs. 13.14–13.16), coagulocytes (cystocytes, COs) (Figs. 13.17, 13.18), spherulocytes (spherule cells, SPs) (Figs. 13.19–13.22), and oenocytoids (OEs) (Figs. 13.23–

13.25). These six cell types form the basis for the observations and terminology throughout this review (see also Chapters 4–10). In some cases the appearance of these cells may alter during the cellular defense reactions, making identification difficult.

Important points to emphasize are that Figs. 13.8–13.25 show freely circulating blood cells and that most previous studies on insect cellular defenses have largely ignored other hemocyte populations. For example, the percentage of sessile hemocytes present in the total hemocyte population is unknown, although Wigglesworth (1965) believed most hemocytes were in fact sessile and not freely circulating. The importance of these "hemocyte reservoirs" (Jones, 1970) is indicated by the great variations that occur in total hemocyte numbers within the life span of any one insect according to its age, sex, developmental stage, or other physiological parameters (Jones, 1962) (see also Chapter 16).

Another group of sessile cells, which may also play an important role in insect immunity and are usually distinguished from sessile hemocytes both by their structure and function, are the pericardial cells (Wigglesworth, 1970; Crossley, 1972). They are found along the dorsal vessel and in various other parts of the body (Nutting, 1951; Wiggles-

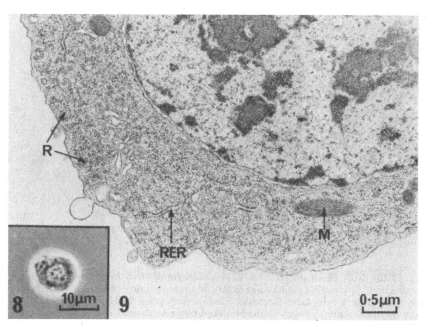

Figs. 13.8, 13.9. Prohemocytes. **13.8.** Phase-contrast micrograph of prohemocyte from *G. mellonella* with characteristic high nuclear/cytoplasmic ratio. **13.9.** Electron micrograph of prohemocyte from *G. mellonella*, showing characteristic poorly developed cytoplasm with many free ribosomes (*R*), little rough endoplasmic reticulum (*RER*), and a few mitochondria (*M*). (**13.8** and **13.9** from Rowley, unpub.)

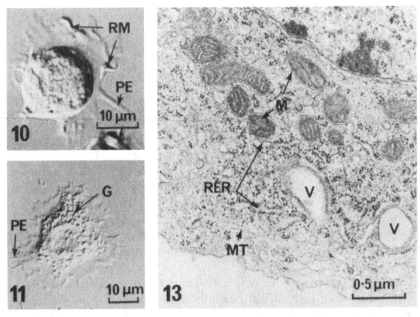

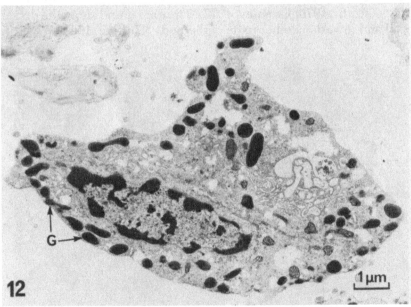

Figs. 13.10–13.13. Plasmatocytes. 13.10. Plasmatocyte of *Pieris brassicae* in process of spreading, with ruffled membranes (*RM*) and protoplasmic extensions (*PE*). Note lack of granules in cytoplasm. Nomarski interference optics. 13.11. Fully spread plasmatocyte of *Clitumnus extradentatus* bordered by protoplasmic extensions (*PE*) and containing large number of cytoplasmic granules (*G*). Nomarski interference optics. 13.12. Electron micrograph of plasmatocyte of *C. extradentatus* with characteristic granules (*G*). 13.13. Electron micrograph showing detail of cytoplasm of plasmatocyte of *G. mellonella*, with vacuoles (*V*), rough endoplasmic reticulum (*RER*), mitochondria (*M*), and peripheral microtubules (*MT*).

worth, 1970). Pericardial cells have been reported to ingest bacteria (Cameron, 1934) but, in general pinocytize colloidal particles from the blood and may be equivalent to the vertebrate reticuloendothelial system (Wigglesworth, 1970).

13.3.2. Phagocytic organs

A population of hemocytes or hemocyte-like cells that probably play an important part in the cellular defenses are found in the so-called phagocytic or hemopoietic organs. Phagocytic organs have been described in several insect orders; they may be diffuse and little more than accumulations of sessile hemocytes or, alternatively, more highly organized and surrounded by a limiting membrane (Jones, 1970) (see also Chapter 2). In orthopteroid insects, they vary in number from two to four pairs, usually arise ventrolaterally in the anterior abdominal segments, originate from the heart wall from which they fan out, and are loosely attached to the alary muscles and diaphragm (Nutting, 1951) (Fig. 13.26). In *Locusta migratoria* they have been shown to contain reticular cells and fully differentiated PLs, GRs, and OEs, and to have a hemopoietic function (Hoffmann et al., 1968). The role of these organs in the removal of bacteria and protozoa from the blood, and hence their importance in the cellular defenses, was recognized originally by Cuénot (1895, 1897) and has recently been confirmed in *L. migratoria* by Hoffmann et al. (1974).

Phagocytic organs have been studied in some detail in the Diptera, Hymenoptera, and Lepidoptera (Jones, 1970) and have been reported to contain PR-like cells, PLs, and OEs (Arvy, 1954; Rizki, 1960; Bairati, 1964). However, although Lange (1932) reported that these cell aggregates are phagocytic, most studies on these structures describe their hemopoietic function (Jones, 1970), not their role in the cellular defenses.

13.3.3. Phagocytosis

Phagocytosis was the first of the host defense reactions of animals to be studied. Metchnikoff's classic work on *Daphnia* sp. (Metchnikoff, 1884) and later on other arthropods, including insects (Metchnikoff, 1892), were the first detailed studies of phagocytosis and subsequently led to modern-day concepts of cellular immunity. What impressed the early authors (e.g., Cuénot, 1895, 1897; Metalnikov, 1908, 1924, 1927; Hollande, 1909, 1930; Iwasaki, 1927; Metalnikov and Chorine, 1928; Chorine, 1931; Cameron, 1934) was that insects could withstand enormous injections of bacteria that were highly pathogenic to man, such as those causing anthrax, tuberculosis, typhoid, pneumonia, cholera, and leprosy.

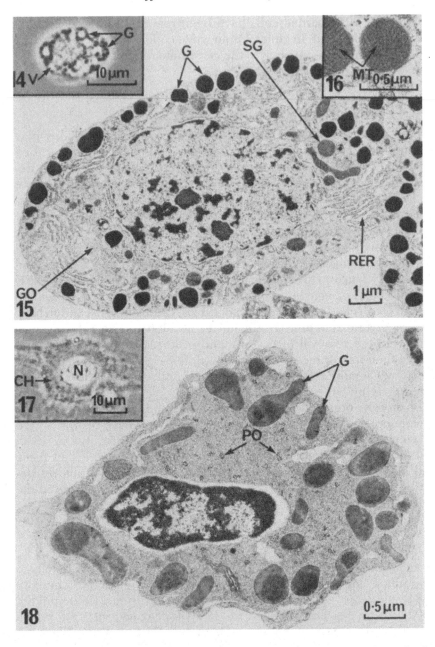

Figs. 13.14–13.16. Granulocytes (granular cells). 13.14. Phase-contrast micrograph of granulocyte from *G. mellonella*. Note granules (*G*) and vacuoles (*V*) formed by granule discharge in vitro. 13.15. Electron micrograph of typical granulocyte from *G. mellonella*. Note swollen rough endoplasmic reticulum (*RER*), Golgi (*GO*), amorphous electron-dense granules (*G*), and structured granule (*SG*). 13.16. High-power view of microtubular substructure (*MT*) present in some granules in a granulocyte of *G. mellonella*. (13.15 from Rowley and Ratcliffe, 1976a)

The majority of work on phagocytosis in insects has been carried out in vivo, and most workers showed that following the injection of test particles such as carmine, polystyrene beads, India ink, erythrocytes, bacteria, fungi, yeast cell walls, mycetomes, and protozoa into the hemocoel, phagocytosis took place within 1 hr (see references in reviews by Jones, 1962; Salt, 1970; Arnold, 1974; Whitcomb et al., 1974). Much confusion, however, has arisen from in vivo work because injections of test particles were variously reported to raise, lower, or have no effect on the numbers and constitution of the circulating hemocytes (Rosenberger and Jones, 1960; Wittig, 1965, 1966; Werner and Jones, 1969; Hoffmann and Brehélin, 1971; Hoffmann et al., 1974; Seryczyńska et al., 1974). This confusion led Arnold (1974), in his review of insect hemocytes, to state that the results of various experimental procedures on insect hemograms are often contradictory and that generalizations can seldom be made.

In vitro techniques
Only with the development of suitable in vitro techniques has it been possible in recent years to study phagocytosis in detail. There are, of course, certain disadvantages to studying phagocytosis in vitro, as conditions are likely to be suboptimal for maximal hemocyte activity. For example, *Galleria mellonella* hemocytes respond to a greater range of test particles in vivo than in vitro (Rabinovitch and De Stefano, 1970). For this reason, results from in vitro studies should, wherever possible, be related back to the in vivo state.

Many early attempts to obtain in vitro phagocytosis apparently failed (Åkesson, 1954; Jones, 1956; Whitten, 1964; Lea and Gilbert, 1966; Scott, 1971a), although possibly, in some of these studies, ingestion may well have occurred but not been recognized owing to the inherent difficulties in identifying intracellular test particles (Smith and Ratcliffe, 1978). More success was obtained with cultured cells (Grace, 1962, 1971; Grace and Day, 1963; Vago, 1964; Vago and Vey, 1970; Vago, 1972; Landureau et al., 1972), although the behavior of cultured cells may not necessarily reflect their in vivo function (Rowley, 1962).

Recently, successes have been reported with in vitro systems for studying phagocytosis (Rabinovitch and De Stefano, 1970; Anderson

Figs. 13.17, 13.18. Coagulocytes (cystocytes). 13.17. Phase-contrast micrograph of coagulocyte from *C. extradentatus* after approximately 10 min in vitro. Note characteristic coagulated hemolymph (*CH*) forming islet of coagulation and prominent nucleus (*N*). 13.18. Electron micrograph of intact coagulocyte from *C. extradentatus* with a cytoplasm devoid of most organelles; only polyribisomes (*PO*) and granules (*G*) are present. Release of these granules causes hemolymph coagulation. Note perinuclear space. (From Rowley, 1977b)

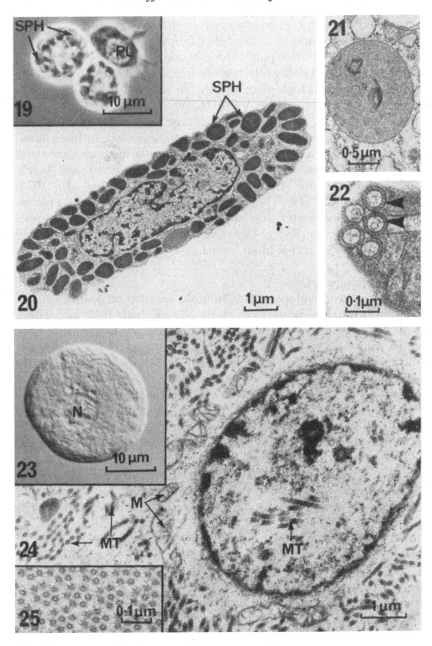

Figs. 13.19–13.22. Spherulocytes (spherule cells). **13.19.** Phase-contrast micrograph showing two spherulocytes and a plasmatocyte (*PL*) from *G. mellonella*. Note prominent spherules (*SPH*), which distend cell periphery and mask nucleus. **13.20.** Typical spherulocyte from *C. extradentatus* with characteristic large unstructured, electron-dense spherules (*SPH*). **13.21.** Portion of spherulocyte of *G. mellonella*, showing spherule with characteristic crystalline substructure. **13.22.** Unusual tubular structures (arrows) found in spherulocytes of *P. americana* as first described by Baerwald and Boush (1970).

et al., 1973a,b; Anderson, 1974, 1975, 1976a,b; Ratcliffe and Rowley, 1975). Rabinovitch and De Stefano (1970) described a simple mono-layer technique for examining phagocytic recognition by *G. mellon-ella* hemocytes and compared the results with those obtained using mouse macrophages. Anderson et al. (1973a,b) and Anderson (1974, 1975, 1976a,b) in excellent studies of the metabolism and bactericidal properties of *Blaberus craniifer* hemocytes and of receptors on insect macrophages used both suspension culture and simple monolayer techniques. Finally, in our studies on the merits of a number of in vitro techniques, we showed that the culture system giving optimal results varied from species to species (Ratcliffe and Rowley, 1975). A roller tube system gave satisfactory results with the hemocytes of *Periplaneta americana* and *Calliphora erythrocephala*, so that workable numbers of cells showing phagocytosis were recovered. However, with *G. mellonella* and *P. brassicae* hemocytes, this system was unsatisfactory because of the low cell recovery resulting from the attachment of many cells to the tube surfaces, even if these were coated with silicon or wax prior to experimentation.

From our limited experience, and the Lepidoptera apart, optimal

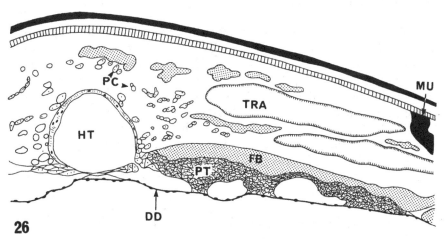

26

Fig. 13.26. Diagram taken from a section through fourth pair of phagocytic organs (*PT*) of *Pterophylla camellifolia*. *HT* = heart; *TRA* = tracheae; *MU* = muscle; *PC* = pericardial cells; *FB* = fat body; *DD* = dorsal diaphragm. (Redrawn from Nutting, 1951)

(facing page)
Figs. 13.23–13.25. Oenocytoids. **13.23.** Typical oenocytoid from *G. mellonella* with characteristic eccentric nucleus (*N*). Nomarski interference optics. **13.24.** Electron micrograph of oenocytoid from *G. mellonella* showing bundles of microtubules (*MT*) present in nucleus and cytoplasm and ring of mitochondria (*M*) around nucleus. **13.25.** High-power view of arrangement of microtubules in oenocytoid of *G. mellon-ella*.

phagocyte–test particle interactions can be obtained by incubation in suspension culture. Quantitative studies, however, can be more conveniently made in simple monolayer culture systems overlaid with the test particles (Rowley and Ratcliffe, 1976b; Ratcliffe et al., 1976a).

Cell types involved

The PL is the predominant cell type involved in phagocytosis in insects (Salt, 1970) both in vivo (Figs. 13.27, 13.28) (Wittig, 1965) and in vitro (Fig. 13.29) (Ratcliffe and Rowley, 1975). Lack of uniformity in hemocyte terminology, however, has led to confusion in identifying the cell types involved. For example, in *G. mellonella*, the phagocytic hemocytes have been called leucocytes, lymphocytes, and spherule cells (Cameron, 1934); plasmatocytoids and adipohemocytes (Werner and Jones, 1969); and microplasmatocytes and macroplasmatocytes (Rabinovitch and De Stefano, 1970). From the descriptions given by these authors, we believe these cells correspond to the PLs and GRs shown in Figs. 13.29 and 13.30. The GRs of *G. mellonella* and *P. brassicae* also have limited phagocytic powers in vitro (Fig. 13.30). However, in *Calpodes ethlius* GRs are the main phagocytic blood cells (Fig. 13.31) and PLs appear to be nonphagocytic (Neuwirth, 1974). Within 15 min of injection of thorium dioxide, Neuwirth observed pinocytotic vesicles enclosing the test particles, and by 17 hr the GRs contained many phagosomes packed with thorium dioxide (Fig. 13.31). Some OEs were also phagocytic; the PLs, although forming numerous pseudopodia, failed to ingest any particles. That the phagocytic cells were indeed GRs is beyond all dispute owing to the presence of characteristic structured granules and the well-developed rough endoplasmic reticulum (RER) (compare Fig. 13.31 with Figs. 13.15, 13.16) (Lai-Fook, 1973; Neuwirth, 1973; Ratcliffe and Price, 1974; Rowley and Ratcliffe, 1976a). Although preliminary work by Neuwirth (1974) indicates that GRs also take up India ink and latex beads, possibly larger test substances would be phagocytized by the *C. ethlius* PLs.

Chemotaxis, attachment, and recognition of foreignness

The first requisite for phagocytosis or any cell-mediated defense reaction is contact between the "foreign" body and the hemocyte surface. Salt (1970) believed that such contacts were likely to be random and that chemotaxis was probably not involved. This view was shared by Jones (1956), who failed to observe any movement of *Sarcophaga bullata* hemocytes toward foreign agents in over 50 coverslip preparations. Salt (1970) based his conclusions on the facts that little evidence for hemocyte chemotaxis had been found and that fewer cells adhered to foreign bodies out of the main blood flow than to ones in it. Presumably, if a chemotactic potential was available, the positioning in the

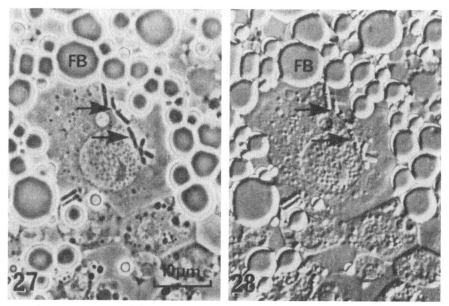

Figs. 13.27, 13.28. In vivo phagocytosis. Plasmatocytes of *Pieris brassicae* containing intracellular *Bacillus cereus* (arrows). Plasmatocytes were found associated with fat body (*FB*). **13.27**, phase contrast; **13.28**, Nomarski interference optics. (From Gagen, unpub.)

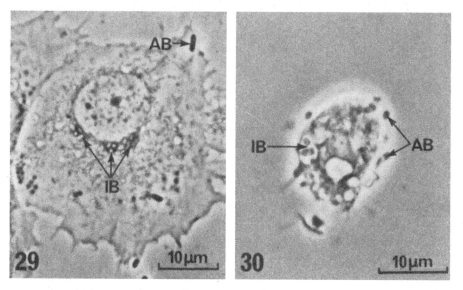

Figs. 13.29, 13.30. In vitro phagocytosis. **13.29**. Plasmatocyte of *P. americana* containing intracellular *Escherichia coli* (*IB*) after 60 min incubation together in monolayer culture. Note darker more refractile attached bacterium (*AB*). **13.30**. Intracellular *E. coli* (*IB*) in granulocyte of *G. mellonella* after 45 min incubation in suspension culture. *AB* = attached bacteria. (From Ratcliffe and Rowley, 1975)

body would make no difference to the rate or extent of encapsulation. Furthermore, in intraspecific organ transplants the blood cells attached only to the cut ends of tracheae and nerves, not to the adjacent surface of the implant. Salt (1970) postulated that if the hemocytes had responded to a chemotactic stimulus, these cells would also have attached to the surrounding unaltered surfaces. Nappi (1974), however, pointed out that the localized aggregation of the hemocytes on the damaged tissues may well have indicated a specific hemocytic response to materials generated from the wounds. Such "wound" or "injury factors," released from damaged cells, have previously been shown to result in the mobilization and attachment of hemocytes at

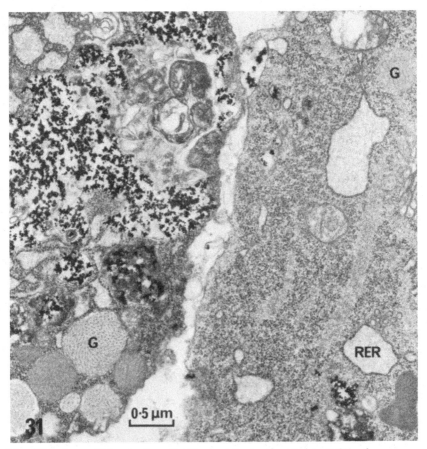

Fig. 13.31. Intracellular thorium dioxide particles in granulocyte (left) and oenocytoid (right) of *Calpodes ethlius* 17 hr after injection. Note granules (G) and sparse, swollen rough endoplasmic reticulum (RER). (From Neuwirth, 1974; reproduced by permission of The National Research Council of Canada from the *Canadian Journal of Zoology*, Vol. 52, pp. 783–4, 1974)

wound sites (Wigglesworth, 1937; Harvey and Williams, 1961; Lea and Gilbert, 1961; Clark and Harvey, 1965; Wyatt and Linzen, 1965; Cherbas, 1973).

Salt (1970) also reasoned that the encapsulation of inert objects, such as pieces of glass or polyfluorocarbon, was additional evidence against the involvement of chemotactic stimuli in hemocyte–foreign body interactions, as chemotactic substances would not diffuse out from such inert bodies. Possibly, however, the cells are responding to a substance produced by interaction between the foreign surface and a specific population of hemocytes (Crossley, 1975; Nappi, 1975) (see also "Nodule Formation" and "Encapsulation," pp. 269 and 380).

Some evidence has now become available to indicate that chemotaxis may well be an important component of the cellular defenses. Again, unfortunately, some of this evidence has come from experiments on encapsulation reactions and not on phagocytosis, but it will be discussed here because the same cell types are probably involved in all the cellular defense reactions (Ratcliffe et al., 1976a). Properties exhibited by the cells during one type of defense reaction are also likely to be important in all the other host defense mechanisms.

Until recently the only "evidence" for a chemotactic function for insect hemocytes was Metchnikoff's (1901) conclusion that the hemocytes of the scarabaeid beetle, *Oryctes* sp., displayed chemotaxis toward anthrax and diphtheria bacteria; Metalnikov's (1927) contention that phagocytes were able to accept or reject foreign substances; and the observations of Vey et al. (1968) and Vey (1969) that hemocytes of *G. mellonella* appeared to be attracted in vitro toward *Aspergillus flavus* conidia.

Nappi and Stoffolano (1972), however, reported that chemotaxis may have been involved during the encapsulation of nematodes by *Musca* spp. hemocytes. They observed that the hemocytes in early *Musca domestica* and *M. autumnalis* larvae were clumped together in masses in the posterior end of the body, but that at pupation these cells became uniformly dispersed throughout the hemocoel. If early larvae, however, were parasitized by the nematode, *Heterotylenchus autumnalis*, the hemocytes were found freely circulating and encapsulating the parasite. These investigators suggested that some chemotactic factor was acting on the hemocytes, or on their control mechanism, so that they prematurely left their posterior position and attached to the parasite.

More direct evidence for the presence of hemocyte chemotaxis has come from recent work in our laboratory (briefly reported in Ratcliffe et al., 1976a). The hemocytes and nodules from *G. mellonella* larvae injected with *Bacillus cereus* (Gagen and Ratcliffe, 1976; Ratcliffe and Gagen, 1976) were placed together on agarose-coated slides. Unidi-

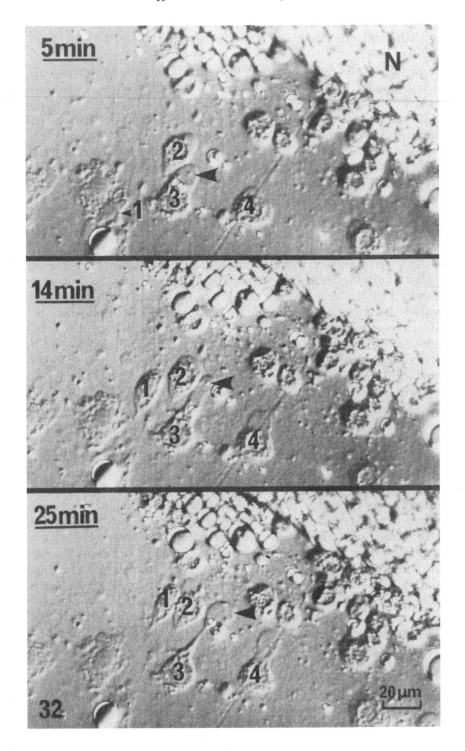

rectional movement of the PLs toward the nodules occurred within 30 min in vitro (Fig. 13.32), providing direct evidence for the chemotactic response in an insect. However, even if insect hemocytes respond to chemotactic stimuli, the recognition mechanism involved may not necessarily be the same as that operating once contact has been made with the foreign body (Wilkinson, 1976).

Recently, a considerable amount has been learned about the nature of the surface receptors and the recognition of foreignness by insect hemocytes, but even so, comparisons with other invertebrates and the much-studied vertebrates are still of great value.

Basically, two forms of particle attachment to phagocytes exist in the animal kingdom: that dependent on and that independent of the presence of serum or hemolymph factors. In the vertebrates, the principal mechanism of recognition of foreignness is mediated by the immunoglobulins. Following exposure to foreign antigen, specific antibody is formed that acts to identify the antigen and assists in its removal by a number of mechanisms, including phagocytosis. In macrophages, and to a lesser extent polymorphonuclear leucocytes, the attachment of particles is often facilitated by antibody (= opsonin), especially in the presence of complement. The role of this opsonin and complement is mediated through specific receptors on the phagocyte surface which have a high affinity for IgG and the C3b component of complement (Roitt, 1974).

Several studies have shown that insect and other invertebrate hemocytes lack the Fc and C3 receptors for immunoglobulin and complement, respectively. Using *G. mellonella* (Rabinovitch and De Stefano, 1970), *P. americana* (Scott, 1971a), and *Spodoptera eridania* (Anderson, 1976a,b) hemocyte monolayers overlaid with either untreated erythrocytes or erythrocytes preincubated with rabbit anti-red cell sera, these workers showed that there was no detectable increase in the attachment and ingestion of the sensitized (i.e., treated) erythrocytes. This result is not surprising, as immunoglobulins have yet to be found in any invertebrate. Anderson (1976a) also looked for complement receptors on the hemocytes of *S. eridania* because there is some evidence for the presence of complement-like substances in insect hemolymph (Anderson et al., 1972; Aston et al., 1976), but failed to obtain increased rosetting capacity of chicken erythrocytes preincubated with antibody plus complement components.

There are, however, numerous reports of serum-dependent phagocytosis in invertebrates (Whitcomb et al., 1974), and in the search for

Fig. 13.32. Chemotaxis of plasmatocytes of *G. mellonella* toward a nodule (*N*) in vitro. Note apparent unidirectional movement of cells *1–4* toward edge of nodule. Protoplasmic extensions in cell 3 are particularly noticeable (arrows). (From Gagen and Ratcliffe, unpub.)

humoral recognition factors the naturally occurring agglutinins present in the body fluids of many invertebrates have been shown to possess opsonic activity. Tripp (1966) was the first to discover that hemagglutinins play an opsonic role and showed that pretreatment of erythrocytes with hemolymph approximately doubled the number of *Crassostrea virginica* hemocytes containing ingested erythrocytes. Further reports (Stuart, 1968; Prowse and Tait, 1969; McKay and Jenkin, 1970; Pauley et al., 1971; Tyson and Jenkin, 1973, 1974; Paterson and Stewart, 1974; Anderson and Good, 1976) on various mollusks and crustaceans have shown similar results, with both bacterial and erythrocyte agglutinins present acting as opsonins.

These invertebrate agglutinins, although playing a role similar to that of vertebrate immunoglobulin, differ considerably from vertebrate antibody (Acton and Weinheimer, 1974). For example, they have different electrophoretic mobilities (Brahmi and Cooper, 1974) and react with the terminal nonreducing N-acetyl-D-galactosamine residues present on the erythrocyte membrane (Uhlenbruck and Prokop, 1966; Prokop et al., 1968; Cann, 1974; Hall and Rowlands, 1974). Indeed, the similarities between these erythrocyte's surface carbohydrate groups and those present on bacteria, viruses, and other cell surfaces drew Renwrantz and Uhlenbruck (1974) to conclude that the anti–red cell agglutinins were part of the primitive antimicrobial defense mechanism of invertebrates.

In insects there is, unfortunately, a dearth of information about the role of agglutinins in the cellular defense reactions. Most studies have been concerned with recording the presence or absence of these substances and whether or not they are inducible (Bernheimer, 1952; Briggs, 1958; Feir and Watz, 1964; Gilliam and Jeter, 1970). Scott (1971a,b, 1972), however, has examined both the physicochemical and biological properties of the naturally occurring hemagglutinin of *P. americana*. He found it was a heat-labile, nondialyzable, euglobulin type of protein that possessed lipid, saccharide, and esterase activity, was noninducible, and on electrophoresis showed no band comparable to vertebrate γ-globulin. Scott (1971a) also attempted to elucidate whether the hemagglutinin in *P. americana* had opsonic activity. He overlaid hemocyte monolayers with erythrocytes preincubated with hemagglutinin and found a decreased adherence of the sensitized red cells. Anderson et al. (1973a) also failed to find an increased killing of bacteria "opsonized" with concentrated hemolymph. More recently (Ratcliffe et al., 1976a; Rowley, 1977a), we have tested the opsonic activity of the naturally occurring hemagglutinins in the hemolymph of *P. americana* and *Clitumnus extradentatus* and shown that incubation of erythrocytes in hemolymph usually causes a reduction in the phagocytic indexes in comparison with the saline controls (Tables 13.1 and 13.2). The reduced phagocytic indexes with sensitized erythrocytes probably resulted from clumping of the erythrocytes on the monolayers, leaving fewer single particles available for ingestion. Thus, in contrast to phagocytosis in many invertebrate groups, phagocytosis in insects appears to be independent of hemolymph factors.

Rabinovitch and De Stefano (1970) have shown that similar patterns of recognition appear to be present, in the absence of hemolymph or serum, for *G. mellonella* hemocytes and mouse macrophages. Both types of phagocytes were capable of recognizing seven different types of modified erythrocytes, so these two phylogenetically distant animals may have common features underlying phagocytic recognition (Rabinovitch and De Stefano, 1970).

Serum-independent attachment by insect hemocytes is probably mediated, at least partially, by interactions with cell surface receptors. Anderson (1976a,b) gave evidence for the presence of these nonspecific surface receptors on the hemocytes of *S. eridania* and *Estig-*

Table 13.1. Effect of hemolymph incubation on phagocytosis of erythrocytes by hemocytes of *Periplaneta americana*

Erythrocyte treatment[a]	Phagocytic index[b]
Hemolymph-incubated	2.1 ± 0.7[c]
Saline-incubated (control)	4.3 ± 2.1

[a] Erythrocytes were incubated for 45 min in either saline or cell-free hemolymph and then rinsed in saline prior to placement on hemocyte monolayers.
[b] Phagocytic index = percent of hemocytes containing one or more intracellular erythrocytes.
[c] Mean values ± standard deviation.

Table 13.2. Effect of hemolymph incubation on phagocytosis of erythrocytes by hemocytes of *Clitumnus extradentatus*

Erythrocyte treatment[a]	Phagocytic index[b]
Hemolymph-incubated	23.3 ± 12.3[c]
Saline-incubated (control)	25.3 ± 9.1

[a] Erythrocytes were incubated for 45 min in either saline or cell-free hemolymph and then rinsed in saline prior to placement on hemocyte monolayers.
[b] Phagocytic index = percent of hemocytes containing one or more intracellular erythrocytes.
[c] Mean values ± standard deviation.

mene acrea, and Scott (1971a) showed that attachment to *P. americana* hemocytes was mainly dependent on trypsin-labile surface receptors. However, as Rabinovitch and De Stefano (1970) showed that *G. mellonella* hemocytes ingested a wider range of modified erythrocytes in vivo than in vitro, additional recognition factors may possibly augment the surface receptors. The only evidence for the presence of such substances comes from studies of the attachment of various test particles to *P. brassicae* hemocytes in vitro (Ratcliffe, 1975) and from observations that the first vital stage in encapsulation reactions and nodule formation in insects is often discharge from and lysis of GRs, COs, or other hemocyte types (Poinar et al., 1968; Crossley, 1975; Rizki and Rizki, 1976; Ratcliffe and Gagen, 1976, 1977; Schmit and Ratcliffe, 1977). Ratcliffe (1975) showed that specific association of formaldehyde-treated sheep erythrocytes, *Escherichia coli,* and *Staphylococcus aureus* occurred with the SPs (= specialized GRs in this species) of *P. brassicae* in vitro (Figs. 13.33, 13.34) and that this binding was probably attributable to a layer of "sticky" acid mucopolysaccharide around this cell type. A similar substance(s) also appears to be discharged onto foreign surfaces by GRs during encapsulation and nodule formation in *G. mellonella* (see Figs. 13.55, 13.77) and, independently of or in combination with material(s) from the foreign bodies, apparently specifically attracts PLs to the alien surfaces (Ratcliffe and Gagen, 1976, 1977; Schmit and Ratcliffe, 1977). We concur with Crossley (1975) and can see no reason why such a recognition process should be confined to encapsulation and nodule formation or

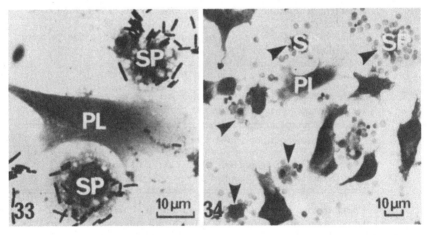

Figs. 13.33, 13.34. Giemsa-stained preparations of monolayer cultures of *P. brassicae* hemocytes incubated with *E. coli* (**13.33**) or formalized erythrocytes (**13.34**). Note specific association of test particles with outside of spherulocytes (*SP*), particularly noticeable with erythrocytes (arrows), giving a rosette-like appearance. *PL* = plasmatocytes. (From Ratcliffe, 1975)

why granular-type hemocytes should not also release recognition factors to stimulate phagocytosis. The recent reports by Amirante (1976) and Valvassori and Amirante (1976) of the synthesis of heteroagglutinins by the GRs and SPs of *Leucophaea maderae* and their localization on the cell periphery of the PLs (Fig. 13.35) are also very interesting in relation to the above observations. But until more is known, further generalizations cannot be made.

Finally, mention should be made of a recent hypothetical model for self–non-self discrimination in invertebrates (Parish, 1977). In this model, 5-glycosyltransferases act as the subunits of the recognition factors and are secreted by the hemocytes into the hemolymph, where they stimulate phagocytosis, regardless of whether they become cell-bound or remain free in the hemolymph. Following secretion into the hemolymph, these transferases randomly polymerize into hexamers, which can react in a multipoint fashion with foreign substances so that their specificity is greatly increased. Evidence in favor of this model is provided by the fact that many invertebrate agglutinins specifically bind to carbohydrates, and to sugars in particular (Renwrantz and Cheng, 1977a,b).

Ingestion

Recognition and attachment of foreign particles to the outside of the blood cells are usually followed by the ingestion phase of phagocytosis. Salt (1970), using the small number of electron micrographs available at the time, proposed that three methods of particle ingestion occurred in insect hemocytes: (1) by formation of pinocytotic ves-

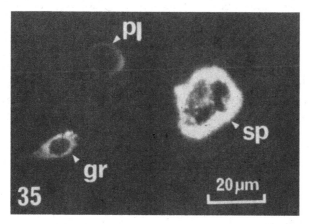

Fig. 13.35. *Leucophaea maderae* hemocytes treated with tagged antiheteroagglutinin serum. Fluorescence is present only at cell periphery of plasmatocyte (*pl*), but is spread throughout cytoplasm of spherulocyte (*sp*) and granulocyte (*gr*). Localization of heteroagglutinin in cytoplasm may well indicate its synthesis in these cells. (From Amirante, 1976)

icles to engulf fluid and small particles such as viruses (Leutenegger, 1967); (2) by encircling particles with pseudopodia, as seen in vertebrate phagocytes (Brewer, 1963; Jones and Hirsch, 1971); and (3) by "close contact" and spreading of the plasma membrane around the particles (Grimstone et al., 1967). This last method of uptake, originally proposed by Paillot (1920), would explain several reports of ingested particles lying directly in the cytoplasm with no sign of phagocytic vacuoles (Salt, 1970).

More recently, Crossley (1968, 1975) working with C. erythrocephala, Vago and Vey (1970) with G. mellonella, Granados (unpub. data cited in Whitcomb et al., 1974) with the saltmarsh caterpillar Amsacta moorei, Neuwirth (1974) with C. ethlius, and ourselves with G. mellonella, C. erythrocephala, and P. americana (Ratcliffe and Rowley, 1974; Rowley and Ratcliffe, 1976c; Ratcliffe et al., 1976a; Rowley, 1977a) have produced convincing light and electron micrographs of the stages in the uptake of foreign substances by insect blood cells.

These micrographs suggest that two methods of absorption of foreign materials seem to be available to insect hemocytes. The first method involves the formation of coated vesicles at the cell surface by indentation of the plasma membrane (Crossley, 1968, 1975). Each vesicle is thickened on its cytoplasmic surface by a layer of bristles projecting into the cytoplasm. In insect hemocytes, as in other animal cells, coated vesicles probably take up soluble proteinaceous material and pass into the cytoplasm to fuse with large vacuoles and so discharge their contents (Crossley, 1975). The second mode of ingestion is true phagocytosis and incorporates Salt's (1970) methods 1 and 2, described above; it is the means by which insect hemocytes incorporate particulate substances such as viruses, bacteria, and some tracer substances. The following description is taken from Ratcliffe and Rowley (1974), Rowley and Ratcliffe (1976c), or Rowley (1977a), unless otherwise stated. Following the particle's attachment to the hemocyte surface, the cell reacts by forming fine pseudopodia, which eventually fuse at their tips or with the cell surface to enclose the particle in a pocket of cytoplasm (Figs. 13.36–13.41). It is not clear whether in every case attachment is a necessary preliminary to filopod formation and ingestion. In experiments with G. mellonella and P. americana hemocytes and bacteria in vitro, uptake was limited to two to four microorganisms per cell, and well-defined phagocytic vacuoles were rarely seen, so that in low-power micrographs the bacteria appeared to lie directly in the cytoplasm (Fig. 13.38a). This close-contact phenomenon probably explains the third method of particle ingestion proposed by Salt (1970) (see above), but high-power micrographs showed that a true phagosome was present in every case (Fig. 13.38b). In similar experiments with C. erythrocephala hemocytes, however, uptake was much more rapid and extensive, so that after 15 min of in-

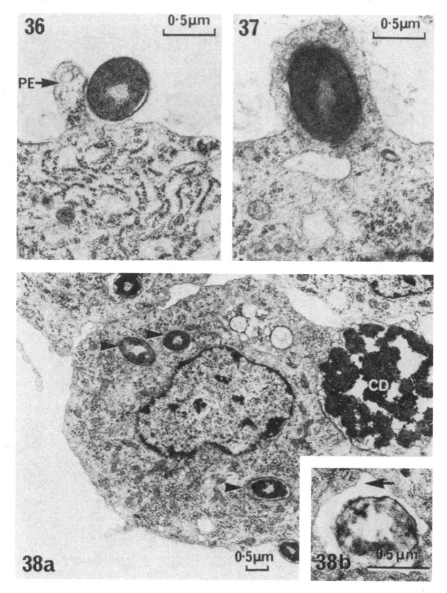

Figs. 13.36, 13.37. Stages in attachment and ingestion of *E. coli* by plasmatocytes of *G. mellonella* in vitro. 13.36. Attachment of *E. coli* to cell membrane. Note small protoplasmic extension (*PE*) enveloping bacterium. 13.37. Plasmatocyte with ingested bacterium projecting from cell surface. (13.36 and 13.37 from Ratcliffe and Rowley, 1974)

Fig. 13.38. Intracellular bacteria in plasmatocytes of *G. mellonella* after incubation in vitro. a. Plasmatocyte with three ingested *E. coli* (arrows). Note lack of a distinct phagocytic vacuole at this magnification. Cell debris (*CD*) is also in process of being ingested by invagination of cell membrane. b. Bacterium within a well-defined phagocytic vacuole (arrow). (From Ratcliffe and Rowley, 1974)

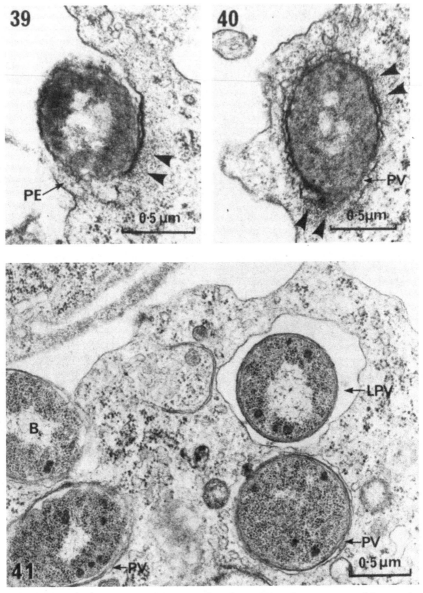

Figs. 13.39, 13.40. Stages in in vitro uptake of *E. coli* by plasmatocytes of *P. americana*. **13.39.** Formation of protoplasmic extension (*PE*) to surround a bacterium. Note increased electron density of cytoplasm at site of attachment (arrows) of bacterium to cell membrane. **13.40.** Intracellular bacterium lying in peripheral cytoplasm and surrounded by a phagocytic vacuole (*PV*). Note increased granularity of cytoplasm around phagosome (arrows). (13.39 and 13.40 from Rowley, 1977a)

Fig. 13.41. Intracellular *E. coli* in peripheral cytoplasm of filamentous plasmatocyte of *Calliphora erythrocephala*. One bacterium is surrounded by a large phagosome (*LPV*). Other phagosomes (*PV*) are, in comparison, much smaller. Bacterium (*B*) appears to be in process of being ingested. (From Rowley and Ratcliffe, 1976c)

cubation numerous bacteria were present in the cytoplasm. Further-more, in the early stages of engulfment the bacteria were surrounded by large phagocytic vacuoles 50–300 nm in diameter, although these rapidly became reduced to 50–100 nm (Fig. 13.41). Ingestion in *P. americana* was characterized by the accumulation of a fine granular material in the cytoplasm, both at the site of bacterial attachment to the cell surface and around the phagosomes within the cell (Figs. 13.39, 13.40). A similar increased cytoplasmic electron density has been recorded in some vertebrate phagocytes during ingestion (Glick et al., 1971) and may perhaps be related to cytoplasmic changes occur-ring during the movement of the phagosomes into the deeper parts of the cell. That this granularity may be associated specifically with the ingestion process is indicated by its loss or reduction around "older" phagosomes.

Additional valuable information about the attachment and ingestion processes in insect hemocytes has recently been provided by Ander-son (1976a,b). He showed that erythrocyte rosetting around *S. eri-dania* hemocytes was reversibly inhibited by the antibiotic cytochala-sin B, promoted by low temperatures, but unaffected by the addition of vinblastine or colchicine to the cultures. Cytochalasin-B-sensitive microfilaments are known to be involved in cell movements, phagocy-tosis, microvilli production, etc; disruption of the microfilaments in the *S. eridania* hemocytes probably interfered with the formation of the surface microvilli (filopods), thus preventing attachment and the initial stages of ingestion (Anderson, 1976b). Since rosetting contin-ued after the disruption of the microtubules with vinblastine and col-chicine, these organelles cannot be involved in the early stages of phagocytosis.

Anderson and his co-workers have also shown that later stages of in-gestion are temperature-sensitive and involve the expenditure of en-ergy (Anderson et al., 1973b; Anderson, 1974), probably as a result of the actin microfilament movements. Using *B. craniifer* hemocytes in-cubated with latex particles or heat-killed *S. aureus*, they showed that uptake was accompanied by stimulated glycogen breakdown, glucose consumption, and lactate production, so that energy for phagocytosis was produced via the glycolytic pathway.

Killing and digestion

This is the final stage of phagocytosis during which ingested mi-croorganisms and other biological agents may be broken down by en-zymes and antimicrobial substances. However, as mentioned above, under "Microbial Pathogens," not all ingested microorganisms are killed by the blood cells, and their fate varies from one microbe spe-cies to another and from one host species to another. For example, Ca-meron (1934) reported that tubercle bacilli remained viable within the hemocytes of *G. mellonella*, whereas Hollande (1930) found that in the blood cells of several other lepidopteran species these bacteria were digested. The phagocyte may also be used as a "haven" in which the pathogen can develop and multiply (see under "Microbial Pathogens"). *Trypanosoma rangeli* in *Rhodnius prolixus* and *Crithidia* spp. in *Drosophila virilis* (Schmittner and McGhee, 1970) use the hemocytes as sites of division and are apparently unaffected by any antimicrobial armory present. A similar situation occurs with *Bacillus thuringiensis* injected into the armyworm, *Pseudaletia unipuncta*: The bacteria are avidly phagocytized, but subsequently released by lysis of the phagocytes to produce a fatal septicemia (Wittig, 1965). Nevertheless, there are a number of reports of the destruction of ingested materials by insect hemocytes, and Salt (1970) concluded that engulfed bacteria are usually destroyed by the phagocyte. Metalnikov (1924), for example, described the dissolution of several bacterial species either directly in the cytoplasm or within vacuoles in the hemocytes. However, apart from the work of Anderson et al. (1973a,b), Ratcliffe and Rowley (1974), Rowley and Ratcliffe (1976b,c), and Kawanishi et al. (1978), few details are known of the killing and digestion of microorganisms by insect hemocytes.

Anderson et al. (1973a) tested the bactericidal activity of *B. craniifer* hemocytes against a range of bacteria (Table 13.3). They found that 25% of the *Staphylococcus aureus* and *Streptococcus faecalis* inocula was killed within 10 min, and their micrographs suggested that the PL was the cell type involved. They showed that the bactericidal capacity was attributable solely to the hemocytes and that no humoral factors were involved. Destruction of *Proteus mirabilis*, *Staphylococcus albus*, and *Serratia marcescens* also occurred, but no killing was recorded with a number of other bacterial species (Table 13.3).

In vertebrates, the antimicrobial systems of the phagocytes have been extensively studied by biochemical methods. In polymorphonuclear (PMN) leucocytes, a number of antimicrobial systems have been identified, including superoxide anion (O_2^-), acid pH, lysozyme, lactoferrin, cationic proteins, and myeloperoxidase-H_2O_2-halide systems (Klebanoff, 1975). The myeloperoxidase system is composed of a halide, usually Cl^- or I^-, hydrogen peroxide, and myeloperoxidase. This complex has been demonstrated cytochemically in the lysosomes of PMN leucocytes, and Baehner et al (1968) showed that following phagocytosis myeloperoxidase was present in the phagolysosomes. Phagocytosis by PMN leucocytes is also accompanied by a respiratory increment in the form of stimulated oxygen consumption, which results in the generation of the hydrogen peroxide (Reed, 1969; Holmes et al., 1970). This process is mediated by flavoprotein oxidase, which, utilizing NADH, reduces nitroblue tetrazolium to produce blue-black formazan deposits in the cytoplasm (Cagan and Karnovsky, 1964).

Anderson et al. (1973a), having illustrated the bactericidal activity of the *B. craniifer* hemocytes, tested for some of the antimicrobial factors mentioned above (Anderson et al., 1973b). Using a special staining technique, they showed that the myeloperoxidase-H_2O_2-halide system was virtually absent, with less than 1% of the cells staining positively. The hemocytes also failed to reduce nitroblue tetrazolium even after the ingestion of test particles. Anderson (1975) concluded that, as with vertebrate macrophages, the bactericidal mechanisms of insect hemocytes are largely unknown. Landureau and Grellet (1975) have since shown that chitinase, which has properties similar to lysozyme, was formed in *P. americana* hemocyte cultures, so this enzyme could possibly play an important role in bacterial killing.

Microbial digestion within vertebrate phagocytes has also been successfully studied by cytochemical and ultrastructural techniques. Early studies showed the fusion of lysosome-like bodies with phagosomes containing bacteria (e.g., Nelson et al., 1962), and later studies, using enzyme cytochemical methods, identified both the primary and secondary lysosomes involved (Horn et al., 1964; Cotran and Litt, 1969). Recently, more attention has been paid to comparing the fate of pathogenic and nonpathogenic bacteria in vertebrate leucocytes. Armstrong and Hart (1971, 1975) showed by quantitative electron microscopy that little fusion of lysosomes and phagosomes containing *Mycobacterium tuberculosis* occurred in cultured macrophages and suggested that the bacteria had the ability to block fusion. In another study, with *Neisseria gonorrheae* and PMN leucocytes, Swanson and Zeligs (1974) found that although lysosomal fusion occurred, killing was impaired owing to leakage of antimicrobial factors through channels formed by the pili attached to the gonococci and leading from the phagolysosome to the outside of the cells. These studies clearly illus-

Table 13.3. Bactericidal activity of *Blaberus craniifer* hemocytes in vitro

Bacterial species	Percent killing after 1 hr
Streptococcus faecalis	45.0
Staphylococcus aureus 502A	51.5
Staphylococcus albus	42.9
Serratia marcescens	36.6
Proteus mirabilis P.O.	32.9
Proteus mirabilis R.Z.	38.3
Pseudomonas aeruginosa	—
Escherichia coli	—
Salmònella typhosa	—
Diplococcus pneumoniae	—

Source: After Anderson et al. (1973a).

trate the potential usefulness of electron microscopy in determining the fate of ingested microorganisms.

Using cytochemistry and electron microscopy, we have attempted to learn something about the intracellular events following phagocytosis of bacteria and other test particles in monolayer and suspension cultures of *G. mellonella*, *C. erythrocephala*, and *P. americana* hemocytes (Ratcliffe and Rowley, 1974; Rowley and Ratcliffe, 1976b,c; Rowley, 1977a).

In *G. mellonella*, PLs taken directly from the insect ingested bacteria after 15–30 min in vitro, but there was little evidence of lysosomal activity in these cells (Fig. 13.38*a*). In longer-term cultures, however, which had been incubated for up to 72 hr, the PLs of both experimental and control cultures contained large numbers of lysosome-like structures termed "multivesicular bodies" (MVBs) (Fig. 13.42). These structures probably arose by budding from the active-looking Golgi, and although they usually lacked acid phosphatase (AcPase) activity, fusion of MVBs with phagosomes containing *E. coli* was often observed (Fig. 13.42). As a result of this fusion, many of the bacteria became surrounded by vesicles derived from the MVBs (Figs. 13.42, 13.43). The intracellular bacteria also often contained a large amount of AcPase activity, and as the MVBs were mainly AcPase-negative, at least some of this material probably originated from small AcPase-positive Golgi-like vesicles, which were sometimes present in the cytoplasm adjacent to the phagosomes. When latex was used as a test particle, all the stages in fusion of these lysosomes with the phagosomes were followed, and later stages were characterized by a "halo" of AcPase reaction product in the phagosomes surrounding the latex beads (Rowley and Ratcliffe, unpub. data). After 12–24 hr in culture, AcPase activity was particularly intense in the bacteria, some of which appeared swollen and surrounded by myelin configurations (Fig. 13.43). Sometimes after incubation, however, the intracellular bacteria appeared normal and either remained entrapped or else rapidly divided within the cell, from which they presumably eventually escaped into the medium. This study shows that under appropriate conditions the PLs have the potential ability to change dramatically. If such changes can occur following microbial challenge in vivo, then the insect cellular defenses may be much more effective than hitherto realized.

In contrast to *G. mellonella*, similar studies with *C. erythrocephala* showed that after only 15 min incubation of *E. coli* with hemocytes taken directly from the insect some of the intracellular bacteria were surrounded by numerous lysosomes, and an electron-dense material, probably resulting from lysosomal discharge, had accumulated in the phagosomes around the bacteria (Figs. 13.44, 13.45). After 120 min incubation, many of the bacteria were swollen and pleomorphic, and there was evidence of cell wall damage (Fig. 13.46).

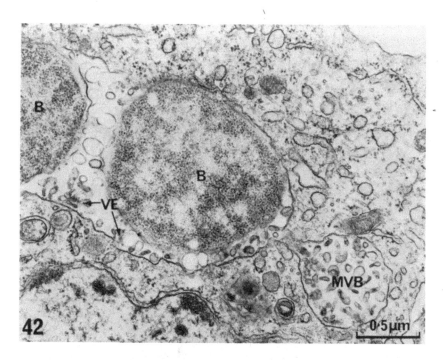

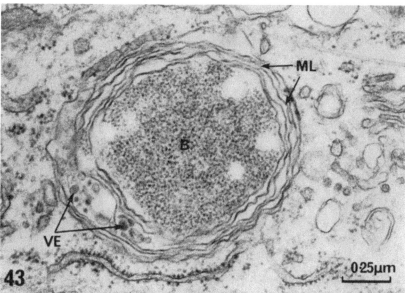

Fig. 13.42. Fusion of multivesicular body (*MVB*), formed in plasmatocyte of *G. mellonella*, with phagosome containing *E. coli* bacteria (*B*). Vesicles (*VE*) derived from MVB are already present in phagosome. Long-term culture. (From Rowley and Ratcliffe, 1976b)

Fig. 13.43. Late stage in degradation of *E. coli* in plasmatocytes of *G. mellonella* after 12 hr in vitro. Bacterium (*B*) is surrounded by myelinlike membranes (*ML*) and a few vesicles (*VE*), presumably derived from multivesicular body. Myelin membranes are characteristic structures present during digestion of various materials. (From Rowley, unpub.)

365

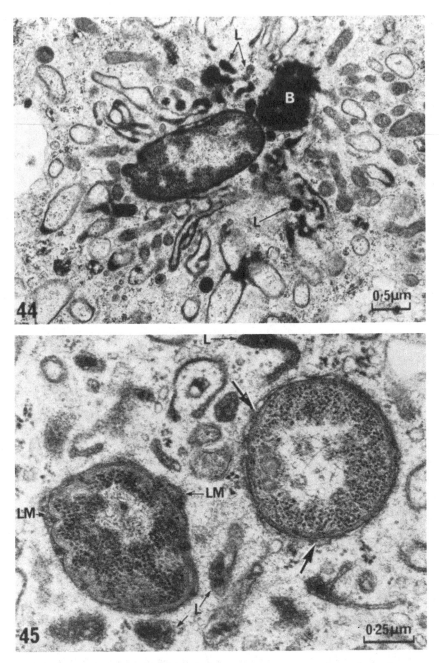

Figs. 13.44, 13.45. Interaction between lysosome-like bodies and *E. coli* in plasmato-cytes of *Calliphora erythrocephala*. **13.44.** Accumulation of electron-dense lysosomes (*L*) and membranous structures around intracellular bacteria. Fusion of lysosomes with bacteria has probably occurred, resulting in an increased electron density espe-cially of bacterium labeled *B*. **13.45.** Accumulation of electron-dense lysosome-like material (*LM*) in phagosome surrounding a bacterium. Arrows also indicate possible lysosomal material. *L* = lysosomes. (**13.44** and **13.45** from Rowley and Ratcliffe, 1976c)

Preliminary work with *P. americana* hemocytes incubated with latex particles has also provided some additional information about intracellular events. During incubation, the PLs of both experimental and control cultures spontaneously degranulated (Figs. 13.47, 13.48) and discharged their contents into the medium. In experimental cultures, some degranulation was also associated with fusion by AcPase-negative granules with the intracellular test particles (Fig. 13.49). After 45 min incubation, however, the latex beads were, as in *G. mellonella*, surrounded by an electron-dense, AcPase-positive reaction product (Fig. 13.50), suggesting that fusion with small AcPase-positive lysosomes present in the cytoplasm had also probably occurred. The spontaneous granule discharge into the medium may correspond to the chitinase release reported by Landureau and Grellet (1975) in *P. americana* hemocyte cultures. The fusion of these granules with the intracellular particulates may thus be highly significant and account for any killing of ingested microorganisms. The concomitant discharge of the AcPase-positive lysosomes may subsequently be responsible for the digestion of the killed microbes.

From these studies we can conclude that many of the events following ingestion and lysosomal fusion in insect blood cells are basically similar to those reported during intracellular killing in vertebrate leucocytes. For example, the degranulation process in *P. americana* is also characteristic of vertebrate phagocytosis (Stossel, 1974), although the hemocyte granules are apparently nonlysosomal. Furthermore,

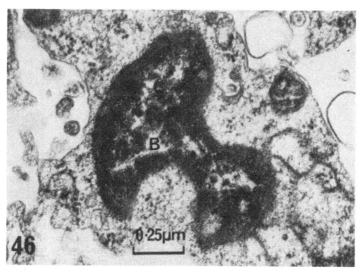

Fig. 13.46. Possible early stage in intracellular breakdown of *E. coli* (*B*) in plasmatocyte of *C. erythrocephala* after 120 min incubation. Note swollen, pleomorphic appearance of bacterium. (From Rowley and Ratcliffe, 1976c)

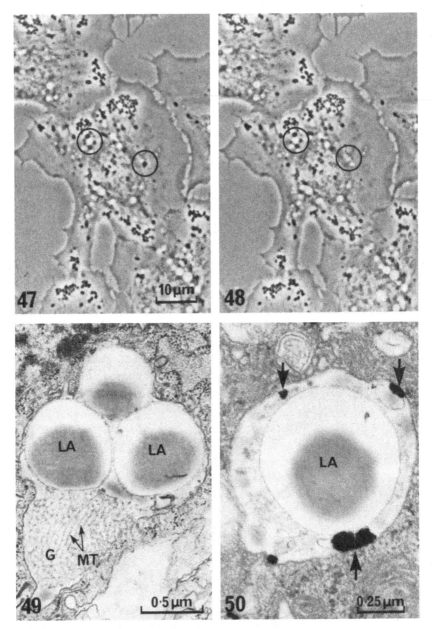

Figs. 13.47, 13.48. Degranulation of plasmatocyte of *P. americana* in vitro. Micrographs, taken 5 sec apart after approximately 75 min in vitro, show lysis and discharge of two granules (enclosed in circles). Such degranulation often leaves a small vacuole, which disappears within a few seconds. (From Rowley, unpub.)

Fig. 13.49. Discharge of granule (*G*) in cytoplasm of plasmatocyte of *P. americana* into phagosome containing latex beads (*LA*). Note microtubule remnants (*MT*) present in granule. Such fusion may well be providing nonlysosomal killing factors and/or digestive enzymes in phagosome. (From Rowley, unpub.)

vertebrate PMN leucocytes enclose a heterogeneous population of granules, and the so-called "secondary granules" (specific granules), which contain alkaline phosphatase and lysozyme, discharge into the phagosome before the "primary granules" (azurophilic granules), which contain the acid hydrolases (Cline, 1975) – events not unlike those recorded during phagocytosis by the *P. americana* hemocytes.

Finally, a recent study of the route and mode of invasion of *Bacillus popilliae* in the larvae of the European chafer, *Amphimallon majalis* (Kawanishi et al., 1978), has shown that many of the in vitro observations on intracellular events in insect blood cells, reported above, also apply to the in vivo situation. Invasion of the gut columnar epithelium by *B. popilliae* resulted in an inflammation reaction by granule-containing hemocytes at the infection focus. These hemocytes formed a capsule on the hemocoelic surfaces and contained numerous phagocytized bacteria (Figs. 13.51–13.54). The phagosomes enclosed not only bacteria but also material that closely resembled the matrix of intact granules (Figs. 13.51–13.53). These findings are similar to the internal degranulation process observed in *P. americana* hemocytes (Fig. 13.49, see above). As with *G. mellonella*, the *A. majalis* hemocytes sometimes overcame the infection, as evidenced by the presence of residual bodies containing bacterial cell wall fragments, but in some cells the bacteria apparently divided and overcame the cells. The bactericidal activity of the hemocytes seemed to be correlated with the tightness of the bacilli within the phagosomes, with maximal degradation occurring in tight vacuoles. Degrading bacteria had the following characteristics: (1) loss of fimbriae, (2) retraction and/or alteration of the cytoplasm, (3) reduction in nuclear material, and (4) alterations of the cell wall (Figs. 13.53, 13.54) (Kawanishi et al., 1978). These changes in bacterial structure are similar to those reported in intracellular bacteria in in vitro experiments (Rowley and Ratcliffe, 1976c).

Another finding of great significance by Kawanishi et al. (1978) was that bacterial endocytosis and degradation were not confined to the hemocytes, but occurred also in the mesenteric cells lining the gut. Phagocytic reactions in insects are not, therefore, the prerogative of the hemocytes, but may also involve other tissues of the body.

13.3.4. Nodule formation

This defense reaction operates against large doses of particulate material that invade the hemocoel and that probably cannot be effec-

Fig. 13.50. Distribution of enzyme acid phosphatase after incubation of latex beads and plasmatocytes of *P. americana* for 45 min in vitro. Enzyme activity is marked by intense lead deposit (arrows) found in phagosome around a latex particle (*LA*). Such deposits indicate that lysosomal fusion with phagosome has occurred. (From Rowley, unpub.)

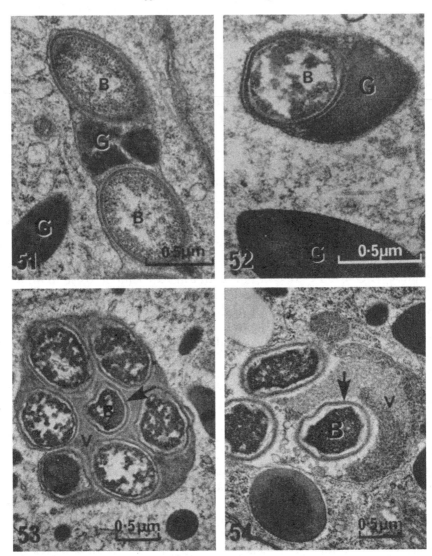

Figs. 13.51–13.54. Stages in breakdown of bacterium, *Bacillus popilliae*, in hemo-cytes of *Amphimallon majalis*. **13.51.** Granules (G) in cytoplasm and in phagosomes containing bacilli (B). **13.52.** Bacillus (B) has a vacuolar appearance characteristic of degradative changes. This material in phagosomes has apparently been derived by fusion with granules (G). **13.53.** Large phagocytic vacuole containing a substance (V) similar to that seen in granules. Arrow indicates doubling of cell wall of bacterium (B). **13.54.** Fibrillar substance (V) present in vacuole containing degrading bacteria (B). Content of vacuole and agglomerated bacterial cytoplasm are atypical in these hemocytes. Arrow indicates apparent doubling of cell wall, characteristic of degraded bacteria in these cells. (**13.51–13.54** from Kawanishi et al., 1978)

tively cleared by phagocytosis alone. Nodules occur in arthropods in response both to naturally occurring infections (Fig. 13.1) and to experimental injections of particulate substances (Metchnikoff, 1884; Bucher, 1959; Salt, 1970). They have been recorded to form around India ink (Durham, 1891), Chinese ink (Marschall, 1966), carmine and Sudan II (Ryan and Nicholas, 1972), celloidin (Iwasaki, 1927), erythrocytes (Ryan and Nicholas, 1972), endotoxin (Schwalbe and Boush, 1971), many different types of bacteria (Metalnikov, 1908, 1924, 1927; Metalnikov and Chorine, 1929; Hollande, 1930; Cameron, 1934; Toumanoff, 1949; Bucher, 1959; Wittig, 1965; Sutter and Raun, 1967; Kurstak et al., 1969; Hoffmann et al., 1974; Gagen and Ratcliffe, 1976), fungi (Speare, 1920; Ermin, 1939; Kawakami, 1965; Vey, 1968; Vey et al., 1968, 1973; Vey and Vago, 1969, 1971; Vey and Quiot, 1975; Vey and Fargues, 1977), and protozoa (Huger, 1960; Henry, 1967; Evans and Elias, 1970; Schmittner and McGhee, 1970). Despite this long history and the recent increase of interest in insect defense reactions, our knowledge about the mechanism of formation of nodules, about the fate of the segregated particles, and about the contribution of nodules to the overall host defenses of insects is small.

Structure and mode of formation

The literature describing nodule formation in insects is confusing, although most authors agree that the end result of this process is an aggregate of blood cells entrapping the particles in a central melanized region surrounded by a sheath of blood cells (Fig. 13.2) (e.g., Hollande, 1930; Vey and Vago, 1971). Frequently, it is difficult to distinguish such structures from capsules (Salt, 1970), and indeed both reactions can occur within one insect in response to invasion by the same parasite (Figs. 13.5, 13.6).

Until recently, it was generally accepted that the first stage in nodule formation was phagocytosis of the particles and that subsequently the phagocytic cells clumped and became ensheathed by fresh hemocytes (Cuénot, 1895; Metalnikov, 1908, 1924, 1927; Iwasaki, 1927; Hollande, 1930; Chorine, 1931). Metalnikov (1924), however, also believed that association of particles and hemocytes could occur without prior phagocytosis, with the bacteria attaching to the outsides of the hemocytes, which then formed giant cells by breakdown of the cell walls. This view was shared by Marschall (1966), who observed that in *Tenebrio molitor* nodule formation began with the entrapment of injected Chinese ink particles in a "network of blood cells," a process much like a coagulation reaction (Grégoire, 1970). But Vey and his coworkers (Vey, 1968; Vey et al., 1968, 1973; Vey and Vago, 1969; Vey and Quiot, 1975; Vey and Fargues, 1977), experimenting mainly with *G. mellonella* larvae in vivo and *G. mellonella* hemocyte cultures injected and inoculated with *Aspergillus* spp. conidia, respectively,

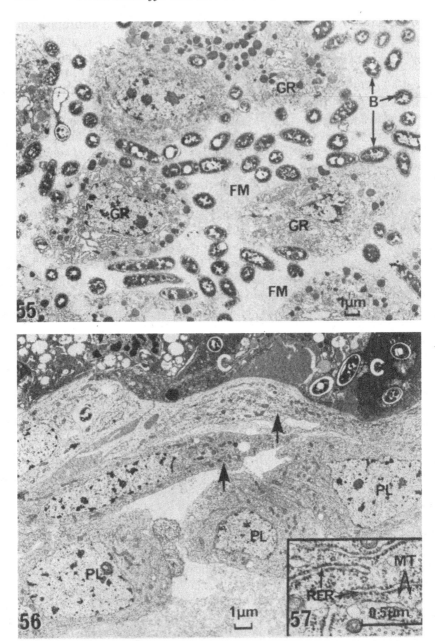

Fig. 13.55. First stage of nodule formation in *G. mellonella* after injection of *Bacillus cereus*. Cell–bacteria aggregate formed 1 min after injection of bacteria. Note large amount of flocculent material (*FM*) and entrapped bacteria (*B*) around granulocytes (*GR*). Flocculent material is formed in hemolymph upon discharge of granules from granulocytes. (From Ratcliffe and Gagen, 1977)

Fig. 13.56. Beginning of second stage of nodule formation in *G. mellonella* 4 hr after injection of bacteria, with attachment of typical plasmatocytes (*PL*) to melanized core

found no prior aggregation of hemocytes or phagocytosis of spores and reported that nodules were formed merely by PLs enveloping melanizing spores to form multicellular sheaths (Fig. 13.2). They also reported that nodule or "granuloma" formation in vitro involved both micro- and macroplasmatocytes (probably GRs and PLs, respectively) and that a chemotactic stimulus emanated from the fungal spores (Vey et al., 1968).

Recently, additional information has become available concerning nodule formation, and unless otherwise stated, the work reported below has been taken from Gagen and Ratcliffe (1976) and Ratcliffe and Gagen (1976, 1977).

In these studies, live or dead *Bacillus cereus, Escherichia coli, Sarcina lutea,* and *Staphylococcus aureus* were injected into final-instar larvae of *G. mellonella* and *P. brassicae,* which were killed and dissected at intervals of from 1 min to 24 hr postinjection. Preliminary experiments have also been carried out on last-instar *T. molitor* larvae and adult *C. extradentatus* injected with killed *B. cereus* and dissected at 5 min, 1 hr, and 24 hr postinoculation (Ratcliffe, Rowley, and Gagen, unpub. data).

Nodules developed within 1–5 min in all insects except those injected with *S. aureus,* and, although varying considerably in detail, nodule formation in all four insect species appeared to be divided into two distinct stages.

In *G. mellonella,* the first stage involved mainly the GRs (Figs. 13.14–13.16). By 1 min postinjection, numerous GRs had clumped together and were discharging their granules to entrap large numbers of bacteria (Fig. 13.55). All stages in granule breakdown were apparent, and the changes occurring in these cells were identical to those described for this cell type during hemolymph coagulation (Rowley and Ratcliffe, 1976a). The clumps formed in *G. mellonella* and the other species were attached to the gut, malpighian tubules, fat body, tracheae, and body wall (Figs. 13.1, 13.58, 13.59). A similar process was observed in *T. molitor* and *C. extradentatus* 5 min after injection, but involved highly degenerate COs in the formation of an extracellular coagulum entrapping the bacteria. In *C. extradentatus,* GRs were also present and degranulating at the periphery of the clumps. The point to emphasize is that at this early stage comparatively few PLs, or any

(continued from facing page)
(C). Some plasmatocytes (arrows) are already flattening onto surface of necrotic core. Core consists of numerous bacteria, an extensive melanized matrix, and necrotic granulocytes. (From Ratcliffe and Gagen, 1977)
Fig. 13.57. Portion of incompletely flattened plasmatocyte from middle region of sheath 24 hr postinjection, showing abundant rough endoplasmic reticulum (*RER*) and numerous microtubules (*MT*). (From Ratcliffe and Gagen, 1977)

other cell type, were observed in nodules from any of the insects. By 5–60 min, in all species, melanization of the degenerating hemocytes and entrapped bacteria had begun, and compaction of the aggregates had taken place. These processes continued until, by 24 hr, the bacteria were enclosed in a central melanized mass of necrotic cells (Figs. 13.59, 13.63). In G. mellonella, these nodules were visible as black specks showing through the cuticle. Upon dissection they were seen to ramify throughout the body cavity (Fig. 13.58).

From approximately 1 to 6 hr postinjection, the second stage in nodule formation had begun, and although this phase varied from one insect species to another, it was generally characterized by the attachment of PLs to the central melanizing core. In G. mellonella and P. brassicae, attachment of PLs began at about 2–6 hr postinjection (Fig. 13.56). These cells contained very few phagocytized bacteria, and they rapidly flattened down until by 24 hr a multicellular sheath had formed (Fig. 13.59), composed of three distinct regions (see "Encapsulation," below). The inner region was composed of flattened, degenerating cells; the middle region contained extremely flattened cells enclosing numerous microtubules, desmosomes, and a well-developed Golgi and RER (Fig. 13.57); the outer region was unremarkable and consisted of newly attached PLs. The main events in nodule

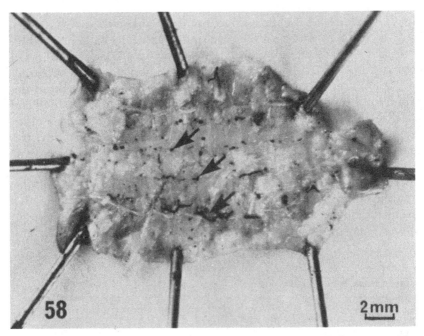

Fig. 13.58. Dissection of G. mellonella 12 hr after injection of killed B. cereus. Note numerous melanized nodules throughout insect (arrows). (From Ratcliffe and Gagen, 1976)

formation in *G. mellonella* are summarized in Fig. 13.60. In *C. extradentatus*, the attaching PLs were observed as early as 1 hr postinjection and, in contrast to the other species, contained numerous phagocytized bacteria (Fig. 13.61), some of which appeared to have fused with intracellular granules (lysosomes?) (Fig. 13.62). By 24 hr, how-

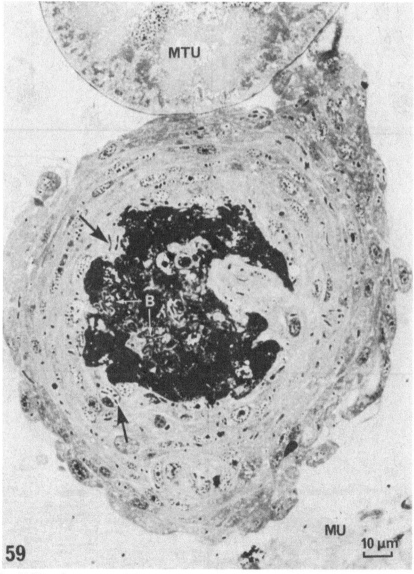

Fig. 13.59. *P. brassicae* nodule 24 hr after injection of *B. cereus*. Note typical structure, with central melanized core containing bacteria (*B*) and sheath of plasmatocytes enclosing a few ingested bacteria (arrows). Nodule is attached to a malpighian tubule (*MTU*). *MU* = muscle. (From Ratcliffe and Gagen, 1976)

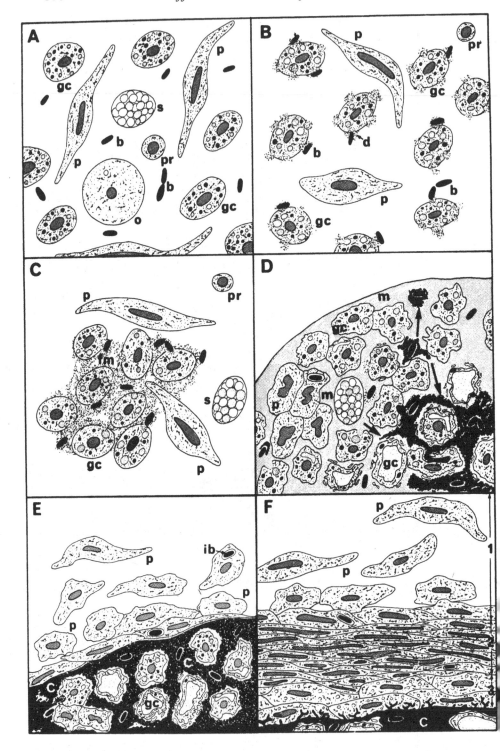

ever, a typical multilayered, flattened sheath was present around a ne-crotic, melanized core. The *T. molitor* nodules also showed some pha-gocytosis by the attaching PLs after only 1 hr, but by 24 hr these structures were very different from those of the other species. A typi-cal necrotic, melanized core was present, but the outer region con-sisted of intact PLs embedded in a matrix of fine flocculent material (Figs. 13.63, 13.64). These cells showed only minimal flattening, and there was no formation of specialized organelles such as desmosomes.

From these studies certain conclusions can be drawn (see Fig. 13.60):

1. Nodule formation apparently occurs in two stages.

2. The first stage is initiated by a granule-containing cell (GR and/or CO) that, in random contact with the bacteria, degranulates to form a sticky substance or a localized coagulum entrapping large numbers of bacteria and other reactive hemocytes. The rapidity of this reaction precludes the involvement of hormonal imbalances or chemotaxis at this stage.

3. Prior to entrapment by the hemocytes, the bacteria are not agglu-tinated by humoral factors; their distribution is clearly random in the cell aggregates.

4. During the first stage, phagocytosis is not an important compo-nent of nodule formation.

5. The second stage involves the specific attachment of large num-bers of PLs to form the outer sheath. These cells contain variable numbers of phagocytized bacteria, depending on the insect species involved. The delay in acquisition of these PLs and the fact that other cell types are rarely involved indicate a chemotactic attraction of these cells to the degenerating central region. This hypothesis has been tested in some preliminary experiments (see "Chemotaxis, Attach-ment, and Recognition of Foreignness," above), which showed, in an in vitro system, the unidirectional movement of *G. mellonella* PLs to-ward nodules (Fig. 13.32). Any chemotactic factors probably origi-

Fig. 13.60. Diagrammatic representation of main events during nodule formation in *G. mellonella*. **A.** Hemocyte types free in hemolymph immediately following injec-tion of bacteria (*b*). *gc* = granulocyte; *p* = plasmatocyte; *pr* = prohemocyte; *s* = spherulocyte; *o* = oenocytoid. **B.** Initial stage in nodule formation. On random contact with bacteria, granulocytes immediately discharge their contents (*d*) to entrap bacteria. **C.** By 1 min postinjection, granulocytes coated with discharged contents co-here to segregate bacteria. *fm* = flocculent material. **D.** At 5–30 min, clumps com-pact and matrix (*m*) is melanizing (arrows), especially in regions of the bacteria. **E.** Beginning of second stage of nodule formation by specific attachment of large num-bers of plasmatocytes, some of which contain intracellular bacteria (*ib*), to inner melanized core (*c*). **F.** Mature nodule is formed by 12–24 hr, and sheath is clearly di-vided into an outer region of newly attached cells (*1*), a middle region of extremely flattened cells (*2*), and an inner region of partially flattened cells containing mela-nized inclusions (*3*). (From Ratcliffe and Gagen, 1977)

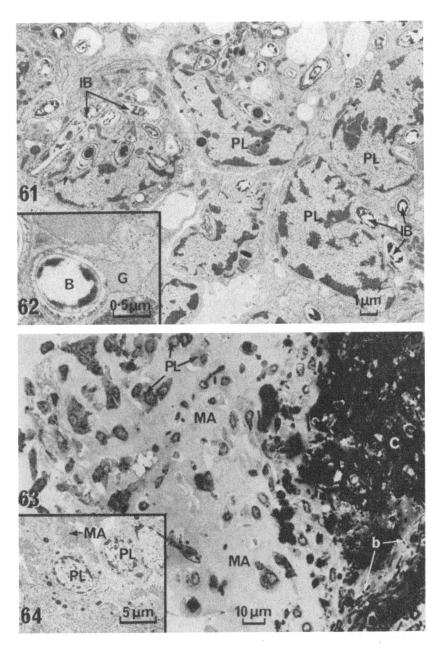

Fig. 13.61. Nodule formed 1 hr after injection of *B. cereus* in *C. extradentatus* composed virtually entirely of plasmatocytes (*PL*). Many bacteria are intracellular (*IB*). (From Ratcliffe, Rowley, and Gagen, unpub.)

Fig. 13.62. *C. extradentatus* nodule showing fusion of granule (*G*) with phagosome containing bacterium (*B*). (From Ratcliffe, Rowley, and Gagen, unpub.)

Fig. 13.63. Light micrograph of 24-hr nodule in *Tenebrio molitor*. Note necrotic core (*C*) containing melanized bacteria (*b*) very much like those seen in *G. mellonella* and *P. brassicae*. However, plasmatocytes (*PL*) in outer sheath are unflattened and surrounded by a large amount of extracellular material (*MA*). (From Ratcliffe, Rowley, and Gagen, unpub.)

378

nated from the GR/CO debris and not from the bacteria (Ratcliffe et al., 1976a).

6. Only the GRs/COs and the PLs are apparently involved in nodule formation.

7. The fine structure of the GRs/COs and the PLs shows that these cells are well adapted to their roles in nodule formation. The GRs/COs contain large numbers of granules synthesized in a cytoplasm rich in RER or polyribosomes (Figs. 13.15–13.18) and are clearly highly sensitive to "non-self." The PLs are more stable cells, capable of phagocytosis and movement by the formation of pseudopodia (Figs. 13.10, 13.11). They frequently become flattened and joined together by desmosomes. The cytoplasm of these flattened cells also often contains an enhanced amount of RER and active-looking Golgi, which may be responsible for synthesizing the numerous microtubules present.

The above description of nodule formation is very similar to that proposed by Metalnikov (1924) and observed by Marschall (1966) following the injection of Chinese ink into *T. molitor*. Our conclusions, however, are based on observations from only a very small number of species; it is hoped they apply to nodule formation in insects in general.

Resistance to segregation in nodules

Very little is known about the avoidance of this cellular defense reaction by pathogenic microorganisms. However, *B. thuringiensis* in *Ephestia kühniella* (Kurstak, 1965), certain pathogenic strains of *S. aureus* in *G. mellonella* (Kurstak et al., 1969; Ratcliffe and Gagen, 1976), and *Metarrhizium anisopliae* in hemocyte cultures of *Oryctes rhinoceros* (Vey and Quiot, 1975) resist segregation into nodules by producing lytic and cytotoxic substances that kill or stress the normally reactive hemocytes. The substances involved may be toxic metabolic products such as lipopolysaccharides (Kurstak et al., 1969), proteolytic enzymes (Bucher, 1960), or other unknown chemicals.

Occurrence and effectiveness as defense reaction

As we have previously discussed (see "Diseases of Insects," above), nodules occur in natural field populations of insects and are not confined to experimental laboratory animals (Speare, 1920; Bucher, 1959; Huger, 1960; Henry, 1967; Sutter and Raun, 1967; Bywater and Ratcliffe, unpub. data). These studies indicate that large numbers of mi-

Fig. 13.64. Electron micrograph showing two plasmatocytes *(PL)* of *T. molitor* and flocculent matrix material *(MA)* surrounding these cells. (From Ratcliffe, Rowley, and Gagen, unpub.)

croorganisms can enter the hemocoel via the gut and elicit this cellular reaction. Virtually nothing, however, is known of the fate of the segregated microorganisms, except that some appear to swell and break down within the melanized cores (Metalnikov, 1924), others can continue dividing and form spores (Huger, 1960), and still others have been observed to resurge and break free of the nodules (Vey and Vago, 1969; Vey and Fargues, 1977). Nodules may persist throughout the remainder of the life of the insect, and both Ermin (1939) and Cameron (1934) reported that in *G. mellonella* nodules formed in the larvae could be recovered from adult insects. Metalnikov (1924) also believed that nodules sometimes attached to the body wall and were discharged to the outside by rupture of the overlying cuticle. Salt (1970) thought that Metalnikov's observations were possibly attributable to the combination of a nodule with a cuticular wound and ecdysis. Schwalbe and Boush (1971), however, in their study of the fate of radioactively labeled endotoxin injected into *G. mellonella*, found that the final deposition of the label was in the insect's cuticle.

The formation of nodules is extremely rapid (Gagen and Ratcliffe, 1976; Ratcliffe and Rowley, unpub. data). In *G. mellonella*, *P. brassicae*, *T. molitor*, and *C. extradentatus* well-developed nodules were present in the hemocoel 5 min after injection of bacteria. In *G. mellonella* and *P. brassicae* nodule formation was accompanied, within 5 min of injection, by approximately an 80% and a 90% drop, respectively, in hemocyte numbers (Figs. 13.65, 13.66). This rapid decline in cells present was immediately followed by a gradual rise in hemocyte numbers (Figs. 13.65, 13.66), as a result either of sedentary cells reentering circulation or of the release of cells from the hemopoietic organs. This fall-off in hemocyte numbers was accompanied by a reduction in circulating bacteria; in *G. mellonella* 95% of a 5×10^5 bacterial injection had been removed from the circulation after only 5 min (Fig. 13.67) (Gagen and Ratcliffe, 1976).

This study clearly shows the potential efficiency of nodule formation in clearing the hemocoel of invading microorganisms. Additional work, however, is required to monitor the fate of the sequestered bacteria and to determine if these are killed or simply prevented from reentering the hemocoel by the physical barriers imposed by the hemocyte sheath. Toumanoff (1949) considered the number of bacteria destroyed in nodules to be related to the degree of melanization, and the role of melanin, if any, should be carefully considered.

13.3.5. Encapsulation

This cellular defense reaction has been studied in over 15 insect orders (Salt, 1963; Hall et al., 1969) and is the method by which insects deal with foreign bodies that enter the hemocoel and that are too

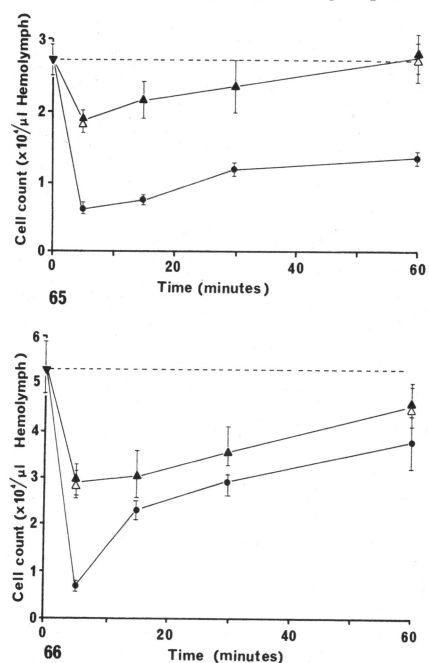

Figs. 13.65, 13.66. Hemocyte counts after injections of 1×10^5 killed *B. cereus* per microliter of hemolymph in *G. mellonella* (**13.65**) and *P. brassicae* (**13.66**). --▼-- = uninjected control level; ▲ = saline-injected controls; △ = pricked controls; ● = experimentals. Bars represent ± 1 standard error of mean. (Redrawn from Gagen and Ratcliffe, 1976)

large to be phagocytized by individual hemocytes. Recently a number of excellent reviews have been published on encapsulation (Shapiro, 1969; Salt, 1970; Nappi, 1974, 1975; Poinar, 1974). The studies to date have given us some idea of capsule morphology and the factors governing capsule formation, but much has yet to be learned about the underlying control mechanisms.

Capsules are formed in insects against a wide range of nonbiological and biological objects, including Araldite (Fig. 13.68) (Grimstone et al., 1967; Reik, 1968; François, 1975; Hillen, 1977), cellophane fragments (Matz, 1965; Brehélin et al., 1975; Zachary et al., 1975), glass rods (Salt, 1956, 1957), latex (Lackie, 1976), nylon fibers (Salt, 1965; Scott, 1971c; Sato et al., 1976; Hillen, 1977), biological implants (Figs. 13.69–13.71) (Scott, 1971c; Lackie, 1976; Schmit and Ratcliffe, 1977), acanthocephalans (Brennan and Cheng, 1975), cestodes (Lackie, 1976), nematodes (Fig. 13.7) (Poinar et al., 1968; Poinar and Leutenegger, 1971; Poinar, 1974; Nappi and Stoffolano, 1971, 1972; Nappi, 1974, 1975), insect parasitoids (Salt, 1963, 1970, 1973, 1975; van den Bosch, 1964; Kitano, 1969a,b, 1974; Nappi, 1970; Vinson, 1972, 1977; Hillen, 1977), large protozoa (Fig. 13.5), and fungi (Salt, 1970; Götz and Vey, 1974) and during tumorigenesis (Rizki and Rizki, 1976) and invasion of the gut by pathogens (Splittstoesser et al., 1978).

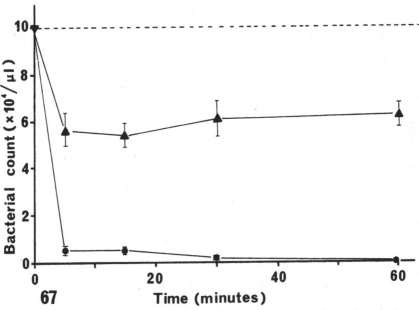

Fig. 13.67. Bacterial counts after injection of *B. cereus*, 1 × 10⁵/μl hemolymph, into *P. brassicae* ▲ and *G. mellonella* ●. --▼-- = injected dose. This drop in number of bacteria is almost certainly caused by their rapid incorporation into nodules. (Redrawn from Gagen and Ratcliffe, 1976)

Structure and mode of formation

Encapsulation usually occurs within 24 hr of parasitization (Shapiro, 1969) and frequently, but not always (see below), involves the blood cells in the formation of a multicellular sheath that may partially or wholly become melanized and kill the enclosed parasite. For the purposes of this synopsis, capsules are classified simply as cellular and humoral, although the cellular forms are sometimes subdivided into simple and melanotic, depending on whether or not melanization of the parasite occurs (Poinar, 1974).

A variety of blood cell types has been reported to take part in cellular encapsulation: amoebocytes (Timberlake, 1912), leucocytes (Hollande, 1920), lymphocytes (Schneider, 1950), macronucleocytes (Paillot, 1928), micronucleocytes (Boese, 1936), proleucocytes (Lartschenko, 1933), SPs (Meyer, 1926), and more recently, GRs (Brehélin et al., 1975) and thrombocytoids (Zachary et al., 1975). Most of these hemocytes probably fall within the PL and/or GR categories (Jones, 1962; Shapiro, 1969; Price and Ratcliffe, 1974), although the thrombocytoid of the Diptera may possibly be regarded as a separate cell type (Zachary et al., 1972). From recent electron microscope studies, it is evident that capsule formation may not be quite as simple as had been suggested by light microscopy, but may involve a well-balanced interplay among different hemocyte types (Sato et al., 1976; Schmit and Ratcliffe, 1977).

Electron microscopy studies of encapsulation (e.g., Grimstone et al., 1967; Poinar et al., 1968; Misko, 1972; Brehélin et al., 1975; François, 1975; Zachary et al., 1975; Schmit and Ratcliffe, 1977) have described capsules in a variety of species and have showed that in many cases, although varying in numbers of cell layers, melanization, etc., the fine structure of the completed capsules resembles that of *E. kühniella*, described by Grimstone et al. (1967). The *E. kühniella* capsule 72 hr after implantation of a piece of Araldite consisted of three distinct layers (Grimstone et al., 1967): (1) an inner region about 10 cells thick, closely adhering to the implant and composed of necrotic PLs that were not appreciably flattened; (2) a middle region of approximately 20 or more cell layers, composed of extensively flattened cells that showed few signs of necrosis; (3) an outer region about 5 cells thick, which adhered loosely to the rest of the capsule and contained cells very similar in fine structure to free blood cells. Plasmatocytes were the only cell type seen in the completed capsule.

The fine structure of a *G. mellonella* capsule formed around *Schistocerca gregaria* nerve cord implant illustrates the characteristic structure described above (Figs. 13.72–13.75) (Schmit and Ratcliffe, 1977). One of the most striking features of this type of capsule is the extreme modification of the cells of the middle region. Not only are they intensely flattened, but they contain numerous microtubules and in-

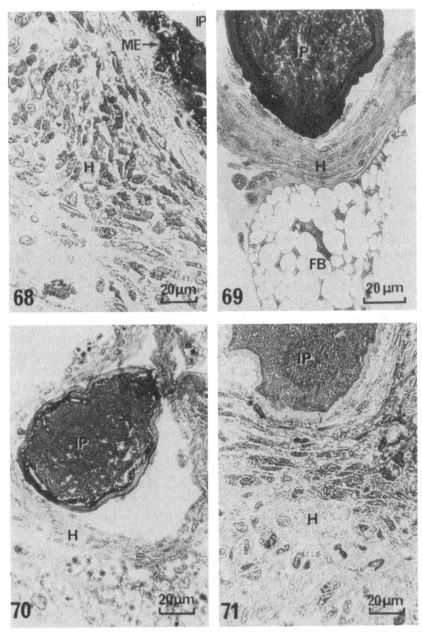

Figs. 13.68–13.71. Appearance of 72-hr-old capsules in four insect species. 13.68. Section through capsule formed in *C. extradentatus* around implanted Araldite (*IP*). Innermost layer is heavily melanized (*ME*) and outer region is composed of hemocytes (*H*), most of which show little sign of flattening. 13.69. Typical capsule formed in *G. mellonella* after implantation of a piece of *Schistocerca gregaria* nerve cord (*IP*). Note highly flattened hemocytes (*H*), which make up bulk of capsule. *FB* = fat body. 13.70. Capsule formed in *T. molitor* around *S. gregaria* nerve cord (*IP*). Like

creased numbers of mitochondria and ribosomes and are joined together by desmosome-like bodies (Fig. 13.76) (François, 1975). Similar modifications have also been reported in the cells of the outer region of capsules (Brehélin et al., 1975). Additional features of this type of capsule include cytolysosomes in the inner necrotic cell layers and a dense intercellular substance that may help to hold the cells together (Grimstone et al., 1967). In *E. kühniella* this substance is composed of acid mucopolysaccharide fused with a basic protein secreted by the cells (Reik, 1968). The degranulation of hemocytes, which has been observed in various regions of capsules, may represent the discharge of this proteinaceous material (Misko, 1972; Schmit unpub. data).

Not all cellular capsules, however, have this flattened, multilayered structure; a number of variations exists. In the phasmids *Carausius morosus* and *C. extradentatus*, for example, the PLs gather around the Araldite implants, but do not adhere closely to one another and are surrounded by a "dense matrix" material (Reik, 1968; Schmit and Ratcliffe, 1978). There is no sign of flattening until several days after implantation (Fig. 13.68), and in many ways these capsules resemble the nodules formed around bacteria in *T. molitor* (Fig. 13.63). Figures 13.68–13.71 show capsules from *C. extradentatus*, *G. mellonella*, *T. molitor*, and *B. craniifer* 72 hr after implantation of Araldite or *S. gregaria* nerve cord; they clearly illustrate the variation that can occur in the extent of hemocyte flattening from one species to another.

The so-called sheath capsules (Salt, 1970) are another variation from the flattened, multilayered, cellular capsules described above. These are much thinner cellular capsules, composed of a melanized sheath overlaid by a few adherent hemocytes. This type of capsule is found in a number of insect species, including *Drosophila melanogaster* parasitized by the wasps, *Pseudeucoila bochei* and *P. mellipes* (Walker, 1959; Nappi and Streams, 1969; Nappi, 1970), and in these species is formed by modified PLs called lamellocytes.

A final variation of the *E. kühniella*-type capsule was reported in *P. americana* by Mercer and Nicholas (1967) in response to the acanthocephalan, *Moniliformis dubius*. The structure of the capsule is almost noncellular, composed mainly of two layers of loosely arranged, nonmelanized vesicles that the authors believed were derived from the surface membranes of hemocytes. This type of capsule is included in the cellular category because of its probable origin from the blood cells. It also illustrates clearly that the type of capsule formed depends

(continued from facing page)
G. mellonella's capsule, it is composed of a few layers of highly flattened hemocytes (*H*). **13.71.** Capsule of *Blaberus craniifer* formed after implantation of a piece of *S. gregaria* nerve cord (*IP*). Very large capsule is composed mainly of flattened and more rounded hemocytes (*H*). (**13.68–13.71** from Schmit, unpub.)

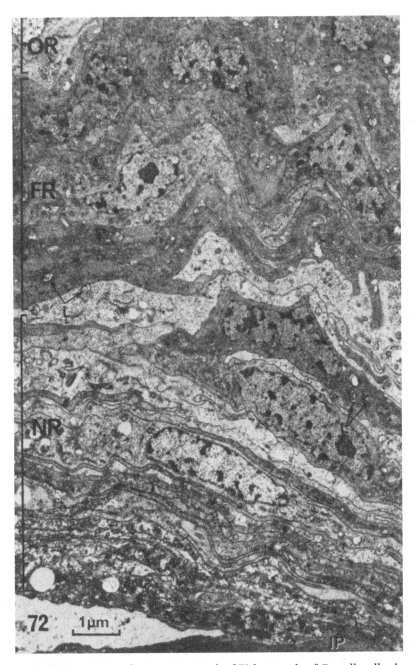

Fig. 13.72. Low-power electron micrograph of 72-hr capsule of *G. mellonella* showing characteristic three regions: an inner necrotic region (*NR*), an electron-dense middle region of extensively flattened plasmatocytes (*FR*), and a less electron-dense outer region (*OR*). *IP* = implant. Cytolysosome-like bodies (*L*) are also present. (From Schmit and Ratcliffe, 1977)

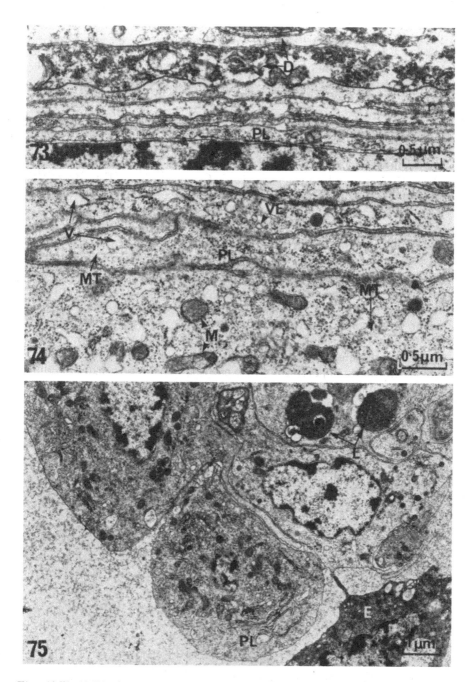

Figs. 13.73–13.75. Characteristic cells in three regions of 72-hr capsule of *G. mellon-ella.* **13.73.** Inner, necrotic region with characteristic granulocyte remnants (*D*) close to implant (just above top of micrograph) and flattening of plasmatocytes (*PL*). **13.74.** Middle region of capsule with extremely flattened plasmatocytes (*PL*). Note microtu-bules (*MT*), vacuoles (*V*), vesicles (*VE*), free ribosomes, and mitochondria (*M*). **13.75.** Outer region of capsule consisting of unflattened plasmatocytes (*PL*), occasionally containing lysosome-like bodies (*L*). Electron-dense cell (*E*) may be melanized. (**13.73–13.75** from Schmit and Ratcliffe, 1977)

greatly upon the type or species of parasite invading the hemocoel, as in response to the nematode, *Caenorhabditis briggsae*, *P. americana* forms a typical multilayered, pigmented cellular capsule (Misko, 1972).

Humoral capsules are formed in insects with few circulating blood cells, apparently often without the participation of the hemocytes (see below, however) and as a result of activation of hemolymph phenoloxidases (Poinar and Leutenegger, 1971; Götz and Vey, 1974). Recently, Götz et al. (1977) examined members of 11 insect orders for their ability to encapsulate nematodes, fungi, Sephadex, etc., and found that only in the Diptera were there any species capable of humoral encapsulation. They also found that there was a direct relationship between the number of circulating hemocytes and the type of encapsulation response. In insects with 500–4,000 cells/mm³, humoral encapsulation occurred; in species with more than 6,000 cells/mm³, cellular encapsulation took place. Humoral encapsulation has also been shown to occur only against living parasites such as nematodes (Poinar and Leutenegger, 1971) and fungi (Fig. 13.3) (Götz and Vey 1974), and not against inanimate objects such as glass fibers and nylon (Götz, 1969). This is similar to observations with cellular capsules, in which melanization is often absent against inanimate objects (Grimstone et al.,

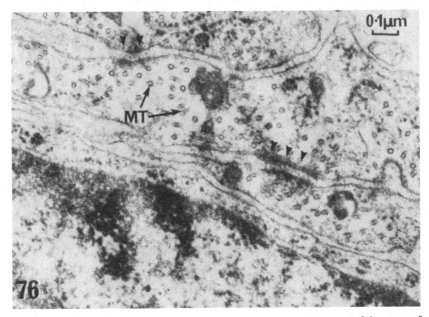

Fig. 13.76. High-power electron micrograph, showing characteristic modifications of plasmatocytes in middle region of 7-day capsule in *Thermobia domestica*. Note numerous microtubules (*MT*) and desmosomes (arrows) connecting adjacent cells. (From François, 1975)

1967), although exceptions have been recorded (François, 1975; Zachary et al., 1975). Furthermore, the inner region of cellular capsules, which covers the parasite with a continuous layer of electron-dense material (Figs. 13.7, 13.68), is very like a humoral capsule. Thus, cellular and humoral capsules have many features in common, and as Götz and Vey (1974) showed that in *Chironomus* spp. disintegrated GRs, which are known to have phenoloxidase activity (Maier, 1970, 1973), were present in the vicinity of encapsulated fungal spores, the possibility exists that humoral encapsulation is an extremely modified form of cellular encapsulation. However, humoral encapsulation in the cuticle of *Chironomus* spp. was shown to take place without any cellular intervention (Götz and Vey, 1974), and Poinar and Leutenegger (1971) reported that true humoral encapsulation occurred around the nematode *Neoplectana carpocapsae* in *Culex pipiens*.

Other factors, besides the nature of the parasite and the numbers of free blood cells, can affect the encapsulation reaction. For example, Blumberg (1976) and Lynn and Vinson (1977) have shown that the eggs of parasitoids are less efficiently encapsulated at extremes of temperature, probably owing to stress induced in the host and perhaps by a reduction in its ability to remove enzymatically the outer fibrous chorion of the parasitoid's eggs (Lynn and Vinson, 1977). Host age can also be an important determinant of encapsulation (van den Bosch, 1964; Salt, 1970; Lynn and Vinson, 1977), with first-instar larvae and adult Diptera and Lepidoptera showing only feeble encapsulation reactions, probably owing in part, to the low numbers of hemocytes present at these stages (Jones, 1962; Salt, 1970). Additional factors that may affect the encapsulation reaction include the sex, health, and nutritional state of the host and the presence of super- and multiparasitism (Shapiro, 1969).

Activation and control of host response

In earlier sections (see "Chemotaxis, Attachment, and Recognition of Foreignness" and "Structure and Mode of Formation of Nodules"), we presented evidence that foreign body–hemocyte contact and the recognition of foreignness involved chemotactic stimuli emanating from lysed granule-containing hemocytes and on the surface receptors of PLs. These studies indicate that the cellular defenses seem to rely on surface phenomena. Additional evidence for this hypothesis is provided by implantation experiments in which hemocyte reactions against implants from the same species are confined to cut or damaged surfaces (Salt, 1970; Scott, 1971c). Similarly, encapsulation usually, but not always (Kitano, 1969a), takes place if the surface of transplants (Salt, 1960) or of tolerated parasites (Salt, 1965, 1966) are abraded, heated, or dissolved with fat solvents. Salt (1970) suggested that because of the wide range of objects encapsulated by a given insect, the recognition of "non-self" was probably attributable to the absence of some factor common to all "self" surfaces. He further suggested that this common factor was a property of connective tissue. However, as Crossley (1975) pointed out, the hemocytes themselves are not surrounded by connective tissue, yet they do not react to each other. Perhaps the mucopolysaccharide material common to many hemocyte inclusions and to the surface of connective tissue (Wigglesworth, 1965) is the important factor determining reactivity. Recent work by Rizki and Rizki (1976) with a tumor-W mutant of D. melanogaster has provided convincing evidence for the importance of the integrity of the basement membrane in the recognition of foreignness. In mid-third-instar tumor-W larvae, the basement membrane over some of the caudal fat body cells is lost and hemocytes are specifically attracted to this region and attach to form melanotic tumors. No such process occurred in the ORE-R wild-type strain.

Attempts have been made to identify the hemocytes involved in the initial recognition reactions of encapsulation by carrying out differential and/or total hemocyte counts soon after implantation or parasitization. Unfortunately, such studies are complicated by changes evoked by the wounding reactions (Walker, 1959; Nappi and Stoffolano, 1971). These studies have been mainly confined to the Diptera, and their chief value has been to provide information about hemocyte transformations during encapsulation (see also Chapter 16).

Clearly the simplest way of identifying the first cells to react to foreign surfaces is by direct ultrastructural observations. Only one such study has been carried out, specifically concentrating on early events in encapsulation during which hemocytes or humoral factors and the foreign object first confront each other (Schmit and Ratcliffe, 1977). In this study, G. mellonella larvae were implanted with S. gregaria nerve cord fragments and examined at intervals from 5 min to 72 hr postimplantation. Within 5 min, contact and lysis of GRs was observed on the implant surface, and the resulting cell debris contained numerous granules (Fig. 13.77). The breakdown of the GRs produced a localized clot formation on the implant surface, and the whole process resembled ordinary hemolymph coagulation (Rowley and Ratcliffe, 1976a). Not until after 20 min were PLs observed attaching, and then solely at sites of GR lysis and discharge. These PLs eventually formed the multilayered sheath shown in Fig. 13.72. Thus, in many respects, encapsulation in G. mellonella is similar to nodule formation in this species and takes place in two stages. The implant is apparently identified as foreign by the GRs, and their breakdown releases some factor(s) (injury factor? Cherbas, 1973), which specifically attracts the PLs to the site of encapsulation. The PLs form the main structural element in the capsule and are the only cell type visible in fully formed capsules. A

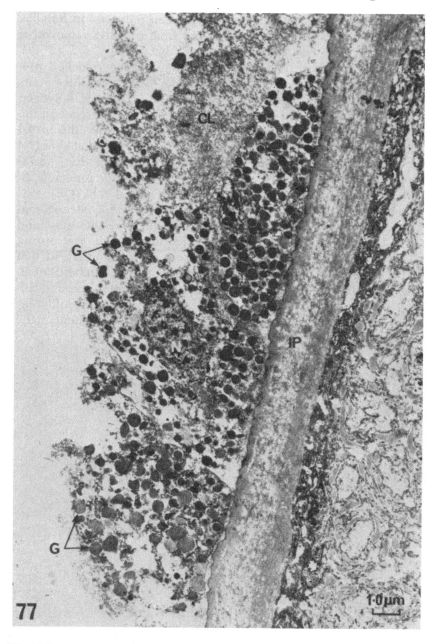

Fig. 13.77. First stage of encapsulation, with granulocytes lysing on surface of nerve cord (*IP*) 5 min after implantation in *G. mellonella*. Resultant debris consists of granules (*G*), isolated nuclei (*N*), and associated clot formation (*CL*). (From Schmit and Ratcliffe, 1977)

similar mechanism for encapsulation has been proposed in *Bombyx mori* (Sato et al., 1976), but was inferred from capsules removed at widely separated times (6 hr, 24 hr, and 72 hr).

The first vital phase in encapsulation has been overlooked previously because most workers examined fully formed capsules hours or even days after implantation or parasitization. There is, however, some considerable evidence in the literature for the presence of the lytic phase in other species. In *C. morosus* (Reik, 1968) the initial reaction to an implant was the contact and lysis of COs; isolated nuclei were found around early capsules in *Diabrotica* spp. (Fig. 13.78) (Poinar et al., 1968) and also, together with cytoplasmic remnants, in *P. americana* (Misko, 1972). The early lysis of crystal cells (OEs) was also reported on the surface of parasites in *D. melanogaster* (Nappi and Streams, 1969), and the disruption of labile hemocytes at wound sites in *Calliphora* sp. led Crossley (1975) to postulate that during parasitization these cells possibly release a recognition signal that may result in the speedy accumulation of blood cells. Brehélin et al. (1975) have described "an immediate plasma coagulation" around cellophane implants in *Locusta migratoria* and *Melolontha melolontha*. Finally, in a recent study of encapsulation of Araldite fragments in *C.*

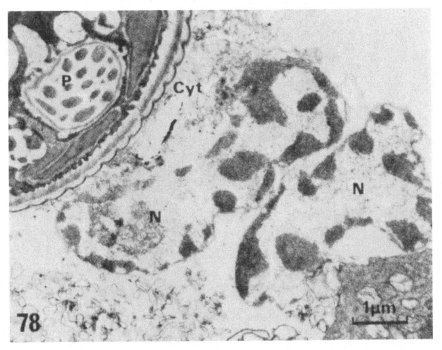

Fig. 13.78. Isolated blood cell nuclei (*N*) coming from lysed blood cells of *Diabrotica* sp. which had made initial contact with parasite, *Filipjevivermis leipsandra* (*P*), 1–2 hr after infection. *Cyt* = fragments of cytoplasm. (From Poinar et al., 1968)

extradentatus, Schmit and Ratcliffe (1978) have observed lysis of COs and GRs on the implant surface within 5 min, which resulted in a localized clot formation and constituted the first phase of encapsulation.

The reason for the termination of the encapsulation process may be the gradual dissipation of the contents of the lysed granule-containing cells with time, or their isolation by continued capsule buildup, so that the presumed chemotactic substance(s) no longer finds its way into the hemolymph.

Two other facets of encapsulation merit consideration: the melanization process and the possible activation and/or control of the cellular reactions by hormonal imbalances.

Nappi (1975) reviewed the role of melanization in host defenses and stated that "very little is known of the role of melanin in the immune reactions of insects." This opinion was influenced, no doubt, by the conflicting results obtained by workers in this field. Melanin deposition is frequently associated with substances released from disrupted hemocytes (Poinar et al., 1968; Nappi and Streams, 1969), and in some insects this pigment may be necessary for completion of the encapsulation reaction. Melanin formation is obviously a requisite of humoral encapsulation, but Nappi (1973) showed that it was also essential for cellular capsule formation. He fed phenylthiourea (PTU) to *Drosophila algonquin* larvae parasitized by *P. bochei* and showed that not only melanin deposition but also cellular encapsulation was prevented. Likewise, Brewer and Vinson (1971) showed a reduction in the encapsulation rate of *Cardiochiles nigriceps* eggs in *Heliothis zea* larvae injected with reduced glutathione or PTU, both known inhibitors of melanization. In contrast, in *C. morosus* Salt (1956) demonstrated that whereas injection of PTU prevented melanization of *Nemeritis canescens* eggs, cellular encapsulation was unaffected.

Despite these conflicting reports, we believe that melanin formation, or release of its precursors, may well be of importance in the initial stages of capsule and nodule formation in insects. This possibility is indicated by the presence of melanin precursors in the GRs of *G. mellonella* (Fig. 13.79) (Schmit et al., 1977), which, as described above, are probably responsible for the initiation of capsule formation, and by the presence of similar precursors in the crystal cells of *D. melanogaster,* which lyse on the surface of *Pseudeucoila mellipes* eggs (Nappi and Streams, 1969) and may well initiate encapsulation. Brewer and Vinson (1971) took things one step further and, based on the results of melanin-inhibition and encapsulation experiments, suggested that tyrosine-containing proteins or polyphenols may be involved in the opsonization of alien surfaces.

Finally, there is some evidence that encapsulation may be controlled in some way by the host's endocrine system (reviewed by

Nappi 1974, 1975). Evidence for this idea comes from observations that changes in the hemocyte populations of *D. melanogaster, Musca domestica,* and *Orthellia caesarion* during encapsulation of parasites closely resemble those occurring during pupation of the normal insects (Walker, 1959; Nappi and Streams, 1969; Nappi and Stoffolano, 1971, 1972). These changes include cellular differentiation, hemocyte mobilization, and lysis of crystal cells (Nappi, 1974). Since pupation is brought about by a low titer of juvenile hormone, the precocious hemocyte changes observed in parasitized larvae may be attributable to decreases in the concentration of this hormone or to increases in the amount of molting hormone (ecdysone) present (Nappi, 1974). Juvenile hormone also seems a likely candidate, as it is known to promote tumorigenesis, a process in which tumors, similar to melanized capsules, are formed (Bryant and Sang, 1969; Madhaven, 1972).

Additional evidence for the endocrine control of encapsulation is provided by Nappi (1975), who reported that capsule formation of *D. algonquin* around the parasitoid *P. bochei* was reduced following ligation of the host larvae to exclude the anterior endocrine center. The problem with this sort of experiment is that unknown stress factors may weaken the host and influence the results.

Despite the evidence in favor of capsule formation being controlled by juvenile and/or molting hormone, recent experiments by Lynn and Vinson (1977) on encapsulation of *C. nigriceps* in *Heliothis virescens* and *H. zea,* showed that neither juvenile hormone analogue nor molting hormone fed to and injected into the hosts, respectively, had any effect on capsule formation. Furthermore, in a study of mermithid

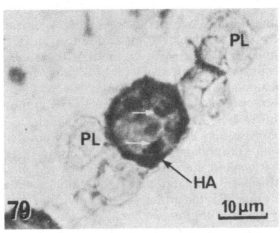

Fig. 13.79. Granulocyte of *G. mellonella,* showing localization of reaction product for polyphenoloxidase, a precursor of melanin, as a "halo" around cell (*HA*) and in vacuoles within cell (arrows). Note absence of any activity in plasmatocytes (*PL*). (From Schmit et al., 1977)

parasitization in the larvae of the blackfly, *Prosimulium mixtum fuscum,* Condon and Gordon (1977) could detect no significant difference in endocrine activity between infected and noninfected hosts.

In conclusion, it is not proposed that any one of the above mechanisms for control and/or activation is solely responsible for encapsulation, which is a complex process probably influenced by an interplay among different factors.

Resistance to encapsulation

Habitual parasites of insects are able to avoid or inactivate the cellular defenses of their hosts in some way and develop unmolested. In recent years, several mechanisms have been proposed by which the parasite can influence host immune reactions (see reviews by Salt, 1968; Kitano, 1969b).

Some of these mechanisms operate directly at the surface of the parasite, where hemocyte recognition and attachment occur. Salt (1965) showed that the eggs of *N. canescens* are resistant to encapsulation in their usual host, *E. kühniella,* as a result of a coating on the surface of the egg which is deposited during its passage through the calyx of the female parasitoid. This coating was subsequently seen to be composed of minute particles (calyx particles) (Rotheram, 1967), which in *E. kühniella* suppressed the reaction of the blood cells to a foreign body, had no general toxic effect on the host, had no inhibitory effect on the defense reactions to other foreign bodies, and acted specifically on certain kinds of blood cells (Salt, 1973). Bedwin (1970) subsequently showed that the *N. canescens* calyx particles had a relatively high ornithine content that because of its chelating properties may prevent hemocyte attachment to the parasitoid's eggs. Salt (1973) confirmed this possibility by showing that in vitro the calyx particles could dissociate the cells of capsules and may therefore act by preventing hemocyte adhesion.

Other workers have also described the presence (Fig. 13.80) and importance of calyx particles in preventing encapsulation. In *C. nigriceps* and *Campoletis sonorensis,* for example, the particles are not attached to the egg surface, but appear to be injected into the host at oviposition (Vinson and Scott, 1974; Norton and Vinson, 1976). In further contrast to the *N. canescens* particles, those in *C. nigriceps* contain DNA (Vinson and Scott, 1975) and are similar to baculovirus (Stoltz et al., 1976).

Recently, an interesting study by Lackie (1976) has shown that other surface-dependent mechanisms may assist the parasite in avoiding recognition and encapsulation. In a series of experiments in which the cysticercoids of the cestode *Hymenolepis diminuta* and host tissues were transplanted in a number of different insect species, she showed that transplanted tissues are tolerated between species in which *H. diminuta* normally survives. The surface of the parasite thus appears similar to the surfaces of tolerant host tissues and therefore escapes recognition as non-self by the host hemocytes.

Another surface-dependent resistance mechanism has been reported by Brennan and Cheng (1975) in the acanthocephalan *Moniliformis dubius* in its host, *P. americana*. The trilaminate envelope surrounding the parasite larvae apparently contains polyanionic mucins that may complex with the divalent cation Cu^{2+} and interfere with the activation of the tyrosinase and melanization reaction of the host.

Other methods of avoiding the host defense mechanisms have also been recorded and are not directly dependent on properties of the parasite surface. In comparative experiments with *D. melanogaster*, parasitized by *P. bochei* (resistant to encapsulation) and *P. mellipes* (susceptible to encapsulation), Nappi and Streams (1969) showed that *P. bochei* had the ability to suppress the immune reactions of the host. The substance(s) responsible was not highly specific, as when both parasites were present in the same host the encapsulation reaction against the normally susceptible *P. mellipes* was also suppressed (Nappi, 1975). The factor inactivating the immune response of *D. melanogaster* may originate within the egg or may be released into the host during oviposition by the female parasitoid. Kitano (1969a,b, 1974) has also reported that some factor(s) is probably released from

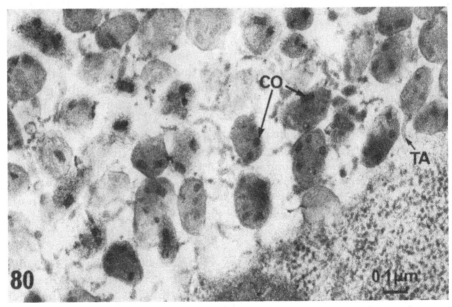

Fig. 13.80. Electron micrograph of calyx particles found in egg reservoir of *Apanteles glomeratus*. These particles have distinct electron-dense core (*CO*) and often have tail-like projections (*TA*). (From Fordy, unpub.)

Apanteles glomeratus eggs present in the larvae of *Pieris rapae cruci-vora*, which suppresses transformation of PRs into PLs or causes reduced adherence of the PLs and so prevents encapsulation. The reduced adherence of the PLs may also have been caused by a nutritional deficiency produced in the host by the presence of numerous giant cells or teratocytes derived from the trophamnion of the egg (Salt, 1971). Alternatively, the teratocytes themselves may have secreted an "encapsulation-inhibiting factor" (Kitano, 1969a).

Occurrence and effectiveness as defense reaction

As mentioned previously (see "Insect Parasitoids," above), as far as we are aware, no quantitative data exist on the occurrence of parasitic invasion and encapsulation reactions in insects under natural conditions. However, Salt (1970) believed that such attacks occur frequently, and Wülker (1961) has reported that in natural habitats dead mermithids are often found encapsulated in the body cavity of *Chironomus* sp. larvae.

Many workers believe that encapsulation probably prevents the development of a wide range of parasites such as trematodes, cestodes, acanthocephalans, nematodes, insect parasitoids, tachinids, and pathogenic fungi (Götz and Vey, 1974). Others take a more conservative view and suggest that encapsulation may merely reflect hemocytic reaction to parasites already killed by some alternative action in the host (van den Bosch, 1964).

If, as seems likely, parasites do succumb within capsules, this may result from asphyxiation (Wigglesworth, 1959; Salt, 1959; Bronskill, 1960) or starvation. Alternatively, it may result from the action of melanin precursors, which may form reactive quinones that have antibiotic effects or protein inactivation properties and form highly resistant nonsoluble substances (Jones, 1958). Indeed, it has been suggested that protein–polyphenol complexes may be of importance in many defense reactions in insects (Taylor, 1969; Maier, 1973). Recently, however, Beresky and Hall (1977) have shown that the phenoloxidase system may be deleterious to the survival of the host.

13.3.6. Hemolymph coagulation including wound repair

Hemolymph coagulation is only briefly mentioned here, as it has been admirably reviewed in recent years (Grégoire, 1970; Crossley, 1975) (see also Chapter 7).

The cells responsible for hemolymph coagulation have been termed "hyaline" hemocytes (Grégoire, 1970) and probably represent the COs of most insect orders and/or the GRs of a few others (Rowley and Ratcliffe, 1976a; Rowley, 1977b).

The role of hemolymph coagulation in insect defenses is to seal off wounds, prevent excessive hemolymph loss, exclude bacteria from access to the hemocoel, and assist in wound repair (Grégoire, 1970). Unfortunately, the events in wound repair have been fully described only in *E. kühniella* and *T. molitor* (Ries, 1932) and in *R. prolixus* (Wigglesworth, 1937, 1973; Locke, 1966; Lai-Fook, 1968, 1970), and many aspects of this process are still only poorly understood. For example, the initial reaction to gut wounding has been reported to be localized blood coagulation in the cockroach, *P. americana* (Day 1952), although this process has not been observed in other insects (Wigglesworth, 1937; Day and Bennetts, 1953). It is also unclear whether the new endocuticle and basement membrane are re-formed by the hemocytes (Lazarenko, 1925; Wigglesworth, 1973) or by the epidermal cells (Wigglesworth, 1937; Lai-Fook, 1968) (see also Chapter 12).

In a recent study of wound repair in *G. mellonella*, which was carried out to clarify the role of the hemocytes, a great similarity between this process and the other defense reactions was noted (Ratcliffe et al., 1976b; Rowley and Ratcliffe, 1978).

Within 10 min of injury, hemolymph coagulation had occurred as a result of degranulation of aggregates of GRs, and this helped seal off

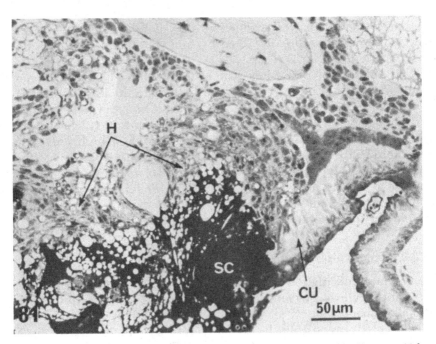

Fig. 13.81. Section through wound made in cuticle (*CU*) of *G. mellonella* some 12 hr previously. Note melanized scab (*SC*) composed of degenerated hemocytes and other tissues and underlying disorganized layers of hemocytes (*H*), forming capsule-like structure. (From Rowley and Ratcliffe, 1978)

the wound. After 60–120 min, the hemocyte aggregates had compacted and melanized and appeared similar in structure to the immature nodules formed in response to the injection of bacteria into the hemocoel. After 6 hr, a second phase of hemocyte involvement in wound healing was apparent and involved a massive influx of PLs into the wound site. These cells flattened down under the wound site and appeared to be aggregating in response to some sort of stimulus, perhaps a "wound factor" produced by the damaged hemocytes or epidermal cells (Harvey and Williams, 1961; Cherbas, 1973; Bohn, 1975). By 12 hr after wounding, the PLs were more flattened and formed a structure around the base of the wound resembling a disorganized capsule (Fig. 13.81).

Thus, in *G. mellonella*, nodule formation, encapsulation, and wound healing appear to involve two-step cellular reactions in which the GRs and PLs play vital roles (see also Chapter 11).

13.4. Concluding remarks

No doubt an efficient host defense system composed of integrated cellular and humoral elements has contributed significantly to the astounding success of the insects. In recent years much has been learned about natural insect diseases and the cellular defense reactions, but few studies have been carried out combining both these aspects of insect pathology. The reason for this is not only the frequent mutual disinterest of ecologists and biological control scientists and cell biologists in each others' work, but also the difficulty of detecting signs of cellular defense reactions in natural insect populations.

Studies of the early stages of parasitic diseases show clearly that the exoskeleton and the gut are all-important in resisting invasion and successfully protect insects from a whole range of would-be pathogens. Once these barriers are crossed, however, experiments show that the host insect is capable of mounting efficient and effective cellular defense reactions. Present evidence indicates that these cellular responses, unlike those in the vertebrates, are nonspecific and that the effector cells lack any degree of "immunological memory." This does not mean that the defense reactions are feeble or impotent, but rather that in insects emphasis is placed on other cell-mediated responses.

Finally, results of research in a very limited number of species indicate that the insect cellular defenses occur in two stages and that hemolymph coagulation and various modifications of this process are responsible for initiating these reactions. The first "recognition" phase appears to be mediated by COs/GRs, or perhaps OEs in the Diptera, which lyse in contact with foreign surfaces and release a recognition factor(s) that subsequently specifically attracts the PLs. Why some foreign surfaces elicit these responses while others fail to do so is unknown, but in many cases results from avoidance mechanisms associated with parasitization.

Clearly, additional work is required on a range of insect species from different habitats to confirm or refute the presence of two-step cellular reactions in insects in general. Even if such confirmation should be obtained, we are still unable to explain why COs/GRs behave so violently toward non-self surfaces.

13.5. Summary

The cellular defense mechanisms of insects are mediated mainly by the hemocytes and include the processes of phagocytosis, nodule formation, encapsulation, and hemolymph coagulation. The type of reaction mounted depends both on the nature and the numbers of the infecting agent. Small doses of particulate material such as viruses, rickettsiae, bacteria, fungal spores, and protozoans are phagocytized; larger doses are usually segregated in nodules. Other foreign materials that enter the hemocoel and are too large to be phagocytized by single hemocytes (e.g., fungal hyphae and metazoan parasites) are, in insects with large numbers of circulating hemocytes, enclosed within multicellular capsules. In species with few free hemocytes, such materials are entrapped in humoral capsules, composed mainly of melanin, which may or may not be derived from the hemocytes. In addition, hemolymph coagulation is of great importance in preventing excessive hemolymph loss and in delaying the entry of infectious agents and parasites following wounding.

Phagocytosis is chiefly carried out, both in vivo and in vitro, by the plasmatocytes. In insects, unlike the case in many other invertebrates, phagocytosis appears to be independent of hemolymph factors and at least partially involves direct interaction of the foreign particle with the hemocyte surface receptors. The release of substances from other hemocyte types may, however, coat the foreign particles and facilitate their uptake. Early stages of phagocytosis are temperature-independent and rely on the presence of cytochalasin-B-sensitive microfilaments; later stages are temperature-dependent and involve the expenditure of energy. Ingested microorganisms are always enclosed within phagosomes and may be killed by unknown antimicrobial factors that are probably released into the phagosomes by fusion with lysosomes and/or granules. Sometimes, however, the microorganisms may survive, divide, and even escape from the hemocyte to invade other tissues.

Nodule formation is a two-step process. The first stage is initiated by the lysis of granule-containing hemocytes (granulocytes and/or coagulocytes) which rapidly entrap other hemocytes and large numbers of bacteria to form melanizing aggregates. The whole process is much like hemolymph coagulation. In the second stage, melanized masses specifically attract large numbers of plasmatocytes, probably as a result of the release of some chemotactic (injury?) factor(s), and these cells form multicellular sheaths around the pigmented cores. Encapsulation also occurs in two stages, with the first stage characterized by a modified hemolymph coagulation on the surface of the foreign body resulting from the lysis of granulocytes and/or coagulocytes. The bulk of the capsule is subsequently formed by the specific attachment of numerous plasmatocytes. The role of endocrine control and melanization in these cellular defense reactions is not fully understood, although neither control nor activation is a simple process, but probably results from an interplay between a number of different factors. The avoidance by parasites of these defense reactions has been recorded, and although little is known about the resistance of microorganisms to segregation in nodules, the resistance of metazoan parasites to encapsulation has been studied in detail, and a number of different mechanisms, including coating of the parasite's eggs with calyx particles and host tissue mimicry, have been identified. The fate of the biological agents isolated in nodules and capsules has rarely been determined, but these structures probably effectively prevent the development of a range of parasites and in some cases may actively kill the invader by the action of reactive melanin precursors or other unidentified factors.

Finally, hemolymph coagulation is an integral part of wound healing. As with nodule formation and encapsulation, two-step cellular reactions appear to be involved, with the attracted plasmatocytes eventually forming a disorganized, capsulelike structure beneath the wound and assisting in the isolation of would-be invaders.

Acknowledgments

We are very grateful to our colleagues, Mr. S. J. Gagen and Mr. A. R. Schmit, who have allowed us to use some of their unpublished results. We are also indebted to the following scientists: Professor G. Amirante, Dr. R. S. Anderson, Miss A. Bywater, Mr. M. Fordy, Dr. J. François, Dr. P. Götz, Dr. C. Y. Kawanishi, Dr. A. Lackie, Dr. M. Neuwirth, Professor W. L. Nutting, Dr. G. O. Poinar, Jr., Dr. C. M. Splittstoesser, and Dr. A. Vey, who have provided both published and unpublished material for this chapter. Thanks also go to Professor E. W. Knight-Jones, in whose department this work was carried out, and to Mrs. M. Colley and Mr. P. Llewellyn for technical assistance. This work was supported by grants from the Royal Society and from the Science Research Council (grant numbers B/RG.5924.3 and GR.A.2286.0).

References

Acton, R. T., and P. F. Weinheimer. 1974. Hemagglutinins: Primitive receptor molecules operative in invertebrate defense mechanisms, pp. 271–82. *In* E. L. Cooper (ed.). *Contemporary Topics in Immunobiology*, Vol. 4, *Invertebrate Immunology*. Plenum Press, New York.

Åkesson, B. 1954. Observations on the haemocytes during the metamorphosis of *Calliphora erythrocephala* (Meig.). *Ark. Zool.* 6:203–11.

Amirante, G. A. 1976. Production of heteroagglutinins in hemocytes of *Leucophaea maderae* L. *Experientia* 32(4):526–8.

Anderson, R. S. 1974. Metabolism of insect hemocytes during phagocytosis, pp. 47–54. In E. L. Cooper (ed.). *Contempory Topics in Immunobiology*, Vol. 4, *Invertebrate Immunology*. Plenum Press, New York.

Anderson, R. S. 1975. Phagocytosis by invertebrate cells *in vitro:* Biochemical events and other characteristics compared with vertebrate phagocytic systems, pp. 152–80. In K. Maramorosch and R. E. Shope (eds.). *Invertebrate Immunity*. Academic Press, New York.

Anderson, R. S. 1976a. Expression of receptors by insect macrophages, pp. 27–34. In R. K. Wright and E. L. Cooper (eds.). *Phylogeny of Thymus and Bone Marrow-Bursa Cells*. Elsevier/North-Holland, Amsterdam.

Anderson, R. S. 1976b. Macrophage function in insects, pp. 215–19. In T. A. Angus, P. Faulkner, and A. Rosenfield (eds.). *Proceedings, First International Colloquium on Invertebrate Pathology*. Queens University, Kingston, Ontario.

Anderson, R. S., N. K. B. Day, and R. A. Good. 1972. Specific hemagglutinin and a modulator of complement in cockroach hemolymph. *Infect. Immun.* 5:55–9.

Anderson, R. S., and R. A. Good. 1976. Opsonic involvement in phagocytosis by mollusk hemocytes. *J. Invertebr. Pathol.* 27(1):57–64.

Anderson, R. S., B. Holmes, and R. A. Good. 1973a. *In vitro* bactericidal capacity of *Blaberus craniifer* hemocytes. *J. Invertebr. Pathol.* 22(1):127–35.

Anderson, R. S., B. Holmes, and R. A. Good. 1973b. Comparative biochemistry of phagocytizing insect hemocytes. *Comp. Biochem. Physiol.* 46B:595–602.

Armstrong, J. A., and P. D. 'A. Hart. 1971. Response of cultured macrophages to *Mycobacterium tuberculosis*, with observations on fusion of lysosomes with phagosomes. *J. Exp. Med.* 134:713–40.

Armstrong, J. A., and P. D. 'A. Hart. 1975. Phagosome-lysosome interactions in cultured macrophages infected with virulent tubercle bacilli: Reversal of the usual nonfusion pattern and observations on bacterial survival. *J. Exp. Med.* 142:1–16.

Arnold, J. W. 1974. The hemocytes of insects, pp. 201–54. In M. Rockstein (ed.). *The Physiology of Insecta*, Vol. 5, 2nd ed. Academic Press, New York.

Arvy, L. 1954. Données sur la leucopoïèse chez *Musca domestica* L. *Proc. R. Entomol. Soc. Lond. A* 29:39–41.

Aston, W. P., J. S. Chadwick, and M. J. Henderson. 1976. Effect of cobra venom factor on the *in vivo* immune response in *Galleria mellonella* to *Pseudomonas aeruginosa. J. Invertebr. Pathol.* 27(2):171–6.

Baehner, R. L., M. J. Karnovsky, and M. L. Karnovsky. 1968. Degranulation of leukocytes in chronic granulomatous disease. *J. Clin. Invest.* 48:187–92.

Baerwald, R., and G. M. Boush. 1970. Fine structure of the hemocytes of *Periplaneta americana* (Orthoptera: Blattidae) with reference to marginal bundles. *J. Ultrastruct. Res.* 31:151–61.

Bairati, A., Jr. 1964. L'ultrastruttura dell'organo dell'emolinfa nella larva di *Drosophila melanogaster. Z. Zellforsch. Mikrosk. Anat.* 61:769–802.

Bedwin, O. R. 1970. The particulate basis of the resistance of a parasitoid to its host. Ph.D. dissertation, University of Cambridge, Cambridge.

Beresky, M. A., and D. W. Hall. 1977. The influence of phenylthiourea on encapsulation, melanization, and survival in larvae of the mosquito *Aedes aegypti* parasitized by the nematode *Neoplectana carpocapsae. J. Invertebr. Pathol.* 29(1): 74–80.

Bernheimer, A. W. 1952. Hemagglutinins in caterpillar blood. *Science* (*Wash., D.C.*) 155:150–1.

Bettini, S. D., S. Sarkaria, and R. L. Patton. 1951. Observations on the fate of vertebrate erythrocytes and hemoglobin injected into the blood of the American cockroach (*Periplaneta americana*). *Science (Wash., D.C.) 113*:9–10.

Blumberg, D. 1976. Extreme temperatures reduce encapsulation of insect parasitoids in their insect hosts. *Experientia 32*:1396–7.

Boese, G. 1936. Der Einfluss tierischen Parasiten auf den organismus der Insekten. *Z. Parasitenk. 8*:243–84.

Bohn, H. 1975. Growth promoting effect of haemocytes on insect epidermis *in vitro. J. Insect Physiol. 21*:1283–93.

Boman, H. G., I. Nilsson, and B. Rasmuson. 1972. Inducible antibacterial defense system in *Drosophila. Nature (Lond.) 237*:232–4.

Boman, H. G., I. Nilsson-Faye, K. Paul, and T. Rasmuson, Jr. 1974. Insect immunity. I. Characteristics of an inducible cell-free antibacterial reaction in hemolymph of *Samia cynthia* pupae. *Infect. Immun. 10*:136–45.

Brahmi, Z., and E. L. Cooper. 1974. Characteristics of the agglutinin in the scorpion, *Androctonus australis*, pp. 261–70. *In* E. L. Cooper (ed.). *Contemporary Topics in Immunobiology*, Vol. 4, *Invertebrate Immunology*. Plenum Press, New York.

Brehélin, M., J. A. Hoffmann, G. Matz, and A. Porte. 1975. Encapsulation of implanted foreign bodies by hemocytes in *Locusta migratoria* and *Melolontha melolontha*. *Cell Tissue Res. 160*:283–9.

Brennan, B. B., and T. C. Cheng. 1975. Resistance of *Moniliformis dubius* to the defense reactions of the American cockroach, *Periplaneta americana. J. Invertebr. Pathol. 26*(1):65–73.

Brewer, D. B. 1963. Electron microscopy of phagocytosis of staphylococci. *J. Pathol. Bacteriol. 86*:299–303.

Brewer, F. D., and S. B. Vinson. 1971. Chemicals affecting the encapsulation of foreign material in an insect. *J. Invertebr. Pathol. 18*:287–9.

Briggs, J. D. 1958. Humoral immunity in lepidopterous larvae. *J. Exp. Zool. 138*:155–88.

Bronskill, J. F. 1960. The capsule and its relation to the embryogenesis of the ichneumonid parasitoid *Mesoleius tenthredinis* Morl. in the larch sawfly, *Pristiphora erichsonii* (Htg.) (Hymenoptera: Tenthredinidae). *Can. J. Zool. 38*:769–75.

Brooks, W. M. 1974. Protozoan infections, pp. 237–300. *In* G. E. Cantwell (ed.). *Insect Diseases*, Vol. I. Dekker, New York.

Bryant, P. J., and J. H. Sang. 1969. Physiological genetics of melanotic tumors in *Drosophila melanogaster*. VI. The tumorigenic effects of juvenile hormone-like substances. *Genetics 62*:321–36.

Bucher, G. E. 1959. Bacteria of grasshoppers of western Canada. III. Frequency of occurrence, pathogenicity. *J. Insect Pathol. 1*:391–405.

Bucher, G. E. 1960. Potential bacterial pathogens of insects and their characteristics. *J. Insect Pathol. 2*:172–95.

Cagan, R. H., and M. C. Karnovsky. 1964. Enzymatic basis of the respiratory stimulation during phagocytosis. *Nature (Lond.) 204*:255–7.

Cameron, G. R. 1934. Inflammation in the caterpillars of Lepidoptera. *J. Pathol. Bacteriol. 38*:441–66.

Cann, G. B. 1974. A laboratory study of the haemagglutinating substance from the snail *Helix aspersa. Med. Lab. Technol. 31*:11–36.

Carter, J. B., and M. L. Luff. 1977. Rickettsia-like organisms infecting *Harpalus rufipes* (Coleoptera: Carabidae). *J. Invertebr. Pathol. 30*(1):99–101.

Chadwick, J. S. 1975. Hemolymph changes with infection or induced immunity in insects and ticks, pp. 241–71. *In* K. Maramorosch and R. E. Shope (eds.). *Invertebrate Immunity*. Academic Press, New York.

Cherbas, L. 1973. The induction of an injury reaction in cultured haemocytes from saturniid pupae. *J. Insect Physiol. 19*:2011–23.

Chorine, V. 1931. Contribution à l'étude de l'immunité chez les insectes. *Bull. Biol. Fr. Belg. 65*:291–393.

404   N. A. Ratcliffe and A. F. Rowley

Clark, R. M., and W. R. Harvey. 1965. Cellular membrane formation by plasmatocytes of diapausing cecropia pupae. *J. Insect Physiol. 11*:161–75.

Cline, M. J. 1975. *The White Cell.* Harvard University Press, Cambridge, Massachusetts.

Condon, W. J., and R. Gordon. 1977. Some effects of mermithid parasitism on the larval blackflies *Prosimulium mixtum fuscum* and *Simulium venustum. J. Invertebr. Pathol. 29*(1):56–62.

Cotran, R. S., and M. Litt. 1969. The entry of granule-associated peroxidase into the phagocytic vacuoles of eosinophils. *J. Exp. Med. 129*:1291–1306.

Crossley, A. C. S. 1968. The fine structure and mechanism of breakdown of larval intersegmental muscles in the blowfly *Calliphora erythrocephala* (Meig.). *J. Insect Physiol. 14*:1389–1407.

Crossley, A. C. S. 1972. The ultrastructure and function of pericardial cells and other nephrocytes in an insect, *Calliphora erythrocephala. Tissue cell 4*(3):529–60.

Crossley, A. C. S. 1975. The cytophysiology of insect blood. *Adv. Insect Physiol. 11*: 117–222.

Cuénot, L. 1895. Etudes physiologiques sur les Orthoptères. *Arch. Biol. Paris 14*:293–341.

Cuénot, L. 1897. Les globules sanguins et les organes lymphoïdes des Invértébrés. (Revue critique et nouvelles recherches.) *Arch. Anat. Microsc. 1*:153–92.

Day, M. F. 1952. Wound healing in the gut of the cockroach *Periplaneta. Aust. J. Sci. Res. (B) 5*:282–9.

Day, M. F., and M. J. Bennetts. 1953. Healing and gut wounds in the mosquito *Aedes aegypti* (L.) and the leafhopper *Orosius argentatus* (EV.). *Aust. J. Biol. Sci. 6*:580–5.

Durham, H. E. 1891. On wandering cells in echinoderms, etc., more especially with regard to excretory functions. *Q. J. Microsc. Sci. 33*:81–121.

Ermin, R. 1939. Ueber Bau und Funktion der Lymphocyten bei Insekten (*Periplaneta americana* L.). *Z. Zellforsch. Mikrosk. Anat. 29*:613–69.

Evans, W. A., and R. G. Elias. 1970. The life cycle of *Malamoeba locustae* (King et Taylor) in *Locusta migratoria migratoides* (R. et F.). *Acta Protozool. 7*:229–41.

Falcon, L. A. 1971. Use of bacteria for microbial control, pp. 67–95. *In* H. D. Burges and N. W. Hussey (eds.). *Microbial Control of Insects and Mites.* Academic Press, New York.

Feir, D., and M. A. Watz. 1964. An agglutinating factor in insect hemolymph. *Ann. Entomol. Soc. Amer. 57*:388.

François, J. 1975. L'Encapsulation hémocytaire expérimentale chez le lépisme *Thermobia domestica. J. Insect Physiol. 21*:1535–46.

Gagen, S. J., and N. A. Ratcliffe. 1976. Studies on the *in vivo* cellular reactions and fate of injected bacteria in *Galleria mellonella* and *Pieris brassicae* larvae. *J. Invertebr. Pathol. 28*(1):17–24.

Gilliam, M., and W. S. Jeter. 1970. Synthesis of agglutinating substances in adult honeybees against *Bacillus larvae. J. Invertebr. Pathol. 16*:69–70.

Glick, A. D., R. A. Gentrick, and R. M. Cole. 1971. Electron microscopy of group A streptococci after phagocytosis by human monocytes. *Infect. Immun. 4*:772–9.

Götz, P. 1969. Die Einkapselung von Parasiten in der Haemolymphe von *Chironomus*-Larven (Diptera). *Zool. Anz. (Suppl.) 33*:610–17.

Götz, P., I. Roettgen, and W. Lingg. 1977. Encapsulement humoral en tant que réaction de défense chez les Diptères. *Ann. Parasitol. 52*(1):95–7.

Götz, P., and A. Vey. 1974. Humoral encapsulation in Diptera (Insecta): Defense reactions of *Chironomus* larvae against fungi. *Parasitology 68*:1–13.

Grace, T. D. C. 1962. The development of a cytoplasmic polyhedrosis in insect cells grown *in vitro. Virology 18*:33–42.

Grace, T. D. C. 1971. The morphology and physiology of cultured invertebrate cells, pp. 171–209. *In* C. Vago (ed.). *Invertebrate Tissue Culture*, Vol. I. Academic Press, New York.

Grace, T. D. C., and M. F. Day. 1963. Film C.S.I.R.O., Canberra, Australia.

Grégoire, Ch. 1970. Haemolymph coagulation in arthropods, pp. 45–74. *In* R. G. Macfarlane (ed.). *The Haemostatic Mechanism in Man and Other Animals*. Symposium, Zoological Society of London, No. 27. Academic Press, New York.

Grimstone, A. V., S. Rotheram, and G. Salt. 1967. An electron-microscope study of capsule formation by insect blood cells. *J. Cell Sci.* 2:281–92.

Hall, J. E., J. E. Weaver, and B. Gomez-Miranda. 1969. Histopathology of lecithodendriid trematode infections in stonefly naiads. *J. Invertebr. Pathol.* 13:91–113.

Hall, J. L., and D. T. Rowlands, Jr. 1974. Heterogeneity of lobster agglutinins. II. Specificity of agglutinin-erythrocyte bindings. *Biochemistry* 13:828–32.

Harvey, W. R., and C. M. Williams. 1961. The injury metabolism of the cecropia silkworm. I. Biological amplification of the effects of localized injury. *J. Insect Physiol.* 7:81–99.

Henry, J. E. 1967. *Nosema acridophagous* sp.n., a microsporidian isolated from grasshoppers. *J. Invertebr. Pathol.* 9:331–41.

Hillen, N. D. 1977. Experimental studies on the reactions of insect haemocytes to artificial implants and habitual parasitoids and on the initiation of wound healing in insects. Ph.D. thesis, University of London, London.

Hoffmann, D., and M. Brehélin. 1971. Evolution de l'image sanguine chez les mâles de *Locusta migratoria* après injection de *Bacillus thuringiensis. C. R. Séances Acad. Sci. Paris* 274D:422–5.

Hoffmann, D., M. Brehélin, and J. A. Hoffmann. 1974. Modifications of the hemogram and of the hemocytopoietic tissue of male adults of *Locusta migratoria* (Orthoptera) after injection of *Bacillus thuringiensis. J. Invertebr. Pathol.* 24:238–47.

Hoffmann, J. A., A. Porte, and P. Joly. 1968. Présence d'un tissue hématopoïétique au niveau du diaphragme dorsal de *Locusta migratoria* (Orthoptère). *C. R. Séances Acad. Sci. Paris* 266D:1882–3.

Hollande, A. C. 1909. Contribution à l'étude du sang des Coléoptères. *Arch. Zool. Exp. Gén.* 42:271–94.

Hollande, A. C. 1920. Réactions des tissus du *Dytiscus marginalis* L. au contact de larves de Distome enkystées et fixées aux parois du tube digestif de l'insecte. *Arch. Zool. Exp. Gén.* 59:543–63.

Hollande, A. C. 1930. La digestion des bacilles tuberculeux par les leucocytes du sang des chenilles. *Arch. Zool. Exp. Gén.* 70:231–80.

Holmes, B., B. H. Park, S. E. Malawista, P. G. Quie, D. L. Nelson, and R. A. Good. 1970. Chronic granulomatous disease in females: A deficiency of leukocyte glutathione peroxidase. *N. Engl. J. Med.* 283:217–21.

Horn, R. G., S. S. Spicer, and B. K. Wetzel. 1964. Phagocytosis of bacteria by heterophil leukocytes. *Amer. J. Pathol.* 45:327–33.

Huger, A. 1960. Untersuchungen zur Pathologie einer Mikrosporidiose von *Agrotis segetum* (Schiff.) (Lepidopt., Noctuidae), verursacht durch *Nosema perezioides* nov. spec. *Z. Pflanzenkr. Pflanzenpathol. Pflanzenschutz.* 67:65–77.

Hurpin, B., and C. Vago. 1958. Les maladies du hanneton commun (*Melolontha melolontha* L.) (Col. Scarabaeidae). *Entomophaga* 3:285–330.

Iwasaki, Y. 1927. Sur quelques phénomènes provoqués chez les chenilles de papillons par l'introduction de corps étrangers. *Arch. Anat. Microsc.* 23:319–46.

Jones, B. 1958. Enzymatic oxidation of proteins as a rate determining step in the formation of highly stable surface membranes. *Proc. R. Soc. B* 149:263–77.

Jones, J. C. 1956. The hemocytes of *Sarcophaga bullata* Parker. *J. Morphol.* 99:233–57.

Jones, J. C. 1962. Current concepts concerning insect hemocytes. *Amer. Zool.* 2:209–46.

Jones, J. C. 1970. Hemocytopoiesis in insects, pp. 7–65. *In* A. S. Gordon (ed.). *Regulation of Hematopoiesis*. Appleton, New York.

Jones, T. C., and J. G. Hirsch. 1971. The interaction *in vitro* of *Mycoplasma pulmonis* with mouse peritoneal macrophages and L-cells. *J. Exp. Med. 133*:231–59.

Kalmakoff, J., B. R. G. Williams, and F. J. Austin. 1977. Antiviral response in insects? *J. Invertebr. Pathol. 29*:44–9.

Kawakami, K. 1965. Phagocytosis in muscadine-diseased larvae of the silkworm, *Bombux mori* (Linnaeus). *J. Invertebr. Pathol. 7*:203–8.

Kawanishi, C. Y., C. M. Splittstoesser, and H. Tashiro. 1978. Infection of the European chafer, *Amphimallon majalis*, by *Bacillus popilliae*. II. Ultrastructure. *J. Invertebr. Pathol. 31*:91–102.

Kitano, H. 1969a. Defensive ability of *Apanteles glomeratus* L. (Hymenoptera: Braconidae) to the hemocytic reaction of *Pieris rapae crucivora* Boisduval (Lepidoptera: Pieridae). *Appl. Entomol. Zool. 4*(1):51–5.

Kitano, H. 1969b. Experimental studies on the parasitism of *Apanteles glomeratus* L. with special reference to its encapsulation-inhibiting capacity. *Bull. Tokyo Gekugei Univ. 21*(4):95–136.

Kitano, H. 1974. Effects of the parasitization of a braconid, *Apanteles*, on the blood of its host, *Pieris*. *J. Insect Physiol. 20*:315–27.

Klebanoff, S. J. 1975. Antimicrobial systems of the polymorphonuclear leukocyte, pp. 45–56. *In* J. Bellanli and D. H. Dayton (eds.). *The Phagocytic Cell in Host Resistance*. Raven Press, New York.

Krieg, A. 1958. Weitere Untersuchungen zur Pathologie der Rickettsiose von *Melolontha* spec. *Z. Naturforsch. 13*:374–9.

Krieg, A. 1963. Rickettsiae and rickettsioses, pp. 577–617. *In* E. A. Steinhaus (ed.). *Insect Pathology*, Vol. I. Academic Press, New York.

Krieg, A. 1971. Possible use of Rickettsiae for microbial control of insects, pp. 173–9. *In* D. H. Burges and N. W. Hussey (eds.). *Microbial Control of Insects and Mites*. Academic Press, New York.

Kurstak, E. 1965. Action de la bactérie *Bacillus thuringiensis* Berl. sur des cellules sanguines d'*Ephestia kühniella* Zell. (Lepidoptera). *C. R. Hebd. Séances Acad. Sci. Paris 260*:2368–70.

Kurstak, E., I. Goring, and C. Vago. 1969. Cellular defense in an arthropod in response to infection with a *Salmonella typhimurium* strain. *Antonie van Leeuwenhoek J. Microbiol. Serol. 35*:45–51.

Lackie, A. M. 1976. Evasion of the haemocytic defence reaction of certain insects by larvae of *Hymenolepis diminuta* (Cestoda). *Parasitology 73*:97–107.

Lafferty, K. J., and R. Crichton. 1973. Immune responses of invertebrates, pp. 300–20. *In* A. J. Gibbs (ed.). *Viruses and Invertebrates*. North-Holland, Amsterdam.

Lai-Fook, J. 1968. The fine structure of wound repair in an insect (*Rhodnius prolixus*). *J. Morphol. 124*:37–78.

Lai-Fook, J. 1970. Haemocytes in the repair of wounds in an insect (*Rhodnius prolixus*). *J. Morphol. 130*:297–314.

Lai-Fook, J. 1973. The structure of the hemocytes of *Calpodes ethlius* (Lepidoptera). *J. Morphol. 139*(1):79–104.

Laudureau, J. C., and P. Grellet. 1975. Obtention de lignées permanentes d'hemocytes de Blatte: Caractéristiques physiologiques et ultrastructurales. *J. Insect Physiol. 21*:137–51.

Landureau, J. C., P. Grellet, and I. Bernier. 1972. Caractérisation, en culture *in vitro*, d'un rôle inconnu des hémocytes d'Insectes: Sa signification physiolgique. *C. R. Séances Acad. Sci. Paris 274D*:2200–3.

Lange, H. H. 1932. Die Phagocytose bei Chironomiden. *Z. Wiss. Biol. B 16*:753–805.

Lartschenko, K. 1933. Die Unempfanglichkeit der Raupen von *Loxostege sticticalis* L. und *Pieris brassicae* L. gegen Parasiten. *Z. Parasitenk. 5*:679–707.

Lazarenko, T. 1925. Beiträge zure vergleichenden Histologie des Blutes und des Bindehewebes. II. Die morphologische. Bedeutung der Blutund Bindegewebe-elemente der Insekten. Z. *Zellforsch. Mikrosk. Anat. 3:409-99.*

Lea, M. S., and L. E. Gilbert. 1961. Cell division in diapausing silkworm pupae. *Amer. Zool. 1:368-9.*

Lea, M. S., and L. E. Gilbert. 1966. The hemocytes of *Hyalophora cecropia* (Lepidoptera). *J. Morphol. 118:197-215.*

Leutenegger, R. 1967. Early events of Sericesthis iridescent virus infection in hemocytes of *Galleria mellonella* (L). *Virology 32:109-16.*

Locke, M. 1966. Cell interactions in the repair of wounds in an insect (*Rhodnius prolixus*). *J. Insect Physiol. 12:389-95.*

Lynn, D. C., and S. B. Vinson. 1977. Effects of temperature, host age, and hormones upon the encapsulation of *Cardiochiles nigriceps* eggs by *Heliothis* spp. *J. Invertebr. Pathol. 29(1):50-5.*

Madhaven, K. 1972. Induction of melanotic pseudotumors in *Drosophila melanogaster* by juvenile hormone. *Wilhelm Roux' Archiv. 169:345-9.*

Maier, W. A. 1970. Die Haemolymphe von *Chironomus* und ihre Beeinflussung durch parasitäre Mermithiden. Dissertation, University of Freiburg, Freiburg.

Maier, W. A. 1973. Die Phenoloxydase von *Chironomus thummi* und ihre Beeinflussung durch parasitäre Mermithiden. *J. Insect Physiol. 19:85-95.*

Marschall, K. J. 1966. Bau und Funktionen der Blutzellen des Mehlkäfers *Tenebrio molitor* L. Z. *Morphol. Oekol. Tiere 58:182-246.*

Matz, G. 1965. Implantation de fragments de cellophane chez *Locusta migratoria* L. (Orthoptère). *Bull. Soc. Zool. Fr. 90:429-33.*

McKay, D., and C. R. Jenkin. 1970. Immunity in the invertebrates: The role of serum factors in phagocytosis of erythrocytes by haemocytes of the fresh-water crayfish (*Parachaeraps bicarinatus*). *Aust. J. Exp. Biol. Med. Sci. 48:139-50.*

McLaughlin, R. E. 1971. Use of protozoans for microbial control of insects, pp. 151-72. *In* H. D. Burges and N. W. Hussey (eds.). *Microbial Control of Insects and Mites.* Academic Press, New York.

Mercer, E. H., and W. L. Nicholas. 1967. The ultrastructure of the capsule of the larval stages of *Moniliformis dubius* (Acanthocephala) in the cockroach *Periplaneta americana. Parasitology 57:169-74.*

Merritt, C. M., G. M. Thomas, and J. Christensen. 1975. A natural epizootic of a coccidian in a population of the Egyptian alfalfa weevil, *Hypera brunneipennis*, and the alfalfa weevil, *H. postica. J. Invertebr. Pathol. 26(3):413-4.*

Metalnikov, S. 1908. Recherches expérimentales sur les chenilles de *Galleria mellonella. Arch Zool. Exp. Gén. 8:489-588.*

Metalnikov, S. 1924. Phagocytose et réactions des cellules dans l'immunité. *Ann. Inst. Pasteur Paris 38:787-826.*

Metalnikov, S. 1927. *L'Infection microbienne et l'immunité chez la mites des abeilles Galleria mellonella.* Monograph, Institut Pasteur. Masson, Paris.

Metalnikov, S., and V. Chorine. 1928. The infectious diseases of *Pyrausta nubilalis* Hb. *Int. Corn Borer Invest. Sci. Rep. 1:41-69.*

Metalnikov, S., and V. Chorine. 1929. On the natural and acquired immunity of *Pyrausta nubilalis* Hb. *Int. Corn Borer Invest. Sci. Rep. 2:22-38.*

Metchnikoff, E. 1884. Ueber eine Sprosspilzkrankheit der Daphnien: Beitrag zur Lehre über den Kampf der Phagocyten gegen Krankheitserreger. *Virchow's Arch. Pathol. Anat. Physiol. 96:177-95.*

Metchnikoff, E. 1892. *Leçons sur la pathologie comparée de l'inflammation.* Masson, Paris.

Metchnikoff, E. 1901. *L'Immunité dans les maladies infectieuses.* Masson, Paris.

Meyer, N. F. 1926. Ueber die Immunität einiger Raupen ihren Parasiten, den Schlupfwespen, gegenuber. Z. *Angew. Entomol. 12:376-84.*

Misko, I. S. 1972. The cellular defense mechanisms of *Periplaneta americana* (L.). Ph.D. thesis, Australian National University, Canberra.

Mohrig, V. W., and B. Messner. 1968. Immunreaktionen bei Insekten. I. Lysozym als grundlegen der antibackterieller Faktor im humoralen abwehrmechanismus der Insekten. *Biol. Zentralbl.* 87:439–70.

Nappi, A. J. 1970. Defense reactions of *Drosophila euronotus* larvae against the hymenopterous parasite *Pseudeucoila bochei. J. Invertebr. Pathol.* 16(3):408–18.

Nappi, A. J. 1973. The role of melanization in the immune reaction of larvae of *Drosophila algonquin* against *Pseudeucoila bochei. Parasitology* 66:23–32.

Nappi, A. J. 1974. Insect hemocytes and the problems of host recognition of foreignness, pp. 207–24. *In* E. L. Cooper (ed.). *Contemporary Topics in Immunobiology*, Vol. 4, *Invertebrate Immunology.* Plenum Press, New York.

Nappi, A. J. 1975. Parasite encapsulation in insects, pp. 293–326. *In* K. Maramorosch and R. E. Shope (ed.). *Invertebrate Immunity.* Academic Press, New York.

Nappi, A. J., and J. G. Stoffolono, Jr. 1971. *Heterotylenchus autumnalis:* Hemocytic reactions and capsule formation in the host, *Musca domestica. Exp. Parasitol.* 29:116–25.

Nappi, A. J., and J. G. Stoffolano, Jr. 1972. Distribution of haemocytes in larvae of *Musca domestica* and *Musca autumnalis* and possible chemotaxis during parasitization. *J. Insect Physiol.* 18:169–79.

Nappi, A. J., and F. A. Streams. 1969. Hemocytic reactions of *Drosophila melanogaster* to the parasites *Pseudeucoila mellipes* and *P. bochei. J. Insect Physiol.* 15:1551–66.

Nelson, E., K. Blinzinger, and H. Hager. 1962. Ultrastructural observations on phagocytosis of bacteria in experimental (*E. coli*) meningitis. *J. Neuropathol. Exp. Neurol.* 21:155–69.

Neuwirth, M. 1973. The structure of the hemocytes of *Galleria mellonella* (Lepidoptera). *J. Morphol.* 139(1):105–24.

Neuwirth, M. 1974. Granular hemocytes, the main phagocytic blood cells in *Calpodes ethlius* (Lepidoptera, Hesperiidae). *Can. J. Zool.* 52:783–4.

Nordin, G. L. 1975. Transovarial transmission of a *Nosema* sp. infecting *Malacosoma americanum. J. Invertebr. Pathol.* 25(2):221–8.

Norton, W. N., and S. B. Vinson. 1976. Egg membrane synthesis of a hymenopteran parasitoid. *Amer. Zool.* 15:828.

Nutting, W. L. 1951. A comparative anatomical study of the heart and accessory structures of the orthopteroid insects. *J. Morphol.* 89:501–98.

Paillot, A. 1920. La phagocytose chez les insectes. *C. R. Séances Soc. Biol.* 83:425–6.

Paillot, A. 1928. On the natural equilibrium of *Pyrausta nubilalis* Hb. *Int. Corn Borer Invest. Sci. Rep.* 1:77–106.

Paillot, A. 1930. *Traité des maladies du ver à soie.* Doin, Paris.

Parish, C. R. 1977. Simple model for self–non-self discrimination in invertebrates. *Nature (Lond.)* 267:711–33.

Paschke, J. D., and M. D. Summers. 1975. Early events in the infection of the arthropod gut by pathogenic insect viruses, pp. 75–112. *In* K. Maramorosch and R. E. Shope (eds.). *Invertebrate Immunity.* Academic Press, New York.

Paterson, W. D., and J. E. Stewart. 1974. *In vitro* phagocytosis by hemocytes of the American lobster (*Homarus americanus*). *J. Fish. Res. Bd. Can.* 31:1051–6.

Pauley, G. B., G. A. Granger, and S. M. Krassner. 1971. Characterization of a natural agglutinin present in the hemolymph of the California sea hare, *Aplysia californica. J. Invertebr. Pathol.* 18:207–18.

Poinar, G. O., Jr. 1969. Arthropod immunity to worms, pp. 173–210. *In* G. J. Jackson, R. Herman, and I. Singer (eds.). *Immunity to Parasitic Animals*, Vol. 1. Appleton, New York.

Poinar, G. O., Jr. 1971. Use of nematodes for microbial control of insects, pp. 181–203. In H. D. Burges and N. W. Hussey (eds.). *Microbial Control of Insects and Mites.* Academic Press, New York.

Poinar, G. O., Jr. 1974. Insect immunity to parasitic nematodes, pp. 167–78. *In* E. L. Cooper (ed.). *Contemporary Topics in Immunobiology,* Vol. 4, *Invertebrate Immunology.* Plenum Press, New York.

Poinar, G. O., Jr., and R. Leutenegger. 1971. Ultrastructural investigation of the melanization process in *Culex pipiens* (Culicidae) in response to a nematode. *J. Ultrastruct. Res. 36:*149–58.

Poinar, G. O., Jr., R. Leutenegger, and P. Götz. 1968. Ultrastructure of the formation of a melanotic capsule in *Diabrotica* (Coleoptera) in response to a parasitic nematode (Mermithidae). *J. Ultrastruct. Res. 25:*293–306.

Price, C. D., and N. A. Ratcliffe. 1974. A reappraisal of insect haemocyte classification by the examination of blood from fifteen insect orders. *Z. Zellforsch. Mikrosk. Anat. 147:*537–49.

Prokop, I., G. Uhlenbruck, and W. Kohler. 1968. A new source of antibody-like substance having anti-blood groups specificity. *Vox Sang. 14:*321–33.

Prowse, R. H., and N. N. Tait. 1969. *In vitro* phagocytosis by amoebocytes from the haemolymph of *Helix aspersa* (Müller). I. Evidence for opsonic factor(s) in serum. *Immunology 17:*437–43.

Rabinovitch, M., and M. J. De Stefano. 1970. Interactions of red cells with phagocytes of the wax moth (*Galleria mellonella,* L.) and mouse. *Exp. Cell Res. 59:*272–82.

Ratcliffe, N. A. 1975. Spherule cell–test particles interactions in monolayer cultures of *Pieris brassicae* hemocytes. *J. Invertebr. Pathol. 26:*217–23.

Ratcliffe, N. A., and S. J. Gagen. 1976. Cellular defense reactions of insect hemocytes *in vivo:* Nodule formation and development in *Galleria mellonella* and *Pieris brassicae* larvae. *J. Invertebr. Pathol. 28(3):*373–82.

Ratcliffe, N. A., and S. J. Gagen. 1977. Studies on the *in vivo* cellular reactions of insects: An ultrastructural analysis of nodule formation in *Galleria mellonella. Tissue Cell 9(1):*73–85.

Ratcliffe, N. A., S. J. Gagen, A. F. Rowley, and A. R. Schmit. 1976a. Studies on insect cellular defense mechanisms and aspects of the recognition of foreignness, pp. 210–14. *In* T. A. Angus, P. Faulkner, and A. Rosenfield (eds.) *Proceedings, First International Colloquium on Invertebrate Pathology.* Queens University, Kingston, Ontario.

Ratcliffe, N. A., S. J. Gagen, A. F. Rowley, and A. R. Schmit. 1976b. The role of granular hemocytes in the cellular defense reactions of the wax moth, *Galleria mellonella,* pp. 295–7. *In* Y. Ben-Shaul (ed.). *Proceedings, Sixth European Congress on Electron Microscopy.* Tal International, Israel.

Ratcliffe, N. A., and C. D. Price. 1974. Correlation of light and electron microscopic hemocyte structure in the Dictyoptera. *J. Morphol. 144(4):*485–98.

Ratcliffe, N. A., and A. F. Rowley. 1974. *In vitro* phagocytosis of bacteria by insect blood cells. *Nature (Lond.) 252(5482):*391–2.

Ratcliffe, N. A., and A. F. Rowley. 1975. Cellular defense reactions of insect hemocytes *in vitro:* Phagocytosis in a new suspension culture system. *J. Invertebr. Pathol. 26:*225–33.

Reed, P. W. 1969. Glutathione and the hexose monophosphate shunt in phagocytizing and hydrogen peroxide-treated rat leukocytes. *J. Biol. Chem. 244:*2459–64.

Reik, L. 1968. Contacts between insect blood cells, with special reference to the structure of the capsules formed about parasites. M.Sc. dissertation, University of Cambridge, Cambridge.

Renwrantz, L. R., and T. C. Cheng. 1977a. Identification of agglutinin receptors on hemocytes of *Helix pomatia. J. Invertebr. Pathol. 29(1):*88–96.

Renwrantz, L. R., and T. C. Cheng. 1977b. Agglutinin-mediated attachment of erythrocytes to hemocytes of Helix pomatia. J. Invertebr. Pathol. 29(1):97–100.

Renwrantz, L. R., and G. Uhlenbruck. 1974. Blood-group-like substances in some marine invertebrates. III. Glycoproteins with blood-group A specificity in the cephalopods Sepia offinalis L. and Loligo vulgaris Lam. J. Exp. Zool. 188:65–70.

Ries, E. 1932. Experimentelle Symbiosestudien. II. Mycetomtransplantationen. Z. Wiss. Biol. A 25:184–234.

Rizki, M. T. M. 1960. Melanotic tumor formation in Drosophila. J. Morphol. 106:147–57.

Rizki, M. T. M., and R. M. Rizki. 1976. Cell interactions in hereditary melanotic tumor formation in Drosophila, pp. 137–41. In T. A. Angus, P. Faulkner, and A. Rosenfield (eds.). Proceedings, First International Colloquium on Invertebrate Pathology. Queens University, Kingston, Ontario.

Roberts, D. W., and W. G. Yendol. 1971. Use of fungi for microbial control of insects, pp. 125–49. In H. D. Burges and N. W. Hussey (eds.). Microbial Control of Insects and Mites. Academic Press, New York.

Roitt, I. 1974. Essential Immunology, 2nd ed. Blackwell, Oxford.

Rosenberger, C. R., and J. C. Jones. 1960. Studies on total blood cell counts of the southern armyworm larva, Prodenia eridania (Lepidoptera). Ann. Entomol. Soc. Amer. 53:351–5.

Rotheram, S. M. 1967. Immune surface of eggs of a parasitic insect. Nature (Lond.) 214:700.

Rowley, A. F. 1977a. Studies on insect cellular defences in vitro. Ph.D. thesis, University of Wales, Swansea.

Rowley, A. F. 1977b. The role of the haemocytes of Clitumnus extradentatus in haemolymph coagulation. Cell Tissue Res. 182:513–24.

Rowley, A. F., and N. A. Ratcliffe. 1976a. The granular cells of Galleria mellonella during clotting and phagocytic reactions in vitro. Tissue Cell 8(3):437–46.

Rowley, A. F., and N. A. Ratcliffe. 1976b. The intracellular fate of bacteria and latex particles in insect blood cells, pp. 301–3. In Y. Ben-Shaul (ed.). Proceedings, Sixth European Congress on Electron Microscopy. Tal International, Israel.

Rowley, A. F., and N. A. Ratcliffe. 1976c. An ultrastructural study of the in vitro phagocytosis of Escherichia coli by the hemocytes of Calliphora erythrocephala. J. Ultrastruct. Res. 55:193–202.

Rowley, A. F., and N. A. Ratcliffe. 1978. A histological study of wound healing and hemocyte function in the wax-moth Galleria mellonella. J. Morphol. 157:181–200.

Rowley, D. 1962. Phagocytosis. Adv. Immunol. 2:241–64.

Ryan, M., and W. L. Nicholas. 1972. The reaction of the cockroach Periplaneta americana to the injection of foreign particulate material. J. Invertebr. Pathol. 19:299–307.

Salt, G. 1956. Experimental studies in insect parasitism. IX. The reactions of a stick insect to an alien parasite. Proc. R. Soc. B 146:93–108.

Salt, G. 1957. Experimental studies in insect parasitism. X. The reactions of some endopterygote insects to an alien parasite. Proc. R. Soc. B 147:167–84.

Salt, G. 1959. The fate of a braconid parasite, Rogas testaceus, in four species of hosts. Biologia (Pakistan) 5(1):84–95.

Salt, G. 1960. Surface of a parasite and the haemocytic reactions of its host. Nature (Lond.) 188:162–3.

Salt, G. 1963. The defence reactions of insects to metazoan parasites. Parasitology. 53:527–642.

Salt, G. 1965. Experimental studies in insect parasitism. XIII. The haemocytic reaction of a caterpillar to eggs of its habitual parasite. Proc. R. Soc. B 162:303–18.

Salt, G. 1966. Experimental studies in insect parasitism. XIV. The haemocytic reaction of a caterpillar to larvae of its habitual parasite. Proc. R. Soc. B 165:155–78.

Salt, G. 1968. The resistance of insect parasitoids to the defence reactions of their hosts. *Biol. Rev. (Cambridge)* 43:200–32.

Salt, G. 1970. *The Cellular Defence Reactions of Insects.* Cambridge Monograph in Experimental Biology, No 16. Cambridge University Press, London.

Salt, G. 1971. Teratocytes as a means of resistance to cellular defence reactions. *Nature (Lond.)* 232:639.

Salt, G. 1973. Experimental studies in insect parasitism. XVI. The mechanism of the resistance of *Nemeritis* to defence reactions. *Proc. R. Soc. Lond.* 183:337–50.

Salt, G. 1975. The fate of an internal parasitoid, *Nemeritis canescens*, in a variety of insects. *Trans. R. Entomol. Lond.* 127(2):141–61.

Sato, S., H. Akai, and H. Sawada. 1976. An ultrastructural study of capsule formation by *Bombyx* hemocytes. *Annot. Zool. Jpn.* 49(3):177–88.

Schmit, A. R., and N. A. Ratcliffe. 1977. The encapsulation of foreign tissue implants in *Galleria mellonella* larvae. *J. Insect Physiol.* 23:175–84.

Schmit, A. R., and N. A. Ratcliffe. 1978. The encapsulation of Araldite implants and recognition of foreignness in *Clitumnus extradentatus. J. Insect Physiol.* 24:511–21.

Schmit, A. R., A. F. Rowley, and N. A. Ratcliffe. 1977. The role of *Galleria mellonella* hemocytes in melanin formation. *J. Invertebr. Pathol.* 29:232–4.

Schmittner, S. M., and R. B. McGhee. 1970. Host specificity of various species of *Crithidia* Léger. *J. Parasitol.* 56(4):684–93.

Schneider, F. 1950. Die Abwehrreaktion des Insektenblutes und ihre Beeinflussung durch die Parasiten. *Vierteljahresschr. Naturforsch Ges. Zur.* 95:22–44.

Schwalbe, C. P., and G. M. Boush. 1971. Clearance of ^{51}Cr-labelled endotoxin from hemolymph of actively immunized *Galleria mellonella. J. Invertebr. Pathol.* 18:85–8.

Scott, M. T. 1971a. Recognition of foreignness in invertebrates. II. *In vitro* studies of cockroach phagocytic haemocytes. *Immunology* 21:817–27.

Scott, M. T. 1971b. A naturally occurring hemagglutinin in the hemolymph of the American cockroach. *Arch. Zool. Exp. Gén.* 112:73–80.

Scott, M. T. 1971c. Recognition of foreignness in invertebrates: Transplantation studies using the American cockroach *(Periplaneta americana). Transplantation* 11: 78–86.

Scott, M. T. 1972. Partial characterization of the hemagglutinating activity in hemolymph of the American cockroach *(Periplaneta americana). J. Invertebr. Pathol.* 19:66–71.

Serycyzyńska, H., M. Kamionek, and H. Sandner. 1974. Defensive reaction of caterpillars of *Galleria mellonella* L. in relation to bacteria *Achromobacter nematophilus* Poinar et Thomas (Eubacteriales: Achromobacteriacae) and bacteria-free nematodes *Neoplectana carpocapsae* Weiser (Nematoda: Steinernematidae). *Bull. Acad. Pol. Sci. Cl. II Sér. Sci. Biol.* 22:193–6.

Shapiro, M. 1969. Immunity of insect hosts to insect parasites, pp. 211–28. *In* G. J. Jackson, R. Herman, and I. Singer (eds.). *Immunity to Parasitic Animals,* Vol. 1. Appleton, New York.

Smith, V. J., and N. A. Ratcliffe. 1978. Host defence reactions of the shore crab, *Carcinus maenas* (L), *in vitro. J. Mar. Biol. Assn. U.K.* 58:367–79.

Speare, A. T. 1920. Further studies of *Sorosporella uvella*, a fungus parasite of noctuid larvae. *J. Agric. Res.* 18:399–440.

Splittstoesser, C. M., C. Y. Kawanishi, and H. Tashiro. 1978. Infection of the European chafer, *Amphimallon majalis*, by *Bacillus popilliae*. I. Light and electron microscope observations. *J. Invertebr. Pathol.* 31:84–90.

Stairs, G. 1972. Pathogenic microorganisms in the regulation of forest insect populations. *Annu. Rev. Entomol.* 17:355–72.

Stossel, T. P. 1974. Phagocytosis. *N. Engl. J. Med.* 290:774–833.

Stoltz, D. B., S. B. Vinson, and E. A. Mackinnon. 1976. Baculovirus-like particles in the reproductive tracts of female parasitoid wasps. *Can. J. Microbiol.* 22:1013–23.

Stuart, A. E. 1968. The reticuloendothelial apparatus of the lesser octopus (*Eledone cirrosa*). *J. Pathol. Bacteriol.* 96:401–12.

Sussman, A. S. 1952. Studies of an insect mycosis. III. Histopathology of an aspergillosis of *Platysamia cecropia* L. *Ann. Entomol. Soc. Amer.* 45:233–45.

Sutter, G. R., and E. S. Raun. 1967. Histopathology of European corn borer larvae treated with *Bacillus thuringiensis*. *J. Invertebr. Pathol.* 9:90–103.

Swanson, J., and B. Zeligs. 1974. Studies on gonococcus infection. VI. Electron microscopic study on *in vitro* phagocytosis of gonococci by human leukocytes. *Infect. Immun.* 10:645–56.

Taylor, R. L. 1969. A suggested role for the polyphenol-phenoloxidase system in invertebrate immunity. *J. Invertebr. Pathol.* 14:427–8.

Timberlake, P. H. 1912. Technical results from the gypsy moth parasite laboratory. VI. Experimental parasitism: A study of the biology of *Limnerium validum* (Cresson). *U.S.D.A. Bureau Entomol. Tech. Ser., No. 19*, 5:71–92.

Tinsley, T. W. 1975. Factors affecting virus infection of insect gut tissue, pp. 55–63. *In* K. Maramorosch and R. E. Shope (eds.). *Invertebrate Immunity*. Academic Press, New York.

Tobie, E. J. 1968. Fate of some culture flagellates in the hemocoel of *Rhodnius prolixus*. *J. Parasitol.* 54(5):1040–6.

Toumanoff, C. 1949. Les maladies microbiennes et l'immunité naturelle chez les Insectes. *Rev. Can. Biol.* 8:343–69.

Triggiani, O., and G. O. Poinar, Jr. 1976. Infection of adult Lepidoptera by *Neoplectana carpocapsae* (Nematoda). *J. Invertebr. Pathol.* 27(3):413–14.

Tripp, M. R. 1966. Hemagglutinin in the blood of the oyster (*Crassostrea virginica*). *J. Invertebr. Pathol.* 8:478–84.

Tyson, C. J., and C. R. Jenkin. 1973. The importance of opsonic factors in the removal of bacteria from the circulation of the crayfish (*Parachaeraps bicarinatus*). *Aust. J. Exp. Biol. Med. Sci.* 51:609–15.

Tyson, C. J., and C. R. Jenkin. 1974. Phagocytosis of bacteria *in vitro* by haemocytes from the crayfish (*Parachaeraps bicarinatus*). *Aust. J. Exp. Biol. Med. Sci.* 52:341–8.

Uhlenbruck, G., and O. Prokop. 1966. An agglutinin from *Helix pomatia* which reacts with terminal N-acetyl-D-galactosamine. *Vox Sang.* 11:519–20.

Vago, C. 1964. *Culture de tissus d'invertébrés*. Service du Film Recherche Scientifique, Paris.

Vago, C. 1972. Invertebrate cell and organ culture in invertebrate pathology, pp. 245–78. *In* C. Vago (ed.). *Invertebrate Tissue Culture*, Vol. 2. Academic Press, New York.

Vago, C., and A. Vey. 1970. *Mycoses d'invertébrés*. Service du Film Recherche Scientifique, Paris.

Valvassori, R., and G. A. Amirante. 1976. Caractéristiques immunochimiques et ultrastructurales des hémocytes de *Leucophaea maderae* L. (Insecta Blattoidea). *Monit. Zool. Ital.* (N.S.) 10:403–12.

van den Bosch, R. 1964. Encapsulation of the eggs of *Bathyplectes curculionis* (Thomson) (Hymenoptera: Ichneumonidae) in larvae of *Hypera brunneipennis* (Boheman) and *Hypera postica* (Gyllenhal) (Coleoptera: Curculionidae). *J. Insect Pathol.* 6:343–67.

van den Bosch, R., and P. S. Messenger. 1973. *Biological Control*. Intext, New York.

Vey, A. 1968. Réactions de défense cellulaire dans les infections de blessures à *Mucor hiemalis* Wehmer. *Ann. Epiphyt.* (Paris) 19:695–702.

Vey, A. 1969. Etude *in vitro* des réactions anticryptogamiques des larves de Lepidoptères: Colloque sur les manifestations inflammatoires et tumorales chez les invertébrés 1967. *Ann. Zool. Ecol. Anim.* 1:93–100.

Vey, A., and J. Fargues. 1977. Histological and ultrastructural studies of *Beauveria bas-*

siana infection in *Leptinotarsa decemlineata* larvae during ecdysis. *J. Invertebr. Pathol.* 30(2):207–15.

Vey, A., and P. Götz. 1974. Humoral encapsulation in Diptera (Insecta): Comparative studies *in vitro*. *Parasitology* 70:77–86.

Vey, A., J. M. Quiot, and C. Vago. 1968. Formation *in vitro* de réactions d'immunité cellulaire chez les insectes, pp. 254–63. *In* C. Barigozzi (ed.). *Proceedings, Second International Colloquium on Invertebrate Tissue Culture*. Istituto Lombardo di Scienze e Lettere, Milan.

Vey, A., J. M. Quiot, and C. Vago. 1973. Mise en évidence et étude de l'action d'une mycotoxine, la beauvericine, sur des cellules d'insectes cultivées *in vitro*. *C. R. Hebd. Séances Acad. Sci. Paris Sér. D* 276:2489–92.

Vey, A., and J. M. Quiot. 1975. Pathologie expérimentale: Effet *in vitro* de substances toxiques produites par le champignon *Metarrhizium anisopliae* (Metsch.) Sorok. sur la réaction hémocytaire du Coléoptère *Oryctes rhinoceros* L. *C. R. Hebd. Séances Acad. Sci. Paris Sér. D* 280:931–4.

Vey, A., and C. Vago. 1969. Recherches sur la guérison dans les infections cryptogamiques d'insectes: Infections á *Aspergillus niger* V. Tiegh. chez *Galleria mellonella* L. *Ann. Zool. Ecol. Anim.* 1:121–6.

Vey, A., and C. Vago. 1971. Réaction anticryptogamique de type granulome chez les insectes. *Ann. Inst. Pasteur (Paris)* 121:527–32.

Vinson, S. B. 1972. Effect of the parasitoid *Campoletis sonorensis* on the growth of its host, *Heliothis virescens*. *J. Insect Physiol.* 18:1501–16.

Vinson, S. B. 1977. *Microplitis croceipes:* Inhibitions of the *Heliothis zea* defense reaction to *Cardiochiles nigriceps*. *Exp. Parasitol.* 41:112–17.

Vinson, S. B., and J. R. Scott. 1974. Parasitoid egg shell changes in a suitable and unsuitable host. *J. Ultrastruct. Res.* 47:1–15.

Vinson, S. B., and J. R. Scott. 1975. Particles containing DNA associated with the oocyte of an insect parasitoid. *J. Invertebr. Pathol.* 25:373–8.

Walker, I. 1959. Die Abwehrreaktion des Wirtes *Drosophila melanogaster* gegen die zoophage Cynipidae *Pseudeucoila bochei* Weld. *Rev. Suisse Zool.* 68:569–632.

Weathersby, A. B. 1975. The hemocoel as barrier to parasitic infection in insects, pp. 273–88. *In* K. Maramorosch and R. E. Shope (eds.). *Invertebrate Immunity*. Academic Press, New York.

Werner, R. A., and J. C. Jones. 1969. Phagocytic haemocytes in unfixed *Galleria mellonella* larvae. *J. Insect Physiol.* 15:425–37.

Whitcomb, R. F., M. Shapiro, and R. R. Granados. 1974. Insect defense mechanisms against microorganisms and parasitoids, pp. 447–536. *In* M. Rockstein (ed.). *The Physiology of Insecta*, Vol. 5, 2nd ed. Academic Press, New York.

Whitten, J. M. 1964. Haemocytes and the metamorphosing tissues in *Sarcophaga bullata, Drosophila melanogaster* and other cyclorrhaphous Diptera. *J. Insect Physiol.* 10:447–69.

Wigglesworth, V. B. 1937. Wound healing in an insect, *Rhodnius prolixus* (Hemiptera). *J. Exp. Biol.* 14:364–81.

Wigglesworth, V. B. 1959. Insect blood cells. *Annu. Rev. Entomol.* 4:1–16.

Wigglesworth, V. B. 1965. *The Principles of Insect Physiology*, 6th ed. Methuen, London.

Wigglesworth, V. B. 1970. The pericardial cells of insects: Analogue of the reticuloendothelial system. *J. Reticuloendothel. Soc.* 7(2):208–16.

Wigglesworth, V. B. 1973. Haemocytes and basement membrane formation in *Rhodnius*. *J. Insect Physiol.* 19:831–44.

Wilkinson, P. C. 1976. Recognition and response in mononuclear and granular phagocytes. *Clin. Exp. Immunol.* 25:355–66.

Williams, C. B. 1960. The range and pattern of insect abundance. *Amer. Nat.* 94:137–51.

Wittig, G. 1965. Phagocytosis by blood cells in healthy and diseased caterpillars. I. Phagocytosis of *Bacillus thuringiensis* Berliner in *Pseudaletia unipuncta* (Haworth). *J. Invertebr. Pathol.* 7:474–88.

Wittig, G. 1966. Phagocytosis by blood cells in healthy and diseased caterpillars. II. A consideration of the methods of making hemocyte counts. *J. Invertebr. Pathol.* 8:461–77.

Wülker, W. 1961. Untersuchungen über die Intersexualität der Chironomiden (Dipt.) nach Paramermis-Infektion. *Ark. Hydrobiol.* (Suppl.) 25:127–81.

Wyatt, G. R., and B. Linzen. 1965. The metabolism of ribonucleic acid in *Cecropia* silkmoth pupae in diapause, during development and after injury. *Biochim. Biophys. Acta* 103:588–600.

Zachary, D., M. Brehélin, and J. A. Hoffmann. 1975. Role of the "thrombocytoids" in capsule formation in the dipteran *Calliphora erythrocephala*. *Cell Tissue Res.* 162:343–8.

Zachary, D., J. A. Hoffmann, and J. A. Porte. 1972. Sur un nouveau type de cellule sanguine (le thrombocytoide) chez *Calliphora erythrocephala*. *C. R. Acad. Sci. Paris* 275:393–5.

Zelazny, B. 1976. Transmission of a baculovirus in populations of *Oryctes rhinoceros*. *J. Invertebr. Pathol.* 27:221–7.

Zeledón, R., and E. De Monge. 1966. Natural immunity of the bug *Triatoma infestans* to the protozoan *Trypanosoma rangeli*. *J. Invertebr. Pathol.* 8:420–4.

14 Cellular and humoral responses to toxic substances

D. FEIR

Biology Department, Saint Louis University, Saint Louis, Missouri, 63103, U.S.A.

Contents

14.1. Introduction

Insects are exposed to the toxins and poisons of their predators as well as to all sorts of chemical poisons. There has been relatively little written on the specific responses of insect hemocytes or hemolymph to poisons. Perry and Agosin (1974) reviewed the physiological responses of insects, resistance in particular, to insecticides. Arnold (1974) has a very short paragraph on detoxication of poisons as a function of hemocytes and concludes that the hemocytes do not function in direct defense against the toxicants. Smith (1962) discussed detoxication mechanisms in insects, but the hemocytes and hemolymph were not included among the mechanisms.

The available literature on this topic is small. I have confined this discussion to responses of hemocytes and hemolymph and have included no other detoxication mechanisms.

14.2. Venoms and toxins of predators

Kamon (1965a–d) has done a number of studies on the responses of locusts to scorpion venom. In one experiment (1965a), he injected scorpion venom into locusts, removed the hemolymph, and stored it at 4 °C. After 24 hr of storage there were considerable sediment and opacity in the hemolymph from the venom-treated locusts, but none in the hemolymph of the controls. He found that the increased sediment and opacity were associated with an increase in the specific

415

gravity of the hemolymph and that these phenomena occurred only in treated females. The specific gravity remained high for 14 days, which was the maximum length of the testing period. There was no increase in specific gravity in treated male hemolymph. Kamon could give no explanation for these findings.

Kamon (1965b) injected sublethal doses of ^{131}I-labeled scorpion venom into *Locusta migratoria* and then measured the radioactivity in various body parts and organs at 5-min intervals for up to 216 hr after injection. He found a low percentage of radioactivity in the hemolymph. At 15–60 min after injection, only 3–8% of all radioactive venom was in the hemolymph. There was an increase in radioactivity in the feces which was proportional to the decrease in radioactivity in the hemolymph with time. The distribution of radioactivity was the same in males and females, and there was a decline of radioactivity in most tissues with time. An age difference was shown in the response of the pericardial cells to the venom. In young adult locusts (2–7 days old), about 12% of the radioactive venom was taken up by the pericardial cells, but in old adult locusts (9–21 days) 25% was absorbed by the pericardial cells. The toxic portion of the scorpion venom is nondialyzable (Kamon, 1965c).

Larvae of *Musca domestica* and *Heliothis virescens* were bitten and envenomized by the brown recluse spider (Eskafi and Norment, 1976). Tissue sections were made in a cryostat after 1–2 hr. Some of the sections were subjected to histochemical procedures for determining enzyme activities, and all were examined with the light microscope. The hemolymph samples were collected on glass slides after puncture of one thoracic leg. The hemocytes examined were classed as oenocytoids and showed lysis, coagulation, and disk-sphere transformation. Hemolymph protein was shown to increase in *H. virescens*, apparently owing to a depletion of proteins in the tissues. Hemolymph protein was not studied in *M. domestica*. A number of morphological changes were also shown in the tissues. Norment and Smith (1968) demonstrated a reduction in hemocyte number in *Acheta domesticus* after treatment with brown recluse venom.

Beard (1963) reviewed insect toxins and venoms from many aspects, including invertebrate responses to venoms. He pointed out that in larvae of *Galleria mellonella* paralyzed by *Bracon hebetor*, the venom is circulated by the blood but there is no apparent defense response of the blood to the venom. He cited research showing that the toxic salivary secretions of asilid flies acted as a general tissue denaturant and that the toxin of reduviid bugs had rather strong proteolytic action.

The exotoxin of *Bacillus thuringiensis* has been used by several investigators for studies on responses of the hemolymph. Vankova (1972) injected the exotoxins into the hemocoel of larval *G. mellonella*. The exotoxins of certain strains caused pathological changes, particularly in the macronucleocytes (= prohemocytes) and the micronu-cleocytes (= plasmatocytes). The cytoplasm became vacuolated and disintegrated, and the nuclear and cytoplasmic membranes were de-scribed as losing their firmness. These are responses of the hemo-cytes, of course, but I do not think they are really defense responses. Hoffmann et al. (1974) found no effect of nonlethal concentrations of the exotoxin on the hemogram of *Locusta migratoria*.

Although the following experiment of Dahlman and Vinson (1977) does not involve a venom or toxin of a predator, it does involve a he-molymph response to a fluid from a parasite (which might be thought of as a kind of predator). These authors injected calyx fluid from the braconid *Microplitis croceipes* into *H. virescens*. This caused an ele-vation of hemolymph trehalose for 3 days after the injection, but the effect was gone by 6 days afterward. The calyx fluid injection also pre-vented the normal increase in hemolymph dry weight seen in the con-trols. The possible relationship of the elevated trehalose levels to de-fense mechanisms was not discussed.

14.3. Insecticides

In a fairly early experiment (Yeager and Munson, 1942), ligated larvae of the southern armyworm were given turnip leaf–cornstarch "sand-wiches" with or without one of the following poisons: nicotine ben-tonite, nicotine peat, rotenone, pyrethrum, phenothiazine, calcium ar-senite, calcium arsenate, arsenic trioxide, Paris green, lead arsenate, sodium fluoride, sodium fluoaluminate, barium fluosilicate, and mer-curic chloride. No hematological changes were caused by nicotine bentonite, nicotine peat, rotenone, pyrethrum, or phenothiazine. The arsenicals, the fluorides, and mercuric chloride did cause marked hematological changes in the fore ends relative to the hind ends of the ligated larvae. The cystocytes (= coagulocytes, COs) and plasmato-cytes (PLs) rounded up; there was marked agglutination, distortion, disintegration, and apparent loss of cells from the blood in the front ends as compared with the hind ends. Yeager and Munson thought there was an increase in mitosis after ingestion of the arsenicals, fluo-rides, and mercuric chloride, but they did not demonstrate it quantita-tively.

Yeager et al. (1942) injected Chinese ink into *Periplaneta ameri-cana*. The Chinese ink physically blocks the hemocytes. The authors emphasized that this morphological blockage may or may not mean in-terference with the normal function of the cell. About 24 hr after the Chinese ink injection, metasodium arsenite or nicotine was applied to the external surface of the roach. Each insect was observed daily for 14 days. The mortality from sodium arsenite was 87% in the ink-in-jected roaches compared with 40% in the controls. The mortality from nicotine was 70% in ink-injected animals compared with 57% in the controls. In a parallel study to the insecticide test, the authors showed that some hemocytes were completely loaded with Chinese ink, some contained none, and most contained intermediate quantities of the phagocytized ink. Very heavy doses of Chinese ink and the subsequent very heavy loading of the hemocytes killed the roaches. The authors recognized the possibility that the loaded hemocytes could not perform normal functions in the insect and that this could have accounted for the increased mortality as well as the interpretation that the hemocytes normally detoxify insecticides.

Larvae of *Tenebrio molitor* crawled on filter paper soaked in 5% DDT in peanut oil for 2 min (Joseph, 1958). Blood was taken 3 days after this exposure. DDT caused no change in protein nitrogen, but did cause a 50% increase in nonprotein nitrogen, a 90% increase in reducing compound, and a 70% increase in amino acid nitrogen. Most amino acids increased, but norleucine and taurine decreased with the poisoning. The author pointed out that the physiological effects of DDT poisoning and starvation were very similar. Jones (1957) reported that hemocytes play no role in the natural defenses of the mealworm (*T. molitor*) against DDT.

Patton (1961) had observed that insects were more susceptible to chemical intoxication when their blood specific gravity was high and that this condition usually foretold ecdysis and a time when the blood count was low. Ethyl alcohol destroys blood cells, and Patton reasoned that this would be a way to reduce the number of hemocytes to determine whether the hemocytes play a role in detoxifying insecticides. He injected 95% ethanol into crickets, and 4 hr later he topically applied parathion to the crickets. The alcohol caused injury to about 30% of the hemocytes in 4 hr, and the LD_{50} for parathion went from 16 $\mu g/g$ body weight to 0.075 $\mu g/g$ body weight. Patton concluded that the hemocytes did detoxify the parathion because a decrease in hemocyte number increased susceptibility to the insecticide. Physical destruction of the hemocytes gave less dramatic results because the LD_{50} was reduced only from 16 $\mu g/g$ body weight to 3 $\mu g/g$ body weight. This suggested to Patton that the solvent used with the insecticide might potentiate the insecticide. The effect of the hemocytes on the susceptibility to parathion was explained as possibly attributable to the presence of high levels of nonspecific esterases in the hemocytes.

Sublethal doses of chlordane were administered to *P. americana*, and eight determinations of total hemocyte counts and differential hemocyte counts were made over the next 30 days (Gupta and Sutherland, 1968). Increases in the total hemocyte counts and in the PLs, granulocytes (GRs), and spherulocytes (SPs) were seen after approximately 15 days. The SPs increased in number right after treatment also, but then declined. The COs showed a slight increase in numbers throughout the 30-day period. No statistical analysis was done, so it is not certain which of these differences in counts are statistically significant. It is also not really clear in the paper whether the same roaches or different roaches were used for the repeated counts, and this might make a difference in interpretation. The authors cite a number of references on changes in total and differential hemocyte counts after the administration of toxic substances.

Rakitin (1974) found that hemocytes adsorbed a considerable proportion of an administered radioactive insecticide, but did not say how this affected the insect's response to the insecticide.

14.4. Soluble immunogens

Although there is a fairly long section on defense reactions of insect blood cells, there is very little on hemocyte responses to toxic substances in Crossley (1975). He cites Chadwick and Vilk's work (1969) showing that endotoxins of *Pseudomonas* in *Galleria* increased the LD_{50} on challenge. Schwalbe and Boush (1971, cited in Crossley, 1975) used labeled polysaccharide derived from *Shigella* as an immunogen in *Galleria*. The label was eventually deposited in the cuticle. No more than 15% of the endotoxin was ever contained by the hemocytes at one time. The immunogen did cause a large increase in the number of circulating hemocytes as monitored by hematocrit determinations.

Kamon (1965d) used various substances to immunize locusts. He found that injected saline, China ink, or charcoal failed to cause immune reactions. Amylose and bovine serum albumin gave a significant immune response when the locusts were challenged with scorpion venom. The response appeared to depend on the number of injections rather than the quantity of injected substance. The mortality to the challenge dose decreased from 100% to 76% in the locusts immunized with two injections and from 100% to 40–50% in those immunized with three injections. Injections of scorpion venom into locusts produced no antibody or antibody-like substances in the hemolymph, and Kamon and Shulov (1965) concluded that the hemolymph does not function in defense against scorpion venom.

14.5. Summary and conclusions

It is fairly easy to see how this topic has interested investigators over the years, but it is not so easy to see why more work has not been done. Although some experiments seem to indicate a detoxifying role for the hemocytes, most of the literature states or implies that the observed hemocyte responses are principally attributable to death or de-

420   D. Feir

generation of the cells. There seems to be little, if any, clear-cut evidence for a defense reaction of hemocytes against toxins. The changes in response to insecticides after blockage of the hemocytes could result from the deleterious effects of the breakdown products of the hemocytes as well as from the loss of a detoxifying role of the hemocytes. In general, I think more critical experiments need to be devised and appropriate poisons need to be used in order to elucidate the role of the hemocytes or hemolymph in defense against toxic substances.

References

Arnold, J. W. 1974. The hemocytes of insects, pp. 201–54. In M. Rockstein (ed.). The Physiology of Insecta, Vol. 5, 2nd ed. Academic Press, New York.

Beard, R. L. 1963. Insect toxins and venoms. Annu. Rev. Entomol. 8:1–18.

Crossley, A. C. 1975. The cytophysiology of insect blood. Adv. Insect Physiol. 11:117–221.

Dahlman, D., and S. Vinson. 1977. Effect of calyx fluid from an insect parasitoid on host hemolymph dry weight and trehalose content. J. Invertebr. Pathol. 29:227–9.

Eskafi, F. M., and B. R. Norment. 1976. Physiological action of Loxosceles reclusa (G. and M.) venom on insect larvae. Toxicon 14:7–13.

Gupta, A. P., and D. J. Sutherland. 1968. Effects of sublethal doses of chlordane on the hemocytes and midgut epithelium of Periplaneta americana. Ann. Entomol. Soc. Amer. 61(4):910–18.

Hoffmann, D., M. Brehélin, and J. A. Hoffmann. 1974. Modifications of the hemogram and of the hemocytopoietic tissue of male adults of Locusta migratoria (Orthoptera) after injection of Bacillus thuringiensis J. Invertebr. Pathol. 24(2):238–47.

Jones, J. C. 1957. DDT and the hemocyte picture of the mealworm, Tenebrio molitor L. J. Cell. Comp. Physiol. 50:423–8.

Joseph, M. T. 1958. Effects of DDT on the blood of Tenebrio molitor Linnaeus. Ann. Entomol. Soc. Amer. 51:554–6.

Kamon, E. 1965a. Effect of scorpion venom on specific gravity of locust hemolymph. Nature (Lond.) 206:640.

Kamon, E. 1965b. Distribution of scorpion venom in locusts. J. Insect Physiol. 11:933–45.

Kamon, E. 1965c. Toxicity of the dialyzable fraction of the venom of the yellow scorpion, Leiurus quinquestriatus, to the migratory locust. Toxicon 2:255–9.

Kamon, E. 1965d. Specificity of the immune response of locust to scorpion venom. J. Invertebr. Pathol. 7:199–202.

Kamon, E., and A. Shulov. 1965. Immune response of locusts to the venom of the scorpion. J. Invertebr. Pathol. 7:192–8.

Norment, B., and O. Smith. 1968. Effects of Loxosceles reclusa Gertsch and Mulaik venom against hemocytes of Acheta domesticus (L.). Toxicon 6:141.

Patton, R. L. 1961. The detoxication function of insect hemocytes. Ann. Entomol. Soc. Amer. 54(5):696–8.

Perry, A. S., and M. Agosin. 1974. The physiology of insecticide resistance by insects, pp. 3–121. In M. Rockstein (ed.). The Physiology of Insecta, Vol. 6, 2nd ed. Academic Press, New York.

Rakitin, A. A. 1974. Adsorption of certain toxicants by various insect tissues. (In Russian, English summary) Zh. Obshch. Biol. 35(1):127–33.

Smith, J. N. 1962. Detoxication mechanisms. Annu. Rev. Entomol. 7:465–80.

Vankova, J. 1972. Changes in the hemolymph of *Galleria mellonella* (L.) caterpillars infected by the exotoxin of *Bacillus thuringiensis*. *Acta Entomol. Bohemosion* 69(5):297–304. (Abstract only.)

Yeager, J. F., E. R. McGovran, S. C. Munson, and E. L. Mayer. 1942. Effects of blocking hemocytes with Chinese ink and staining nephrocytes with trypan blue upon resistance of the cockroach *Periplaneta americana* (L.) to sodium arsenite and nicotine. *Ann. Entomol. Soc. Amer.* 35(1):23–40.

Yeager, J. F., and S. C. Munson. 1942. Changes induced in the blood cells of the southern armyworm (*Prodenia eridania*) by the administration of poisons. *J. Agric. Res.* 64(6):307–32.

15 Biochemical and ultrastructural aspects of synthesis, storage, and secretion in hemocytes

A. C. CROSSLEY

School of Biological Sciences, University of Sydney, Sydney, N.S.W. 2006, Australia

Contents

15.1. Introduction

Mesodermal stem cells (i.e., cells migrating inward at the gastral groove during gastrulation of the embryo) differentiate into a host of specialized tissues including hemocytes (Anderson, 1972a,b) (see also Chapter 1). Specialization is recognized by particular synthetic abilities, and ultimately differentiation might be regarded as derepression of certain genes leading to synthesis of particular proteins. A particular gene may be derepressed briefly or more permanently, when we come to be able to recognize the cell as differentiated using biochemical or ultrastructural criteria. The wide variety of possible protein syntheses makes it difficult not only to classify differentiated hemocytes, but also to distinguish hemocytes from other mesodermal cells, for if hemocytes are considered simply as a mobilization or muster of mesodermal cells, then this muster at times includes transient migrations of mesodermal cells such as myoblasts (Crossley, 1965, 1972a). Conversely, if circulation is made an exclusive parameter, then hemocytes in hemocytopoietic organs and in hemocytic reservoirs (Jones, 1970) are excluded artificially. The importance of sessile hemocytic reservoirs, noted by Wigglesworth (1965), is emphasized by comparison of hemocyte counts before and after heat treatment, which prevents clumping and attachment of hemocytes (Wheeler, 1963).

The activities and hence appearance of sessile hemocytes may not differ greatly from those of connective tissue cells such as fibroblasts. Indeed, the embryological origin of fibroblasts is not clear, and they may, like hemocytes, arise from median mesoderm. Their relation to hemocytes is considered below, under "Hemocytes, Fibroblasts, and Collagen Secretion." Cells believed to arise from lateral mesodermal somites are excluded from this review; this excludes myoblasts, cardioblasts, fat body cells, and pericardial cells. A scheme showing possible pathways of differentiation of mesodermal cells, based on synthetic activities, is shown in Fig. 15.1. This chapter considers the synthetic activities of all forms derived from median mesoderm, with the exception of phagocytes and thrombocytoids, which have recently been reviewed by the author (Crossley, 1975).

15.2. Hemocytes and hemolymph homeostasis

15.2.1. Small molecule homeostasis

The hemolymph is both a controlled environment and a transport interface for all insect cells and is regulated with respect to ionic and

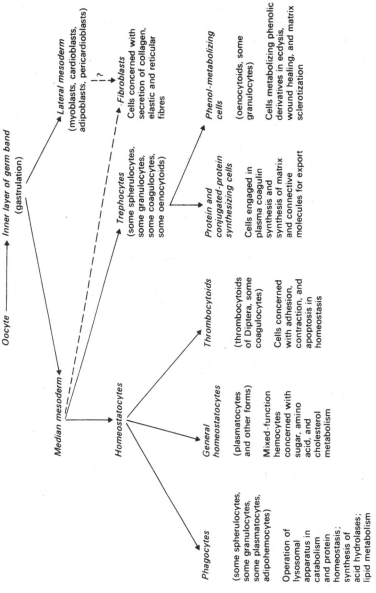

Fig. 15.1. Possible differentiation pathways of mesodermal cells, based on synthetic activities.

Oocyte ⟶ Inner layer of germ band (gastrulation)

Lateral mesoderm (myoblasts, cardioblasts, adipoblasts, pericardioblasts)

Fibroblasts
Cells concerned with secretion of collagen, elastic and reticular fibres

Median mesoderm

Trephocytes (some spherulocytes, some granulocytes, some coagulocytes, some oenocytoids)

Protein and conjugated-protein synthesizing cells
Cells engaged in plasma coagulin synthesis and synthesis of matrix and connective molecules for export

Phenol-metabolizing cells
(oenocytoids, some granulocytes)
Cells metabolizing phenolic derivatives in ecdysis, wound healing, and matrix sclerotization

Homeostatocytes

Thrombocytoids
(thrombocytoids of Diptera, some coagulocytes)
Cells concerned with adhesion, contraction, and apoptosis in homeostasis

General homeostatocytes
(plasmatocytes and other forms)
Mixed-function hemocytes concerned with sugar, amino acid, and cholesterol metabolism

Phagocytes
(some spherulocytes, some granulocytes, some plasmatocytes, adipohemocytes)
Operation of lysosomal apparatus in catabolism and protein homeostasis; synthesis of acid hydrolases; lipid metabolism

425

molecular composition. Regulation is the outcome of the coordinated activities of many kinds of cells, including hemocytes, but the gut, fat body, salivary glands, and nephrocyte complex are also involved. In this section we shall consider the regulation by hemocytes of small molecules such as sugars and amino acids. Significant peculiarities of insect hemolymph include a trend toward replacement of ionic osmotic effectors by organic molecules; a cationic composition low in Na^+; an unusual blood sugar, trehalose, in addition to glucose; and the presence of organic phosphates and a wide variety of enzymes (Florkin and Jeuniaux, 1974).

Most insects can regulate the osmotic pressure of their hemolymph, and it lies between 270 and 490 mOsm (Sutcliffe, 1963). In *Bombyx*, the osmotic pressure remains at a nearly constant level throughout development, largely as a result of regulation of free amino acid concentration (Jeuniaux et al., 1961). Both amino acid and sugar levels are strongly influenced by hemocytes, as discussed below. Sutcliffe (1963) has pointed out that the participation of inorganic ions in the osmotic pressure of the hemolymph tends to decrease with the phylogenetic level of the insect, so that in Lepidoptera, Hymenoptera, and Coleoptera organic molecules are the main osmotic effectors. These orders also have low Na^+ and high Mg^{2+} and K^+ levels in the hemolymph (Duchâteau et al., 1953). The hemolymph ionic patterns in different orders at different stages in their life histories are compared in a useful table by Florkin and Jeuniaux (1974). The role of hemocytes in ionic regulation has not been studied, and there is no information on the relative composition of hemocytes and plasma.

Large amounts of organic acids, especially citrate, are present in insect hemolymph (Levenbook and Hollis, 1961), and the equilibrium concentrations of calcium must be low. Organic phosphates such as α-glycerophosphate and glucose-6-phosphate are present in high concentrations in insect hemolymph, and as chloride concentrations are generally low, the phosphates make a substantial contribution to the ionic fraction. Insects appear to carry dissolved in their plasma substantial levels of a number of metabolites that other groups of animals keep typically within their cells (Wyatt et al., 1963).

There is at present no information on the types of hemocytes involved in small molecule homeostasis, although this could readily be obtained by tracer studies. Ultrastructural specialization of homeostatocytes might be confined to membrane systems without the development of special organelles, and such cells would then resemble plasmatocytes (PLs). Other cell types, and especially phagocytes replete with lytic enzymes, are also likely to have ancillary homeostatic functions (Fig. 15.1).

Biochemical analysis of hemocytes has been greatly hampered by the lack of suitable cell separation techniques. Density gradient centrifugation methods developed for mammalian cells are unsuitable because of the high density of insect hemocytes (< 1.12 g/cm^3). Some success has been achieved recently with Ficoll gradients for buoyant density analysis of *Calliphora* hemocytes (Peake, 1979).

15.2.2. Trehalose metabolism

Insect hemolymph contains trehalose, a nonreducing dimer of α-glucose, in high concentration (Wyatt and Kalf, 1956, 1957), and fluctuations in the level of trehalose reflect the state of glycemia during exercise (Wyatt, 1967). The level of hemolymph trehalose is regulated by a hyperglycemic factor (Steele, 1961) originating in the corpus cardiacum (Steele, 1969; Normann and Duve, 1969; Vejbjerg and Normann, 1974). The hyperglycemic factor of *Manduca* appears to be a glucagon-like peptide (Tager et al., 1975); furthermore, in this insect there appears to be an opposing insulin-like peptide hormone derived from the corpus cardiacum–corpus allatum complex that decreases the level of trehalose in the hemolymph (Tager et al., 1976). Although the fat body appears to be a primary target tissue of these hormones, their effect on the hemocytes has not been determined, but is probably important because of their role in metabolism of trehalose.

Two α-glucosidases, one of which is a trehalase and specifically catalyzes the hydrolysis of trehalose, have been demonstrated in *Periplaneta* hemolymph by Matthews et al. (1976). The trehalase is present both in hemocyte and serum fractions of hemolymph. The enzyme activities in the two fractions share identical electrophoretic mobilities, have similar pH sensitivities, and are inhibited to the same extent by EDTA, indicating the same enzyme. The relative contribution of hemocyte and serum trehalases to the overall trehalase activity of the whole hemolymph was calculated by Matthews et al. (1976). The average hemocyte count of *Periplaneta* is 3×10^4 cells/μl hemolymph (Brady, 1967), and the hemocyte trehalase activity on calcium-activated EDTA-prepared cells has been found to be 0.70 μ-mol glucose/hr per 10^6 cells. Thus the hemolymph activity attributable to hemocytes is 21 nmol glucose/hr per μl hemolymph. As the trehalase activity of the whole hemolymph is 32 nmol glucose/hr per μl hemolymph, the hemocytes contribute two-thirds of the activity. As soluble trehalase may well be released from cells during preparation of serum fractions, this is a minimal figure for hemocyte enzyme content.

Friedman (1960) demonstrated that hemolymph trehalase in the blowfly, *Phormia*, is inhibited in vivo and that activity is increased upon dilution of the hemolymph with water. A similar activation is obtained when *Periplaneta* hemolymph is diluted with buffer and is attributed to lysis of hemocytes by Matthews et al. (1976). There are indications that *Periplaneta* trehalase activity may be modulated in vivo both by calcium ion concentration and pH. The latter is significant in view of the fact that the pH of *Periplaneta* drops from a resting level between pH 7.5 and 7.8 to pH 6.3 after 2 min of flight. Thus the hemolymph trehalase activity will increase in response to the decrease in pH caused by exercise (Matthews et al., 1976). Support for this concept also comes from observation of a marked increase in hemolymph glucose levels after 5 min of exercise, and it is possible that hemolymph trehalose activity is regulated through the influence of CO_2 from muscle upon the pH. At the present time there is no information on the distribution of trehaloses in different types of hemocytes or on the influence of insulin-like hormones on the structure of hemocytes.

15.2.3. Glycogen and glucose metabolism

As in Crustacea (Johnston et al., 1971), insect blood plasma contains little glycogen (Wyatt, 1961), although in both groups glycogen is abundant in the hemocytes. Babers (1941) assayed the distribution of glycogen between hemocytes and plasma in *Prodenia* larvae fed on a glucose-starch paste. He found 16.5 mg glycogen per 100 ml blood in the cell fraction, but only 0.92 mg glycogen per 100 ml blood in the plasma, and concluded that most of the glycogen was deposited in the hemocytes. Glycogen has also been reported in *Galleria* hemocytes by Ashhurst and Richards (1964b), who used PAS reagent with diastase controls. Costin (1975), using the PAS-diastase method, failed to detect glycogen in fifth-instar *Locusta* hemocytes, but as glycogen is abundant in granulocytes of fifth-instar *Locusta*, as evidenced by electron microscopy (Brehélin et al., 1975), it appears that abundant mucopolysaccharide or mucoprotein masks glycogen in histochemical procedures. Glycogen, especially in its β form, has a distinctive ultrastructure (Crossley, 1975) and is frequently encountered in hemocytes (e.g., in *Rhodnius;* Lai-Fook, 1968; in *Calliphora:* Crossley, 1968; in *Blaberus:* Moran, 1971; in *Antheraea:* Beaulaton and Monpeyssin, 1976), although its presence may be transient, as discussed below.

The phosphatase involved in the breakdown of glycogen to glucose-1-phosphate has been reported in *Bombyx* hemolymph, although its location has not been determined (Faulkner, 1955). Phosphoglucomutase, responsible for the interconversion of glucose-1-phosphate and glucose-6-phosphate, which is involved not only in glycolysis but also in metabolism of oligo- and polysaccharides, is present in the midgut of *Bombyx* (Newburgh and Cheldelin, 1955), although it has not been detected in the hemolymph (Faulkner, 1955).

Glucose is present in large amounts in the hemolymph of *Phormia* (Evans and Dethier, 1957) and *Agria* larvae when reared on defined diets rich in glucose (Barlow and House, 1960). In another dipteran,

Drosophila, the larva can regulate both the glucose concentration and the osmotic pressure of the hemolymph, but the contribution of the hemocytes to this homeostasis is not known (Zwicky, 1954). Hemolymph glucose is known to be incorporated into the chitin of growing cuticle in several insect orders (Bade and Wyatt, 1962). Flight muscle stores and metabolizes its own glycogen by control of Ca^{2+} flux and regulation of the phosphate potential (Sacktor, 1976). Florkin and Jeuniaux (1974) have suggested that high hemolymph sugar levels are restricted to a number of species of Hymenoptera and Diptera that are adapted for sustained flight. However, the role of tissues and especially hemocytes in sugar homeostasis needs further investigation.

15.2.4. Lipid metabolism

Lipid droplets are formed in *Prodenia* hemocytes as glycogen disappears from the cells (Munson and Yeager, 1944). In both *Ephestia* (Arnold, 1952) and *Galleria* (Ashhurst and Richards, 1964b) larval hemocytes accumulate stored lipid before pupation, indicating that some hemocytes parallel the lipid metabolism of fat body cells. Homogeneous electron-lucent or gray droplets of smooth, rounded profile are commonly encountered in electron micrographs of hemocytes and are usually interpreted as lipid (e.g., in *Ephestia:* Grimstone et al., 1967; Smith, 1968; in *Calliphora:* Crossley, 1968, 1975; Zachary and Hoffmann, 1973). In *Pectinophora* larvae phagocytic granulocytes progressively accumulate lipid droplets, giving rise to adipohemocytes (Raina, 1976).

15.2.5. Amino acid metabolism

Insect blood contains high levels of free amino acids. For example, in *Phormia* third-instar larvae, 60% of the total free amino acid is found in the hemolymph (Levenbook, 1966). The distribution of amino acids between hemocytes and plasma was not investigated by earlier workers, but has recently assumed importance as a result of the proposal that glutamate acts as a transmitter at the insect neuromuscular junction (Kerkut et al., 1965; Usherwood and Cull-Candy, 1975). Sufficient L-glutamate is apparently present in the hemolymph to reduce or abolish neurally evoked contractions (Murdock and Chapman, 1974; Clements and May, 1974). Indeed, if L-glutamate is injected into blowfly larval hemolymph, reversible paralysis ensues (Evans and Crossley, 1974; Irving et al., 1976). Enclosure of free amino acids within hemocytes would keep glutamate away from receptors at the neuromuscular junction, but measurements on cockroaches indicated only 12% of the total free amino acid pool bound to hemocytes (Holden, 1973). However, in experiments on *Calliphora,* where particular precautions were taken to prevent hemocyte disruption during preparation of samples, over 60% of the dicarboxylic amino acids glutamate and aspartate appeared to be sequestered in the hemocyte fraction (Evans and Crossley, 1974). In *Lucilia*, glutamate is also present in hemocytes, from which it is released on clotting, and aged hemolymph thus gains pharmacological activity, even though L-glutamate is virtually absent from fresh hemolymph (Irving et al., 1976). When glutamate was injected into *Calliphora* hemolymph, the concentration of this amino acid within hemocytes rapidly increased, indicating that hemocytes were capable of selectively accumulating and retaining amino acids against a concentration gradient. Thus, hemocytes apparently have a homeostatic function with respect to plasma amino acids (Evans and Crossley, 1974).

Hemolymph amino acids are, of course, in a pool drawn on by many tissues engaged in protein synthesis. In adult *Phormia* males, for example, it has been estimated that the entire pool of free lysine and alanine is replaced within 1 hr (Levenbook and Krishna, 1971). Although (^{14}C)-glycine is rapidly incorporated into larval *Bombyx* fat body and silk gland, slower incorporation into gut and hemolymph proteins, after a long period, has been noted (Faulkner and Bheemeswar, 1960). In lepidopteran larvae, hemolymph amino acid homeostasis can be disturbed by extirpation of silk glands, an operation that causes accumulation of massive amounts of amino acids such as glutamine, threonine, and glycine, which are metabolic precursors of silk. A delicate balance normally exists between secretion and withdrawal of hemolymph amino acids (Wyatt, 1963).

In the tsetse fly, *Glossina*, the composition of the hemolymph free amino acid pool shows changes related to feeding and flight, and proline appears to be a readily mobilizable energy reserve (Bursell, 1963, 1966). Both proline and alanine are present in large quantities in *Glossina* hemolymph (Knight, 1960; Petit, 1968). During each pregnancy cycle in *Glossina*, there is a switch from leucine to tyrosine utilization in hemolymph protein synthesis, but whereas leucine is continuously removed from the hemolymph, tyrosine is at times released and builds up in the hemolymph (Tobe and Davey, 1975). The metabolism of tyrosine is considered below in connection with phenol metabolism.

15.2.6. Peptide–amino acid conversion

Small peptides form an important and highly accessible form of amino acid storage in the hemolymph, hydrolysis being controlled by hemocyte enzymes. Collett (1976a) has shown that during the period of intensive posteclosion feeding in adult *Calliphora*, the concentration of free amino acids remains unchanged, but the total amount of peptide increases dramatically, sometimes to amounts equivalent to 10% of the total protein of the fly. These peptides contain, at most, four to five amino acid residues. Collett (1976b) suggests that peptides emerge from nonhemolymph cells and pass into the hemolymph for hydrolysis, although he ignores the role of the cellular compartment of the hemolymph, which is included with the plasma in his samples. A homeostatic control was indicated in experiments in which the hemolymph was diluted with saline in vivo. The total amount of peptide released in response to dilution approximates the amount of amino acid needed to restore the hemolymph to normal concentration. Five out of the six molecular species of peptidases detected in *Calliphora* are present in the whole hemolymph, although the hemocyte/plasma ratio appears not to have been investigated.

Collett (1976b) suggests that a major function of one or more of the types of cells in the hemolymph may be the production of peptidases that mediate the controlled release of amino acids from peptides. This control system would provide not only metabolites, but also osmotic stabilization. In *Drosophila*, Begg and Cruikshank (1962) estimated that about half the osmotic pressure of the hemolymph was accounted for by amino acids. Further analysis of the distribution of both peptides and peptidases in the cells and plasma of the hemolymph is urgently needed in order to develop a comprehensive picture of amino acid turnover in insects. In *Schistocera* hemolymph, Miller et al. (1973) found most of the glutamate in the plasma, but in a pharmacologically inactive form. Pharmacological activity develops on standing and is enhanced by an unknown factor present in the hemocytes. With our recently gained knowledge of peptide turnover, these results could be interpreted as inactive peptide hydrolysis by a peptidase from hemocytes to produce the neurotransmitter glutamate.

15.2.7. Hemocyte trephocytosis

The mobile nature of hemocytes would make localized release of metabolites on receptive tissues feasible, a phenomenon known as trephocytosis. The first indication of this phenomenon came from a study of development in the pupa of *Anagasta*, where hemocytes were observed releasing lipid droplets in the developing wing (Zeller, 1938). It is usually not obvious whether hemocytes are involved in release or uptake of metabolites unless detailed ultrastructural and histochemical studies are performed. In a study of the developing (*Lucilia*) blowfly wing, Seligman et al. (1975) interpret their micrographs to indicate hemocyte phagocytosis of fragments including lipid droplets from disintegrating wing hypodermis rather than trephocytosis. In *Sarcophaga*, Whitten (1969) described the digestion of fragments of larval cells in hemocytes within the puparium, followed by release of osmiophilic globules into the extracellular space. Similar globules appear in adjacent epidermal cells, providing cir- cumstantial evidence of trephocytosis. For *Calliphora* there is ultra- structural evidence of lipid production by hemocytes and release into the hemolymph by exocytosis (Crossley, 1968), but not of localized re- lease.

15.3. Hemocytes in protein synthesis

15.3.1. Qualitative importance of hemocytes in plasma protein synthesis

Insect hemolymph contains abundant protein, both in the cell fraction and in the plasma. In this section we shall consider the relative impor- tance of hemocytes in synthesis of hemolymph proteins, the turnover and fate of hemolymph proteins, and the changes in the system that result from injury. The relationship of ultrastructural to biochemical studies will be discussed.

The possibility that hemocytes were involved in plasma protein me- tabolism was first envisaged by Cuénot (1891), but quantitative assess- ment has indicated that hemocytes have a minor role in bulk synthesis of hemolymph protein. In *Diatraea,* for example, Chippendale (1970a) measured the relative capacity of larval hemolymph and mid- gut to synthesize protein. The rate of ^{14}C-leucine incorporation into hemolymph (consisting of both hemocytes and plasma) and into mid- gut proteins was measured under identical conditions for 13-day lar- vae. The rate of protein synthesis by hemolymph was one-tenth that by midgut. The fat body in *Diatraea* synthesizes hemolymph protein even more rapidly than the midgut, and in terms of bulk protein syn- thesis the hemocytes are of minor importance (Chippendale, 1970b). Qualitatively, however, proteins synthesized by hemocytes are likely to be of great significance, as evidenced by studies of injury and of cu- ticle and connective tissue formation.

The synthesis of hemolymph proteins has been extensively studied in a variety of silk moths. It was suggested by Sissakjan and Kuvajeva (1957) that in the *Bombyx* pupa the hemocytes synthesize hemolymph protein. Soon afterward, however, Shigematsu (1958) demonstrated that *Bombyx* fat body incubated in vitro was able to synthesize and secrete two globulin proteins and that electrophoretic comparison in- dicated that these were identical to hemolymph proteins. Protein syn- thesis by fat body cells in vitro has also been demonstrated for *Hyalo- phora* by Stevenson and Wyatt (1962).

Laufer (1960) used electrophoretic zymograms to follow changes in *Hyalophora* and *Samia* blood proteins during development and de- tected esterase, phosphatase, lipase, glucosidase, galactosidase, tyrosinase, and chymotrypsin activity in blood plasma. Pupal *Hyalophora* blood plasma contains six esterases, one of which is sensitive to eserine and is probably an acetylcholinesterase. Circumstantial evidence suggests passage of protein molecules from fat body to blood to oocyte. However, the proteins of the hemocyte fraction were not surveyed, and no direct evidence of sites of protein synthesis was obtained. Several of the enzymes found in plasma have also been detected in hemocytes, including phosphatase, lipase, and tyrosinase, and the cell compartment of the blood is probably important in synthesis.

Methods used by several workers fail to distinguish between or separate proteins present in hemocytes from those present in the plasma. Thus, the method of Lysenko (1972) in which whole hemolymph is frozen and thawed before centrifugation is certain to disrupt hemocytes and lead to a mixture of cellular and plasma proteins. It is not surprising, therefore, that 19 protein bands were obtained on acrylamide gels for *Galleria*. As a caution it should be borne in mind that 55 proteins can be detected in a single cellular organelle, the ribosome (Garrett and Wittmann, 1973). In their study of *Oncopeltus* hemolymph proteins, Feir and Krzywda (1969) do not report separation of cells from plasma at all, and the biological significance of their results is obscure.

15.3.2. Hemolymph as a protein transit system

That the hemolymph is an important marshaling locus for proteins is illustrated by two systems: the oocyte and the salivary gland. In *Hyalophora*, seven proteins of the oocyte have antigenic counterparts in the blood. Even noninsectan proteins injected into the hemocoel have been detected in the oocyte, but quantitative measurements indicate selective protein uptake (Telfer, 1960, 1961). Telfer and Melius (1963) used ^{3}H-leucine labeling of blood proteins in *Hyalophora* to demonstrate uptake into yolk spheres of the oocyte.

In disk electrophoresis of *Chironomus* blood plasma 17 protein bands were detected, 9 of which have been tentatively identified as hemoglobin; 10 of the plasma protein bands are shared by salivary gland extracts, including 4 hemoglobins, and their identity has been confirmed by immunoelectrophoresis and Ouchterlony gel diffusion (Doyle and Laufer, 1969). All the antigens detectable in salivary gland secretion are also present in the hemolymph (Laufer and Nakase, 1965). These results, together with studies on amino acid uptake, provide convincing evidence that the mucoprotein secretion of the salivary gland is not synthesized in the gland, but in other tissues, with the hemolymph acting as a transit system. Whereas the work of Orr (1964), Price (1966), and Martin et al. (1969) have demonstrated that the fat body represents a major site of protein synthesis in Diptera, the role of hemocytes in salivary and oocyte protein synthesis remains largely unexplored.

15.3.3. Hemocytes in synthesis of cuticle proteins

In one developing system, the cuticle, the role of the hemocytes in protein synthesis is better documented. During larval development in several species, protein levels rise (Denucé, 1958; Chen and Levenbook, 1966; Chippendale and Beck, 1966), reach a peak concentration during the premolt stage, then decline (Fox and Mills, 1969). During apolysis, a specific blood protein is synthesized, which disappears after apolysis (Steinhauer and Stephen, 1959). These results support Ludwig's (1954) suggestion that there is a relationship between hemolymph and cuticle proteins. About 14 electrophoretic bands of protein are found in *Periplaneta* cuticle, and about 11 in the hemolymph, with at least 6 protein bands appearing from both sources (Fox and Mills, 1969). Electrophoretic migration (ratio of distance traveled by a protein band to that traveled by the tracking dye) fails to prove that proteins having the same mobility are identical, but this datum is further supported by serological evidence. Antibodies prepared against partially purified cockroach cuticular proteins cross reacted with at least one blood protein, indicating similarity of antigenic sites (Fox et al., 1972). (It should be borne in mind that immunoelectrophoresis of blood proteins demonstrates in most instances that single electrophoretic peaks are composed of several antigenically different proteins: Nelstrop et al., 1970.) In *Locusta* fifth-instar nymphs, hemolymph proteins labeled with ^{3}H-leucine were vigorously taken up by epidermal cells and were deposited in the cuticle, although the possibility of breakdown to amino acids before incorporation remained (Tobe and Loughton, 1969). Evidence that degradation to amino acids does not occur before uptake into the cuticle of *Periplaneta* is provided by Fox and Mills (1969) and Koeppe and Mills (1972). Diphenols bound to blood proteins are translocated to the cuticle, where the ^{14}C label becomes incorporated into the matrix (see below, under "Hemocyte, Hemolymph, and Sclerotization of the Integument"). Immunochemical studies indicate that *Manduca* cuticle and blood proteins have identical antigenic sites (Koeppe and Gilbert, 1973).

In a series of experiments to test the hypothesis that cuticle proteins are made in hemocytes, Geiger et al. (1977) injected ^{14}C-leucine (ul) into the hemocoel of *Periplaneta* and observed first that label became incorporated into protein both of plasma (termed serum by these authors) and hemocytes. Plasma proteins were purified by gel filtration, concentrated, and reinjected into other freshly ecdysed insects. After incubation of 1 hr, radioactivity was detected in serum, hemocyte, and cuticle proteins. The data suggest that the same proteins pass rapidly among hemocytes, plasma, and cuticle, although there is no direct evidence that synthesis of the proteins takes place in the hemocytes.

15.3.4. Hemocytes in synthesis of injury proteins

When glycine-1-^{14}C was injected into the hemocoel of diapausing pupal and developing adult *Hyalophora*, incorporation into the proteins of the blood plasma was found to be accelerated by prior injury to the integument (Telfer and Williams, 1955, 1960). Diapausing *Hyalophora* pupae proved useful experimental subjects for studies on injury protein synthesis. Berry et al. (1964) found that injection of 0.36 mg/g of puromycin at the time of an experimental injury prevented the appearance of injury protein and the increase in concentration of the other blood proteins, suggesting that injury stimulated the actual synthesis of protein. However, threshold concentrations (0.5 μg/g) of actinomycin D prevented the synthesis of new blood protein, but did not prevent the injury-induced stimulation of synthesis of proteins already in production. Pupae injected with threshold amounts of actinomycin D to prevent synthesis of injury protein were used as hosts, and various tissues of previously injured but uninjected pupae were used as implants. The host blood was then checked for injury protein. This experiment revealed that the hemocytes of injured pupae are one source of injury protein and that hemocytes synthesize and release injury protein into the blood of actinomycin-treated hosts, presumably calling forth the mRNA they had made before they were exposed to the actinomycin in the blood of the host. The injury protein has the same electrophoretic mobility as a protein that normally appears during adult development and seems to be identical with it. Thus, by wounding a pupa one induces the precocious synthesis of a protein that normally appears later in development.

Coles (1965) found for *Rhodnius* that most hemolymph proteins corresponded to fat body proteins in electrophoresis gels, but some of the quantitatively minor bands did not. The level of non-fat-body proteins increased after injury, and it was suggested that these could be formed by hemocytes.

Marek (1969) reported that *Galleria* tissues isolated in vitro released proteins into an incubation medium containing centrifuged hemolymph. That proteins were synthesized rather than simply released by dying cells was evidenced by addition of actinomycin-D to the nutrient medium, which resulted in the absence of certain bands in electrophoretic gels. Unfortunately, no electrophoretic data on the unincubated medium were provided, and this hampers interpretation of the results. It appears that two proteins, both esterases, are synthesized de novo by midgut cells and by hemocytes, but not by fat body cells. The most abundant protein is identified as an esterase apoenzyme and is termed X_2 "injury protein"; a second X_1 "companion protein" is inhibited by eserine and is considered an acetylcholinesterase. In later papers, Marek (1970a,b) amplifies his earlier results using in vivo techniques and a different form of stress. Cooling pupae of *Galleria* without their cocoons for 3–7 days at 4 °C induced synthesis of a "cooling protein," a globulin that appears in the hemolymph. The cooling protein is again an esterase apoenzyme. It is assumed by Marek (1970a) that cooling protein is synthesized by hemocytes, but this does not follow directly from the published results of his in vivo experiments, and other tissues such as midgut could contribute, as indicated for injury protein by his in vitro results. It is clear that plasma alone, without hemocytes is not active in protein synthesis in *Galleria*.

There are analogies between the results of Laufer (1960) and Berry et al. (1964) for *Hyalophora;* Coles (1965) for *Rhodnius;* and Marek (1969, 1970a,) for *Galleria*. In *Hyalophora* and *Galleria*, injury proteins that have esterase activity appear, like the *Rhodnius* injury protein, to derive from hemocytes. Marek (1970a,b) suggests that injury proteins are the result of induced hemocyte gene activation and invokes the Kroëger (1963) hypothesis to explain the induction on a basis of changed Na^+/K^+ ionic ratio. This hypothesis has recently been critically reviewed by Ashburner and Cherbas (1976).

The globulin induced by cooling of *Galleria* is termed an "immunoglobulin" by Marek (1970b) because its electric charge determined by electrophoresis and its molecular weight of 470,000 determined by gel filtration are within the range of variation of vertebrate γ-globulins. This may not be a valid induction, as was pointed out by Nelstrop et al. (1970). They made a comparison of *Locusta* serum proteins with other invertebrate and vertebrate blood proteins and concluded that although proteins with the same electrophoretic mobility as the γ-globulins were detected in an insect, it was not possible to characterize a protein as an immunoglobulin merely on the basis of its electrophoretic mobility or its molecular size alone. Furthermore, there is no evidence that the *Galleria* cooling protein produces an acquired immune response.

Day et al. (1970) have shown the existence of complement component 3 proactivator in the primitive arthropod *Limulus*, and a similar humoral factor is present in *Blaberus* (Anderson et al., 1972), indicating that at least some analogues of the vertebrate complement system may be present in insects. A natural, heat-labile, hemagglutinin with activity against vertebrate erythrocytes is also present in *Blaberus* hemolymph (Anderson et al., 1972). The origin and molecular nature of these components remain to be investigated, but it is probable that they are proteins derived from hemocytes.

Changes in the protein spectrum of insect hemolymph caused by active immunization were observed by Stephens (1959), Sauerländer and Erhardt (1961), Crowley (1964), Gingrich (1964), and Kamp (1968). Kamp (1968) demonstrated that actinomycin C did not significantly inhibit the synthesis of immuno-factor in the larvae of *Galleria*, thus establishing a distinction between this factor and the cooling protein described by Marek (1970a).

One protein that is readily identifiable in the hemolymph of injured *Galleria* is lysozyme. If the insect is stressed by injection of gram-positive bacteria, the natural lysozyme values of 25–500 μg/ml hemolymph rise to 9,000 μg/ml within 24 hr. This protein may be largely responsible for humoral immunity in insects and is certainly a major bactericidal blood protein (Möhrig and Messner, 1968). Lysozyme is also produced by *Calliphora* larvae in response to bacterial infection and has been shown to be produced by pericardial cells. In uninjured *Calliphora*, the maximal titer of lysozyme in the blood occurs shortly after puparium formation, at the time of pericardial cell degranulation (Crossley, 1972b). In *Locusta*, an enzyme active on *Micrococcus* (probably lysozyme) is present in the plasma and in greater quantity in the hemocytes. If hemolymph is separated from hemocytes after blood coagulation, the serum contains four times as much activity as the plasma (Hoffmann et al., 1977). These authors report that lysozyme levels rise on injection of *Bacillus thuringiensis*, but rapidly fall again, and are not responsible for retained immunity. In Tettigonioidea and Grylloidea, a phagocytic organ of sessile hemocytes or reticular cells is present (Cuénot, 1897; Nutting, 1951). Ultrastructural evidence for a causal relationship between injection of living bacteria and protein synthesis in these cells is provided by Hoffmann and Porte (1973). *Locusta* adults injected with a weak dose of living *B. thuringiensis*, or of culture medium cleared of bacteria, show not only proliferation and liberation of phagocytic hemocytes, but also changes of the reticular cells from the hemopoietic organ. Marked dilation of cisternae of the rough endoplasmic reticulum, with the formation of crystalloid inclusions, suggests that reticular cells are engaged in protein synthesis and storage. Further ultrastructural work is needed to correlate changes in hemocytes with synthesis of specific proteins.

15.4. Hemocytes in mucopolysaccharide and conjugated protein synthesis

Mucopolysaccharides and conjugated proteins, including both glycoproteins and lipoproteins, form matrix components of blood coagulum, antibacterial nodules, and connective tissues. Hemocytes of certain types in most insects contain mucopolysaccharides or conjugated proteins that are apparently precursors of matrix components, and the secretion of these molecules has been studied in an effort to understand blood coagulation, wound healing, bacteriostasis, and connective tissue formation.

15.4.1. Plasma coagulation

The plasma coagulum is the most accessible and hence most studied matrix, but even so we have only a vague idea of its composition. Brehélin (1972) approached the problem by asking what proportion of hemolymph conjugated proteins formed part of the coagulum in *Locusta*. He found that $18 \pm 5.5\%$ of the total blood protein was incorporated into the coagulum and that this protein was represented by a single electrophoretically negative band of material. Postulating that the titer of protein might correlate with the rate of coagulation, he assayed the protein band during the fifth instar, when rapidity of coagulation increased threefold. However, there was no significant change in the protein titer during this period. Instead, the increased rate of coagulation correlated with increases in number and proportion of both granulocytes (GRs) and coagulocytes (COs) (Brehélin, 1972). More recently, Brehélin (1977) has reported that in *Locusta* the coagulum is formed from a glycolipoprotein (MW 500,000–1,000,000) and that he has evidence that a nondialyzable (i.e., small molecular fraction) thermolabile factor, which provokes coagulation, can be isolated from GRs in *Locusta*. In the fresh cell-free hemolymph of *Periplaneta* six conjugated proteins have been detected by electrophoresis (Siakotos, 1960a). The concentration of these proteins changed during developmental cycles and as the blood coagulated. Comparison of cell-free fresh hemolymph (plasma) and cell-free clotted hemolymph (serum) revealed a marked redistribution of neutral lipid, the appearance of a new protein fraction, and increases in existing protein levels, but a decrease in glycoproteins (Siakotos, 1960b). No simple functional interpretation of these complex changes is possible.

If insect hemolymph is centrifuged as soon as possible after removal, using conditions that remove intact hemocytes, a normal clot fails to form. This is not surprising for *Calliphora* (Åkesson, 1953), where cellular agglutination occurs (Crossley, 1975). It is, however, indicative of hemocyte involvement in plasma gelation in *Locusta* (Brehélin, 1972) and *Periplaneta* (Franke, 1960a,b), where coagulation of soluble components of the plasma, rather than cellular agglutination, is the dominant clotting event. Similar results have been obtained for crustaceans and for the chelicerate, *Limulus* (Levin and Bang, 1964a). These results indicate that certain hemocytes form part of the chain of events leading to plasma gelation, but do not indicate that hemocytes themselves release the coagulable molecules. Whole-insect X-irradiation has also been shown to disrupt clotting changes in

Periplaneta, *Carausius*, and *Locusta* (Grégoire, 1955b; Hoffmann, 1972). However, demonstrable damage to a wide variety of cells, including hemocytes, again tells us no more than that some unknown link in the blood coagulation chain has been broken (see also Chapter 7).

15.4.2. Hemocytes in localized coagulation

It has been known for many years that changes in certain hemocytes show spatial and temporal relationships with the onset of clotting. For crustaceans, "explosive corpuscles" were described by Hardy (1892) and their role in clotting explored by Tait (1911) and later workers. For insects, similar unstable hemocytes were first described by Yeager and Knight (1933), and a class of hyaline hemocytes termed "coagulocytes" has been advanced, on the basis of phase-contrast microscopy of many species of insects, as the type of hemocyte efficient in the inception of coagulation (Grégoire and Florkin, 1950a; Grégoire, 1955a). The significance of the term CO has been discussed by the author elsewhere (Crossley, 1975). Observation of cellular changes in COs are not invariably followed by plasma gelation (Grégoire, 1955a), and a causal relationship has not been established by phase-contrast microscopy. It is possible that changes in labile cells may be the effect rather than the cause of coagulation, as has been suggested by Gupta and Sutherland (1966). If COs are instrumental in plasma gelation, we might expect a positive correlation between strong plasma gelation and the titer of COs. It is difficult to assay coagulation quantitatively, but certain insects do show strong plasma gelation. Within this group there is a wide variation in the percentage of COs; the range is from 8–36% in *Gryllotalpa* and *Carausius* to 95–100% in *Meloe* and *Heliocopris* (Grégoire, 1964, p. 171). On the other hand, there is a correlation between the developmental titer of COs in differential hemocyte counts of *Periplaneta* (Wheeler, 1963) and *Locusta* (Brehélin, 1971) and the ability of the blood to clot.

The formulation of granular precipitate, presumably plasma coagulum, around COs is visible in electron microscope preparations of *Locusta* blood, but the timing of events suggests that changes in the ultrastructure of the COs, such as disappearance of electron-dense granules, follow plasma gelation and are not causal (Hoffmann and Stoekel, 1968; Hoffmann et al., 1968). In *Carausius*, Goffinet and Grégoire (1975) report that COs directly eject nuclear and cytoplasmic substances through microruptures of the cytoplasmic membrane (see Chapter 7). Although other types of hemocytes undergo structural alterations, these take place without breakage of the cytoplasmic membrane and when plasma coagulation is already under way. In this species, the organelle complement of COs is not distinctive, and the observed functional peculiarity, which Grégoire (1964) has called a "differential sensitiveness," is presumed to result from differences in the permeability of the plasma membrane. Support for this idea comes from the observations of Belden and Cowden (1971) on *Periplaneta*, where the CO gives indications of becoming more hydrophobic as the blood clots.

It is important to emphasize that COs are not the only reactive cells that change during the clotting reaction. Others include the oenocytoids (OEs), which are discussed below in connection with phenol metabolism, the GRs or spherulocytes (SPs) discussed next, and the thrombocytoids of Diptera (Zachary and Hoffmann, 1973).

There is evidence for several species that functional and structural relationships exist between GRs, SPs, and COs. In Dictyoptera, for example, both cystocytes (= coagulocytes) and SPs are labile, and in both types there are granules that show tubular substructure (Ratcliffe and Price, 1974). The SPs of several species of Dictyoptera, including *Periplaneta* and *Leucophaea*, contain spherules that react positively in histochemical tests for acid mucopolysaccharides (Gupta and Sutherland, 1967). Spherules are released into the hemolymph by rupture of the hemocytes. *Bombyx* SPs also contain mucopolysaccharide and glycoprotein (Akai and Sato, 1973). The adipohemocytes (= GRs) of *Galleria* contain acidic mucopolysaccharides (Ashhurst and Richards, 1964b), and the granules appear to be composed of hexose-containing nonsulfated acidic mucosubstances with *vic*-glycol and carboxyl groups (Neuwirth, 1973). In the electron microscope, the granules vary in electron density, with granules of least density in close association with well-developed Golgi complexes. The granules of least density are substructured, with 10- to 15-nm tubules packed together in swirls (see Chapter 4). Electron histochemistry with PTA at low pH, and light microscope histochemistry, indicate that it is the substructured granules that contain acidic mucosubstances. The same type of GR contains distinct lipid droplets, which become more numerous as the animal nears pupation, and also acid phosphatase–positive, lysosome-like vacuoles. The SPs, in contrast, contain sulfated, weakly acidic mucosubstances; and the ultrastructure of the spherules is a regular latticework of globules (Neuwirth, 1973). These ultrastructural results for *Galleria* were confirmed by Rowley and Ratcliffe (1976), who also studied the formation and release of granules. Their electron micrographs indicated that an initial swelling of the rough endoplasmic reticulum was followed by association of the reticulum with Golgi complexes, where condensation of vesicles to larger vacuoles occurred and where tubular substructure became apparent. During incubation for short terms in vitro, these cells showed progressive degranulation as material derived from the granules was discharged into the medium. Some of the granules released from the hemocytes contained remnants of tubular substructure, but this disappeared as a result of interaction of granule contents and plasma to form a floccular precipitate. This precipitate is interpreted as hemolymph coagulum by Rowley and Ratcliffe (1976).

In the thysanuran, *Thermobia*, studied by François (1974, 1975a), GRs and SPs can be distinguished in light microscope preparations, but in the electron microscope, no separation of these types emerged from a study of organelles. The granules of GRs appeared homogeneous and electron-dense, and when numerous, gave the cell the morula form typical of an SP. Histochemical tests indicated an acidic mucoprotein content (alcian-blue-positive at pH 2.5, and hyaluronidase-resistant). A subclass of GRs was distinguished, on the basis of rapid cytological change on contact with a glass slide and on ultrastructural grounds, and termed a CO. COs were found to contain heterogeneous granules, some containing 15-nm-diameter tubules, others containing granular material. In spite of the ultrastructural differences, a histochemical relationship between GRs and COs is evident. Both show reactions indicative of the presence of acidic mucopolysaccharides and glycoproteins.

15.4.3. Capsule and nodule matrixes

When François (1975b) introduced foreign objects into *Thermobia*, the hemocytes surrounding the objects were predominantly COs. At 8 hr after initiation, the capsule was composed of 60% COs, 30% PLs, and 10% GRs. The granules of GRs appeared to be extruded into the intercellular space by exocytosis. The intercellular matrix, like the hemocyte granules, reacted positively in a number of histochemical tests for nonsulfated acidic mucopolysaccharides, but contained no collagen. Eventually, the hemocytes forming the capsule became degranulated, stretched, and linked together by desmosomes in concentric layers shaped like an onion.

In *Pieris*, test particles such as the bacteria *Escherichia coli* and *Staphylococcus* became associated with SPs in greater numbers than with other cell types and adhered firmly to granules released from cells into the in vitro test medium. All the hemocyte types in *Pieris* contained acidic mucosubstances identified by alcian blue (pH not specified) and toluidine blue staining (Ratcliffe, 1975). Capsule formation in *Pieris* is also a function of body region, fewer hemocytes being deposited on capsules forming around nylon implants in the anterior part of the abdomen than in distal regions (Carton, 1977).

According to Brehélin et al. (1975), introduction of sterilized cellophane fragments into *Locusta* induces an immediate but limited plasma coagulation around the implant involving only the COs. Following this reaction, the implants become surrounded by GRs, but the

PLs, although occasionally embedded in the capsule, do not participate in its formation. Typical GRs also surround experimentally severed abdominal muscle and aggregate beneath integumental wounds in *Locusta* (Brehélin et al., 1976). Electron microscopy of *Locusta* hemocytes had indicated that two ultrastructural classes of GRs could be recognized, one of which was a CO (Hoffmann, 1966; Hoffmann and Stoekel, 1968). Histochemistry on *Locusta* hemocytes, using the periodic acid–Schiff (PAS) test and alcian blue (at pH 3.0), had indicated that all cell types except prohemocytes (PRs) contained mucopolysaccharide (Hoffmann, 1967). In a more detailed histochemical study of *Locusta*, Costin (1975) found mucosubstances in all hemocyte types. The PL and CO granules were of four types, containing (1) glycoprotein and neutral mucopolysaccharide, (2) nonsulfated, periodate-unreactive sialomucin, (3) nonsulfated, periodate-reactive sialomucin, or (4) sulfated, periodate-reactive sialomucin. Spherules of SPs were also found to contain nonsulfated sialomucins or glycoprotein and mucopolysaccharides, and were strongly PAS-positive. The granules of the GRs were all found to be similar, containing periodate-reactive and sulfated sialomucin. OEs reacted for glycoprotein and neutral mucopolysaccharide and for sulfated, periodate-reactive sialomucins.

Cellular aggregates formed rapidly around killed *Bacillus cereus* introduced into the hemolymph of *Galleria*. Initially, adhesive flocculent material was released by GRs in contact with the bacteria. Granule discharge occurred by localized breakdown of the cell membrane, not by exocytosis. Later stages involved PL encapsulation and melanization (Ratcliffe and Gagen, 1976, 1977). As mentioned above, the granules in GRs of this insect are composed of acidic mucosubstances (Ashhurst and Richards, 1964b; Neuwirth, 1973), and there is an obvious relationship between the nodule matrix and plasma coagulum. We cannot, however, be certain that the hemocyte secretions involved are identical in both systems, in view of the heterogeneity of granules in these cells.

15.4.4. Connective tissue matrixes

Secretion of basement membrane by circulating cells was postulated by Lazarenko (1925a,b), and evidence for hemocyte involvement in noncollagenous connective tissue deposition in *Rhodnius* is provided by Wigglesworth (1933, 1956, 1973). This evidence is based on histochemistry, on electron microscopy, and on the temporal and spatial relationship of hemocytes and developing matrixes beneath the epidermis. Ultrastructural evidence for the involvement of hemocytes in the secretion of the connective matrix around the prothoracic glands in *Leucophaea* has been obtained by both Beaulaton (1968) and Scharrer (1972). Micrographs show a similarity between the vacuolar contents of hemocytes at extrusion and the material making up the tunica propria, both of which contain 2-nm microfibrillae (Beaulaton, 1968). The Golgi apparatus is involved in the packaging of this secretion in a fashion comparable to that of vertebrate goblet cells secreting mucopolysaccharide (Neutra and Leblond, 1966). Vacuoles replete with bundles of tubular elements 15 nm in diameter, derived from the Golgi apparatus, are implicated in the genesis of the 2-nm microfibrillae (Beaulaton, 1968) (see also Chapters 11, 12).

Histochemical comparison of hemocyte contents and connective matrix rests heavily, although not exclusively, on the results of the PAS reaction for both *Bombyx* (Beaulaton, 1968) and *Rhodnius* (Wigglesworth, 1956, 1973). PAS-positive compounds include polysaccharides, glycoproteins, glycolipids, and unsaturated lipids and phospholipids (Pearse, 1968, p. 311); PAS reactivity thus does not provide definitive correlations. It is also important to bear in mind that PAS-positive secretions may be largely protein. The salivary gland of *Chironomus* gives a strongly positive PAS reaction, but nevertheless the secretion contains less than 4% carbohydrate (Kato et al., 1963; Doyle and Laufer, 1969). In spite of this, the protein is still classified as a mucoprotein.

Beaulaton and Monpeyssin (1976) remark that, in *Antheraea*, the GR has both secretory and phagocytic properties. Heterophagic vacuoles appear alongside vacuoles with tubular substructure in electron micrographs. The latter react positively in the periodic acid–thiocarbohydrazide test for α-glycol groups and appear to contain a glycoprotein. Presumptive glycoprotein vacuoles appear electron-dense and tubular in the vicinity of the Golgi apparatus, but become less dense and granular near the plasma membrane, where profiles indicate release by exocytosis. Similar substructured vacuoles have been described for *Calpodes* (Lai-Fook, 1973) and for *Galleria* (Neuwirth, 1973), and are analogous to glycoprotein granules in diverse cells as listed by Beaulaton and Monpeyssin (1976). Costin (1975) has reported that all six hemocyte types recognized in *Locusta* contained PAS-positive material, as did the connective tissues examined. However, on the basis of an extended range of histochemical tests, a distinction can be made between hemocyte granules, which contain sialomucins (Costin, 1975), and the fibrous connective tissue, which contains glycosaminoglycans, chondroitin and dermatan sulfates, and keratan sulfate (Ashhurst and Costin, 1971). The relative contributions of the hemocyte and the ensheathed cell in the elaboration of the matrix have not been determined. Several types of connective tissue in *Galleria* have been investigated by ruthenium red staining coupled with enzyme digestion. The basement membrane of the ovarian sheath apparently attracts ruthenium red by the presence of negative charges on sialic acid residues complexed with proteins and is predominantly glycoprotein (Przelecka and Dutkowski, 1973). The outer surface of the neural lamella, the connective tissue ensheathing the brain, shows neuraminidase-, hyaluronidase-, and to some extent phospholipase-C-sensitive ruthenium red staining. This suggests that the negative charge in this tissue depends on the presence of the anionic groups of sialic and hyaluronic acids and phospholipids, probably as conjugated proteins (Dybowska and Dutkowski, 1977). Although we do not know which, if any, components of *Galleria* matrixes are secreted by hemocytes, we do know that the granules in GRs of this insect are composed of acidic mucosubstances and could contribute to the matrix (Ashhurst and Richards, 1964a; Neuwirth, 1973). Jones (1976) argues that on the basis of examination of histological serial sections of *Galleria* and other species there is no evidence for the involvement of hemocytes in basement membrane formation. It is probable that many insect cells can secrete their own basement membranes without the involvement of accessory cells (Edwards et al., 1958). Lai-Fook (1968) suggests that epidermal cells of *Rhodnius* secrete basement membranes without the participation of hemocytes. Simple, nonfibrous matrixes containing glycoproteins but no glycosaminoglycans could be formed by the cells they underlie; other more complex matrixes may involve accessory cells such as hemocytes or fibroblasts (Ashhurst and Costin, 1974).

15.5. Hemocytes, fibroblasts, and collagen secretion

The status of fibroblasts in relation to hemocytes is ambiguous, partly because we know nothing about the embryological origin of fibroblasts in insects, and partly because the fibrous components of insect connective matrixes are little studied. In vertebrates, collagenous, elastic, and probably reticular fibers are secreted by fibroblasts, but in insects only collagen has been reliably demonstrated. Histochemical evidence for collagen in insect nerve has been provided for mantids by Rudall (1955) and for *Locusta* by Ashhurst (1959), and collagen was identified in electron micrographs of neural lamellae of *Periplaneta* (Hess, 1958), *Rhodnius* (Smith and Wigglesworth, 1959), and *Locusta* (Ashhurst and Chapman, 1961). Cells in *Thermobia* (François, 1974) and *Locusta* (Ashhurst and Costin, 1974) have been called fibroblasts on the basis of their ultrastructure and function in secretion of collagen matrixes. Fibroblasts appear from undifferentiated cells recognizable as stable populations in fixed locations in second-instar *Locusta* nymphs, and it is possible that they arise from the same stem cell population as migratory hemocytes (Ashhurst and Costin, 1974). The ultrastructure of fibroblasts, and in particular the presence of distended cisternae of rough endoplasmic reticulum, distinguish them from GRs and SPs (Ashhurst and Costin, 1976). However, fibroblasts and PLs resemble each other in having extensive development of ribosome-studded endoplasmic reticulum (François 1975b). Collagen substructure has been identified within vacuoles of cells identified as hemocytes associated with the prothoracic gland in *Leucophaea* (Scharrer, 1972), but we cannot exclude the possibility that such vacuoles form part of the lysosomal apparatus engaged in endocytosis. Ashhurst and Costin (1974) suggest that collagen-containing cells in both *Locusta* and *Leucophaea* are permanently sessile and should be called fibroblasts. The collagen of the neural lamella is secreted by glial cells (Ashhurst and Richards, 1964a; Ashhurst and Costin, 1971). In *Thermobia*, fibroblasts differentiate from stem cells associated with the hemopoietic organ, which is located between pericardial cells and dorsal diaphragm in thorax and abdomen. These cells are associated with hemocytes and with reticular cells, and the complex thus formed secretes a fibrous connective tissue. Three types of fibers are present: collagen and reticular fibers, both with 66-nm periodicity, and thicker elastic fibers that show a positive reaction to paraldehyde-fuchsin and Owen's blue stains (François, 1975a). The distribution of elastic fibres in invertebrates has been reviewed by Elder (1973). In an ultrastructural study of *Calpodes* larvae, four connective tissue components were distinguished by selective staining. These were matrix, collagen, fibers less than 6 nm in diameter, and 40 nm fibres, which are most abundant in elastic tissue (Locke and Huie, 1972). Three components can also be seen in electron micrographs of noncollagenous elastic ligaments that anchor pericardial cells in *Calliphora* làrvae (Crossley, 1972b). Further research is needed to clarify the contribution of hemocytes, fibroblasts and supported tissues in the synthesis of this multicomponent system (see also Chapter 12).

15.6. Hemocytes in phenol metabolism

15.6.1. Phenoloxidases in hemocytes

Hemocytes containing enzymes capable of accelerating the oxidation of phenolic compounds were first described by Dennell (1947) for *Sarcophaga*, and subsequent reports for *Drosophila* (Rizki and Rizki, 1959), *Calliphora* (Crossley, 1964, 1975), *Antheraea* (Evans, 1967, 1968), *Periplaneta* (Mills et al., 1968), *Locusta* (Hoffmann et al., 1970), *Leucophaea* (Preston and Taylor, 1970), *Chironomus* (Maier, 1973), *Galleria* (Schmit et al., 1977), and *Aedes* (Andreadis and Hall, 1976) confirm the occurrence of phenoloxidases (or tyrosinases) in hemocytes of many insects. It has also been established that insect blood plasma usually contains phenolic substrates such as tyrosine (Finlayson and Hamer, 1949) and dihydroxyphenylalanine (= DOPA) (Pryor,

1955). A dipeptide containing tyrosine in stored form is present in the hemolymph of *Sarcophaga* (Bodnaryk and Levenbook, 1969; Levenbook et al., 1970), *Musca* (Bodnaryk, 1970), and *Celerio* (Wilinska and Piechowska, 1975), and the availability of peptide tyrosine is in all probability controlled by hemocytes (see above, under "Peptide–Amino Acid Conversion"). Tyrosine is also released from the fat body into the hemolymph of *Calliphora* under humoral control, and most of this tyrosine remains in the plasma (Price, 1972). Nevertheless, generalized melanization of the hemolymph does not occur in healthy insects, and the reasons for localization of interaction of enzyme and substrate, both spatial and temporal, have been reviewed in detail by Brunet (1963, 1965) and, in connection with hemocytes, by Crossley (1975). Two types of explanation are currently significant. The first involves the localized activation of an inactive proenzyme; the second involves the physical separation of enzyme from its substrates by cellular membranes.

15.6.2. Activation of phenoloxidases

Proenzymes

Activation of an inert prophenoloxidase to form an active phenol-oxidizing system was first recognized in *Melanoplus* eggs (Bodine et al., 1937; Bodine and Allen, 1938; Bodine, 1945), but similar hemolymph systems were discovered later in *Drosophila* (Ohnishi, 1953, 1954; Horowitz and Fling, 1955), *Musca* (Ohnishi, 1958, 1959; Inaba and Funatsu, 1964), *Tenebrio* (Heyneman, 1965), *Antheraea* (Evans, 1968), *Leucophaea* (Preston and Taylor, 1970), *Bombyx* (Ashida, 1971), and *Sarcophaga* (Hughes and Price, 1975a). The activator of *Antheraea* hemolymph prophenoloxidase appears to be a protein from the hemocytes (Evans, 1968). Prophenoloxidase from *Musca* can be artificially activated by proteolytic enzymes such as chymotrypsin (Ohnishi et al., 1970). In *Bombyx*, a natural prophenoloxidase activator can be isolated from cuticle, and this activates hemolymph propro-phenoloxidase by release of a single peptide (Ashida et al., 1974).

The first step of the conversion of tyrosine to quinones by phenoloxidases is the hydroxylation of tyrosine to DOPA, and an enzyme with tyrosine hydroxylase activity is present in inactive form in *Ostrinia* (Lepidoptera) larval hemolymph. Activation of this proenzyme involves a small thermolabile molecule, itself derived from hemolymph, in a system responding to trauma of hemorrhage (Brennan and Beck, 1972). Activation proceeds with a similar time course whether or not the blood is centrifuged before assay, but damaged hemocytes may well release an activator during collection of hemolymph, and it is possible that this is a peptide. A protein activator of *Sarcophaga* hemolymph prophenoloxidase is present both in cuticle and in hemolymph, but the site of synthesis of the hemolymph activator is unknown (Hughes and Price, 1976). Both proenzyme and activator are present in the hemolymph several days before pupation occurs, but enzyme activity does not develop in the hemolymph until pupariation has begun. The mechanism that prevents premature activation of prophenoloxidase may be an inhibitor, such as that described in the fat body and plasma of *Antheraea* by Evans (1968) or that secreted into the plasma by salivary glands in *Calliphora* (Thomson and Sin, 1970). On the other hand, the inhibitory mechanism may be based on the ionic strength of the plasma (Hughes and Price, 1976). In *Drosophila*, the "crystal cells" are believed to contain both the enzyme and substrate necessary for melanin formation (Rizki and Rizki, 1959). If this is so, the enzyme may be an inactive prophenoloxidase or the substrate may be rendered inaccessible by crystallization.

Membrane barriers

A concept of physical separation of phenoloxidase from its substrates is supported by the observations of Mills and Whitehead (1970) on *Periplaneta* and of Post (1972) on *Pieris* hemocytes. In *Periplaneta*, permeability of hemocytes toward tyrosine changed at ecdysis, indicating that a hormone, such as bursicon, could allow access of tyrosine to enzymes in the hemocytes that convert it to DOPA and dopamine. In *Pieris*, the in vitro conversion of tyrosine to DOPA and dopamine by hemolymph is induced by bursicon (Post, 1972; Post and DeJong, 1973). Alternatively, overall blood cell permeability could be increased by hormone, with the result that swelling and eventual lysis of cells would liberate tyrosine-metabolizing enzymes.

There is evidence that the phenol-oxidizing enzymes in plasma are formed and released by hemocytes. This was first demonstrated for certain crustacean hemocytes, which release a phenoloxidase during explosive contact interaction with foreign surfaces (Pinhey, 1930; Bhagvat and Richter, 1938). Insect hemocytes that contain phenoloxidases in *Sarcophaga* were termed OEs (Dennell, 1947); analogous (type B) hemocytes in *Calliphora*, which contained both phenoloxidase and peroxidase activity, were found to be labile, becoming blistered as the blood clotted and accumulating at wound sites (Crossley, 1964, 1975). Phenol-oxidizing enzymes are also present in the COs in *Locusta* (Hoffmann et al., 1970). In *Galleria*, phenol oxidation occurs in OEs (Neuwirth, 1973) and possibly in GRs (Schmit et al., 1977). The very large number of hemocytes in *Galleria* that are associated with phenolic reaction product (half the total hemocyte population in the preparations of Schmit et al., 1977) may be misleading. Long incubation times at high temperature in unstated concentrations of phenolic substrates may have led to reaction product dispersal and secondary association with nonsynthetic hemocytes. In *Leucophaea*, diphenoloxidase was detected in granular PLs, the granules being similar to a vertebrate premelanosome (Preston and Taylor, 1970). In addition to histochemical evidence for the synthesis of prophenoloxidase in hemocytes, there is biochemical evidence that hemocytes in *Sarcophaga* synthesize prophenoloxidase and transport it into the plasma immediately after synthesis (Hughes and Price, 1975a,b, 1976). There is genetic evidence that phenoloxidase molecular structure is controlled by α-locus at 52.4 on chromosome II of *Drosophila*. Although the whole-insect extracts used in this work yield a proenzyme and a protein activator, the cellular origins of the molecules have not been traced (Lewis and Lewis, 1963).

15.6.3. Nature of hemolymph and hemocyte phenoloxidases

Preparations of diphenoloxidase from the hemolymph of *Calliphora* have been purified and separated from calliphorin and other contaminating proteins with comparable sedimentation and diffusion coefficients by Munn and Bufton (1973). They reported that the purified inactive proenzyme had a molecular weight, determined by gel filtration, of 115,000 and a sedimentation coefficient of 9.4 S. The phenoloxidase is not identical with calliphorin, which has a sedimentation coefficient of 18 S and which probably contaminated earlier preparations of the enzyme, such as those of Karlson and Liebau (1961). The active diphenoloxidase is relatively insoluble, has a higher sedimentation coefficient (16 S or more), and is immunologically distinct from the proenzyme (Munn and Bufton, 1973). Hughes and Price (1974) have obtained an insoluble tyrosinase fraction from *Sarcophaga* larvae by dilution of centrifuged blood. This enzyme, which has an estimated molecular weight of 400,000 appears to be a lipoprotein complex. The active form of the enzyme binds to subcellular particles (Sekeris and Mergenhagen, 1964), coprecipitates with lipoprotein (Hughes and Price, 1974), or undergoes a process of polymerization or association (Munn and Bufton, 1973).

The procedures for obtaining hemolymph by pooling ice-cold samples, used by all workers preparing phenoloxidase, would be expected to include molecules released from labile hemocytes in response to trauma. Subsequent centrifugation will not, therefore, invariably distinguish between enzymes, activators, and substrates of hemocyte and plasma origins.

The hemolymph phenoloxidases isolated from *Calliphora* (Munn and Bufton, 1973) and *Sarcophaga* (Hughes and Price, 1976) both have *o*-diphenoloxidase activity, and the *Sarcophaga* enzyme has tyrosine hydroxylase activity, indicating that both are tyrosinases (EC 1.10.3.1, *o*-diphenol:O_2-oxidoreductases). The comparative biochemistry of the phenoloxidase complex has been reviewed by Mason (1955). Hemolymph diphenoloxidase from *Leucophaea* activated with methanol, acetone, or α-chymotrypsin catalyzed the oxidation of a variety of diphenols besides DOPA, but not the monophenol tyrosine, even in the presence of a catalytic amount of DOPA (Preston and Taylor, 1970). The native enzyme, upon natural activation, may show higher monophenolase activity when bound to mitochondria, allowing some coupling with the hydrogen transport system (Karlson and Sekeris, 1964; Sekeris and Mergenhagen, 1964; Brunet, 1965). Various other enzymes active in phenol metabolism have been detected in insect hemocytes. *Calliphora* hemocytes contain mono- and *o*-diphenoloxidase activity and peroxidase (donor: H_2O_2-oxidoreductase) activity (Crossley, 1964, 1975). *Periplaneta* hemocytes contain monoamine oxidase, which is involved in the conversion of tyramine to protocatechuic acid (Whitehead, 1970b). *Glossina* hemocytes contain decarboxylase active on tyrosine and possibly DOPA (Whitehead, 1973).

The presence of these enzymes in the hemocytes, rather than in the plasma or integument, offers options in the control of metabolic pathways leading to melanin and tanning agents. Control of hemocyte synthesis and permeability in turn controls interaction of enzymes with phenolic substrates in the hemolymph, with implications in wound healing, capsule formation, and bacteriostasis. These activities, together with the possibility of hemocyte involvement in production of quinones responsible for hardening and darkening the cuticle after ecdysis, will be discussed next. It is also worth recalling that in many insects, as in crayfish (Unestam, 1976), phenoloxidase activity is very closely linked with blood clotting.

15.6.4. Hemocytes, hemolymph, and sclerotization of the integument

Sclerotization scenarios

In 1964, Karlson and Sekeris suggested that during puparium formation in *Calliphora* a prophenolase moves from the hemolymph into the cuticle, where it is activated, and brings about sclerotization of the cuticle by conversion of N-acetyldopamine to quinones. The quinones react with free amino groups in cuticle proteins (Mason and Peterson, 1965), generating covalent linkages that convert soft proteins into hard insoluble materials, the sclerotins (Pryor, 1940). Evidence about the nature of sclerotins is fragmentary (Brunet, 1967). Quinone polymers also form melanin pigment. Insect hemocytes have been implicated in two scenarios leading to sclerotization of the cuticle. In the first it is proposed that hemocytes synthesize and release into the hemolymph proteins that are transported to the cuticle. In the second scenario, the hemocytes are supposed to gain access to tyrosine or its derivatives and to produce DOPA, dopamine, or other precursors of quinones, and the integument is supposed to take up and utilize this pool of substrate, either for sclerotization or for melanin formation. In support of the first scenario, there is evidence (discussed above, under "Hemocytes in synthesis of cuticle proteins") that some cuticle proteins are very similar to hemolymph proteins and, furthermore, that transfer of hemocyte-made proteins to the cuticle occurs. An alternative explanation for these results is that the epidermal cells synthesize proteins, which are transferred both to hemolymph and to cuticle. There is no doubt that the epidermis itself synthesizes protein and passes this to the cuticle (Locke and Krishnan, 1971) in developmental systems involving the Golgi complex and plasma membrane plaques (Locke, 1976). Further biochemical and ultrastructural analysis is, however, needed to clarify the form in which cuticle components pass from hemocyte to cuticle. Sclerotin precursors and phenoloxidizing enzymes as well as small metabolites are possible candidates for transfer. In *Manduca* (Koeppe and Gilbert, 1973), a dopamine-carrying protein appears to be transported from hemolymph to cuticle, traversing the epidermis, as evidenced by both radiotracer studies and electrophoretic analysis.

In many insects (e.g., *Sarcophaga:* Lai-Fook, 1966; *Lucilia:* Hackman and Goldberg, 1967), both hemocyte and cuticular enzymes include *o*-diphenoloxidases, but in *Bombyx* and *Drosophila* the cuticle-bound phenoloxidase is reported to be a laccase (EC1.10.3.2,*p*-diphenol: O_2-oxidoreductase), with specificity for *para* phenols (Yamazaki, 1972). In both these species, the larval hemolymph phenoloxidase is an *ortho* diphenoloxidase (Ohnishi, 1953, 1954; Nakamura and Sho, 1964) so we can conclude that some, at least, of the cuticular enzyme does not have a hemocyte origin. The prophenoloxidase-activating enzyme from the cuticle does, however, activate the hemolymph prophenoloxidase in *Calliphora* (Karlson et al., 1964) and in *Bombyx* (Dohke, 1973), indicating a functional affinity between the two systems.

Quinone precursor synthesis

There is abundant evidence that hemocytes can metabolize precursors of quinones, as discussed above. Significantly, this aspect of hemocyte metabolism is synchronized with ecdysis. Thus, bursicon, the cuticle-hardening hormone (Fraenkel and Hsiao, 1965), causes greatly increased uptake of radioactively labeled tyrosine into hemocytes (Mills and Whitehead, 1970), allowing tyrosine conversion to quinone precursors to take place inside the hemocytes (Post, 1972). Changes in hemolymph tyrosine levels during sclerotization in *Calliphora* (Price,

1971) and in *Pieris* (Post and DeJong, 1973) are consistent with the proposed hemocyte system, but do not exclude other possibilities. In *Leucophaea*, bursicon is released into the hemolymph as the old cuticle splits at the thorax, and hormone activity remains high throughout ecdysis and for 1.5 hr thereafter (Srivastava and Hopkins, 1975). In Diptera (Fraenkel and Hsiao, 1965) and Lepidoptera (Truman, 1973), bursicon is released at variable intervals after ecdysis, as evidenced by fly ligation assays. Further work is needed to correlate the syntheses taking place in hemocytes with bursicon activity and, further, with sclerotization and melanization of the cuticle.

In adult Diptera, tyrosine hydroxylation rather than a subsequent reaction may be the rate-limiting step in producing sclerotizing molecules (Seligman et al., 1969). This limit may also apply to *Periplaneta* (Hopkins and Wirtz, 1976) because the essential decarboxylation reaction leading to dopamine derivatives depends on a DOPA decarboxylase that has little activity with tyrosine as substrate. In *Periplaneta* larval hemolymph DOPA decarboxylase activity increases to reach a peak 6 hr after ecdysis, coinciding with the main period of cuticular sclerotization. DOPA decarboxylase (combined cell and plasma) activity is about four times greater in hemolymph than in the integument, based on tissue dry weights, but the integument may also serve as a major site for dopamine biosynthesis because of its greater mass (Hopkins and Wirtz, 1976). L-DOPA decarboxylase has been detected within the hemocytes of *Periplaneta* (Whitehead, 1969) and *Sarcophaga* (Whitehead, 1970a). *Sarcophaga* hemocytes in vitro preparations can synthesize DOPA and dopamine from tyrosine, but these are oxidized by an *o*-diphenoloxidase and do not accumulate in the medium. Although the hemocyte enzymes are capable of converting tyrosine to dopamine at eclosion, the rate of synthesis is not accelerated until after the release of bursicon, 20–30 min later. The most likely explanation for the effect of bursicon is not that it activates the enzymes, but that it overcomes a barrier to tyrosine in the appropriate hemocytes, as discussed above, under "Membrane Barriers" (Whitehead, 1970a).

An alternative pathway to dopamine by direct decarboxylation of tyrosine to form tyramine is also known and was detected in *Periplaneta* hemocytes by Whitehead (1969). The tyrosine decarboxylase activity of the total hemolymph of this insect is, however, relatively unimportant (Hopkins and Wirtz, 1976), and the tyrosine is in all probability first converted to DOPA by tyrosine dehydroxylase.

Although there is abundant evidence that quinone precursors, including DOPA and dopamine, are synthesized in hemocytes concurrently with humoral changes at ecdysis, proof of transfer to the integument is still lacking.

15.6.5. Phenol-oxidizing hemocytes in wound repair and capsule and nodule formation

Wound repair

Blackening of wound sites in the integument often follows hemostasis in insects, and there is evidence that the epidermal phenoloxidase involved in wound repair differs from enzymes associated with sclerotization of the integument (Lai-Fook, 1966). A dual role for hemocyte phenol-oxidizing systems can also be envisaged, and we can next review the evidence for Taylor's (1969) suggestion that the hemocyte phenol-oxidizing system plays a major part in sealing lesions and combating invading organisms.

Wounding dramatically increases the number of hemocytes in circulation (Harvey and Williams, 1961; Lea and Gilbert, 1961; Jones, 1967; Hoffmann, 1969), and preferential attachment of hemocytes at wound sites has often been reported (Lazarenko, 1925b; Wigglesworth, 1937; Harvey and Williams, 1961). It is likely that an "injury factor" is generated at the margin of the wound (Clark and Harvey, 1965), and the influence of this factor on hemocytes has been discussed elsewhere (Crossley, 1975). In an ultrastructural study of *Rhodnius*, Lai-Fook (1968, 1970) showed that a form of GR (termed a PL), with dense, homogeneous granules, was active in wound repair. GRs accumulated at incisions in the integument, and these underwent changes. Of particular interest was the observation that hemocyte granules, which were at first homogeneous, subsequently developed tubular substructure (15 nm in diameter), analogous to cells involved in mucoprotein synthesis (as discussed above, under "Hemocytes in Localized Coagulation"). Such hemocytes are likely to be involved in the local coagulation response of homeostasis, but are they also involved in melanization?

It has been demonstrated that certain crustacean hemocytes release phenolase during an explosive contact interaction with foreign surfaces (Pinhey, 1930; Bhagvat and Richter, 1938). In *Calliphora*, hemocytes containing phenol-oxidizing enzymes accumulated at wound sites and appeared also to release enzymes by breaking down on contact with foreign surfaces (Crossley, 1975). Hemocyte population studies also implicate phenol-oxidizing hemocytes in bacteriostasis. When *Galleria* larvae were injected with bacteria (*Serratia* or *Enterobacter*), the percentage of hemocytes containing phenoloxidase increased when compared with control larvae injected with saline (Pye and Yendol, 1972).

Encapsulation and nodule formation

Encapsulated parasites are frequently melanized (Salt, 1970). The significance of melanin in the defense reaction is particularly apparent in *Strongylogaster* parasitized by *Mesolius*, where the parasite egg is invariably encapsulated, but the parasite is able to develop only when melanin fails to form in the capsule (Adam, 1966). In cellular encapsulation, melanin precursors are formed by disintegration of the innermost cells of the hemocyte envelope (Poinar et al., 1968).

For *Drosophila*, Nappi (1970) suggests that deposition of pigment on nonencapsulated hymenopterous parasites accounts well for observed mortality. The parasitization is associated with increased numbers of OEs. Similarly in *Musca*, aggregation and fusion of OEs form a pigmented layer that adheres to a parasitic nematode (Nappi and Stoffolano. 1971).

Encapsulation of mermithid nematodes by *Chironomus* larvae involves the deposition of a capsule from precursors present in hemolymph, without the direct participation of hemocytes and has been called "humoral encapsulation" (Götz, 1969). The humoral encapsulation reaction is not elicited by inert substances such as nylon, but is effective against pathogenic fungi. Parts of disintegrated GRs have been found in the immediate neighborhood of encapsulated fungal spores, and cell material appears to be involved in the formation of the capsule. The capsule reacts positively in several tests for melanin (Götz and Vey, 1974). *Chironomus* GRs contain phenoloxidase (Maier, 1973) and may be involved indirectly in synthesis of capsule components. In *Culex* infected with a nematode, a similar noncellular capsule has also been observed (Poinar and Leutenegger, 1971). Inhibition of melanin formation in *Heliothis* larvae by injection of phenoloxidase inhibitors such as phenylthiourea reduced encapsulation of parasite eggs, not simply melanization, the inference being that melanization accompanies or precedes capsule formation (Brewer and Vinson, 1971).

Quantitative study of the *Drosophila* hemocyte picture during infection with eggs of the parasite *Pseudeucoila*, allowed Walker (1959) to show not only an increase in the total number of hemocytes, but also a transformation of PLs to lamellocytes, the latter forming a capsule around parasite eggs. Later, parasitized *Drosophila* larvae were shown to undergo lysis of "crystal cells" (= oenocytoids?) with the release of phenolic substrates (Nappi and Streams, 1969; Nappi, 1970). Nappi (1973, 1974) suggests that some stimulus in infected larvae results in precocious timing of changes to the phenol-oxidizing hemocytes that normally occur at metamorphosis.

In *Galleria*, GRs have been implicated in the formation both of capsules around heterografts (Schmit and Ratcliffe, 1977) and of nodules around bacterial aggregates (Gagen and Ratcliffe, 1976; Ratcliffe and Gagen, 1976) (see also Chapter 13). If fragments of *Schistocerca* nerve cord are implanted into the hemolymph of *Galleria*, a local clot is formed by contact lysis of GRs. Later, PLs adhere and a capsule forms.

Similar responses of hemocytes bring injected bacteria into an aggregate or nodule, which is later enclosed within a multicellular PL sheath. Injections of killed *Bacillus cereus* elicit changes in, and eventual lysis of, GRs to form a melanized capsule (Ratcliffe and Gagen, 1977). Both GRs and OEs of *Galleria* reportedly contain enzymes that lead to melanin production (Schmit et al., 1977), although some reservations about these results are discussed above (under "Membrane Barriers"). There is no evidence that phenol-oxidizing enzyme is located within intact GRs of *Galleria* (Neuwirth, 1973). Perhaps the most plausible interpretation is that GRs provide a conjugated protein matrix that is converted to melanin by the action of phenol-oxidizing enzymes derived in all probability from OEs on phenolic substrates present in the hemolymph. The presence of OEs in small numbers is consistent with a catalytic function, and the detection of prophenoloxidases in this type of hemocyte implicates them in melanin formation (Crossley, 1975).

Peroxide bacteriostasis

In *Calliphora* hemocytes, phenol-oxidizing enzymes of o-diphenolase character (= tyrosinase, = o-diphenol:O_2-oxidoreductase) and peroxidase character (= donor:H_2O_2-oxidoreductase) have been reported to accumulate at wound sites (Crossley, 1975). Peroxidases are also present in the integument of *Calliphora* and *Calpodes* (Locke, 1969). The presence of peroxidase is of particular interest in view of its implication in a bactericidal system in vertebrate leucocytes (Klebanoff, 1967a,b, 1968). In the vertebrate system, hydrogen peroxide is probably produced as a result of oxidation of NADH or NADPH and forms the substrate for peroxidase in conversion of halides to halogens, as well as having direct bactericidal properties (Pincus and Klebanoff, 1971; Klebanoff, 1971; Homan-Müller et al., 1974). In hemocytes of insects, including *Galleria*, NADH or NADPH oxidase activity has been demonstrated histochemically and is enhanced by challenge with injected bacteria (Messner, 1976). Thus, by analogy with the vertebrate system, two components of the peroxide antibacterial defense system have been demonstrated in insects. Peroxidase has also been detected in 2% of *Blaberus* hemocytes by Anderson (1974), but little or no iodination reaction was detected. Although microbodies containing catalase and urate oxidase have been identified in insect fat body cells (Locke and McMahon, 1970), their occurrence in hemocytes has yet to be confirmed by histochemistry. Microbody-like elements have been described in *Leucophaea* hemocytes by Hagopian (1971). Further work on insects, such as *Calliphora*, which have abundant peroxidase is needed to resolve the peroxide-linked activities of hemocytes.

15.7. Synthesis by hemocytes in vitro

15.7.1. Nature of cultured cells

Early attempts to culture hemocytes in vitro were reasonably success-ful in that adhesion and locomotion of hemocytes were observed when these were simply mounted in a drop of saline or plasma on a slide (Glaser, 1917). Although most cells disintegrated, a few report-edly survived for weeks and showed some mitoses (Glaser, 1918; Tay-lor, 1935). Many cultures of hemocytes failed to divide (review in Jones, 1970), but *Peridroma* hemocytes formed monolayer cultures showing limited mitosis and survived for 2 weeks when ascorbic acid and a high [nitrogen] atmosphere were used to delay oxidation of phe-nols (Martignoni and Scallion, 1961). Mitosis in cultured hemocytes from *Drosophila* was reported by Horikawa and Kuroda (1959), and division continued for 2 weeks in medium changed twice weekly. It appears that limitations in cell growth result either from exhaustion of metabolites or from the accumulation of toxic growth inhibitors. At-tempts to formulate culture media were at first empirical (Trager, 1935), but following the work of Wyatt's group on hemolymph compo-sition, formulations approached the native environment (Wyatt et al., 1956; Wyatt, 1956).

The success of Grace's (1958) techniques for establishment of sub-cultured lines of heterogeneous cells derived from lepidopteran ovar-ian tissue encouraged many workers to attempt to establish insect cell clones. As Wigglesworth (1959) has suggested, some of these cloned cells may be hemocytes. Although this is difficult to prove, the macro-phage-like habit of some cultured cells (e.g., Shields and Sang, 1970) is suggestive of hemocyte origin. Some cells cultured from embryos of *Periplaneta* contained granules with 17-nm tubular substructure, very similar to hemocyte granules of other species. Cells in these *Periplaneta* cultures produced an enzyme that broke down *Micrococcus* cell walls, interpreted as a chitinase (Deutsch and Landureau, 1970). Dif-ficulties in this interpretation arise from the absence of 17-nm tubular substructure in *Periplaneta* hemocytes (Baerwald and Boush, 1970) and the possibility that the *Micrococcus* assay may detect lysozyme rather than chitinase. Grace has not succeeded in obtaining subcul-tured hemocyte lines, although he has set up primary hemocyte cul-tures from several lepidopteran and dipteran species (Jones, 1970). In fact, the circulating hemocytes may provide a less favorable primary explant than hemocytopoietic tissues, because the latter contain a higher proportion of PRs in active mitosis and fewer terminally dif-ferentiated nonmitotic forms. The lymph glands of *Drosophila* have been cultured and show active mitosis (Castiglioni and Rezzonico, 1961, 1963), but further work in this direction is needed.

Hemocytes from larvae of *Chilo* (Lepidoptera) have been repeatedly subcultured by Mitsuhashi (1966), and although some differentiated hemocyte types degenerated, the mitotic index of stem cell types increased after a lag period lasting several months. In all, eight generations of *Chilo* hemocytes grew in vitro, and multiplication in the subcultures was stimulated when numbers were depleted by cell removal (Mitsuhashi, 1966). This observation suggests a chalone-like hemocytopoietic control, akin to that described for vertebrate blood cell populations (Bullough and Laurence, 1964; Rytomaa and Kiviniemi, 1968). The change in hemocyte population dynamics during culture in vitro has been documented for *Malacosoma* by Arnold and Sohi (1974). Of the five cell types distinguished in fresh blood, only three survived in vitro: PRs, PLs, and GRs. These were all able to divide, but the cultured cells were both larger and more polyploid than those in fresh blood, indicating some degree of transformation. GRs were less numerous in culture than in fresh *Malacosoma* blood, in line with the observation that *Tenebrio, Periplaneta* (Gupta and Sutherland, 1967), and *Galleria* (Rowley and Ratcliffe, 1974) GRs also underwent degeneration in vitro. Mazzone (1971) reported that following colchicine metaphase block, both diploid and polyploid chromosome numbers were observed in cultured *Lymantria* hemocytes. In virus-infected cultures, the hemocyte protein synthetic pathway was diverted to synthesis of viral molecules. These results indicate that the activities of some specialized cell types may not be represented by the in vitro culture method. On the one hand, this reduces the usefulness of the method as a technique for studying cell synthesis; on the other hand, it provides a selection system for functional analysis of certain cell types. The extent of metabolic modification that accompanies cell transformation in culture is unknown. PLs from diapausing saturniid pupae collected without contact with a wound appeared in vitro as rounded or spindle-shaped cells. An injury reaction, consisting of cellular transformation to ameboid and adhesive forms was induced in vitro by a component from injured epidermis. This component, named hemokinin, has been partially purified and has a molecular weight of about 50,000 (Cherbas, 1973). As most cultured hemocytes derive from blood collected at wounds, it is likely that injury reactions are normally triggered in hemocyte cultures.

15.7.2. Analysis of synthesis in vitro

Analysis of changes wrought in the in vitro medium by metabolism of insect cells is possible (Grace and Brzostowski, 1966; Clements and Grace, 1967), and opportunities for analysis are increased by the development of hemolymph-free defined media for insect cell culture

(Yunker et al., 1967). It is, therefore, surprising how little attention has been paid to the synthetic capacity of hemocytes in vitro. There are opportunities, in particular, to study the influence of hormones on in vitro synthesis (Oberlander, 1976), including the induction by ecdysone of specific protein synthesis (Best-Belpomme and Courgeon, 1976a,b). Blood from the cockroach, *Gromphadorhina*, has been cultivated in vitro for periods in excess of a year, and the cells secrete a PAS-positive material interpreted as chitin (Ritter and Bray, 1968). Because epidermal-type cells also appear in these cultures, both the source and the nature of this secretion are doubtful. The requirement of hemocytes in vitro for oxygen is generally ignored in favor of reducing the most spectacular synthetic activity of hemocytes in hemolymph – the production of quinones and melanin. Thus, Rowley and Ratcliffe (1974) in their study of phagocytosis of bacteria in *Galleria* in vitro set up their cultures in a nitrogen atmosphere. However, oxygen starvation has been shown to lead to hemocyte degeneration in vitro for *Hyalophora* (Calvet and Clark, 1974). As the physiological state of the culture is reflected in the decrease in cytoplasmic viscosity that accompanies cell degeneration (Bessis, 1963), immersion refractometry can be used to monitor the physiological state of cultured hemocytes. Using this technique Calvet and Clark (1974) showed that *Hyalophora* hemocytes could be maintained in a modified Grace's medium for 3 days without a significant change in cytoplasmic viscosity, provided the medium had been aerated prior to cell culture. However, a significant change in viscosity, indicative of changed hemocyte solute concentration and cell decay, occurred in unaerated medium.

15.8. Summary

Hemocytes are implicated in the maintenance of unusual levels of certain amino acids and sugars in insect blood homeostasis. A trehalase and glycolytic enzymes present in hemocytes influence hemolymph glucose levels. Levels of hemolymph amino acids, particularly the neurotransmitter glutamate, are controlled by selective uptake into hemocytes, and the pharmacological activity of the hemolymph relates to the integrity of the hemocytes. Hemocyte peptidases also control the level of free amino acids through action on the bulk-stored peptides of the hemolymph. There is at present no information on the types of hemocytes involved in small molecule homeostasis.

The hemolymph is an important pool of stored and transient proteins, some of which are synthesized in hemocytes. Available data suggest that some proteins pass among hemocytes, plasma, and integument. Injury or stress elicits production of injury proteins by hemocytes, and these include several esterases, which pass to the hemolymph. Other injury proteins of the hemolymph, such as lysozyme and possibly globulins, are synthesized in noncirculating cells such as reticular cells and pericardial cells, and the contribution of hemocytes to the injury protein pool remains to be defined.

In many orders of insects, hemocytes are active in the synthesis of conjugated proteins. In species that have been comprehensively studied by histochemical methods, glycoproteins or mucopolysaccharides are present in most categories of hemocytes. Conjugated protein synthesis involves both ribosome-studded endoplasmic reticulum and Golgi complexes, and granules of distinctive ultrastructure are formed. Release of secretion granules is a local phenomenon, and effective release stimuli include wounding and penetration by foreign objects. In some insects one consequence of hemocyte secretion is hemostasis, but certain capsule or nodule matrixes and connective tissue matrixes composed of sialomucins may also be produced by hemocytes. The role of hemocytes in the formation of multicomponent fibrous connective tissues is less clear, because other cell types such as fibroblasts, glial cells, and the ensheathed cells themselves may contribute to the matrix. Cells of unknown embryological origin with some of the properties of vertebrate fibroblasts are present in insects and are associated with collagen production.

Hemocytes synthesize and secrete several enzymes concerned with the oxidation and decarboxylation of phenolic precursors of sclerotization agents and melanin. Phenoloxidase is present in hemocytes as an inert proenzyme, and local activation is controlled by diverse activators and inhibitors. The permeability of phenoloxidase-containing hemocytes to substrates such as tyrosine forms another component of the control system, and this in turn relates to humoral changes at ecdysis. Hemolymph diphenoloxidase, which is the best known insect hemocyte secretion product, has a molecular weight of 115,000 and a sedimentation coefficient of 9.4 S. Other hemocyte enzymes involved in phenol metabolism include both tyrosine and DOPA decarboxylases, peroxidase, and peptidases.

Precursors of cuticle sclerotization agents appear to be produced by hemocyte enzymes under humoral control, but proof of transfer to the integument is lacking. There is some evidence that diphenols bound to blood proteins are transferred to the cuticle.

Hemocyte phenol-oxidizing systems located in oenocytoids (or possibly granulocytes) play a major part in sealing lesions and in bacteriostasis through local sclerotization and melanin production. Injury markedly affects the population dynamics of these hemocytes. Quinones or melanin serve as bacteriostatic agents and inhibit development of eukaryotic parasites, although the hemocytes are not always directly associated with encapsulation. Central components of a peroxide bacteriostatic mechanism are also present in insect hemocytes.

Improvements in techniques for culture of insect hemocytes in vitro have opened up avenues for exploration of hemocyte secretory function, particularly under humoral influences, but this exploration has hardly begun.

Acknowledgments

I would like to acknowledge with gratitude the assistance of Ms. Ann Smith and the financial support of the Australian Research Grants Committee during the preparation of this review.

References

Adam, H. 1966. Die hämocytaren Abwehrreaktionen des Blutes von *Strongylogaster xanthoceros* (Stephens) und *Strongylogaster lineata* (Christ) gegen die endoparasitische Ichneumonida *Mesolius niger* (Gravenhorst). *Beitr. Entomol. 15:*893–965.

Akai, H., and S. Sato. 1973. Ultrastructure of the larval hemocytes of the silkworm *Bombyx mori. Int. J. Insect Morphol. Embryol. 2:*207–31.

Åkesson, B. 1953. Observations on the haemocytes during the metamorphosis of *Calliphora erythrocephala. Ark. Zool. 6:*203–11.

Anderson, D. T. 1972a. The development of hemimetabolous insects, pp. 95–164. *In* S. J. Counce and C. H. Waddington (eds.). *Developmental Systems: Insects*, Vol. 1. Academic Press, New York.

Anderson, D. T. 1972b. The development of holometabolous insects, pp. 165–242. *In* S. J. Counce and C. H. Waddington (eds.). *Developmental Systems: Insects*, Vol. 1. Academic Press, New York.

Anderson, R. S. 1974. Metabolism of insect hemocytes during phagocytosis. *Contemp. Top. Immunobiol. 4:*47–54.

Anderson, R. S., N. K. Day, and R. A. Good. 1972. Specific hemagglutinin and a modulator of complement in cockroach hemolymph. *Infect. Immun. 5:*55–9.

Andreadis, T. G., and D. W. Hall. 1976. *Neoaplectana carpocapsae:* Encapsulation in *Aedes aegypti* and changes in host haemocytes and hemolymph proteins. *Exp. Parasitol. 39:*252–61.

Arnold, J. W. 1952. The haemocytes of the Mediterranean flour moth, *Ephestia kühniella* Zell. (Lepidoptera: Pyralidae). *Can. J. Zool. 30:*352–64.

Arnold, J. W., and S. S. Sohi. 1974. Hemocytes of *Malacosoma disstria* Hübner (Lepidoptera: Lasiocampidae): Morphology of the cells in fresh blood and after cultivation *in vitro. Can. J. Zool. 52:*481–5.

Ashburner, M., and P. Cherbas. 1976. The control of puffing by ions–the Kroeger hypothesis: A critical review. *Mol. Cell. Endocrinol. 5:*89–107.

Ashhurst, D. E. 1959. The connective tissue sheath of the locust nervous system: A histochemical study. *Q. J. Microsc. Sci. 100:*401–12.

Ashhurst, D. E., and J. A. Chapman. 1961. The connective tissue sheath of the nervous system of *Locusta migratoria:* An electron microscope study. *Q. J. Microsc. Sci. 102:*463–7.

Ashhurst, D. E., and N. M. Costin. 1971. Insect mucosubstances. III. Some mucosubstances of the nervous systems of the wax-moth *Galleria mellonella* and the stick insect *Carausius morosus. Histochem. J. 3:*379–87.

Ashhurst, D. E., and N. M. Costin. 1974. The development of a collagenous connective tissue in the locust *Locusta migratoria. Tissue Cell 6:*279–300.

Ashhurst, D. E., and N. M. Costin. 1976. The secretion of collagen by insects: Uptake of [³H]proline by collagen-synthesizing cells in *Locusta migratoria* and *Galleria mellonella. J. Cell Sci. 20:*377–403.

Ashhurst, D. E., and A. G. Richards. 1964a. The histochemistry of the connective tissue associated with the central nervous system of the pupa of the wax-moth *Galleria mellonella* L. *J. Morphol. 114:*237–46.

Ashhurst, D. E., and A. G. Richards. 1964b. Some histochemical observations on the blood cells of the wax-moth *Galleria mellonella* L. *J. Morphol. 114:*247–53.

Ashida, M. 1971. Purification and characterization of pre-phenol oxidase from haemolymph of the silkworm *Bombyx mori. Arch. Biochem. Biophys. 144:*749–62.

Ashida, M., K. Dohke, and E. Ohnishi. 1974. Activation of prephenoloxidase. III. Release of a peptide from prephenoloxidase by the activating enzymes. *Biochem. Biophys. Res. Commun. 57:*1089–95.

Babers, F. H. 1941. Glycogen in *Prodenia eridania*, with special reference to the ingestion of glucose. *J. Agric. Res. 62:*509–30.

Bade, M., and G. R. Wyatt. 1962. Metabolic conversions during pupation of the *Cecropia* silkworm. I. Deposition and utilization of nutrient reserves. *Biochem. J. 83:*470–8.

Baerwald, R. J., and G. M. Boush. 1970. Fine-structure of the hemocytes of *Periplaneta americana* (Orthoptera: Blattidae) with particular reference to marginal bundles. *J. Ultrastruct. Res. 31:*151–61.

Barlow, J. S., and H. L. House. 1960. Effects of dietary glucose on haemolymph carbohydrates of *Agria affinis* (Fall). *J. Insect Physiol. 5:*181–9.

Beaulaton, J. 1968. Etude ultrastructurale et cytochimique des glandes prothoraciques de vers à soie aux quatrième et cinquième âges larvaires. I. La *tunica propria* et ses relations avec les fibres conjunctives et les hémocytes. *J. Ultrastruct. Res. 23:*474–98.

Beaulaton, J., and M. Monpeyssin. 1976. Ultrastructure and cytochemistry of hemocytes of *Antheraea pernyi* Guer. (Lepidoptera, Attacidae) during the fifth larval stage. I. Prohaemocytes, plasmatocytes and granulocytes. *J. Ultrastruct. Res. 55:*143–56.

Begg, M., and W. J. Cruickshank. 1962. A partial analysis of *Drosophila* larval haemolymph. *Proc. R. Soc. Edin. 68:*215–36.

Belden, D. A., and R. R. Cowden. 1971. Detection of early changes in cockroach haemocytes during coagulation with 8-anilino-1-naphthalene sulfonic acid. *Experientia 27:*448–9.

Berry, S. J., A. Krishnakumaran, and H. A. Schneiderman. 1964. Control of synthesis of RNA and protein in diapausing and injured *Cecropia* pupae. *Science (Wash., D.C.) 146:*928–40.

Bessis, M. 1963. Studies on cell agony and death: An attempt at classification, pp. 287–315. *In* A. V. S. de Renck and J. Knight (eds.). *CIBA Foundation Symposium on Cellular Injury.* Little Brown. London.

Best-Belpomme, M., and A.-M. Courgeon. 1976a. Inductions protéiques par l'ecdystérone dans des clones cellulaires de *Drosophila melanogaster* cultivés *in vitro.* *C. R. Acad. Sci. Paris 282 D:*469–71.

Best-Belpomme, M., and A.-M. Courgeon. 1976b. Etude comparative des effects del'α et de la β-ecdysone sur un clone de cellules diploides de *Drosophila melanogaster* en culture *in vitro:* Induction protéique et modifications morphologiques. *C. R. Acad. Sci. Paris 283 D:*155–8.

Bhagvat, K., and D. Richter, 1938. Animal phenolases and adrenaline. *Biochem. J. 32:*1397–1406.

Bodine, J. H. 1945. Tyrosinase and phenols: Action of diversely activated tyrosinase on monohydric and o-dihydric phenols. *Proc. Soc. Exp. Biol. Med. N.Y. 58:*205–9.

Bodine, J. H., and T. H. Allen. 1938. Enzymes in ontogenesis (Orthoptera). IV. Natural and artificial conditions governing the action of tyrosinase. *J. Cell. Comp. Physiol. 11:*409–23.

Bodine, J. H., T. H. Allen, and E. J. Boell. 1937. Enzymes in ontogenesis (Orthoptera).

III. Activation of naturally occurring enzymes (tyrosinase). *Proc. Soc. Exp. Biol. Med.* 37:450–3.

Bodnaryk, R. P. 1970. Effect of dopa decarboxylase inhibition on the metabolism of β-alanyl-L-tyrosine during puparium formation of the fleshfly *Sarcophaga bullata*. *Comp. Biochem. Physiol.* 35:221–7.

Bodnaryk, R. P., and L. Levenbook. 1969. The role of β-alanyl-L-tyrosine (sarcophagine) in puparium formation in the fleshfly *Sarcophaga bullata*. *Comp. Biochem. Physiol.* 30:909–21.

Brady, J. 1967. The relationship between blood ions and blood-cell density in insects. *J. Exp. Biol* 47:313–26.

Brehélin, M. M. 1971. La coagulation de l'hémolymphe et ses modalités chez les larves de l'orthoptère *Locusta migratoria* (phase grégaire). *C. R. Acad. Sci. Paris* 273:1598–1601.

Brehélin, M. M. 1972. Etude du mecanisme de la coagulation de l'hémolymph d'un acridien, *Locusta migratoria migratoroides* (R and F). *Acrida* 1:167–75.

Brehélin, M. 1977. Sur la coagulation de l'hémolymph chez *Locusta migratoria*. *Ann. Parasitol. Hum. Comp.* 52:98–9.

Brehélin, M., J. A. Hoffmann, G. Matz, and A. Porte. 1975. Encapsulation of implanted foreign bodies by haemocytes in *Locusta migratoria* and *Melolontha melolontha*. *Cell Tissue Res.* 160:283–9.

Brehélin, M., D. Zachary, And J. A. Hoffmann. 1976. Functions of typical granulocytes in healing of the orthopteran *Locusta migratoria*. *J. Microsc. Biol. Cell.* 25:133–6.

Brennan, J. J., and S. D. Beck. 1972. Activation of tyrosine hydroxylase in haemolymph from diapausing larvae of the European corn-borer *Ostrinia nubilalis*. *Insect Biochem.* 2:451–9.

Brewer, F. D., and S. B. Vinson. 1971. Chemicals affecting the encapsulation of foreign material in an insect. *J. Invertebr. Pathol.* 18:287–9.

Brunet, P. C. J. 1963. Tyrosine metabolism in insects. *Ann. N.Y. Acad. Sci.* 100:1020–34.

Brunet, P. C. J. 1965. The metabolism of aromatic compounds. *Biochem. Soc. Symp.* 25:49–77.

Brunet, P. C. J. 1967. Sclerotins. *Endeavour* 98:68–74.

Bullough, W. S., and E. B. Laurence. 1964. Mitotic control by internal secretion: The role of the chalone-adrenalin complex. *Exp. Cell Res.* 33:176–94.

Bursell, E. 1963. Aspects of the metabolism of amino acids in the tse-tse fly *Glossina* (Diptera). *J. Insect Physiol.* 9:439–52.

Bursell, E. 1966. Aspects of the flight metabolism of tse-tse flies (*Glossina*). *Comp. Biochem. Physiol.* 19:809–18.

Calvet, J. P., and R. M. Clark. 1974. Immersion refractometric analysis of cultured *Hyalophora cecropia* hemocytes. *Protoplasma* 80:29–40.

Carton, Y. 1977. La réaction d'immunité cellulaire (capsule hémocytaire) chez un insecte, en function de la région du corps. *Ann. Parasitol. Hum. Comp.* 52:59–62.

Castiglioni, M. C., and R. G. Rezzonico. 1961. First results of tissue culture in *Drosophila*. *Experientia* 17:88–90.

Castiglioni, M. C., and R. G. Rezzonico. 1963. Genotypical differences between stocks of *D. melanogaster* revealed from culturing *in vitro*. *Experientia* 19:527–9.

Chen, P. S., and L. Levenbook. 1966. Studies on the haemolymph proteins of the blowfly *Phormia regina*. II. Synthesis and breakdown as revealed by isotopic labelling. *J. Insect Physiol.* 12:1611–27.

Cherbas, L. 1973. The induction of injury reaction in cultured haemocytes from saturniid pupae. *J. Insect Physiol.* 19:2011–24.

Chippendale, G. M. 1970a. Metamorphic changes in haemolymph and midgut proteins of the southwestern corn borer *Diatraea grandiosella*. *J. Insect Physiol.* 16:1909–20.

Chippendale, G. M. 1970b. Metamorphic changes in fat body proteins of the southwestern corn borer Diatraea grandiosella. J. Insect Physiol. 16:1057–68.

Chippendale, G. M., and S. D. Beck. 1966. Haemolymph proteins of Ostrinia nubilalis during diapause and prepupal differentiation. J. Insect Physiol. 12:1629–38.

Clark, R. M., and W. R. Harvey. 1965. Cellular membrane formation by plasmatocytes of diapausing Cecropia pupae. J. Insect Physiol. 11:161–76.

Clements, A. N., and T. D. C. Grace. 1967. The utilization of sugars by insect cells in culture. J. Insect Physiol. 13:1327–32.

Clements, B., and T. May. 1974. Studies on locust neuromuscular physiology in relation to glutamic acid. J. Exp. Biol. 60:673–705.

Coles, G. C. 1965. The haemolymph and moulting in Rhodnius prolixus Stål. J. Insect Physiol. 11:1317–23.

Collett, J. I. 1976a. Peptidase-mediated storage of amino acids in small peptides. Insect Biochem. 6:179–85.

Collett, J. I. 1976b. Small peptides, a life-long store of amino acid in adult Drosophila and Calliphora. J. Insect Physiol. 22:1433–40.

Costin, N. M. 1975. Histochemical observations of the haemocytes of Locusta migratoria. Histochem. J. 7:21–43.

Crossley, A. C. 1964. An experimental analysis of the origins and physiology of haemocytes in the blue blow-fly Calliphora erythrocephala (Meig.). J. Exp. Zool. 157:375–98.

Crossley, A. C. 1965. Transformations in the abdominal muscles of the blue blow-fly Calliphora erythrocephala (Meig.), during metamorphosis. J. Embryol. Exp. Morphol. 14:89–110.

Crossley, A. C. 1968. The fine-structure and mechanism of breakdown of larval intersegmental muscles in the blowfly Calliphora erythrocephala. J. Insect Physiol. 14:1389–1407.

Crossley, A. C. 1972a. Ultrastructural changes during transition of larval to adult intersegmental muscle at metamorphosis in the blowfly Calliphora erythrocephala. I. Dedifferentiation and myoblast fusion. J. Embryol. Exp. Morphol. 27:43–74.

Crossley, A. C. 1972b. The ultrastructure and function of pericardial cells and other nephrocytes in an insect: Calliphora erythrocephala. Tissue Cell. 4:529–60.

Crossley, A. C. 1975. The cytophysiology of insect blood. Adv. Insect Physiol. 11:117–222.

Crowley, G. J. 1964. Studies in arthropod serology. II. An invertebrate response to injected antigenic materials. Wasmann J. Biol. 22:185–224.

Cuénot, L. 1891. Etudes sur le sang et les glandes lymphatiques dans la série animale. Arch. Zool. Exp. Gén. (Ser. 2) 9:365–475, 592–670.

Cuénot, L. 1897. Les globules sanguines et les organes lymphöides des invertébrés. Arch. Anat. Microsc. 1:153–92.

Day, N. K. B., H. Gewurz, R. Johannsen, and J. Finstad, 1970. Complement and complement-like activity in lower vertebrates and invertebrates. J. Exp. Med. 132:941–50.

Dennell, R. 1947. A study of the insect cuticle: The formation of the puparium of Sarcophaga falculata. Proc. R. Soc. B. 134:79–110.

Denucé, J. M. 1958. Zonenolekphoretische Untersuchungen der Hamolymphe-proteins von Insekten in Verschiedenen Studien der Larven-Entwieklung. Z. Naturf. 136:215–18.

Deutsch, V., and J. C. Landureau. 1970. Caractéristiques ultrastructurales de cellules d'insectes produisant une chitinase in vitro. C. R. Acad. Sci. Paris D270:1491–4.

Dohke, K. 1973. Studies on prephenoloxidase-activating enzyme from cuticle of the silkworm Bombyx mori. I. Activation reaction by the enzyme. Arch. Biochem. Biophys. 157:203–9.

Doyle, D., and H. Laufer. 1969. Sources of larval salivary gland secretion in the dipteran Chironomus tentans. J. Cell Biol. 40:61–78.

Duchâteau, G. N., M. Florkin, and J. Leclercq. 1953. Concentrations des bases fixes et types de composition de la base totale de l'hémolymph des insectes. *Arch. Int. Physiol. Biochim.* *61*:518–49.

Dybowska, H. E., and A. B. Dutkowski. 1977. Ruthenium red staining of the neural lamella of the brain of *Galleria mellonella*. *Cell Tissue Res.* *176*:275–84.

Edwards, G. A., H. Ruska, and E. DeHarven. 1958. Electron microscopy of peripheral nerves and neuromuscular junctions in the wasp leg. *J. Biophys. Biochem. Cytol.* *4*:107–14.

Elder, H. Y. 1973. Distribution and functions of elastic fibers in the invertebrates. *Biol. Bull.* (*Woods Hole*) *144*:43–63.

Evans, D. R., and V. G. Dethier. 1957. The regulation of taste thresholds for sugars in the blowfly. *J. Insect Physiol.* *1*:3–17.

Evans, J. J. T. 1967. The activation of prophenoloxidase during melanization of the pupal blood of the Chinese oak silkworm *Antheraea pernyi*. *J. Insect Physiol.* *13*:1699–1711.

Evans, J. J. T. 1968. Natural substrates of the phenoloxidase in the pupal blood of the silkmoths, *Antheraea pernyi* and *A. eucalypti*. *J. Insect Physiol.* *14*:277–91.

Evans, P. D., and A. C. Crossley. 1974. Free amino acids in the haemocytes and plasma of the larva of *Calliphora vicina*. *J. Exp. Biol.* *61*:463–72.

Faulkner, P. 1955. A hexose-1-phosphatase in silkworm blood. *Biochem. J.* *60*:590–6.

Faulkner, P., and B. Bheemeswar. 1960. Studies on the biosynthesis of proteins in the silkworm *Bombyx mori* L. *Biochem. J.* *76*:71–8.

Feir, D., and L. Krzywda. 1969. Concentration of insect haemolymph proteins under various experimental conditions. *Comp. Biochem. Physiol.* *31*:197–204.

Finlayson, L. H., and D. Hamer. 1949. Free amino acids in the haemolymph of *Calliphora erythrocephala*, Meigen. *Nature* (*Lond.*) *163*:843–4.

Florkin, M., and C. Jeuniaux. 1974. Haemolymph composition, pp. 255–308. *In* M. Rockstein (ed.). *The Physiology of Insecta*, Vol. 5, 2nd ed. Academic Press, New York.

Fox, F. R., and R. R. Mills. 1969. Changes in haemolymph and cuticle protein during the moulting process in the American cockroach. *Comp. Biochem. Physiol.* *29*:1187–95.

Fox, F. R., J. R. Seed, and R. R. Mills. 1972. Cuticle sclerotization by the American cockroach: Immunological evidence for the incorporation of blood proteins into the cuticle. *J. Insect Physiol.* *18*:2065–70.

Fraenkel, G. S., and C. Hsiao. 1965. Bursicon, a hormone which mediates tanning of the cuticle in the adult fly and other insects. *J. Insect Physiol.* *11*:513–56.

François, J. 1974. Etude ultrastructural des hémocytes du Thysanoure *Thermobia domestica* (Insecte, Aptérygote). *Pedobiologia* *14*:157–62.

François, J. 1975a. Hemocytes and the hematopoietic organ of *Thermobia domestica* (Thysanura: Lepismatidae). *Int. J. Insect Morphol. Embryol.* *4*:477–94.

François, J. 1975b. L'encapsulation hemocytaire experimentale chez le lépisme *Thermobia domestica*. *J. Insect Physiol.* *21*:1535–46.

Franke, H. 1960a. Licht- und elektronenmikroskopische Untersuchungen über die Blutgerinnung bei *Periplaneta orientalis*. *Zool. Jahrb. Abt. Allg. Zool. Physiol. Tiere* *68*:499–518.

Franke, H. 1960b. Licht- und elektronenmikroskopische Untersuchungen uber die Blutgerinnung bei *Periplaneta orientalis*. *Zool. Jahrb. Abt. Allg. Zool. Physiol. Tiere.* *69*:131–2.

Friedman, S. 1960. The purification and properties of trehalase isolated from *Phormia regina* Meig. *Arch. Biochem. Biophys.* *87*:252–8.

Gagen, S. J., and N. A. Ratcliffe. 1976. Studies on the *in vivo* cellular reactions and fate of injected bacteria in *Galleria mellonella* and *Pieris brassica* larvae. *J. Invertebr. Pathol.* *28*:17–24.

464   A. C. Crossley

Garrett, R. A., and H.-G. Wittmann. 1973. Structure and function of the ribosome. *Endeavour* 32:8–14.

Geiger, J. G., J. M. Krolak, and R. R. Mills. 1977. Possible involvement of cockroach haemocytes in the storage and synthesis of cuticle proteins. *J. Insect Physiol.* 23:227–30.

Gingrich, R. E. 1964. Acquired humoral immune response of the large milkweed bug, *Oncopeltus fasciatus* (Dallas) to injected materials. *J. Insect Physiol.* 10:179–94.

Glaser, R. W. 1917. The growth of insect blood cells *in vitro. Psyche* 24:1–17.

Glaser, R. W. 1918. On the structure of immunity principles in insects. *Psyche* 25:39–46.

Goffinet, G., and Ch. Grégoire, 1975. Coagulocyte alterations in clotting haemolymph of *Carausius morosus* L. *Arch. Int. Physiol. Biochim.* 83:707–22.

Götz, P. 1969. Die Einkapselung von Parasiten in der Hämolymph von *Chironomus*-larven. *Zool. Anz. Suppl. BD 33.* (*Verh. Zol. Ges. S.*):610–17.

Götz, P., and A. Vey. 1974. Humoral encapsulation in Diptera (Insecta): Defence reactions of *Chironomus* larvae against fungi. *Parasitology* 68:193–205.

Grace, T. D. C. 1958. The prolonged growth and survival of ovarian tissue of the promethea moth *Callosamia promethea in vitro. J. Gen. Physiol.* 41:1027–34.

Grace, T. D. C., and H. W. Brzostowski. 1966. Analysis of the amino acids and sugars in an insect cell culture medium during cell growth. *J. Insect Physiol.* 12:625–33.

Grégoire, Çh. 1955a. Blood coagulation in Arthropods. V. Studies on haemolymph coagulation in 420 species of insects. *Arch. Biol.* 66:103–43.

Grégoire, Ch. 1955b. Coagulation de l'hémolymph chez les insectes irradiés par les rayons X. *Arch. Int. Physiol. Biochim.* 68:246–8.

Grégoire, Ch. 1964. Haemolymph coagulation, pp. 153–88. In M. Rockstein (ed.). *The Physiology of Insecta*, Vol. 3. Academic Press, New York.

Grégoire, Ch., and M. Florkin. 1950a. Etude au microscope à contraste de phase du coagulocyte du image granulaire et de la coagulation plasmatique dans la sang des insectes. *Experientia* 6:297–8.

Grimstone, A. V., S. Rotherham, and G. Salt. 1967. An electron microscope study of capsule formation by insect blood cells. *J. Cell Sci.* 2:281–92.

Gupta, A. P., and D. J. Sutherland. 1966. *In vitro* transformations of the insect plasmatocyte in some insects. *J. Insect Physiol.* 12:1369–75.

Gupta, A. P., and D. J. Sutherland. 1967. Phase contrast and histochemical studies of spherule cells in cockroaches (Dictyoptera). *Ann. Entomol. Soc. Amer.* 60:557–65.

Hackman, R. H., and M. Goldberg. 1967. The o-diphenoloxidases of fly larvae. *J. Insect Physiol.* 13:531–44.

Hagopian, M. 1971. Unique structures in the insect granular haemocytes. *J. Ultrastruct. Res.* 36:646–58.

Hardy, W. B. 1892. The blood corpuscles of the Crustacea, together with a suggestion as to the origin of the crustacean fibrinferment. *J. Physiol.* 13:165–90.

Harvey, W. R., and C. M. Williams. 1961. The injury metabolism of the *Cecropia* silkworm. I. Biological amplification of the effects of localized injury. *J. Insect Physiol.* 7:81–99.

Hess, A. 1958. The fine structure of nerve cells and fibers, neuroglia, and sheaths of the ganglion chain the cockroach *Periplaneta americana. J. Biophys. Biochem. Cytol.* 4:731–42.

Heyneman, R. A. 1965. Final purification of a latent phenolase with mono- and diphenolase activity from *Tenebrio molitor. Biochem. Biophys. Res. Comm.* 21:162–9.

Hoffmann, D., and A. Porte. 1973. Secretory differentiation of reticular cells in the hematopoietic organ of *Locusta migratoria C. R. Acad. Sci. Paris D* 276:677–80.

Hoffmann, D., M. Brehélin, and J. A. Hoffmann. 1977. Premiers résultats sur less réactions de défense antibactériennes de larves et d'imagos de *Locusta migratoria. Ann. Parasitol. Hum. Comp.* 52:87–8.

Hoffmann, J. A. 1966. Etude ultrastructurale de deux hémocytes à granules de *Locusta migratoria. C. R. Acad. Sci. Paris 263*:521–4.

Hoffmann, J. A. 1967. Etude des hémocytes de *Locusta migratoria* L. Orthoptère. *Arch. Zool. Exp. Gén. 108*:251–91.

Hoffmann, J. A. 1969. Etude de la récupération hémocytaire après hémorragies expérimentales chez l'orthoptere *Locusta migratoria. J. Insect Physiol. 15*: 1375–84.

Hoffmann, J. A. 1972. Modifications of the haemogramme of larval and adult *Locusta migratoria* after selective X-irradiations of the haemocytopoietic tissue. *J. Insect Physiol. 18*:1639–52.

Hoffmann, J. A., A. Porte, and P. Joly. 1970. On the localization of phenoloxidase activity in coagulocytes of *Locusta migratoria* L. (Orthoptera). *C. R. Acad. Sci. D. Paris 270*:629–31.

Hoffmann, J. A., and M-E. Stoeckel. 1968. Sur les modifications ultrastructurales des coagulocytes au cours de la coagulation de l'hémolymph chez un insecte orthoptéröide: *Locusta migratoria. C. R. Soc. Biol. (Strasburg) 162*:2257–9.

Hoffmann, J. A., M.-E. Stoeckel, A. Porte, and P. Joly. 1968. Ultrastructure des hémocytes de *Locusta migratoria* (Orthoptère). *C. R. Acad. Sci. Paris. 266*:503–5.

Holden, J. S. 1973. Free amino acid levels in the cockroach *Periplaneta americana. J. Physiol. 232*:61P–62P.

Homan-Müller, J. W. T., R. S. Weening, and D. Roos. 1974. Production of hydrogen peroxide by phagocytizing human granulocytes. *J. Lab. Clin. Med. 85*:198–207.

Hopkins, T. L., and R. A. Wirtz. 1976. DOPA and tyrosine decarboxylase activity in tissues of *Periplaneta americana* in relation to cuticle formation and ecdysis. *J. Insect Physiol. 22*:1167–71.

Horikawa, M., and K. Kuroda. 1959. *In vitro* cultivation of blood cells of *Drosophila melanogaster* in a synthetic medium. *Nature (Lond.) 184*:2017.

Horowitz, N. H., and M. Fling. 1955. The autocatalytic production of tyrosinase in extracts of *Drosophila melanogaster*, pp. 207–18. In W. D. McElroy and B. Glass (eds.). *Amino Acid Metabolism*. Johns Hopkins Press, Baltimore.

Hughes, L., and G. M. Price. 1974. The isolation and properties of a lipoprotein fraction possessing tyrosinase activity from the haemolymph of the larva of the fleshfly *Sarcophaga barbarta. Trans. Biochem. Soc. 2*:336–8.

Hughes, L., and G. M. Price. 1975a. Tyrosinase activity in larvae of the fleshfly *Sarcophaga barbarta. Trans. Biochem. Soc. 3*:72–5.

Hughes, L., and G. M. Price. 1975b. The inhibition of catecholase activity by cuticle extracts in larvae of the fleshfly *Sarcophaga barbarta. Trans. Biochem. Soc. 3*:549–51.

Hughes, L., and G. M. Price. 1976. Hemolymphal activation of protyrosinase and the site of synthesis of hemolymph protyrosinase in larvae of the fleshfly *Sarcophaga barbarta. J. Insect Physiol. 22*:1005–11.

Inaba, T., and M. Funatsu. 1964. Studies on tyrosinase in the housefly. III. Activation of protyrosinase by natural activator. *Agric. Biol. Chem. 28*:206–15.

Irving, S. N., M. P. Osborne, and R. G. Wilson. 1976. Virtual absence of L-glutamate from the haemoplasm of arthropod blood. *Nature (Lond.) 263*:431–3.

Jeuniaux, Ch., Gh. Duchâteau-Bosson, and M. Florkin. 1961. Contributions à la biochemie du ver à soie. XII. Modifications de l'aminoacidémie et de la pression osmotique de l'hémolymphe au cours du développement de *Bombyx mori* L. *Arch. Int. Physiol. Biochim. 69*:617–27.

Johnston, M. A., P. S. Davies, and H. Y. Elder. 1971. Possible hepatic function for crustacean blood cells. *Nature (Lond.) 230*:471–2.

Jones, J. C. 1967. Effects of repeated haemolymph withdrawals and of ligating the head on differential haemocyte counts of *Rhodnius prolixus* (Stål). *J. Insect Physiol. 13*:1351–60.

Jones, J. C. 1970. Haemocytopoiesis in insects, pp. 7–65. In A. S. Gordon (ed.). Regulation of Haematopoiesis, Vol. 1. Appleton, New York.

Jones, J. C. 1976. Do insect hemocytes normally transform into basement membrane or fat body? Amer. Zool. 16:220.

Kamp, H. 1968. Untersuchungen zur humoraline Immunitat bei Pyrrhocoris apterus und Galleria mellonella. Z. Vgl. Physiol. 58:441–64.

Karlson, P., and H. Liebau. 1961. Zum Tyrosinstoffwechsel der Insekten V. Reindaestellung, Kristallisation und Substratspezifität der o-Diphenoloxydase aus Calliphora erythrocephala. Hoppe-Seyler's Z. Physiol. Chem. 326:135–43.

Karlson, P., D. Mergenhagen, and C. E. Sekeris. 1964. Zum Tyrosinstoffwechsel der Insekten. XV. Weitere Untersuchungen über das o-Diphenoloxidase-system von Calliphora erythrocephala. Hoppe-Seyler's Z. Physiol. Chem. 338:42–50.

Karlson, P., and C. E. Sekeris. 1964. Biochemistry of insect metamorphosis. pp. 221–241. In M. Florkin and H. S. Mason (eds.). Comparative Biochemistry, Vol. 6. Academic Press, New York.

Kato, K. I., E. Perkowska, and J. L. Sirlin. 1963. Electro and immunoelectrophoretic patterns in the larval salivary secretion of Chironomus thummi. J. Histochem. Cytochem. 11:484–8.

Kerkut, G. A., A. Shapira, and R. G. Walker. 1965. The effect of acetylcholine, glutamic acid and GABA on contractions of the perfused cockroach leg. Comp. Biochem. Physiol. 16:37–48.

Klebanoff, S. J. 1967a. Iodination of bacteria. A bactericidal mechanism. J. Exp. Med. 126:1063–79.

Klebanoff, S. J. 1967b. A peroxidase-mediated antimicrobial system in leucocytes. J. Clin. Invest. 46:1078.

Klebanoff, S. J. 1968. Myeloperoxidase-halide-hydrogen peroxide antibacterial system. J. Bacteriol. 95:2131–8.

Klebanoff, S. J. 1971. Intraleukocytic microbicidal defects. Annu. Rev. Med. 22:39–62.

Knight, R. H. 1960. Free amino acids in the haemolymph of Glossina species. E. Afr. Trypanosomiasis Res. Organ. Rep. 1960:22–3.

Koeppe, J. K., and L. E. Gilbert. 1973. Immunochemical evidence for the transport of haemolymph protein into the cuticle of Manduca sexta (Lep., Sphingidae). J. Insect Physiol. 19:615–24.

Koeppe, J. K., and R. R. Mills. 1972. Hormonal control of tanning by the American cockroach: Probable bursicon mediated translocation of protein-bound phenols. J. Insect Physiol. 18:465–9.

Kroëger, H. 1963. Chemical nature of the system controlling gene activities in insect cells. Nature (Lond.) 200:1234–5.

Lai-Fook, J. 1966. The repair of wounds in the integument of insects. J. Insect Physiol. 12:195–226.

Lai-Fook, J. 1968. The fine-structure of wound repair in an insect Rhodnius prolixus. J. Morphol. 124:37–78.

Lai-Fook, J. 1970. Haemocytes in the repair of wounds in an insect Rhodnius prolixus. J. Morphol. 130:297–314.

Lai-Fook, J. 1973. The structure of the haemocytes of Calpodes ethlius (Hesperiidae: Lepidoptera). J. Morphol. 139:79–104.

Laufer, H. 1960. Blood proteins in insect development. Ann. N.Y. Acad. Sci. 89:490–515.

Laufer, H., and Y. Nakase. 1965. Salivary gland secretion and its relation to chromosomal puffing in the dipteran Chironomus thummi. Proc. Natl. Acad. Sci. U.S.A. 53:511–6.

Lazarenko, T. 1925a. Beiträge zur vergeichenden Histologie des Blutes und des Bindegewebes. II. Die morphologische Bedeutung der Blut- und Bindegewebeelemente der Insekten. Z. Mikrosk. Anat. Forsch. 3:409–99.

Lazarenko, T. 1925b. Histological observations on healing of integument wounds in insects. *Bull. Biol. Res. Perm. Univ.* 2:287–398.

Lea, M. S., and L. I. Gilbert. 1961. Cell division in diapausing silkworm pupae. *Amer. Zool.* 1:368–9.

Levenbook, L. 1966. Hemolymph amino acids and peptides during larval growth of the blowfly *Phormia regina*. *Comp. Biochem. Physiol.* 18:341–51.

Levenbook, L., R. P. Bodnaryk, and T. F. Spande. 1970. Levels of free glutamic acid, phenylalanine and γ-glutamyl-L-phenylalanine during pupal sclerotization. *Comp. Biochem. Physiol.* 35:499–502.

Levenbook, L., and V. W. Hollis. 1961. Organic acid in insects. 1. Citric acid. *J. Insect Physiol.* 6:52–61.

Levenbook, L., and I. Krishna. 1971. Effect of ageing on amino acid turnover and rate of protein synthesis in the blowfly *Phormia regina*. *J. Insect Physiol.* 17:9–12.

Levin, J., and F. B. Bang. 1964. A description of cellular coagulation in *Limulus*. *Bull. Johns Hopkins Hosp.* 115:265–74.

Lewis, H. W., and H. S. Lewis. 1963. Genetic regulation of dopa oxidase activity in *Drosophila*. *Ann. N.Y. Acad. Sci.* 100:827–39.

Locke, M. 1969. The localization of a peroxidase associated with hard cuticle formation in an insect, *Calpodes ethlius* Stoll, Lepidoptera, Hesperiidae. *Tissue Cell* 1:555–74.

Locke, M. 1976. The role of plasma membrane plaques and Golgi complex vesicles in cuticle deposition during the moult/intermoult cycle, pp. 237–58. *In* H. R. Hepburn (ed.). *The Insect Integument.* Elsevier, New York.

Locke, M., and P. Huie. 1972. The fiber components of insect connective tissue. *Tissue Cell* 4:601–12.

Locke, M., and N. Krishnan. 1971. The distribution of phenoloxidases and polyphenols during cuticle formation. *Tissue Cell* 3:103–26.

Locke, M., and J. T. McMahon. 1970. The origin and fate of microbodies in the fat body of an insect. *J. Cell Biol.* 48:61–78.

Ludwig, D. 1954. Changes in distribution of nitrogen in blood of Japanese beetle (*Popillia japonica* Newman) during growth and metamorphosis. *Physiol. Zool.* 27:325–34.

Lysenko, O. 1972. Some characteristics of *Galleria mellonella* hemolymph proteins. *J. Invertebr. Pathol.* 19:335–41.

Maier, W. A. 1973. Die Phenoloxidase von *Chironomus thummi* und ihre Beeinflussung durch Parasitare Mermithiden. *J. Insect Physiol.* 19:85–95.

Marek, M. 1969. *In vitro* synthesis of injury protein and a companion protein in organs of prepupa of *Galleria mellonella*. *Comp. Biochem. Physiol.* 29:1231–7.

Marek, M. 1970a. Effect of actinomycin D on synthesis of cooling protein in haemolymph of pupae of *Galleria mellonella*. *Comp. Biochem. Physiol.* 34:221–9.

Marek, M. 1970b. Effect of stress of cooling on the synthesis of immunoglobulins in haemolymph of *Galleria mellonella* L. *Comp. Biochem. Physiol.* 35:615–22.

Martignoni, M. E., and R. J. Scallion. 1961. Preparation and uses of insect haemocyte monolayers *in vitro*. *Biol. Bull.* (*Woods Hole*) 121:507–20.

Martin, M.-D., J. F. Kinnear, and J. A. Thompson. 1969. Developmental changes in the late larva of *Calliphora stygia*. II. Protein synthesis. *Aust. J. Biol. Sci.* 22:935–46.

Mason, H. S. 1955. Comparative biochemistry of the phenolase complex. *Adv. Enzymol.* 16:105–84.

Mason, H. S., and E. W. Peterson. 1965. Melanoproteins. I. Reactions between enzyme-generated quinones and amino acids. *Biochim. Biophys. Acta* 111:134–46.

Matthews, J. R., R. G. H. Downer, and P. E. Morrison. 1976. γ-Glucosidase activity in haemolymph of the American cockroach *Periplaneta americana*. *J. Insect Physiol.* 22:157–63.

Mazzone, H. M. 1971. Lepidoptera cell culture and its application to the study of plant viruses and animal parasites. I. Cultivation of gypsy moth haemocytes. *Curr. Top. Microbiol. Immunol.* 55:196–200.

Messner, B. 1976. NADH oxidase NADPH oxidase activity in the blood cells of hemimetabolous and holometabolous insects. *Acta Histochem.* 56:261–9.

Miller, R., G. Leaf, and P. N. R. Usherwood. 1973. Blood glutamate in arthropods. *Comp. Biochem. Physiol.* 44A:991–6.

Mills, R. R., S. Androuny, and F. R. Fox. 1968. Correlation of phenoloxidase activity with ecdysis and tanning hormone release in the American cockroach. *J. Insect Physiol.* 14:603–11.

Mills, R. R., and D. L. Whitehead. 1970. Hormonal control of tanning in the American cockroach: Changes in blood cell permeability during ecdysis. *J. Insect Physiol.* 16:331–40.

Mitsuhashi, J. 1966. Tissue culture of the rice stem borer *Chilo suppressalis* Walker (Lepid. Pyralidae). II. Morphology and *in vitro* cultivation of haemocytes. *Appl. Entomol. Zool.* 1:5–20.

Möhrig, W., and B. Messner. 1968. Immunreaktionen bei Insekten. I. Lysozymals grundlegender antibakterieller Faktor im humoralen Abwehrmechanisms der Insekten. *Biol. Zentralb.* 87:439–70.

Moran, D. T. 1971. The fine structure of cockroach blood cells. *Tissue Cell* 3:413–22.

Munn, E. A., and S. F. Bufton. 1973. Purification and properties of a phenol oxidase from the blowfly *Calliphora erythrocephala*. *Eur. J. Biochem.* 35:3–10.

Munson, S. C., and J. F. Yeager. 1944. Fat inclusion in the blood cells of the southern armyworm *Prodenia eridania*. *Ann. Entomol. Soc. Amer.* 37:396–400.

Murdock, L., and G. Chapman. 1974. L-glutamate in arthropod blood plasma: Physiological implication. *J. Exp. Biol.* 60:783–94.

Nakamura, T., and S. Sho. 1964. Studies on silkworm tyrosinase. *J. Biochem. (Tokyo)* 55:510–15.

Nappi, A. J. 1970. Defence reactions of *Drosophila euronotus* larvae against the hymenopterous parasite *Pseudocoila bochei*. *J. Invertebr. Pathol.* 16:408–18.

Nappi, A. J. 1973. Hemocytic changes associated with the encapsulation and melanization of some insect parasites. *Exp. Parasitol.* 33:285–302.

Nappi, A. J. 1974. Insect hemocytes and the problem of host recognition of foreignness. *Contemp. Top. Immunobiol.* 4:207–24.

Nappi, A. J., and J. G. Stoffolano. 1971. *Heterotylenchus autumnalis* haemocytic reactions and capsule formation in the host, *Musca domestica*. *Exp. Parasitol.* 29:116–25.

Nappi, A. J., and F. A. Streams. 1969. Haemocytic reactions of *Drosophila melanogaster* to the parasites *Pseudocoila mellipes* and *P. bochei*. *J. Insect Physiol.* 15:1551–68.

Nelstrop, A. E., G. Taylor, and P. Collard. 1970. Comparative studies of serum and haemolymph proteins. *Comp. Biochem. Physiol.* 35:191–9.

Neutra, M., and C. P. Leblond. 1966. Synthesis of the carbohydrate of mucus in the Golgi complex as shown by electron microscope radioautography of goblet cells from rats injected with glucose-H^3. *J. Cell. Biol.* 30:119–36.

Neuwirth, M. 1973. The structure of the haemocytes of *Galleria mellonella* (Lepidoptera). *J. Morphol.* 139:105–24.

Newburgh, R. W., and V. H. Cheldelin. 1955. Oxidation of carbohydrates by the pea aphid *Macrosiphum pisi* (KLTB). *J. Biol. Chem.* 214:37–4.

Normann, T. C., and H. Duve. 1969. Experimentally induced release of a neurohormone influencing hemolymph trehalase level in *Calliphora erythrocephala* (Diptera). *Gen. Comp. Endocrinol.* 12:449–59.

Nutting, W. L. 1951. A comparative anatomical study of the heart and accessory structures orthopteroid insects. *J. Morphol.* 89:501–98.

Oberlander, H. 1976. Hormonal control of growth and differentiation of insect tissues cultured *in vitro*. *In Vitro* 12:225–35.

Ohnishi, E. 1953. Tyrosinase activity during puparium formation in *Drosophila melanogaster. Jpn. J. Zool.* 11:69–74.

Ohnishi, E. 1954. Activation of tyrosinase in *Drosophila virilis. Annot. Zool. Jpn.* 27:188–93.

Ohnishi, E. 1958. Tyrosinase activation in the pupae of the housefly *Musca vicina. Jpn. J. Zool.* 12:179–88.

Ohnishi, E. 1959. Studies on the mechanism of tyrosinase activity in the housefly, *Musca vicina. J. Insect Physiol.* 3:219–29.

Ohnishi, E., K. Dohke, and M. Ashida. 1970. Activation of pre-phenoloxidase. II. Activation by α-chymotrypsin. *Arch. Biochem. Biophys.* 139:143–8.

Orr, G. W. 1964. The influence of nutritional and hormonal factors on the chemistry of the fat body, blood and ovaries of the blowfly *Phormia regina* (Meig.). *J. Insect Physiol.* 10:103–19.

Peake, P. W. 1979. Isolation and characterization of the haemocytes of *Calliphora vicina* on density grandients of Ficoll. (In preparation)

Pearse, A. G. E. 1968. *Histochemistry Theoretical and Applied*, Vol. 1, 3rd ed., Churchill-Livingstone, Edinburgh.

Petit, J. P. 1968. Hémolymphe de glossines: Récolte et analyse. *Rev. Elevage Méd. Vet. Pays Trop.* 21:493–8.

Pincus, S. H., and S. J. Klebanoff. 1971. Quantitative leucocyte iodination. *N. Engl. J. Med.* 284:744–50.

Pinhey, K. G. 1930. Inhibitors of tyrosinase. *J. Exp. Biol.* 32:468–84.

Poinar, G. O., and R. Leutenegger. 1971. Ultrastructural investigation of the melanization process in *Culex pipiens* (Culicidae) in response to a nematode. *J. Ultrastruct. Res.* 36:149–58.

Poinar, G. O., R. Leutenegger, and P. Götz. 1968. Ultrastructure of the formation of a melanotic capsule in *Diabrotica* (Coleoptera) in response to a parasitic nematode. (Mermithidae). *J. Ultrastruct. Res.* 25:292–306.

Post, L. C. 1972. Bursicon: Its effect on tyrosine permeation into insect haemocytes. *Biochim. Biophys. Acta* 290:424–8.

Post, L. C., and B. J. DeJong. 1973. Bursicon and the metabolism of tyrosine in the moulting cycle of *Pieris* larvae. *J. Insect Physiol.* 19:1541–6.

Preston, J. W., and R. L. Taylor. 1970. Observations on the phenoloxidase system in the haemolymph of the cockroach *Leucophaea maderae. J. Insect Physiol.* 16:1729–44.

Price, G. M. 1966. The *in vitro* incorporation of (U = ¹⁴C) valine into fat body protein of the larva of the blowfly *Calliphora erythrocephala. J. Insect Physiol.* 12:731–40.

Price, G. M. 1971. Tyrosine metabolism in the blowfly larva *Calliphora erythrocephala. Biochem. J.* 121:28P–29P.

Price, G. M. 1972. Tyrosine metabolism in the larva of the blowfly *Calliphora erythrocephala. Insect Biochem.* 2:175–85.

Pryor, M. G. M. 1940. On the hardening of the ootheca of *Blatta orientalis. Proc. R. Soc. B.* 128:378–93.

Pryor, M. G. M. 1955. Tanning of blowfly puparia. *Nature (Lond.)* 175:600.

Przelecka, A., and A. B. Dutkowski. 1973. The structure of the ovariole wall in *Galleria mellonella. Folia Histochem. Cytochem.* 11:269–74.

Pye, A. E., and W. G. Yendol. 1972. Haemocytes containing prophenoloxidase in *Galleria* larvae after injection of bacteria. *J. Invertebr. Pathol.* 19:166–70.

Raina, A. K. 1976. Ultrastructure of the larval haemocytes of the pink bollworm *Pectinophora gossypiella* (Lepidoptera: Gelechiidae). *Int. J. Insect Morphol. Embryol.* 5:187–95.

Ratcliffe, N. A. 1975. Spherule cell–test particle interactions in monolayer cultures of *Pieris brassicae* hemocytes. *J. Invertebr. Pathol.* 26:217–23.

Ratcliffe, N. A., and S. J. Gagen. 1976. Cellular defence reactions of insect haemocytes *in vivo:* Nodule formation and development in *Galleria mellonella* and *Pieris brassicae* larvae. *J. Invertebr. Pathol.* 28:373–82.

Ratcliffe, N. A., and S. J. Gagen. 1977. Studies on the *in vivo* cellular reactions of insects: An ultrastructural analysis of nodule formation in *Galleria mellonella. Tissue Cell. 9:*73–85.

Ratcliffe, N. A., and C. D. Price. 1974. Correlation of light and electron microscopic haemocyte structure in the Dictyoptera. *J. Morphol. 144:*485–97.

Ritter, H., and M. Bray. 1968. Chitin synthesis in cultivated cockroach blood. *J. Insect Physiol. 14:*361–6.

Rizki, M. T. M., and R. M. Rizki. 1959. Functional significance of crystal cells in the larva of *Drosophila melanogaster. J. Biophys. Biochem. Cytol. 5:*235–40.

Rowley, A. F., and N. A. Ratcliffe. 1974. Studies on *in vitro* phagocytosis in insects. *Soc. Invertebr. Pathol. Abstract.* (Arizona Mtg.).

Rowley, A. F., and N. A. Ratcliffe. 1976. The granular cells of *Galleria mellonella* during clotting and phagocytic reactions *in vitro. Tissue Cell 8:*437–46.

Rudall, K. M. 1955. The distribution of collagen and chitin. *Symp. Soc. Exp. Biol. 9:*49–71.

Rytomaa, T., and K. Kiviniemi. 1968. Control of granulocyte production. 1. Chalone and antichalone, two specific humoral regulators. *Cell Tissue Kinet. 1:*329–41.

Sacktor, B. 1976. Biochemical adaptations for flight in the insect. *Biochem. Soc. Symp. 41:*111–31.

Salt, G. 1970. *The Cellular Defence Reactions of Insects.* Cambridge Monogr. in Experimental Biology, No. 16. Cambridge University Press, Cambridge.

Sauerländer, S., and P. Erhardt. 1961. Reduzierung der Proteinfraktionen in der Hämolymphe bakterienkrander Blattiden. *Naturwissenschaften 48:*674–5.

Scharrer, B. 1972. Cytophysiological features of hemocytes in cockroaches. *Z. Zellforsch. Mikrosk. Anat. 129:*301–19.

Schmit, A. R., and N. A. Ratcliffe. 1977. The encapsulation of foreign tissue implants in *Galleria mellonella* larvae. *J. Insect Physiol. 23:*175–84.

Schmit, A. R., A. F. Rowley, and N. A. Ratcliffe. 1977. The role of *Galleria mellonella* hemocytes in melanin formation. *J. Invertebr. Pathol. 29:*232–4.

Sekeris, C. E., and D. Mergenhagen. 1964. Phenoloxidase system of the blowfly *Calliphora erythrocephala. Science* (Wash., D.C.) *145:*68–9.

Seligman, I. M., B. K. Filshie, F. A. Doy, and A. C. Crossley. 1975. Hormonal control of morphogenetic cell death of the wing hypodermis in *Lucilia cuprina. Tissue Cell 7:*281–96.

Seligman, I. M., S. Friedman, and G. S. Fraenkel. 1969. Bursicon mediation of tyrosine hydroxylation during tanning of the adult cuticle of the fly *Sarcophaga bullata. J. Insect Physiol. 15:*553–61.

Shigematsu, H. 1958. Synthesis of blood protein by the fat body in the silkworm *Bombyx mori. Nature* (Lond.) *182:*880–2.

Siakotos, A. N. 1960a. The conjugated plasma proteins of the American cockroach. *J. Gen. Physiol. 43:*999–1013.

Siakotos, A. N. 1960b. The conjugated plasma proteins of the American cockroach. II. Changes during moulting and clotting process. *J. Gen. Physiol. 43:*1015–30.

Sissakjan, N. M., and E. B. Kuvajeva. 1957. On the peculiarities of protein synthesis in the coelomic fluid of the silkworm *Bombyx mori. Biochimija. 22:*686–94.

Shields, G., and J. H. Sang. 1970. Characteristics of five cell types appearing during *in vitro* culture of embryonic material from *Drosophila melanogaster. J. Embryol. Exp. Morphol. 23:*53–69.

Smith, D. S. 1968. *Insect Cells: Their Structure and Function.* Oliver & Boyd, Edinburgh.

Smith, D. S., and V. B. Wigglesworth. 1959. Collagen in the perilemma of insect nerve. *Nature* (Lond.) *183:*127–8.

Srivastava, B. B. L., and T. L. Hopkins. 1975. Bursicon release and activity in haemolymph during metamorphosis of the cockroach *Leucophaea maderae. J. Insect Physiol. 21:*1985–93.

Steele, J. E. 1961. Occurrence of a hyperglycaemic factor in the corpus cardiacum of an insect. *Nature (Lond.) 192*:680–1.

Steele, J. E. 1969. The hyperglycaemic activity of the corpus allatum in an insect. *J. Insect Physiol. 15*:421–3.

Steinhauer, A. L., and W. P. Stephen. 1959. Changes in blood proteins during the development of the American cockroach *Periplaneta americana*. *Ann. Entomol. Soc. Amer. 52*:733–8.

Stephens, J. M. 1959. Immune responses of some insects to some bacterial antigens. *Can. J. Microbiol. 5*:203–28.

Stevenson, E., and G. R. Wyatt. 1962. The metabolism of silkmoth tissues. I. Incorporation of leucine into protein. *Arch. Biochem. Biophys. 99*:65–71.

Sutcliffe, D. W. 1963. The chemical composition of haemolymph in insects and some other arthropods, in relation to their phylogeny. *Comp. Biochem. Physiol. 9*:121–35.

Tager, H. S., J. Markese, R. D. Speirs, and K. J. Kramer. 1975. Glucagon-like immunoreactivity in insect corpus cardiacum. *Nature (Lond.) 254*:707–8.

Tager, H. S., J. Markese, K. J. Kramer, R. D. Speirs, and C. N. Childs. 1976. Glucagon-like and insulin-like hormones of the insect neurosecretory system. *Biochem. J. 156*:515–20.

Tait, J. 1911. Types of crustacean blood coagulation. *J. Mar. Biol. Assoc. 9*:191–8.

Taylor, A. 1935. Experimentally induced changes in the cell complex of the blood of *Periplaneta*. *Ann. Entomol. Soc. Amer. 28*:135.

Taylor, R. L. 1969. A suggested rôle for the polyphenol-phenoloxidase system in invertebrate immunity. *J. Invertebr. Pathol. 14*:427–8.

Telfer, W. H. 1960. A selective accumulation of blood proteins by the oocytes of saturniid moths. *Biol. Bull. (Woods Hole) 118*:338–51.

Telfer, W. H. 1961. The route of entry and localization of blood proteins in the oocytes of saturniid moths. *J. Biophys. Biochem. Cytol. 9*:747–59.

Telfer, W. H., and M. E. Melius. 1963. The mechanism of blood protein uptake by insect oocytes. *Amer. Zool. 3*:185–91.

Telfer, W. H., and C. M. Williams. 1955. Incorporation of radioactive glycine into the blood proteins of *Cecropia* silkworm. *Anat. Rec. 122*:441–2.

Telfer, W. H., and C. M. Williams. 1960. The effect of diapause, development and injury on the incorporation of radioactive glycine into the blood proteins of the *Cecropia* silkworm. *J. Insect Physiol. 5*:61–72.

Thomson, J. A., and Y. T. Sin. 1970. The control of prophenoloxidase activation in larval haemolymph of *Calliphora stygia*. *J. Insect Physiol. 16*:2063–74.

Tobe, S. S., and K. G. Davey. 1975. Synthesis and turnover of haemolymph proteins during the reproductive cycle of *Glossina austeni*. *Can. J. Zool. 53*:614–29.

Tobe, S. S., and B. G. Loughton. 1969. An autoradiographic study of haemolymph protein uptake by the tissues of the fifth instar locust. *J. Insect Physiol. 15*:1331–46.

Trager, W. 1935. Cultivation of the virus of Grasserie in silkworm tissue cultures. *J. Exp. Med. 61*:501–14.

Truman, J. W. 1973. Physiology of insect ecdysis. III. Relationships between hormonal control of eclosion and of tanning in the tobacco hornworm *Manduca sexta*. *J. Exp. Biol. 58*:821–9.

Unestam, T. 1976. Measurement of phenol oxidase activity in rapidly clotting arthropod blood. *J. Invertebr. Pathol. 27*:391–3.

Usherwood, P. N. R., and S. G. Cull-Candy. 1975. Pharmacology of somatic nerve-muscle synapses, pp. 201–280. *In* P. N. R. Usherwood (ed.). *Insect Muscle*. Academic Press, New York.

Vejbjerg, K., and T. C. Normann. 1974. Secretion of hyperglycaemic hormone from the corpus cardiacum of flying blowflies. *Calliphora erythrocephala*. *J. Insect Physiol. 20*:1189–92.

Walker, I. 1959. Die Abwehrreaktion des Wirtes *Drosophila melanogaster* gegen die zoophage Cynipide *Pseudeucoila bochei* Weld. *Rev. Suisse Zool.* 66:569–632.

Wheeler, R. E. 1963. Studies on the total haemocyte count and haemolymph volume in *Periplaneta americana* (L.) with special reference to the last moulting cycle. *J. Insect Physiol.* 9:223–35.

Whitehead, D. L. 1969. New evidence for the control mechanism of sclerotization in insects. *Nature (Lond.)* 224:721–3.

Whitehead, D. L. 1970a. L-dopa decarboxylase in the haemocytes of Diptera. *FEBS Lett.* 7:263–6.

Whitehead, D. L. 1970b. The role of haemocytes in the biosynthesis of protocatechuate in the cockroach colleterial system. *Biochem. J.* 119:65–6.

Whitehead, D. L. 1973. The events preceding formation of the puparium in *Glossina* larvae. *Trans. R. Soc. Trop. Med. Hyg.* 67:300–1.

Whitten, J. M. 1969. Haemocyte activity in relation to epidermal cell growth, cuticle secretion and cell death in a metamorphosing cyclorrhaphan pupa. *J. Insect Physiol.* 15:763–78.

Wigglesworth, V. B. 1933. The physiology of the cuticle and of ecdysis in *Rhodnius prolixus* with special reference to the function of oenocytes and dermal glands. *Q. J. Microsc. Sci.* 76:269–318.

Wigglesworth, V. B. 1937. Wound healing in an insect, *Rhodnius prolixus* (Hemiptera). *J. Exp. Biol.* 14:364–81.

Wigglesworth, V. B. 1956. The haemocytes and connective tissue formation in an insect *Rhodnius prolixus* (Hemiptera). *Q. J. Microsc. Sci.* 97:89–98.

Wigglesworth, V. B. 1959. Insect blood cells. *Annu. Rev. Entomol.* 4:1–16.

Wigglesworth, V. B. 1965. *The Principles of Insect Physiology*, 6th ed. Dutton, New York.

Wigglesworth, V. B. 1973. Haemocytes and basement membrane formation in *Rhodnius. J. Insect Physiol.* 19:831–44.

Wilinska, L., and M. J. Piechowska. 1975. Metabolism of free tyrosine in hemolymph and fat body of caterpillar of *Celerio euphorbiae* (Lepidoptera). *Bull. Acad. Pol. Sci. Ser. Sci. Biol.* 23:735–8.

Wyatt, G. R. 1961. The biochemistry of insect haemolymph. *Annu. Rev. Entomol.* 6:75–102.

Wyatt, G. R. 1963. Biochemistry of diapause, development and injury in silkmoth pupae, pp. 23–41. *In* V. J. Brookes (ed.). *Insect Physiology*. Oregon State University Press, Corvallis.

Wyatt, G. R. 1967. The biochemistry of sugars and polysaccharides in insects. *Adv. Insect Physiol.* 4:287–360.

Wyatt, G. R., and G. F. Kalf. 1956. Trehalose in insects. *Fed. Proc.* 15:388.

Wyatt, G. R., and G. F. Kalf. 1957. The chemistry of insect haemolymph. II. Trehalose and other carbohydrates. *J. Gen. Physiol.* 40:833–47.

Wyatt, G. R., R. B. Kropf, and F. G. Carey. 1963. The chemistry of insect haemolymph. IV. Acid-soluble phosphates. *J. Insect Physiol.* 9:137–52.

Wyatt, G. R., T. C. Lougheed, and S. S. Wyatt. 1956. The chemistry of insect haemolymph: Organic components of the haemolymph of the silkworm *Bombyx mori* and two other species. *J. Gen. Physiol.* 39:853–68.

Wyatt, S. S. 1956. Culture in vitro of tissue from *Bombyx mori* L. *J. Gen. Physiol.* 39:841–52.

Yamazaki, H. I. 1972. Cuticular phenoloxidase from the silkworm, *Bombyx mori:* Properties, solubilization and purification. *Insect Biochem.* 2:431–44.

Yeager, J. F., and H. H. Knight. 1933. Microscopic observations on blood coagulation in several different species of insects. *Ann. Entomol. Soc. Amer.* 26:591–602.

Yunker, C., J. Vaughn, and J. Cory. 1967. Adaptation of an insect cell line (Grace's *Antheraea* cells) to a medium free of insect haemolymph. *Science (Wash., D.C.)* 155:1565–6.

Zachary, D., and J. A. Hoffmann. 1973. The haemocytes of *Calliphora erythrocephala* (Meig.) (Diptera). Z. *Zellforsch. Mikrosk. Anat. 141*:55–73.

Zeller, H. 1938. Blut und Feltkörper im Flugel der Mehlmotte *Ephestia kuhniella*. Z. *Morphol. Oekol. Tiere 34*:663–738.

Zwicky, K. T. 1954. Osmoregulatorische Reaktionen der Larve von *Drosophila melanogaster*. Z. *Vgl. Physiol. 36*:367–90.

16 Changes in hemocyte populations

M. SHAPIRO

Gypsy Moth Methods Development Laboratory, U.S. Department of Agriculture, Otis Air Force Base, Massachusetts 02542, U.S.A.

Contents

16.1. Introduction

The life of an insect may be uneventful, progressing from the immature to the adult with no trauma. On the other hand, the insect may face the trauma of injury, the lack of food, or the challenge of natural enemies. In this chapter I hope to discuss quantitative and qualitative changes in hemocyte populations during the life of an insect undergoing growth and development or exposed to injury or insult. Population changes will be determined primarily by changes in numbers (total hemocyte counts), in types of hemocytes (differential hemocyte counts), and in blood volume. Some emphasis will be placed on the wax moth, *Galleria mellonella* (L.) (Lepidoptera), on which much information has been obtained. In vivo studies will be emphasized.

16.2. Changes during development

Salt (1970), in his excellent review, *The Cellular Defense Reactions of Insects*, cautioned about the total hemocyte count (THC). He felt the THC must be used with caution because of inherent changes in the ratio of circulating to sedentary hemocytes and the changes owing to sampling procedures.

16.2.1. Orthoptera

Tauber and Yeager (1935) studied the THC in Orthoptera, Odonata, Hemiptera, and Homoptera. THCs from *Blatta* and *Periplaneta* were higher in females carrying oothecae than in "normal" females. Adults of *Udeopsylla* and *Melanoplus* (crickets and grasshoppers), *Plathermis* (a dragonfly), and *Euschistus* (a stink bug) had higher cell populations than nymphs. The authors felt that a gradual increase in hemocytes occurred during development.

In *Periplaneta*, the variation in cell counts was great (Smith, 1938). The lowest average count was 34,000/mm³ and the highest was 158,000/mm³. Although counts from nymphs were higher (96,000/mm³) than from adults (85,000/mm³), the author felt these differences were minimal owing to great variations in counts among individuals. Analyses of counts showed a bimodal frequency with peaks of 70,000/mm³ and 120,000/mm³. Counts of the field cricket, *Gryllus*, showed a trimodal distribution of cell counts, with peaks of 50,000/mm³, 75,000/mm³, and 125,000/mm³. In *Schistocerca* (Mathur and Soni, 1937) and in *Locusta* (Webley, 1951), the THC was lower in nymphs than in adults.

Patton and Flint (1959) noted that counts from *Periplaneta* decreased at the molt and later increased postmolt. The THC from insects known to be between molts was 50,000/mm³, with a range between 21,000/mm³ and 99,000/mm³. These counts were in general agreement with those of Tauber and Yeager (1935). Gupta and Sutherland (1968) utilized only adult males. The average count was quite low (ca. 18,000/mm³) in comparison with values found by Tauber and Yeager (1935) (ca. 51,000/mm³) and Smith (1938) (93,000/mm³).

The great differences in counts may be attributable to the acetic acid vapor treatment (Fisher, 1935; Smith, 1938), which was shown to increase the number of circulating cells (Fisher, 1935). Tauber and Yeager (1935) used glacial acetic acid; Gupta and Sutherland (1968) used versene in their diluting fluid.

16.2.2. Hemiptera

Tauber and Yeager (1935, 1936) determined THCs in 62 species of insects, representing eight orders. Salt (1970) summarized the data in Orthoptera, Hemiptera, and Homoptera and concluded that, in general, the hemolymph of immatures and adults contains "some tens of thousands of hemocytes per mm³."

Feir and O'Connor (1969) determined the total number of hemocytes within fifth-stage nymphs of *Oncopeltus*. Counts increased slightly during the first 3 days postecdysis, from 35,500/mm³ to 40,500/mm³. At this time, cell populations decreased during the next 3 days to 26,000/mm³ and then increased to 40,000/mm³. The data, which conflicted with those from a previous study (Feir, 1964), were felt to be attributable to variations of blood volume and/or differences in the adherence of hemocytes to tissues or to the wound. The authors felt that more information is needed on adhesiveness and blood volumes in relationship to the availability of circulating hemocytes.

In *Rhodnius,* as in the orthopterans *Locusta* (Webley, 1951) and *Periplaneta* (Patton and Flint, 1959), the THC increased prior to the molt, decreased at ecdysis, and increased postmolt (Wigglesworth, 1955). In the hemipteran *Halys dentata* cell numbers decreased prior to ecdysis. After ecdysis, numbers continued to decrease and later increased in midinstar (Bahadur and Pathak, 1971).

The THCs from fourth-stage *Rhodnius* nymphs were quite low (800–2,000/mm³) (Jones, 1962). Hemocyte populations were higher in fifth-stage nymphs than during the fourth stage (Jones, 1967c). Jones and Liu (1961) followed hemocyte changes in *Rhodnius*. Prior to the molt, prohemocytes (PRs) decreased but adipohemocytes (ADs) increased. At the molt, PRs and granulocytes (GRs) increased, whereas plasmatocytes (PLs), ADs, and oenocytoids (OEs) decreased. Following the molt, PLs and OEs increased, and GRs decreased.

Jones (1967c) studied hemocyte population changes in *Rhodnius* and concluded that great differences occurred between fed and unfed nymphs on the fifth day after ecdysis. In fed nymphs, the percentages of PLs and OEs increased greatly. In adult females, following a blood meal, PRs increased and lysing GRs decreased.

16.2.3. Neuroptera and Plecoptera

In *Chrysops* (Neuroptera), the THC was higher in larvae than in adults (Tauber and Yeager, 1936). In the stonefly adult, *Acroneuria* (Plecoptera), progressively more hemocytes adhered to the walls of wing veins, resulting in a decline in the number of circulating cells. Arnold (1966) felt that changes in hemocyte populations during adult life were not caused by increases, but by the removal of certain types from circulation.

Differential hemocyte counts (DHCs) were made in two ways: (1) by scanning a vein and identifying all hemocytes, both sedentary and in circulation; and (2) by identifying only those hemocytes flowing across a fixed field of view in a vein. The proportions of hemocytes observed in method 1 were: PRs (0.5%), PLs (40.5%), GRs (53%), and hyaline hemocytes (6%). In method 2, PLs were greatly reduced (18%) owing to their propensity for adhering to vein walls (Arnold, 1966).

16.2.4. Coleoptera

Tauber and Yeager (1936) found that the THC was higher in *Phyllophaga* and *Osmorderma* larvae than in adults. The mean THC from third-stage *Melolontha* larvae was 4,000/mm³ (Collin, 1963). Cell counts from *Tenebrio* larvae varied from 9,200/mm³ to 128,000/mm³, with an overall average of 48,000/mm³. No differences in THC were noted, with an increase in length from 20 to 33 mm (Jones, 1950). The mean THC for *Popillia japonica* larvae was 25,000/mm³ (Schwartz and Townshend, 1968). No differences in THCs were observed among larvae, pupae, and adult *Leptinotarsa* (Arvy et al., 1948).

DHCs of *Tenebrio* larvae were obtained by Jones (1950). PLs and GRs (= coagulocytes, COs) were the principal hemocytes. The DHC was not changed materially as larvae developed within an instar (i.e., as larvae increased in weight). Collin (1963) studied the hemocytes of *Melolontha* larvae and found great individual variability of different types of cells. He felt that this variability made it difficult to use the DHC as a measure of health.

16.2.5. Diptera

PLs and crystal cells were found throughout the larval life of *Drosophila melanogaster* (Rizki, 1957). PLs accounted for 90–95% of the hemocyte population. During the early stages of pupation, podocytes

(POs) were observed. These cells increased in number as PLs decreased. During pupation, lamellocytes (LAs) also appeared, which the author (Rizki, 1962) observed transforming from PLs. In a subsequent study, Nappi and Streams (1969) demonstrated that crystal cells increased up to the last larval instar, decreased during the prepupal stage, and then disappeared. LAs were initially observed in mid-second-instar larvae. These cells increased up to the early third instar and then decreased. This decrease, the authors felt, was attributable to adhesion to tissue surfaces. During pupation, the LAs became mobile and became circulating cells.

PLs (including PRs) were present throughout the larval life of *Drosophila euronotus* (Nappi, 1970). In the first-instar larvae these cells were the only hemocytes present; in second-stage larvae OEs and LAs were observed. As the third stage progressed, PLs decreased to about 50%, as both OEs (25%) and LAs (17%) increased. In the prepupal stage, PLs (35%), OEs (32%), and LAs (34%) were present in nearly equal proportions.

Hemocytes were initially seen at the end of the first larval instar in *Calliphora erythrocephala* (Crossley, 1964). The number of hemocytes continued to increase during the larval stage. Very few hemocytes were in circulation in 2-day-old *Orthellia caesarion* larvae. During larval development, the number of hemocytes increased (Nappi and Stoffolano, 1972a). PLs (including PRs) increased during the experimental period, while the proportion of OEs decreased.

As in the two previous reports, hemocytes of *Musca domestica* and *M. autumnalis* larvae were located in the posterior part of the body (Nappi and Stoffolano, 1972b). During larval development, the number of hemocytes increased within the posterior region. At the time of pupation, hemocytes were distributed throughout the hemocoel (Nappi and Stoffolano, 1972b).

The hemocytes of the blowfly, *Sarcophaga bullata,* were studied in detail by Jones (1956, 1967b). Hemolymph was examined during larval development to pupation. PRs were found in all stages examined, but were rare. PLs were observed in all stages and were the only hemocytes seen in adults. These cells tended to clump rapidly in fresh, unfixed coverslip films. In general, the proportion of PLs decreased during larval development, whereas GRs increased. Spherulocytes (SPs) increased in proportion during the first 6 days of larval life, then decreased abruptly. In in vitro preparations, these cells degenerated to "empty" SPs. During the last larval stage, the THC increased from about 8,000/mm³ in the crop-full stage to about 34,000/mm³ in the brown-spiracled stage, just prior to pupation. The proportion of PLs decreased, while the GRs increased. Both types, however, increased numerically as larvae approached pupation. During the experimental period, blood volume decreased slightly, despite dramatic shifts in

THC. From these data, Jones concluded that the changes in blood cell populations could not be accounted for by changes in blood volume (Jones, 1956). During the time from prepupal to brown pupal stage, the THC decreased from 34,000/mm³ to 12,000/mm³. GRs disappeared prior to adult emergence.

In a subsequent study, Jones (1967b) calculated the numbers of different hemocytes within the entire hemocoel. These values were obtained by multiplying the blood volume (in microliters) and the number of hemocytes (per microliter). The absolute number of hemocytes increased from about 500,000 in feeding larvae to about 2,300,000 in brown-spiracled prepupae. From this time, the numbers decreased to about 540,000 in black pupae. Jones showed that large numbers of PLs and GRs came into circulation between the crop-full stage and the brown-spiracled stage. No correlations were found between changes in the populations of PLs, GRs, and SPs at any time. From these data, Jones concluded that PLs did not transform to GRs or SPs as Gupta and Sutherland (1966) claimed.

16.2.6. Lepidoptera and Hymenoptera

A fair amount of literature dealing with changes in hemocyte populations involves the Lepidoptera. Emphasis will be placed on the wax moth, *Galleria mellonella*, because of the relatively large amount of information available. In *Prodenia eridania*, Rosenberger and Jones (1960) showed that the THC did not vary significantly in sixth-instar larvae as they approached pupation. Counts from heat-fixed larvae ($\bar{x}$ 27,700) were higher than those from unfixed larvae ($\bar{x}$ 17,200). Yeager (1945) described 10 cell classes and 32 hemocyte types and followed changes from first-instar larvae to adults. Especially noted were changes in the form of the hemocyte from passive to active; for example, "the disappearance of spindle ends and a rounding up or spreading of the cell." The passive-active transformation (F/N) increased until the fifth and sixth instars, then decreased in the prepupal stage. At pupation, the F/N ratio increased rapidly. The development of *Prodenia*, as evidenced by changes in the blood picture, could be divided into early larval, late larval, metamorphic, and imaginal phases. Cell numbers were higher in larvae than in adults.

In *Trichoplusia ni* larvae, counts varied from about 14,000/mm³ to 25,000/mm³ with no apparent trend (Laigo and Paschke, 1966). Little change in cell numbers occurred in 6- to 10-day-old *Heliothis zea* larvae. Counts varied from 25,000/mm³ to 31,000/mm³ (Shapiro et al., 1969). SPs increased in numbers in 7- to 9-day-old larvae; then decreased. THCs increased in larvae of *H. virescens* from 12,000/mm³ to about 31,000/mm³ (Vinson, 1971). SPs increased from 38% in 5-day-old larvae to 59% at day 8 and then decreased. PRs and PLs decreased from day 5 to day 8 and then increased up to pupation; OEs remained fairly stable at 1–2%. THCs in the armyworm, *Pseudaletia unipuncta*, were high in molting fifth-instar larvae (ca. 40,000/mm³). After the molt, the cell population decreased to 20,000/mm³ and continued to fall until prepupation (Wittig, 1965b). GRs sometimes reached a peak at the sixth instar and then declined. PLs, on the other hand, increased during the sixth instar and during pupation. SPs showed a peak in early sixth instar and then declined. The author concluded that THCs were a more reliable index than DHCs of developmental changes during the sixth instar.

Arnold and Hinks (1976) studied the multiplication of circulating hemocytes in the noctuid *Euxoa declarata*. In general, cell numbers increased from 6,000/mm³ to 20,000/mm³ in the second to sixth instars. All cell types (PRs, PLs, GRs, SPs, and OEs) occurred in higher numbers in the sixth-instar larvae. Changes in proportions of cell types were also followed; these did not always parallel changes in numbers.

It is difficult to generalize concerning the data obtained from *Prodenia, Trichoplusia, Pseudaletia,* and *Euxoa,* although all the genera belong to a single family, Noctuidae. In most cases, the values cited were from control insects during a specified experimental period. Changes in the proportions of hemocyte types also occurred and were often dramatic (Yeager, 1945; Wittig, 1965b; Arnold and Hinks, 1976).

THCs and DHCs of nondiapause larvae and 5-day-old pupae of the pink bollworm, *Pectinophora gossypiella,* were studied by Clark and Chadbourne (1960). No significant differences were found between larval counts ($\bar{x}$ ca. 19,000/mm³) and pupal counts ($\bar{x}$ ca. 22,800/mm³). The proportions of PRs, PLs, ADs, and COs varied from stage to stage; this, the authors felt, was a reflection of hemocyte functions in food transport, storage, and metabolism.

Cell populations increased during larval development of the Mediterranean flour moth, *Anagasta kühniella,* and reached a peak during the prepupal period (73,500/mm³). During pupation, the numbers decreased (4,000/mm³) and increased slightly (6,000/mm³) during the first day of adult life. PR, PL, AD, and OE populations were followed during the life of the insect (from first-instar larvae to adults). PRs were present in high proportions (ca. 35%) in neonates, decreased during the second instar (6.4%), and remained few thereafter. PLs increased to a high level (78%) in fourth-instar larvae and then gradually declined. Subsequently (e.g., day 2 of pupation), their numbers increased throughout adulthood. ADs were "approximately the numerical reciprocal of PLs." OEs were observed rarely. Fusiform calls increased during the first four instars and fluctuated during the last two instars at about 30%. A decrease occurred during pupation and was high in newly eclosed adults. Arnold (1952a) felt that fusiform hemocytes were indicative of cell activity.

Cell populations from early fifth-instar *Pieris rapae crucivora* females appeared to be higher than those of later fifth-instar females (Kitano, 1969). PLs (40%), GRs (54%), PRs (1%), and OEs (4.4%) were identified in fifth-instar *Pieris* (Takada and Kitano, 1971).

The THC of the silkworm, *Bombyx mori*, reached its peak at every molt. The highest cell density (ca. 8,000/mm^3) was reached during the fifth instar and subsequently declined. After adult emergence, the THC reached its lowest level (ca. 200–300/mm^3) (Nittono, 1960). PLs occurred in high numbers during active growth at each instar. These cells decreased during the fifth instar, increased before adult eclosion, and attained a maximal value (60–70%) in adults. GRs (including both ADs and GRs) reached a peak at each molt (60–70%). Minimal values (10%) were observed in adults. Larval PRs and SPs were not observed after the midstage of pupation. Adult SPs, however, were recognized. OEs usually decreased after pupation.

Nittono (1960) examined the larval blood cells of 301 silkworm strains. In 26 strains, SPs were not present, and the author concluded that these strains produced less silk than those strains containing SPs. Depending on the insect strain, the SP-lacking character was recessive (H-5 minor) or dominant (Shisen-sanmin).

Large variations in cell numbers occurred from the fourth instar to adulthood in *Hyalophora cecropia* silkworms (Lea, 1964). Counts decreased at pupation, remained low in newly emerged adults (4,000/mm^3), and were lowest (500–1,500/mm^3) in older adults. PLs and GRs made up more than 90% of the total hemocyte population. PLs were predominant in late fifth-instar larvae and in adults; GRs were predominant in the fourth and early fifth instars. OEs were relatively scarce, and SPs occurred at the time of cocoon spinning. In a subsequent study, Lea and Gilbert (1966) further classified cells as PRs and ADs. All the cell types, including vermiform cells (VEs), were also observed in *Samia cynthia* and *Antheraea polyphemus*.

Stephens (1963), using last-instar larvae of *Galleria*, obtained a mean value of 32,200 circulating hemocytes per cubic millimeter from unfixed larvae. The variation was great, as the standard error of 1,200 indicates. Jones (1967a) followed hemocyte populations from the eleventh through the twenty-first day of larval life by measuring THCs, DHCs, and hemolymph volumes. In both heat-fixed and unfixed larvae, the THCs increased during larval development, although counts from heat-fixed ($\bar{x}$ 51,500/mm^3) were significantly higher than those from unfixed ($\bar{x}$ 28,200/mm^3) larvae. Counts from unfixed 20- to 22-day-old larvae (32,000–37,700/mm^3) were comparable to the 33,200/mm^3 obtained by Stephens (1963). On the basis of his data, Jones (1967a) disagreed with Stephens's statement that heat fixation did not affect the counts. Preliminary data suggested that cell populations continued to increase during cocoon spinning and decreased in the prepupae. THCs were determined in fifth-, sixth-, and seventh-instar larvae of *G. mellonella* (Shapiro, 1966) (Table 16.1). The THC increased during larval development from 20,724/mm³ among fifth-instar larvae (11–20 mg) to a high of 47,220/mm³ among seventh-instar larvae (301–360 mg). In this study, larvae were randomly selected from rearing colonies, weighed, and their head width measurements taken. Because a high positive correlation ($r = \cdot + 0.798$) was found between larval weight and head capsule width, and because head width increases in a linear fashion during larval development and the larval weight of *Galleria* increases with age in a linear fashion (Beck, 1960), it was considered justified to utilize larval weight as a criterion for larval development and larval age. However, this criterion is valid only when rearing conditions are optimal.

There have been numerous studies on the DHC of the wax moth. The main aim of the work in the 1920s and 1930s was to elucidate

Table 16.1. Total hemocyte counts in *G. mellonella* larvae during development

Weight class (mg)	Instar	No. of larvae	THC (cells/mm³) Mean	Standard deviation
11– 20	Fifth	25	20,724	2,378
21– 30	Sixth	27	23,400	1,982
31– 40		25	22,487	3,278
41– 50		30	32,036	3,512
51– 60		31	27,362	3,744
61– 70		25	29,315	3,239
71– 80	Seventh	27	32,513	3,765
81– 90		25	31,719	3,840
91–100		48	27,542	3,028
101–110		25	30,925	4,296
111–120		25	26,351	3,088
121–130		38	32,873	3,628
131–140		36	28,770	3,712
141–150		25	28,285	3,479
151–160		25	36,799	2,783
161–170		25	33,968	3,075
171–180		26	35,382	3,075
181–190		25	32,539	3,242
191–200		25	28,975	4,389
201–210		25	38,130	3,381
211–220		32	35,742	2,733
221–230		29	38,563	3,360
231–250		34	41,015	3,642
251–270		30	42,079	4,105
271–300		27	41,561	3,919
301–360		15	47,220	3,553

changes in the blood formula during bacterial infection and immunization. An examination of the literature revealed that much of this work was done by Serge Metalnikov, a colleague of Elie Metchnikoff, and co-workers at the Pasteur Institute in Paris. In spite of the vast amount of data collected, it is difficult to appraise the results owing to the absence of controls and the small sample sizes used. Subsequent to Metalnikov's work (1908, 1922, 1927), *Galleria* larvae were used for blood pathology studies, but it was not until the 1960s that the wax moth was used in hematological studies without reference to immunological aspects (Ashhurst and Richards, 1964; Jones, 1964, 1967a; Shrivastava and Richards, 1964, 1965; Jones and Liu, 1968).

A listing of the hemocyte classification systems of Metalnikov and other investigators is shown in Table 16.2. It can be seen that confusion may occur when one classification system is compared with another.

Jones (1967a) recognized five classes of hemocytes in unfixed hemolymph: PRs, PLs, SPs, ADs, and OEs. GRs, COs, POs, and VEs were not seen. During the active feeding phase, PRs and PLs were the predominant hemocytes (95%). During the nonfeeding, pre-cocoon-spinning period, ADs were observed. As spinning continued, the proportion of ADs increased, reached a peak in lightly cocooned larvae (ca. 16%), and declined. Mature ADs increased to a maximum of 57% in newly formed pupae. SPs decreased during prepupation and were not observed in pupae. As larvae transformed into pupae, the percentage of PLs decreased to about 40%.

Shapiro (1966), utilizing May-Grünwald-Giemsa-stained smears from heat-fixed larvae, recognized PRs, PLs, GRs, ADs, SPs, OEs, and POs. The DHCs from sixth- and seventh-instar larvae are summarized in Table 16.3. Note that the highest percentage of PRs occurs from the middle to the end of the sixth instar (weight classes 41–70 mg) and the period prior to pupation (weight classes 211 and over). At no time did

Table 16.2. Hemocyte types in *G. mellonella* larvae described by Metalnikov (1927), Cameron (1934), and in present study, using Yeager-Jones system

Metalnikov (1927)	Cameron (1934)	Yeager-Jones
Lymphocytes	Lymphocytes	Prohemocytes
Lymphocytes, proleucocytes	Lymphocytes, leucocytes	Plasmatocytes
Leucocytes	Leucocytes	Granular hemocytes
Spherule cells, empty spherule cells	Leucocytes, spherule cells	Adipohemocytes
Spherule cells	Spherule cells	Spherule cells
Oenocytes	Oenocytes	Oenocytoids
?	?	Podocytes

Table 16.3. Mean differential hemocyte counts in *G. mellonella* larvae during development

Weight class (mg)	No. of insects	PRs	Round PLs	Fusiform PLs	Irregular PLs	GRs	ADs	SPs	POs	OEs	Unclassified cells	Degenerating cells	Cells in mitosis
21– 30	32	0.6	66.1	23.9	0.7	4.4	0.02	2.2	0.05	0.05	0.05	0.4	0.15
31– 40	20	1.0	63.9	27.1	0.6	6.7	0.30	2.5	0.03	0.15	0.10	0.4	0.10
41– 50	24	2.8	53.6	31.1	0.6	7.7	1.50	1.4	0.04	0.04	0.25	0.7	0.15
51– 60	18	1.5	53.6	30.1	0.8	8.1	1.40	1.0	0.10	0.10	0.10	0.6	0.15
61– 70	29	2.2	56.1	30.8	0.7	7.8	1.00	0.8	0.00	0.05	0.10	0.7	0.10
71– 80	21	1.1	55.2	33.6	0.6	6.8	0.45	1.3	0.00	0.05	0.10	0.5	0.15
81– 90	24	0.5	54.2	35.5	0.6	7.1	0.45	0.5	0.00	0.05	0.15	0.5	0.15
91–100	45	0.4	54.1	32.0	0.5	10.8	0.05	1.6	0.03	0.10	0.05	0.5	0.10
101–110	23	0.2	50.5	35.0	0.4	11.5	0.05	1.7	0.05	0.05	0.05	0.5	0.15
111–120	21	0.2	53.8	32.3	0.4	10.7	0.05	1.8	0.00	0.05	0.05	0.5	0.10
121–130	40	0.2	49.6	37.7	0.5	10.0	0.10	1.3	0.04	0.10	0.10	0.5	0.10
131–140	33	0.3	46.6	36.7	0.4	14.2	0.80	1.1	0.05	0.15	0.05	0.4	0.10
141–150	20	0.2	47.2	38.4	0.4	12.7	0.40	0.6	0.00	0.05	0.05	0.4	0.10
151–160	22	0.2	43.9	40.3	0.5	12.8	0.85	1.1	0.03	0.10	0.05	0.4	0.10
161–170	29	0.2	45.0	43.8	0.4	11.1	1.55	1.4	0.04	0.20	0.05	0.4	0.10
171–180	23	0.3	42.6	42.7	0.4	11.3	1.55	1.7	0.00	0.15	0.05	0.3	0.15
181–190	26	0.2	40.6	42.2	0.3	12.9	1.75	1.5	0.02	0.25	0.10	0.4	0.10
191–200	20	0.2	41.6	42.5	0.4	10.8	2.10	1.9	0.00	0.35	0.05	0.4	0.05
201–210	18	0.2	38.2	47.1	0.5	9.4	1.15	1.2	0.05	0.40	0.10	0.4	0.10
211–220	30	0.5	35.2	50.5	1.0	8.2	1.20	0.7	0.00	0.20	0.15	0.5	0.10
221–240	25	0.1	32.9	57.4	0.7	10.3	2.35	0.3	0.05	0.25	0.15	0.5	0.05
241–260	25	0.1	25.6	53.3	0.5	13.5	3.35	0.0	0.15	0.10	0.10	0.7	0.10
261–290	11	0.8	25.5	50.0	1.0	12.5	4.10	0.0	0.20	0.20	0.10	0.6	0.05

the proportion of these cells exceed 2.8%, and the proportion varied only from 0.20 to 0.30% among larvae weighing 101–210 mg. The proportion of irregular PLs varied from 0.3% to 1.0% during larval development. As the larva matured, the proportion of round PLs decreased; the reverse was true for the fusiform PLs. A point is reached (weight class 171–180 mg) when the ratio of round to fusiform PLs is equal to 1 (Table 16.4). Hollande and Aghar (1928) also noted that hemocytes can become fusiform as lepidopterous larvae mature. The percentage of GRs increased until the mid-seventh instar (weight class 131–140 mg). From that time on, the proportion of these cells varied between 8.2% and 13.5%. The ADs increased in proportion during larval development, attaining a high of 4.1% in larvae weighing 261–290 mg. The highest percentage of SPs (2.5) occurred at the beginning of the sixth larval instar (weight classes 21–40 mg). In larvae weighing 41–220 mg, the proportion varied between 0.5% and 1.9%. At that time, the number of SPs decreased until none could be found in larvae weighing 241 mg. The percentage of OEs was greater in seventh-instar larvae (0.05–0.40) than in sixth-instar larvae (0.04–0.15). In general, the proportion of OEs increased as the larva in-

Table 16.4. Ratio of round to fusiform plasmatocytes in *G. mellonella* larvae during development

Weight class (mg)	No. of insects	Round/fusiform plasmatocytes
21– 30	32	2.77
31– 40	20	2.36
41– 50	24	1.72
51– 60	18	1.78
61– 70	29	1.82
71– 80	21	1.65
81– 90	24	1.53
91–100	45	1.69
101–110	23	1.44
111–120	21	1.66
121–130	40	1.31
131–140	33	1.27
141–150	20	1.22
151–160	22	1.09
161–170	29	1.03
171–180	23	1.00
181–190	26	0.96
191–200	20	0.98
201–210	18	0.78
211–220	30	0.70
221–240	25	0.64
241–260	25	0.48
261–290	11	0.48

creased in weight. The proportion of POs was very low, ranging from 0 to 0.2%, with the highest value occurring in larvae weighing 261–290 mg. Degenerating cells were not limited to any particular class of hemocytes, but were found among all types. The percentage of these cells varied little throughout the sixth and seventh instars, ranging from 0.3 to 0.7. A low proportion (0.05–0.25%) of cells could not be classified with certainty during the last two instars.

It is difficult to compare the results obtained by Jones (1967a) and Shapiro (1966) because different methods were used (e.g., unfixed vs. heat-fixed preparations; unstained vs. stained material; age in days vs. weight). These differences led to (1) combining PRs and PLs and (2) nonobservance of GRs by Jones. When the same hemocytes were recognized (e.g., ADs and SPs), the trends were similar. Moreover, both workers noted an increase in the number of circulating cells per cubic millimeter during larval development. Neuwirth (1973) studied the ultrastructure of *Galleria* hemocytes and identified four cell types: plasmatocytoids, GRs, SPs, and OEs.

At this time, it would be of interest to compare results obtained from *Galleria* and other insects. In *Galleria* (Shapiro, 1966) and in *Prodenia* (Yeager, 1945), a peak in PRs occurred in the middle of the penultimate instar. The proportion of PRs then decreased and subsequently increased as the last-instar larvae approached prepupation. In Nittono's (1960) study of the silkworm, peaks occurred 24 hr after each molt. The percentage of PRs in *Euxoa* reached a maximum during the third instar and thereafter decreased (Arnold and Hinks, 1976). In *Galleria*, *Prodenia*, and *Bombyx* a reduction in PRs occurred during the last instar up to prepupation, when an increase occurred. Hemocyte changes in *Euxoa* were followed only during larval development.

The significance of the increase in fusiform PLs and reciprocal decrease in round PLs during larval development, under optimal conditions, is not known. It would be interesting to determine the ratio of round to fusiform PLs under various environmental and physiological conditions to ascertain whether PLs respond by changes in the ratio. Such changes could serve as an index of "health."

Yeager (1945) also obtained an increase in GRs in *P. eridania* up to the middle of the last instar. In this insect, the proportion of GRs then decreased, whereas in *Galleria* a plateau was reached (Shapiro, 1966), and in *B. mori* the proportion of GRs increased during the last instar (Nittono, 1960). In the larvae of *Euxoa*, these cells occurred in greater proportions than any other hemocyte type. These cells decreased from 49% to 31% in second-instar larvae and then increased during larval development.

In both *Anagasta* (Arnold, 1952a) and *Prodenia* (Yeager, 1945), the greatest percentage of ADs occurred in pupae. In *Galleria*, a spectacular increase in the proportion of ADs, from 0 to 69%, occurred when larvae began to spin cocoons (Jones, 1964). In a subsequent study (Jones, 1967a), ADs were initially observed in 17-day-old larvae. In general, the proportion of these cells increased during cocoon formation. Shapiro (1966) observed an increase in ADs during larval development. In *Hyalophora cecropia*, ADs included diverse cells. Even though the subclasses were often indistinguishable when fixed and stained, most cells were easily categorized under the phase-contrast microscope. Adipohemocytes of subclass I, most numerous in larvae, were found in all stages; those of subclass II occurred only in pre-pupae and newly molted pupae through the first day of adulthood (Lea and Gilbert, 1966).

Trends similar to those obtained in *Galleria* (Shapiro, 1966; Jones, 1967a) were found in *Sarcophaga* (Jones, 1956), *Bombyx* (Nittono, 1960), and *Pseudaletia* (Wittig, 1965b) for SPs; in *Prodenia* (Yeager, 1945), *Tenebrio* (Jones, 1950), *Bombyx* (Nittono, 1960), and *Euxoa* (Arnold and Hinks, 1976) for OEs; and in *Prodenia* (Yeager, 1945) for POs. However, the functions of these hemocytes are not well known, and it is difficult to explain the results.

Wille (1974) investigated changes in hemocytes of *Apis* during bacterial septicemia, rickettsiosis, nosematosis, and mixed infections. No correlations could be made between changes in cell morphology and disease, but the author concluded that changes in hemocyte morphology might be the result of extreme physiological changes. Zapol'skikh (1976) classified hemocytes of larvae, pupae, and adult workers of *A. mellifera* as PRs, PLs, OEs, and SPs (= ADs?). The PRs divided and developed into PLs, OEs, and SPs. PLs were differentiated into larval, larval and pupal, and pupal and adult types. SPs were described in pupae and adults.

16.3. Blood volume

Florkin (1937) investigated changes in blood volume (BV) of *Bombyx* from the start of spinning to eclosion. Up to the eleventh day after spinning, the BV was nearly constant. It then fell and remained at a constant level from the twelfth day up to the time of eclosion, when another reduction of volume occurred. The THC increased greatly at each molt without any decrease in BV (Nittono, 1960). At the fifth instar or during an early stage of cocoon spinning, depending on the strain of *Bombyx* studied, the THC reached its peak. Nittono felt that the increase in THC was attributable to a real increase of circulating hemocytes as well as to a reduction of BV.

During the development of *Sarcophaga*, the BV decreased (Jones, 1956, 1967b). The BV was expressed as percent of body weight (Jones, 1956) or as volume (in microliters) (Jones, 1967b). The amount of hemolymph in fifth-stage *Rhodnius* decreased from 3–4 μl on the day of ecdysis to less than 1 μl on day 7 (Jones, 1967c). The BV increased in *Euxoa* during larval development from 0.69 mm³ (second instar) to 127 mm³ (Arnold and Hinks, 1976). In 1961, Lee examined the variation of BV with age in the desert locust, *Schistocerca gregaria*, using the dye method. The BV per unit weight fell during the period following ecdysis and rose thereafter. In 1962, Wheeler, using the dye method, showed that the increase in BV of *Periplaneta* at ecdysis was statistically significant.

In *Galleria*, the BV (as percent of body weight) did not change significantly in seventh-instar larvae, whereas the amount of blood (in microliters) increased as the larvae gained weight (Shapiro, 1966). Jones (1967a) found that the BV (as percent of body weight) remained unchanged (ca. 34%) during the feeding period and then gradually decreased in newly formed pupae (ca. 16%). BVs from both unfixed and heat-fixed larvae were identical.

Two earlier reports on the BV of *Galleria* larvae were published (Richardson et al., 1931; Zielinska and Wroniszewska, 1957). In neither case was the dye method used; instead, the amount of hemolymph was determined by subtracting the weight of the larva after bleeding from the weight of the intact larva. Richardson et al. (1931) obtained a mean value of 41% for 16 larvae ranging in weight from 76.5 to 221 mg. The variation was great: from a low of 20.4% to a high of 58.8%. Zielinska and Wroniszewska (1957), using last-instar larvae with a mean weight of 180 mg, obtained a BV of 28.3 ± 2.3%.

16.4. Interrelationships among total hemocyte counts, differential hemocyte counts, and blood volume

To appreciate the changes in hemocyte populations and hemocyte types, it is necessary to combine the THC and/or DHC and/or BV measurements. By using the THC and DHC, the numbers of particular hemocytes per cubic millimeter during larval development can be approximated (Shapiro, 1966; Arnold and Hinks, 1976).

In *Galleria*, several phenomena would not have been revealed by the sole use of the DHC data (Tables 16.3, 16.5): (1) peak of PRs at the end of the sixth larval instar; (2) increase in GRs through larval development; (3) sudden decrease in SPs; (4) increase in POs, OEs, irregular PLs, and degenerating cells; and (5) increase in cells undergoing mitosis in larvae weighing 51–120 mg (i.e., from late sixth to early seventh instar) (Shapiro, 1966). These observations make clear that more meaningful results may be obtained by considering both DHC and THC, as a change in the proportion to a given cell type present may or may not be associated with a parallel change in total numbers of these hemocytes.

The absolute number of circulating hemocytes per cubic millimeter

Table 16.5. Total number of hemocytes per cubic millimeter in G. mellonella larvae during development

Weight class (mg)	No. of insects	Cell classes											
		PRs	Round PLs	Fusiform PLs	Irregular PLs	GRs	ADs	SPs	POs	OEs	Unclassified cells	Degenerating cells	Cells in mitosis
21– 30	32	124	17,062	4,968	189	828	51	414	14	14	14	84	41
31– 40	20	225	14,440	6,075	135	1,575	68	563	7	34	23	94	23
41– 50	24	396	17,280	9,920	192	2,560	160	448	13	13	80	158	34
51– 60	18	411	14,696	8,220	219	2,192	384	274	27	27	27	192	48
61– 70	29	635	16,408	9,083	205	2,344	293	234	0	15	29	192	27
71– 80	21	358	17,875	11,050	195	2,175	166	413	0	16	33	147	44
81– 90	23	159	17,118	11,412	190	2,219	143	159	8	16	48	163	49
91–100	45	110	14,850	8,800	138	3,025	14	360	16	28	16	159	48
101–110	23	62	15,759	10,815	124	3,708	16	525	0	16	16	155	48
111–120	21	53	14,256	8,448	106	2,904	13	475	13	66	13	132	40
121–130	40	66	16,450	12,502	165	3,290	33	420	14	33	33	165	33
131–140	33	86	13,536	10,656	115	4,032	230	317	0	43	14	115	29
141–150	20	57	13,201	10,754	113	3,679	113	170	11	14	14	113	28
151–160	22	74	16,192	14,720	184	4,784	313	405	14	37	18	147	37
161–170	29	68	15,300	14,960	140	3,740	427	476	0	68	17	136	34
171–180	23	106	16,222	16,222	142	3,894	539	682	7	18	18	102	17
181–190	26	65	13,325	13,650	98	4,125	569	488	0	81	33	142	33
191–200	20	60	12,180	12,470	130	3,190	609	551	0	92	16	130	16
201–210	18	76	14,478	17,907	145	3,429	438	457	0	152	29	116	29
211–220	30	178	12,495	18,207	381	2,956	428	250	19	71	57	191	38
221–240	25	39	12,738	22,002	250	3,860	907	116	57	97	54	179	18
241–260	25	41	10,660	21,730	193	5,740	1,196	0	19	36	39	270	39
261–290	11	337	11,946	21,050	410	5,473	1,681	0	82	84	41	246	21

can be calculated from the THC and BV (Smith, 1938; Shapiro, 1966; Jones, 1967b,c; Arnold and Hinks, 1976). The total blood cell content of an average American cockroach was estimated to be about 16.5 million cells (Smith, 1938). Wheeler (1963) calculated that nymphs and adults of *Periplaneta* contain between 9 million and 13 million cells. The total number of hemocytes in *Sarcophaga* increased from 415,000 in feeding-stage larvae to 2.3 million in brown-spiracled larvae; subsequently, the hemocyte population decreased to 540,000 during pupation (Jones, 1967b). In *Galleria,* the absolute number of hemocytes per cubic millimeter increased from 2.2 million in larvae weighing 151–160 mg to 3.97 million in larvae weighing 231–240 mg (Shapiro, 1966) (Table 16.6). Jones (1967a) calculated that 831,000–2,448,000 cells are available in the hemolymph of 15- 21-day-old unfixed larvae. Subsequently, Werner and Jones (1969) estimated that 113,000–1,570,000 cells are in circulation within last-instar larvae. In *Euxoa,* the total number of hemocytes in circulation increased from 4,457 in second-instar larvae to 2.45 million in sixth-instar larvae (Arnold and Hinks, 1976).

Webley (1951) examined the THC and BV of *Locusta migratoria migratorioides* and attributed the higher concentration of blood cells in older adults to a decrease in BV rather than to an increase in hemocytes. Wheeler (1962, 1963) showed that an increase in BV of *Periplaneta* was accompanied by a decrease in the number of circulating hemocytes per cubic millimeter at ecdysis. By multiplying the THC and the BV, it was found that the absolute number of circulating hemocytes was relatively constant.

It has been reported that an inverse relationship exists between the THC and the BV in *Bombyx* larvae (Nittono, 1960), in adults of *Locusta* (Webley, 1951), and in last-stage larvae and adults at ecdysis of

Table 16.6. *Absolute number of circulating hemocytes in seventh-instar G. mellonella larvae*

Weight class (mg)	Mean blood volume (% body weight)	Mean THC (cells/mm³)	Absolute number of circulating hemocytes (cells/mm³)
151–160	38.5	36,800	2,196,040
161–170	40.5	34,000	2,272,050
171–180	38.9	35,400	2,409,155
181–190	37.2	32,500	2,296,650
191–200	37.5	29,000	2,123,525
201–210	37.1	38,000	2,890,090
211–220	37.4	35,700	2,870,637
221–230	37.1	38,600	3,222,135
231–240	35.6	41,528	3,970,077

Periplaneta (Wheeler, 1962, 1963). In *Sarcophaga*, however, Jones (1956) concluded that changes in the THCs could not be accounted for by the slight changes in BV. In *Bombyx*, Nittono (1960) felt that an increase in THC at the start of cocoon spinning was attributable to both increases in the number of blood cells in circulation and decreases in the BV. In *Galleria* larvae, the THC increased without a significant decrease in BV. Thus it appears that an increase in the THC is not merely a reflection of a decrease in BV. In addition, the absolute number of circulating hemocytes per cubic millimeter did not remain constant in spite of changes in the THC and BV, but increased steadily from 2.2 million to 4 million (Shapiro, 1966).

As it has been shown that hemocyte populations may change quantitatively and qualitatively during growth and development, I will endeavor to analyze the factors involved in these changes.

16.5. Cyclical changes

16.5.1. Ecdysis

Changes in THC, DHC, BV, and mitotic index (MI) have been associated with the molt. Patton and Flint (1959) noted that the THC decreased significantly at the molt in *Periplaneta;* this finding was contrary to the report of Tauber and Yeager (1935). The decrease in THC began about 48 hr prior to the molt and became more striking during the next 24 hr (Patton and Flint, 1959). Following the molt, cell population changes were also noted in *Locusta* (Webley, 1951), *Rhodnius* (Wigglesworth, 1955), *Sarcophaga* (Jones, 1956), *Bombyx* (Nittono, 1960), and *Periplaneta* (Wheeler, 1963). Yeager (1945) noted ill-defined morphological changes in the hemocytes at the larval molts, especially into the fourth, fifth, and sixth larval instars. No consistent changes in the proportions of PRs, PLs, GRs, and OEs were observed in *Rhodnius* at ecdysis. Following the molt, however, PLs increased and GRs decreased, and the BV declined in the fasting insects (Jones, 1967c). The proportion of mitotically dividing cells in the cockroach, *Blatta orientalis*, decreased prior to and during molting and rose after ecdysis. The maximum percentage usually occurred on the third day after ecdysis (Tauber, 1935, 1937).

During the period following ecdysis, the BV per unit weight fell and rose thereafter in *Schistocerca* (Lee, 1961). The decrease in THC at the molt was attributed to an increase in BV. At the time of the molt, the coagulability of the blood increased. Although changes in the number of circulating hemocytes might be associated with changes in BV, this relationship may not be clear-cut. Jones (1956) and Arnold (1966) felt that decreases in THC were attributable to the adherence of hemocytes to tissue surfaces. Nonetheless, changes in hemocyte populations during ecdysis must be recognized and taken into account.

16.5.2. Periodicity

Arnold (1969) studied changes in hemocyte populations of the giant cockroach, *Blaberus giganteus*, during the light-dark cycle. The proportions of SPs and GRs changed in association with the light-dark cycle in 25% of the insects tested. Proportions of PRs and PLs did not change during the light-dark cycle. Although clear-cut evidence for a diurnal response could not be demonstrated, Arnold cautioned that "there is a potential for periodicity in the haemocyte complex and a need to consider it in the planning of experiments."

Fluctuations of the THC were examined in a group of sixth-instar *Pseudaletia* larvae during the third day after molt (Wittig, 1966). One group of control larvae showed two minima at 12:15 P.M. and 5:25 P.M. (PST); in other control groups minima were observed at about 1,5,8, and 11 P.M. (PST). These minima could not be explained, but the author felt they were associated with the rhythm of feeding. The blood picture of *Rhodnius*, following a blood meal, was different from that of unfed bugs: The proportion of PLs decreased, whereas GRs and OEs increased. In addition, the BV increased. On the basis of these observations, Jones (1967c) felt that the proportions of hemocytes were affected by ecdysis and by the nutritional state of the insect.

Although differences in the MI of hemocytes of fifth-instar *Oncopeltus* were observed throughout the instar, Feir and McClain (1968a) could find no diurnal rhythm of activity. Moreover, great variations in the MI were found from hour to hour, which might have been caused by the use of different insects for each determination.

16.6. Changes under other conditions

16.6.1. Mitosis

It is generally agreed that not all types of hemocytes divide mitotically. In *Prodenia*, Yeager (1945) found that most of the mitotically dividing figures were observed among the PLs, smooth-contour chromophils, and ADs, and only occasionally among GRs and PRs. Arnold (1952a) obtained similar results in *Anagasta* larvae. In *Sarcophaga*, mitosis occurred primarily in PRs (Jones, 1956). In *Galleria*, mitosis was observed in PRs, PLs, and GRs (Shapiro, 1966, 1968a); Jones and Liu (1968) found mitosis only among plasmatocyte-like cells. In

Euxoa, PRs, GRs, and SPs were observed to divide mitotically (Arnold and Hinks, 1976). PLs rarely divided, and OEs were never seen to divide.

Some authors have included the percentage of mitotically dividing cells. Yeager and Tauber (1932) reported that about 5 out of 1,000 hemocytes divided in the roach, *Periplaneta fuliginosa*. In a subsequent study (1933), the same authors reported that 0.2–1.0% of the hemocytes divided mitotically in nymphs of *P. orientalis*. In *P. americana*, no cell division was observed in some roaches (Taylor, 1935). An average of 0.2% of the hemocytes divided mitotically (range 0.00–0.50%) in nymphs and adults of *Blatta orientalis* (Tauber, 1937). Tauber's values were not affected by age, sex, or fluctuations in laboratory temperatures between 17 °C and 32 °C. At the molt, the mitotic count decreased and tended to remain low (0.00–0.15%) for about a day. Then the mitotic activity reached a maximum on the third day postmolt (0.65–1.00%), and decreased on day 4. Feir and McClain (1968a) observed that the mitotic activity in *Oncopeltus* was low following the molt, began to increase after about a day, and reached a peak in the second day postmolt. The index remained high until the third day, then declined steadily until the end of the stadium.

Rooseboom (1937) noted that mitotic activity in the stick insect, *Carausius*, was most active between 3 A.M. and 9 A.M. No significant increases in the mitotic activity of *Galleria* hemocytes were observed during the hours of 8 A.M. to 4 P.M., owing to great individual variations (Jones and Liu, 1968).

In general, the normal MI is less than 10 cells per thousand (= 1%). Paillot (1924a) observed that in the caterpillar, *Euproctis chrysorrhoea*, the MI was 0.3–0.4%. Rooseboom (1937) observed a mean MI of 0.40% in *Carausius*. Yeager (1945) obtained values of 0.00–1.70% in *Prodenia* larvae. Among second- to sixth-instar larvae, the mean percentage was 0.481, with a range of 0.00–1.70%. In *Anagasta*, a mean value of 0.1–0.2% was obtained by Arnold (1952a), and this varied during larval development. In *Euxoa*, the MI varied from 0.4% to 0.87% during larval development, with no apparent trends. The maximum daily production of hemocytes, however, increased during larval development (Arnold and Hinks, 1976).

In the blowfly, *Sarcophaga*, mitotic figures were seen only in 1- and 2-day-old larvae (Jones, 1956); Crossley (1964) observed mitosis in third-instar *Calliphora* larvae. The MI in *Sarcophaga* ranged from 0.4% to 0.8% and from 0.8% to 1.0% in *Calliphora*.

Several reports have been made on the mitotic counts of circulating hemocytes in *Galleria*. Iwasaki (1927) stated that the normal percentage was 0.1–0.2; Tateiwa (1928) was unable to observe any mitosis of hemocytes. Jones (1962) found that mitotic divisions of circulating hemocytes may be absent in larvae. Shapiro (1966) examined the MI from 547 larvae reared in colonies and observed 127 mitotic figures. The PRs had the highest percentage of mitosis, followed by GRs and PLs. The proportion of cells in mitosis varied from 0.05 to 0.15% during the sixth and seventh instars, and the highest number of cell divisions occurred in late sixth- and early seventh-instar larvae (weighing 51–120 mg). Jones (1967a) noted that cells divided from day 15 to day 21 (i.e., from feeding to prepupation). The greatest numbers of mitotically dividing cells were observed in 17-day-old *Galleria*. In a subsequent study, Jones and Liu (1968) found that between the nineteenth and twenty-second days, more larvae (27%) had no mitotically dividing hemocytes than from day 14 to day 18 (7%). These authors found no difference in the MI among 14- to 17-day-old heat-fixed (0.27%) and unfixed larvae (0.21%).

The percentage of mitotically dividing cells of *Periplaneta orientalis* was 0.071 when only anaphase and telophase were included; the value increased to 0.5%, when prophase, metaphase, anaphase, and telophase were counted (Yeager and Tauber, 1933). Tauber (1935) calculated that the duration of telophase was about 6 times that of metaphase, about 1.7 times anaphase (which is 3.5 times metaphase), and about 1.2 times prophase (5 times metaphase). Telophase was the longest and metaphase the shortest phase in the mitotic cycle. In this study, Tauber assumed that there was a direct relationship between the relative duration of a phase of mitosis and the actual frequency of the stage and further concluded that the duration of the different phases was identical to the relation between their frequencies.

All stages of mitosis were observed by Yeager and Tauber (1933), Tauber (1935), Yeager (1945), and Feir and McClain (1968a,b). It was noted that prophase was the most difficult stage to identify (Yeager, 1945), and numerous prophase figures are undoubtedly not recognized from the mitotic counts (Feir and McClain, 1968a). Thus, prophase could not be determined with certainty and was omitted from the MI by Jones (1950), Clark and Harvey (1965), Lea and Gilbert (1966), Shapiro (1966, 1968a,b), and Jones and Liu (1968). Jones (1954) observed only metaphases from the hemolymph of unfixed *Tenebrio* larvae and metaphases, anaphases, and telophases from heat-fixed larvae. Metaphases accounted for 70% of divisions in *Galleria* larvae; anaphase and telophase occurred in equal proportions (Jones and Liu, 1968). Yeager (1945) noted that the MI did not become very high even when the majority of hemocytes were young or transitional. Yeager felt this might indicate that the hemocytes were derived from sessile cells or that too few hemocytes divided. In *Galleria*, Jones and Liu (1968) concluded that the low MI was sufficient to account for "the normal maintenance of the haemocytes." Moreover, the authors felt mitoses of circulating or sessile cells were the only means for hemocyte replenishment in *Galleria*. Yeager (1945) stated that the balance between degeneration and mitotic division, accounting for the absolute number of hemocytes, was not known in *Prodenia*. In *Galleria*, the number of cells in mitosis per cubic millimeter was compared with the number of degenerating cells per cubic millimeter during the last three larval instars, and it was found that there were from 2.0 to 10.7 more degenerating cells than cells in mitosis (Shapiro, 1966). At no time were there more cells in mitosis than degenerating cells. If mitoses of circulating hemocytes were sufficient to account for the increase in cell numbers, there would have to be a higher incidence of mitosis than of degeneration, for mitosis of circulating hemocytes would have to be great enough to replace the dying cells and greater still to account for the new cells. In addition, in larvae weighing 151–290 mg, when the absolute number of circulating hemocytes per cubic millimeter increased from 2.2 million to 4 million, the ratio of degenerating to mitotically dividing hemocytes per cubic millimeter increased from 4.0 to 10.2. These data suggest it is unlikely that mitosis alone could account for the increase in the number of hemocytes per cubic millimeter.

A limitation in this line of reasoning is that not all stages of mitosis were seen. As prophase was not distinguished, the mitotic counts were lower than would be the case if prophase were included. But, from the data available, it is still reasonable to conclude that even if prophase were included, mitosis would not be great enough to be solely responsible for the increase in circulating hemocytes (Shapiro, 1966). In *Euxoa*, Arnold and Hinks (1976) studied the multiplication of circulating hemocytes and concluded that most of the hemocyte production could be accounted for by mitotic division among circulating hemocytes. The authors further concluded that "(1) mitosis produces like cells; (2) a high mitotic index among like cells results in a large population of similar cells, unless they degenerate rapidly or are lost through differentiation to cells of another class; and (3) a low mitotic index or the absence of division among like cells implies that cells of that form originate from cells of another form or from a separate source." On the basis of their studies on mitosis, Feir and McClain (1968b) suggested that mitotic activity was under hormonal control, which set the time at which hemocytes responded to different treatments by an increase in mitosis (see also Chapters 3, 8).

16.6.2. Sex

In general, differences in either DHC or THC between sexes were not significant in *Periplaneta* (Smith, 1938; Ogel, 1955; Wheeler, 1963), *Tenebrio* (Jones, 1950), *Bombyx* (Nittono, 1960), *Hyalophora* (Lea, 1964), or *Pectinophora* (Clark and Chadbourne, 1960). In their studies on THCs in Orthoptera, Odonata, Hemiptera, and Homoptera

(1935) and in Neuroptera, Coleoptera, Lepidoptera, and Hymenoptera (1936), Tauber and Yeager found that the average count from females was higher than that from males. Some of the high female counts appeared to be associated with periods of oviposition. Webley (1951) observed no significant differences in THCs from male and female nymphs of *Locusta*, but adult males had higher cell counts than adult females.

As male larvae of *Anagasta* can easily be recognized after the third instar because of a pigmented testis, Arnold (1952a) used only males for his studies. In *Euxoa*, small differences occurred: fewer PLs and a greater proportion of GRs in the females. These differences, however, were not consistent throughout the test period.

16.6.3. Diapause

The majority of research on the hemocytes of diapausing insects, principally the cecropia moth, has dealt with wounding (see below). Aside from these reports, few data compare hemocyte populations from diapausing and nondiapausing insects. Clark and Chadbourne (1960), studying the hemocytes of the pink bollworm, *Pectinophora gossypiella*, found few differences in the THCs of nondiapausing and diapausing larvae and pupae. The proportions of PRs and PLs were lower, and those of ADs and GRs higher, in diapausing larvae. Raina and Bell (1974), also studying the hemocytes of the pink bollworm, distinguished seven classes of hemocytes: PRs, PLs, GRs, SPs, ADs, OEs, and POs. The authors noted that all cell types decreased in number during diapause. As diapause was terminated, the hemocyte numbers increased. The number of ADs in pharate pupae that developed from diapausing pupae was three times greater than in pharate pupae developing from nondiapausing larvae. The authors concluded that decreases in cell numbers during diapause were probably attributable to the low metabolic rate during diapause.

16.6.4. Starvation

Jones (1950) carried out a series of experiments on the effects of starvation on the hemocytes of last-instar *Tenebrio* larvae. Larvae starved for periods ranging from 1 to 30 days showed THCs within the normal range. There was a tendency for GRs to increase and PLs to decrease in number. After 120 days of starvation, there was an increase in PRs and a marked apparent hemocytopenia. In 1952, Jones and Tauber noted decreases in THCs and in the relative proportions of PLs after prolonged starvation.

In both *Leptinotarsa* (Arvy et al., 1948) and *Prodenia* (Rosenberger and Jones, 1960), THC increased during starvation. In both cases, the authors suggested that changes in BV were responsible for the changes in THC. Among fourth-instar *Anagasta* larvae starved for 24 hr, Arnold (1952a) found a significant increase in the relative numbers of ADs and degenerating cells, but a decrease in the number of fusiform PLs. The MI of ADs was also found to increase. Arnold concluded that the blood picture was similar to that found during prepupation.

Ovanesyan (1951) found that starvation reduced the number of fusiform hemocytes in *Bombyx* larvae, and Nittono (1960) reported that the THC in starved silkworm larvae decreased gradually and became lower than in the controls. The relative proportion of SPs increased as the starvation period increased, and sometimes these cells accounted for more than 50% of the total hemocyte population. In most of the SPs there was a decrease in the number of spherules but an increase in the size of each spherule. In *Galleria*, the THC among control insects increased, whereas that of starved insects decreased (Shapiro, 1966). Significantly more round PLs were found in starved larvae during the first 3 days. From this time on, however, more of these hemocytes were observed in control larvae. Both the percentage and the number of fusiform PLs were significantly higher in control than in starved larvae, owing to a greater decrease of these cells in starved larvae. Even though the percentage of GRs increased in both groups, a smaller number of these hemocytes occurred in starved insects owing to the significant decrease in THC. Other changes that occurred were (1) decrease in ADs, SPs, and cells in mitosis in starved insects; and (2) increase in degenerating cells in starved insects. During the course of the experiment (10 days), there were no significant changes in BV in the control larvae or in the starved larvae. When starved and control larvae were compared, no significant differences in BV were found.

These results suggest that changes in THC are not simply a consequence of changes in BV. The data indicate that the wax moth possesses some means of regulating its BV, and that this ability is maintained even under conditions of starvation. During starvation of *P. americana*, the BV did not change, and it was concluded that the regulation of the BV involved a transfer of water from tissues to blood (Yeager and Munson, 1950). It is possible that the same mechanism exists in starved *Galleria*.

16.6.5. Wounding

"The influence of a common lesion, e.g., in an injection or burn, has never been studied, but must be examined" (Rooseboom, 1937). Rooseboom found that the MI of *Carausius* was greatest between 3 and 9 hr after bleeding. In the same year, a classic study of wound healing in an insect (*Rhodnius*) was published (Wigglesworth, 1937). Hemocytes accumulated at the margin of the incision within a few hours. This accumulation was followed by the formation of "wound" tissue and continued until the wound was plugged and until any matter that had escaped into the hemocoel was completely surrounded (Day, 1952). Whereas Wigglesworth (1937) found frequent mitoses among the circulating hemocytes, Day (1952) found no such increase either 5 or 24 hr after wounding adult *Periplaneta*. Day hypothesized that hemocytes were probably mobilized from a previously sedentary state and that replacement occurred slowly by mitosis of hemocytes in the hemolymph. Ries (1932) observed that accumulation of blood cells did not take place until 2–3 days after wounding in *Tenebrio* larvae.

In the leafhopper, *Orosius argentatus*, the early stages of the wound-healing process are attributable to hemocytes; in *Periplaneta* healing of gut wound is caused by the combined action of hemocytes and the regeneration of midgut epithelium. In *Aedes aegypti*, only the simple epithelium and the gut musculature are involved (Day and Bennetts, 1953). In her study of the injury reaction of saturniid pupae, Cherbas (1973) concluded that an injury factor was released at the site of incision and that this factor was responsible for increased motility and adhesiveness among the hemocytes. This material was termed "haemokinin."

An increase of mitotic activity in the circulating hemocytes was observed following injury to diapausing *H. cecropia* pupae (Davis and Schneiderman, 1960; Harvey and Williams, 1961; Lea and Gilbert, 1961; Bowers and Williams, 1964; Lea, 1964; Clark and Harvey, 1965). Lea (1964) reported a decrease of 72% in the THC within the first 30 min after the initial wound. In the next 30-min period the count rose and continued to rise in succeeding intervals (4 hr, 24 hr, 48 hr). The THC in pupae injured repeatedly was significantly higher than in pupae injured only once, and the THC was higher in pupae in which a fresh incision was made each time than in pupae bled from the same incision each time (Harvey and Williams, 1961). After wounding, the circulating blood cells increased in number. Mitoses were not observed within the first 24 hr, nor was there any evidence of peak mitotic activity on any given day (Clark and Harvey, 1965).

In *Calliphora*, a small wound led to a smaller increase in mitotic activity than did a large wound (Crossley, 1964). At 18 hr after wounding, a marked increase in the MI was observed. *Tenebrio* larvae were subjected to multiple hemorrhages (Jones, 1954). After the first hemorrhage, the proportion of PLs decreased. Subsequently, the level of these cells returned to prehemorrhage levels. The proportions of OEs and GRs remained unchanged. An increase in the MI of *Oncopeltus* was noted following (1) a transverse incision in the midabdominal dorsum, (2) a longitudinal dorsal incision extending the length of the insect, or (3) small holes burned in the abdomen (Feir and McClain,

1968b). On the other hand, the blood picture of unfed or fed fifth-instar *Rhodnius* was unaffected even after 5 hemorrhages (Jones, 1967d).

The first study devoted to wound healing in *Galleria* was made by Cameron (1934). He reported an increase in PRs; a decrease in PLs, GRs, and ADs; and no change in the number of SPs 24 hr after hemorrhage of last-instar larvae. At this time, there were frequent mitoses in the PRs. Shrivastava and Richards (1965) stated that the number of circulating hemocytes could be approximately doubled within 6–24 hr by bleeding last-instar larvae. These authors reported that a greater increase occurred when last-instar larvae were bled a small drop (1 µl) rather than a large drop (ca. 5 µl), and that in both cases stimulation of mitotic activity occurred.

At 24 hr after wounding, a significant increase in the number of circulating hemocytes per cubic millimeter was observed (Shapiro, 1966). The MI reached its peak 48 hr later and fell during the next 2 days. The THC increased throughout the 5-day test period among controls, and no differences in THCs were noted at day 5 between the groups (Shapiro, 1968a). Wittig (1966) also noted an increase in THC when the cuticle of *Pseudaletia* was pierced. In *Bombyx*, the THC increased 20–30 hr after wounding and returned to normal after 2 days (Nittono, 1960).

In *Galleria*, the proportion and number of mitotically dividing cells and round PLs increased, whereas the fusiform PLs decreased within 24 hr after wounding. GR levels remained unchanged; ADs decreased; and PRs and degenerating cells initially increased. After 72 hr, higher numbers of ADs, SPs, and cells in mitosis were observed. Within the next 2 days (i.e., at 120 hr after wounding) mitotic activity declined, and no differences were evident in the blood pictures of control and wounded larvae (Shapiro, 1968a). When BV measurements were taken, no significant differences were observed between control and wounded larvae during the experimental period. In *Popillia*, however, severe hemorrhage involved a loss of more than 50% of the BV, without an appreciable change in the THC (Beard, 1949).

When *Periplaneta* was subjected to slight hemorrhage, the THC and BV were maintained. When insects were bled repeatedly, "each hemorrhage but the last was followed by a readjustment involving (1) replacement of lost plasma by fluid, and (2) failure to replace all of the lost blood cells" (Yeager and Tauber, 1933). In *Locusta*, the normal BV level was restored within 24 hr (Hoffmann, 1969). Brehélin et al. (1976) investigated the functions of GRs during wound healing. They concluded that the GRs were involved in wound healing and showed no ultrastructural changes during the healing process (i.e., within 24 hr).

The effects of the wound response should never be overlooked.

Jones (1950) concluded that most of the cytopathological changes observed in *Tenebrio* hemocytes, except in the case of bacterial infection and cauterization, were probably the result of the wound itself.

16.6.6. Other injuries

The effects of ligation have been studied by several workers. Arnold (1952a) noted that metamorphosis in *Anagasta* was inhibited. Results from ligated larvae were almost identical to those from starved larvae (e.g., a significant increase in ADs). When fourth-instar *Bombyx* larvae were ligated between head and thorax, the THC was nearly the same as that from fifth-instar larvae. The THC remained at this high level until the larva developed into a precocious pupa. These larvae did not spin cocoons (Nittono, 1960). The THC of *Prodenia* larvae increased from 28,700/mm³ 17 hr postligation to 38,800/mm³ 31 hr later (Rosenberger and Jones, 1960).

A detailed study of the effects of ligation on *Rhodnius* was made by Jones (1967c). Insects were ligated 1 hr, 4 days, 8 days, and 14 days postfeeding. Significant differences were found in groups ligated before molting hormones were fully secreted (1 hr to 4 days postfeeding) and those ligated after the critical period (8 to 14 days postfeeding). During the initial period, BV was high, the number of hemocytes was high, and the blood picture differed from that of the controls. After the molting hormones were fully secreted, the blood picture of ligated insects was similar to that of control insects. These studies showed, Jones (1967c) felt, that hormones played a role in the regulation of hemolymph available from an appendage and in regulation of cyclic changes involving numbers and types of hemocytes.

Whereas unligated *Galleria* pupate on day 20 or 22 (30–36 °C), ligated larvae lived for a much longer time. The site of ligation had an effect upon hemocyte population counts. Ligation of the head did not result in a significant change in cell numbers, but ligation of the thorax resulted in a rapid decrease in cell numbers posterior to the ligature. Significantly greater numbers of hemocytes were found in the front half when a larva was ligated in the middle of its body. The mitotic rate was much higher in the anterior half, but Jones and Liu (1969) could not correlate the mitotic rate with specific changes in THCs in either half.

When *Tenebrio* larvae were cauterized, hemocytopenia occurred, and numerous abnormal hemocytes were observed (Jones and Tauber, 1952). Low temperature (5 °C) resulted in a decrease in mitotically dividing cells in *Blatta orientalis* (Tauber, 1935). When roaches were returned to normal temperatures, the mitotic rate suddenly increased. Moderate temperatures (15–17 °C) did not cause any change in the blood picture of *Anagasta* larvae (Arnold, 1952a). No change in the THC of *Prodenia* was observed when larvae were exposed to 4 °C (Rosenberger and Jones, 1960).

The mitotic rate of *Blaberus* hemocytes slowly increased at 37 °C, reaching a peak a few days before death (Tauber, 1935). A small increase in the number of degenerating cells was observed when *Anagasta* larvae were maintained at 34 °C for 48 hr (Arnold, 1952a). The THC increased in *Prodenia* larvae exposed to drying conditions at 25 °C (Rosenberger and Jones, 1960). Presumably, the increase might be attributable to desiccation and a loss of body fluids.

Anoxia produced no apparent change in the blood picture of *Anagasta* larvae (Arnold, 1952a). Gamma irradiation (3,000–9,000 rads) of potato tuberworm larvae, *Gnorimoschema operculella,* caused a decrease in the THC, regardless of whether the larvae were irradiated as young or mature (Elbadry, 1964). DHCs, on the other hand, were not altered. Changes in hemocyte populations of *Oncopeltus* were followed after gamma irradiations with ^{60}Co (Feir and McClain, 1968b). The 1,000-rad dose had no effect on the hemocytes, but the 10,000-rad dose did. In the latter treatment, MIs were very high, and most of the mitotic figures were in metaphase. The authors concluded that irradiation resulted in mitotic arrest.

16.6.7. Injection of materials

Nonparticulates and solutions

"It should not be forgotten that these leucocytic variations, as well as the changes in the cells themselves, depend, not only on a microbe, an excitant, but also on a whole series of other causes: the temperature, the age of the animal, its physiological state, the quality and the quantity of injected substances, etc" (Metalnikov, 1927). These words of the great pathologist and immunologist, Serge Metalnikov, recognized the complexity of the hemocytic response. In 1966, Wittig noted that the hemocytic response is complex and depends upon the response to the wound, to the properties and volume of the diluent used, and to the pathogen injected. The wound response has been discussed above; this section will deal with the injection of nonparticulates and solutions.

Distilled water. Rooseboom (1937) observed vacuolation of *Carasius* hemocytes after injection of distilled water. Injection of distilled water into *Tenebrio* larvae did not affect the THC. Mitotic counts were increased, but the percentages of fusiform cells and PRs were reduced (Jones, 1950). In a subsequent study, Jones and Tauber (1952) injected 40 mealworm larvae with nonsterile distilled water. Many abnormal hemocytes were observed in some of the larvae, but no correlation could be made between the cytopathologies produced and the injection of water. At 1–12 hr postinjection, the THC was increased in *Pseudaletia* larvae (Wittig, 1966). The increase was not attributable to the wounding response alone, and Wittig assumed that the volume of water injected produced a further rise in the THC. The DHC, on the other hand, did not change following the injection of water.

Salts and other substances. After injection of saline, the THC increased in *Euproctis* (Paillot, 1923) and in *Schistocerca* (Lee, 1961), but was reduced in *Galleria* (Jones, 1962). Sterile salt solutions produced no change in the proportion of mitotic figures in *Blatta* (Tauber, 1935) or in *Tenebrio* (Jones, 1950). PRs and PLs increased, while ADs decreased, in *Tenebrio* larvae (Jones, 1950). In larvae, the greatest increase in PRs occurred 24 hr postinjection; no increase in this cell type occurred in adults. Fusiform cells increased in adults, however (Jones, 1950). Jones and Tauber (1952) observed few abnormal hemocytes following injection of Yeager's saline. After injection of acetone saline (10^{-3}), abnormal hemocytes were observed after 24 hr.

THCs were reduced in *Prodenia* following injection of trypan blue (Rosenberger and Jones, 1960). Injections of sterile nutrient broth, beef extract, and peptone resulted in increased mitotic activity in *Blatta* (Tauber, 1935). Aspartic acid, glutamic acid, glycine, alanine, and tryptophan did not alter the incidence of mitosis, but injection of cysteine enhanced the MI (Tauber, 1935).

Injection of insulin, both concentrated (40 units) and diluted, caused an increase in the incidence of mitotically dividing cells (Tauber, 1935). Adrenaline (10^{-3} g/ml) increased the MI in *Oncopeltus* (Feir and McClain, 1968b). Two carcinogens, 20-methylcholanthrene and dimethylbenzanthracene, were also tested in milkweed bugs by Feir and McClain (1968b). In one experiment, the MI among bugs treated with 20-methylcholanthrene significantly increased 48 hr after molting. In the second experiment, the diluent itself (Tween-80) resulted in a higher MI than did the carcinogen. No changes in mitotic activity of the hemocytes were noted after exposure to dimethylbenzanthracene.

Particulate materials

Much work on the injection of particulates has centered around the phagocytic response (Hollande, 1909; Metalnikov, 1927; Toumanoff, 1930; Cameron, 1934; Rooseboom, 1937; Arnold, 1952a; Jones, 1956; Wittig, 1966; Werner and Jones, 1969) and has primarily involved the use of ink or carmine. In some cases, the injected particulates were detrimental to the hosts. Cameron (1934) noted that India ink was toxic to *Galleria* larvae when volumes greater than 0.02–0.05 ml

(1/150–1/200 dilution) were injected. Lea and Gilbert (1966) noted the possible toxicity of Chinese ink (in Ringer's) following injection into *Cecropia*. Jones (1956) noted that carmine seriously interfered with the development of *Sarcophaga*. After injection of sheep erythrocytes (2.5 million and 3.5 million/ml) and latex particles (0.005, 0.01, and 0.1% in saline), *Galleria* larvae died within 24–48 hr (Werner and Jones, 1969).

No change occurred in the hemocyte picture of *Carausius* after injection of Chinese ink (Rooseboom, 1937), but changes were found in *Apis* (Toumanoff, 1930) and in *Galleria* (Cameron, 1934). Wittig (1966) observed an increase in GRs and a decrease in PLs following injection in *Galleria*. The incidence of cytopathology among *Tenebrio* hemocytes was greater following injection of India ink. Although abnormal hemocytes (e.g., degeneration, vacuolization) were seen among lampblack-treated insects, the incidence was lower than among insects treated with India ink (Jones and Tauber, 1952). The THC was increased in *Rhodnius* (Wigglesworth, 1956), but cell counts were reduced in *Galleria* (Werner and Jones, 1969) following injection of ink. PLs and ADs increased; the proportions of SPs, OEs, and mitotically dividing hemocytes did not change (Werner and Jones, 1969).

Injection of colloidal gold or colloidal iron into *Tenebrio* larvae resulted in abnormalities in 10–30% of the hemocytes (Jones and Tauber, 1952). The percentage of ADs increased in *Galleria* after injection of starch or sheep erythrocytes. No changes were observed in SPs, OEs, and mitotically dividing cells in either case. Injection of starch resulted in an increase in THC; injection of sheep erythrocytes led to a decrease (Werner and Jones, 1969). Takada and Kitano (1971) followed changes in THC and DHC in *Pieris rapae crucivora* following injection of India ink. GRs increased and PLs decreased in proportion. PRs increased, reached a peak 24 hr postinjection, and then decreased. The THC decreased for the first 24 hr and then rose during the next 48 hr.

Although India ink particles have been commonly used, there are great variations in both size and shape. These variations are eliminated by the use of polystyrene latex particles (Wittig, 1966). When 6.6 million or 13.3 million particles were injected, the larvae appeared normal. After injection of 66.3 million particles, the segment at the site of injection became constricted, and posterior segments showed signs of paralysis. Low doses caused an increase in the THC; high doses led to a decrease. GRs increased and PLs decreased in *Galleria*, but no decrease in PLs was observed in *Pseudaletia* (Wittig, 1966). Wittig's data (1966) may be summarized as follows: (1) low doses resulted in an increase in THC, which may have been caused by the volume of water introduced into the hemocoel; high doses resulted in a decrease in THC; (2) the proportions of GRs and PLs were not altered by low doses of latex particles; at high doses, however, the GRs increased and the PLs decreased.

Following injection of latex, *Galleria* larvae died within 24–48 hr without noticeable changes in the proportions of PLs, SPs, OEs, and mitotically dividing cells. ADs and the number of circulating cells decreased (Werner and Jones, 1969).

16.6.8. Phagocytosis

Although the study of phagocytosis in insects has an important place in insect immunology and insect pathology, it is not the intent of this chapter to review its history and evaluate its significance. Before changes in hemocyte populations during phagocytosis are described, it is important to define certain terms. "*Phagocytic response* of the blood is the totality of the reactions of the hemocytes to small foreign particles present in the hemocoel . . . *Phagocytic capacity* of the blood is the maximum amount of phagocytosis that can be carried out in a given situation" (Wittig, 1966).

Phagocytosis of inert materials

Changes in the blood picture were noted in *Galleria* by early workers (Metalnikov, 1927; Cameron, 1934) and more recently by Werner and Jones (1969); in *Apis* by Toumanoff (1930); and in *Carausius* by Rooseboom (1937). A decrease in THC was noted following injection of erythrocytes (Bettini et al., 1951; Werner and Jones, 1969), latex (Wittig, 1966; Werner and Jones, 1969), and ink (Werner and Jones, 1969); cell populations increased following injections of starch (Werner and Jones, 1969) and latex (Wittig, 1966). Following injection of India ink, GRs increased in percentage, and PLs decreased. The THC decreased for the first 24 hr postinjection, but greatly increased during the next 48 hr in *Pieris rapae crucivora* (Takada and Kitano, 1971). The phagocytic index increased from 26% (2 hr) to 75% (72 hr) during the experimental period. During this time, GRs increased to represent 92% of all circulating hemocytes. About 69% of all GRs and 33% of all PLs were found to be phagocytic. When latex particles were injected, PLs were the primary phagocytic hemocytes, but GRs were also phagocytic (Takada and Kitano, 1971).

Studies on phagocytosis in *Galleria* (Werner and Jones, 1969) are difficult to appraise, as mortality occurred 24–48 hr postinjection of sheep erythrocytes and latex particles. No mortality occurred, however, following injection of starch and Chinese ink. The phagocytic response to Chinese ink was greater than that to latex, red blood cells, and starch (Werner and Jones, 1969).

Heat-killed *Bacillus thuringiensis* was injected into *Pseudaletia* larvae. The THCs of injected larvae were dependent on the doses of bacilli. THCs were close to normal following injection of low doses, but were only one-fifth to one-fourth of these values after injection of high numbers of heat-killed bacilli (Wittig, 1965b). A high dose (16.8 million per larva) resulted in an increase in the proportion of GRs and a decrease in PLs. The author noted that injection of a high dose exceeded the phagocytic capacity of the blood. Phagocytized bacilli were digested within the phagocytes. As these observations involved only circulating cells, Wittig stressed that the phagocytic capacity of an insect can only be determined when the relation between circulating and sessile hemocytes is determined. Killed *Bacillus cereus* and other bacteria were injected into *Galleria* and *Pieris brassicae*. A hemocytic response occurred following injection with nonpathogens, but no response occurred after a pathogen (*Staphyloccus aureus*) was injected (Ratcliffe and Gagen, 1976).

Rooseboom (1937) found that Norit powder (a medicinal carbon) particles less than 10 μm in diameter were phagocytized in hemocytes of *Carausius*, whereas particles larger than 10 μm remained free. In *Pseudaletia*, large ink particles were not phagocytized within individual hemocytes, but were surrounded by clusters of hemocytes (= nodules) (Wittig, 1966). When low doses of ink were injected, GRs were much more active than PLs as phagocytes. As the concentration of ink increased, however, the proportion of phagocytic hemocytes in larval blood was greater than in adult blood.

The phagocytic capacity of *Galleria* blood was higher than that of *Pseudaletia* blood. For a given incitant, one-third of armyworm hemocytes were active as phagocytes, whereas one-half of the wax moth cells responded. *Galleria* may be more active because of the higher proportions of phagocytic GRs occurring in the blood during the first 5 days of the last instar. At the beginning of prepupation, the reverse is true: *Pseudaletia* blood contains more GRs than PLs (Wittig, 1966).

Phagocytosis of biologically active materials

A great deal of our information on the hemocytic response to bacteria is still based on the early studies of Hollande, Metalnikov and his co-workers, Cameron, and Paillot (Wittig, 1966). Metalnikov (1927) noted that phagocytic hemocytes elongated in *Galleria*, but tended to become round in *Carausius*, following injection of *Bacterium tumefaciens* (Toumanoff, 1949). During bacterial challenge, phagocytosis can (1) occur from the beginning to the end of infection (Metalnikov and Chorine, 1929); (2) occur at the beginning of infection, but be diminished or absent at the end (Tateiwa, 1928; Toumanoff, 1930); (3) be absent or weak at the beginning of infection, but increase during infection; or (4) be absent throughout the infection cycle (Stephens, 1963; Schwartz and Townshend, 1968).

Even assuming that phagocytosis does occur, the end result may be (1) phagocytosis, leading to the destruction of the bacterium; (2) growth of the bacterium, leading to death of the host, despite an active phagocytic response; or (3) phagocytosis, leading to the destruction of most bacilli. The small number of active bacteria continue to grow and eventually kill the host (Cameron, 1934).

Bacillus thuringiensis was injected into the hemocoel of *Pseudaletia* larvae. Digestion of the bacterial cells occurred following uptake of the cells. Within 3 hr, the hemocytes began to degenerate following challenge with a heavy dose (Wittig, 1965a). Differences in the phagocytic response were dependent on the bacterial concentration. After injection of a heavy dose (16 million to 33 million per larva), many free bacilli were observed, indicating that the phagocytic capacity was exceeded. After injection of a light dose (1/100th of the heavy dose), phagocytosis occurred within 30 min but did not prevent the reappearance of bacilli in the hemolymph (Wittig, 1965b). The high concentration resulted in death within 3–4 hr postinjection; larvae receiving the light dose died within 28 hr. It is obvious that both doses exceeded the phagocytic capacity of the hemocytes much more than comparable doses of heat-killed bacilli (Wittig, 1966). *Sarcina lutea* and *Staphylococcus epidermidis* were phagocytized within 1 hr postinjection in *Galleria* larvae. The phagocytic response to *S. epidermidis* was greater than those to India ink, starch, latex particles, sheep red blood cells, and *Sarcina* (Werner and Jones, 1969).

Phagocytosis of insect viruses can occur (1) when free nonoccluded virions are found in the hemocoel, and (2) after tissue lysis has occurred, liberating viral inclusion bodies into the hemocoel (Wittig, 1968). Most PLs phagocytize polyhedra in virus-infected larvae of *Malacosoma alpicola* (Benz, 1963) and *Thaumetopoea wilkinsoni* (Harpaz et al., 1965). In *Pseudaletia* larvae, more than 50% of the total hemocyte population was phagocytic in granulosis-infected larvae, whereas more than 20% had similar activity in nucleopolyhedrosis-infected larvae (Wittig, 1965b). In both instances, phagocytic activity was directed against viral inclusion bodies. During granulosis, nearly all virus capsules were engulfed within 30 min. In all cases, GRs had a greater phagocytic capacity than PLs. No changes in DHC and THC were noted in fifth-stage larvae. When polyhedra were injected, the polyhedral inclusion bodies (PIBs) were engulfed within 30 min. As in the case of virus capsules, GRs showed greater phagocytic capacity than PLs (Wittig, 1968).

During nucleopolyhedrosis in the cotton bollworm, *Heliothis zea*, the average number of phagocytic cells decreased in both control (31%) and virus-infected (63%) larvae from the second to the third day postinfection. Little difference in the percentages of phagocytic hemocytes occurred during viral infection, however. Latex particles

(1.3 μm) were injected (4.8 million per larva) into larvae previously exposed to nucleopolyhedrosis virus. The presence of inclusion bodies in the nucleus and latex particles in the cytoplasm indicated that infected hemocytes were still phagocytic (Shapiro et al., 1969).

Galleria larvae previously injected with India ink were more severely infected by nucleopolyhedrosis virus than uninjected larvae. (Stairs, 1964). Stairs suggested that the phagocytic hemocytes were blocked by India ink. Prior injection of India ink into *Pseudaletia* larvae did not lessen the phagocytic capacities of hemocytes to engulf virus capsules (Wittig, 1968).

Phagocytosis in protozoan (e.g., microsporidian) infections can occur (1) early in infection, when planonts penetrate gut epithelium and enter the hemocoel, and (2) late in infection, when cells and tissues lyse, liberating all stages of the pathogen into the hemocoel (Weiser, 1963). Whereas phagocytosis of *Nosema* occurred in *Trichoplusia* larvae (Laigo and Paschke, 1966), phagocytosis decreased in *Ecdyonurus* during infection (Weiser, 1956). In a subsequent study, Weiser (1961) showed that *Nosema* spores were phagocytized by hemocytes, which fused to form giant polynucleated cells. Similar cells or nodules were observed during nosematosis of *Manduca sexta* (Brooks, 1971).

Speare (1920) investigated the effects of an entomopathogenic fungus, *Sorosporella uvella*, in both nonsusceptible and susceptible hosts. When low doses were injected into nonsusceptible hosts, *Bombyx* and *Lachnosterna*, phagocytosis occurred. No phagocytosis occurred following injection of a large dose. In a susceptible host, *Prodenia*, phagocytosis was not observed within 48 hr. A day later, however, blastocysts were engulfed within phagocytic hemocytes. Phagocytosis was more intense against *Aspergillus flavus* (a pathogen) than against *A. niger* (a nonpathogen of *Hyalophora cecropia*) (Sussman, 1952). Numerous changes in the ultrastructure of *Galleria* hemocytes were observed in response to the fungus *Paecilomyces farinosus* and the parasitic nematode *Neoaplectana carpocapsae*. These changes included intense vacuolization, delamination of the nuclear membrane, the appearance of lysosomal vacuoles (Kamionek and Seryczyńska, 1976).

16.6.9. Poisons

Several studies have been made on lesions of hemocytes (Pilat, 1935; Lepesme, 1937; Yeager and Munson, 1942; Yeager et al., 1942; Jones, 1950, 1957; Toumanoff and Lapied, 1950; Arnold, 1952b; Jones and Tauber, 1954; Gupta and Sutherland, 1968), and an excellent review on the pathology of blood cells was written by Wittig (1962).

Yeager and Munson (1942) observed that calcium arsenate, calcium arsenite, arsenic trioxide, sodium fluoride, and mercuric chloride caused cytological changes in the hemocytes of *Prodenia*. These changes were divided into three categories: (1) passive-active trans-formation (rounding up or spreading of fusiform cells) and cell aggluti-nation; (2) possible regenerative changes (increased MI); and (3) de-generative changes. These changes were also seen in *Anagasta* larvae treated with dichloroethyl ether, chloroform, and methylbromide (Ar-nold, 1952b). Sublethal doses of chlordane did not cause any passive-active transformations in the fusiform PLs (Gupta and Sutherland, 1968).

Both dichloroethyl ether and methylbromide caused a significant decrease in the THC, whereas an increase in THC occurred in chloro-form-fed *Anagasta* larvae (Arnold, 1952b). Jones and Tauber (1954) observed a significant increase in the THCs of last-instar *Tenebrio* lar-vae following fumigation with nicotine. In a later study, Jones (1957) showed that THCs in unfixed and heat-fixed *Tenebrio* larvae treated with pure, finely powdered p,p'DDT were within the normal range prior to the moribund stage. On the basis of this study, the author felt that hemocytes played a very small role, if any, in the defense of the insect.

A decrease in THCs was also observed by Pilat (1935), Fisher (1936), and Trehan and Pajni (1961); increases were noted by Arvy et al. (1950) and Gupta and Sutherland (1968). Following exposure to chlordane, PLs, GRs, SPs, and COs increased in percentages (Gupta and Sutherland, 1968). The proportions of PRs and OEs did not change greatly following fumigation, but the PL/AD relationship was markedly changed (Arnold, 1952b). These changes were indicative of starvation. Arnold felt that ADs were important in the recovery of the insect because of their ability to synthesize fats and contribute to the insect's energy pool.

Although Toumanoff and Lapied (1950) noted increases in the num-ber of PRs and GRs in *Galleria*, Arvy et al. (1950) and Jones (1957) observed increases of these cells in *Chrysomela* and *Tenebrio*, respec-tively. Although carbon disulfide and diethyl ether did not appreci-ably alter the THC in *Blatta*, diethyl ether reduced the coagulability of the hemolymph (Fisher, 1936). When *Tenebrio* larvae were ex-posed to sodium fluoride dust, none died or showed signs of intoxica-tion. Neither THC nor DHC showed any changes from the "norm," and abnormal hemocytes were not observed (Jones and Tauber, 1954). Adult mealworms, on the other hand, were more sensitive to sodium fluoride, as abnormal hemocytes were observed. Fusiform cells were found in greater numbers among treated worms than among controls. Nicotine intoxication resulted in a fourfold increase of the THC among *Tenebrio* larvae (Jones and Tauber, 1954).

The BV of *Periplaneta* was decreased following poisoning with carbon disulfide (Shull et al., 1932) and arsenic (O'Kane and Glover, 1935). A similar decrease occurred in *Periplaneta* following TEPP poisoning (Roan et al., 1950). No marked change occurred in the BV of *Tenebrio* larvae exposed to DDT, although the insecticide-treated insects lost twice as much weight as the untreated, starved controls (Jones, 1957a). Arnold (1952b) reported that the response of *Anagasta* hemocytes went through three stages, similar to the ones reported by Yeager and Munson (1942): (1) changes involving passive-active transformation; (2) pathologic changes, leading to degeneration and a reduction in cell numbers; and (3) regenerative changes, resulting in an increased MI and an increase in cell numbers, especially the ADs. Rotenone (Tauber, 1935) and nicotine and sodium fluoride (Jones and Tauber, 1954) had no effect on the number of mitotically dividing cells in *Blatta* and *Tenebrio*, respectively. In *Anagasta* larvae, changes in the MI were noted (Arnold, 1952b). The MI increased after dichloroethyl ether fumigation, especially among the ADs. At the end of fumigation, the MI increased as a result of carbon tetrachloride fumigation, especially among the ADs. Within 5 min of methylbromide fumigation, the MI increased suddenly. In each case, few mitotically dividing cells were seen in moribund larvae (Arnold, 1952b) (see also Chapter 14).

16.6.10. Various diseases

"The great sensitivity of the leukocytes against microbial species permits the supposition that these cells have an importance in the fight of the organism against these poisons. Their injection generally provokes a pronounced hyperleucocytosis of the blood" (Metchnikoff, 1901). These words served as an impetus for insect pathology, insect immunology, and, indirectly, insect hematology.

Bacterial diseases

André Paillot (1920) described a reaction, caryocinetosis, which was evidenced by an abnormal multiplication of hemocytes, as a response to microbial infection. Moreover, he demonstrated that this reaction could be provoked by all microbes under certain conditions. Tauber (1940) also noted an increase in the number of mitotically dividing cells, as a response to two pathogens, *Staphylococcus albus* and *Bacterium* sp. An increase in the THC was observed in the hemolymph of insects with naturally occurring bacterial infections (Tauber and Yeager, 1935, 1936).

No reduction in the THC was observed in *Popillia* larvae infected with *Bacillus popillae* prior to sporulation. At sporulation, the number of circulating hemocytes increased (Beard, 1945). Similar results were obtained by Schwartz and Townshend (1968), but St. Julian et al. (1970) noted a decrease in THC during "milky disease." Many authors have observed drastic reductions in the numbers of hemocytes during bacterial infections (Kostritsky et al., 1924; Metalnikov, 1927; Toumanoff, 1930, 1949; Chorine, 1931; Babers, 1938; Wittig, 1965b; Werner and Jones, 1969). THCs in bacteria-infected *Prodenia* larvae either remained unchanged or declined markedly. No increase in the MI was noted, and the authors concluded that hemocytes were not very effective in protecting the insect (Rosenberger and Jones, 1960). The THC in *Galleria* increased up to 48 hr following inoculation with *Staphylococcus epidermidis*, and then decreased rapidly. Following injection of *Sarcina lutea*, the BV increased (Werner and Jones, 1969).

Changes in the DHC of *Galleria* were followed by Metalnikov (1922, 1927), Tateiwa (1928), Chorine (1931), Cameron (1934), and Werner and Jones (1969). In general, PLs decreased, and GRs, PRs, and especially SPs increased in percentages. Metalnikov (1927) concluded from his extensive studies that each microbe or excitant provoked a specific reaction among the hemocytes. Arnold (1952a) noted unusually large numbers of ADs associated with bacterial infections in *Anagasta*. Following inoculation of *Sarcina lutea*, PLs decreased and ADs increased. Inoculation of *Staphylococcus epidermidis* resulted in an increase of PLs. In both cases, SPs and OEs did not change in proportion (Werner and Jones, 1969). Toumanoff (1930) noted changes in the THC in *Apis* infected by *Staphylococcus*. In a subsequent study on *Carausius* infected with *B. tumefaciens*, Toumanoff (1949) noted the passive-active transformation.

High doses (28 million to 33.5 million per *Pseudaletia* larva) of *B. thuringiensis* resulted in a drastic reduction in PLs, a great increase in SPs and PRs, and little change in GRs. Low doses (0.26 million per larva) had a different effect on the blood picture. For the first 6.5 hr, bacilli could not be observed in the blood. Although GRs increased, bacilli reappeared in the blood. GRs decreased greatly, PLs decreased slightly, and SPs and PRs increased dramatically (Wittig, 1966). In *Apis*, no relationship was established between changes in hemocyte morphology and bacterial septicemias, rickettsioses, nosematoses, and mixed infections (Wille, 1974). From his studies on the mealworm, *Tenebrio molitor*, Jones (1950) concluded that changes in the blood picture "may indicate a physiological response to some general excitatory agent whether it be hormone, poison, or bacteria."

Fungal diseases

Few studies have been devoted to hemocyte changes during fungal development. In the case of a true entomopathogenic fungus, when phagocytosis occurs, it is ineffective. In some instances, the hemocytes do not appear to be adversely affected (Speare, 1920: *Soro-sporella uvella* in *Prodenia*). In other cases, the hemocytes: form giant cells (Boczkowska, 1935: *Beauveria bassiana, Metarrhizium anisopliae* in *Galleria*); round up (= passive-active transformation) (Cermakova and Samsinakova, 1960: *B. bassiana* in *Leptinotarsa*); or show degenerative changes and eventually lyse (Sussman, 1952: *Aspergillus flavus* in *Hyalophora;* Sirotina, 1961: *B. bassiana* in *Leptinotarsa*).

The blood picture of *Leptinotarsa*, during infection by *Beauveria*, showed profound pathological changes. The number of pathologic forms and degenerating cells increased markedly, while the numbers of phagocytic hemocytes decreased drastically (Sirotina, 1961). From these studies, Sirotina concluded that hematological assays could be utilized effectively to determine the health of the insect population and could be utilized as "a valuable basis for microbiological measures against the Colorado beetle." The number of blood cells of *Cecropia* increased upon infection by *Aspergillus flavus* (a pathogen) but decreased as the fungus multiplied throughout the host. In the case of *A. niger* (a nonpathogen), the number of circulating cells increased for 2 days and then decreased by the third day after challenge.

Protozoan diseases

Paillot (1918) noted the association of giant blood cells with microsporidian infections in larvae of *Pieris brassicae*. Weiser (1956), on the other hand, observed no change in hemocyte morphology in *Ecdyonurus venosis* infected by *Nosema baetis*. No active phagocytosis occurred, and the hemocytes did not increase in number. The blood picture of potato tuberworm larvae, *Gnorismoschema operculella*, was altered during infection by *Nosema destructor* (Steinhaus and Hughes, 1949).

No significant differences in THC were observed in *Anagasta* larvae infected by a microsporidian (Arnold, 1952a). Decreases in the THC were observed in *Trichoplusia ni* (Laigo and Paschke, 1966) and in *Apis mellifera* (Gilliam and Shimanuki, 1967) following infection by *Nosema* sp. and *N. apis*, respectively, and an increase was observed in *Sericesthis* infected by *Adelina sericesthis* (Weiser and Beard, 1959). Following inoculation with the trypanosome, *Leptomonas pyrrhocoris*, the blood picture in *Galleria* was changed, as the flagellates multiplied in the hemocoel (Zotta and Teodoresco, 1933). At an elevated temperature (39 °C), the protozoa were lysed, and the hemocytes multiplied in great numbers. The authors concluded that the hemocytic response was not specific to the protozoan, but was similar to that obtained utilizing different bacteria and inert materials as Chinese ink (Zotta and Teodoresco, 1933).

Rickettsial diseases

No microscopic changes were noted in the blood of the Japanese beetle, *Popillia japonica,* in the early stages of the blue disease, caused by *Coxiella popillae.* Moreover, the blood picture continued to show no pathologic changes, even late in infection (Dutky and Gooden, 1952). During rickettsiosis caused by *Rickettsia melolonthae,* the PLs of *Melolontha* decreased to one-sixth their normal number, and ADs declined to one-third their normal value. Little or no melanization occurred in infected hemolymph, owing to a decrease or complete loss of tyrosine (Krieg, 1958). Vago (1959) demonstrated that the fat body and hemocytes underwent pathologic changes. Niklas (1960) confirmed Krieg's (1958) results and demonstrated a decrease in the number of PLs and ADs and a decrease in melanization. In addition, coagulation of the blood decreased during infection.

Viral diseases

It is well known that nucleopolyhedrosis viruses multiply in the tracheal matrix, fat body, hypodermis, and hemocytes of Lepidoptera. For a detailed review on the pathology of insect blood cells during viral infection, the reader is referred to the review by Wittig (1962). At an early stage of viral infection, chromatin aggregates and small granules showing brownian movement appear at the periphery of the nucleus of infected cells (i.e., nuclear ring zone or "anneau mirotant") (Paillot, 1924a,b, 1926, 1933). This is one of the earlier microscopic signs of the disease and has also been reported by later authors (Steinhaus, 1948, in *Colias eurytheme;* Bergold, 1952, in *Bombyx mori;* Jaques, 1962, in *Trichoplusia ni;* Smirnoff, 1962, in *Erranis tiliaria*).

In other Lepidoptera, polyhedra have not been observed in blood cells until late in infection (Aizawa et al., 1957, in *Euproctis flava* and *E. pseudoconspersa;* Drake and McEwen, 1959, in *Trichoplusia;* Smirnoff, 1964, in *Operophthera bruceata*). In *Trichoplusia,* only the GRs, PLs, and PRs were infected. From the initial appearance of polyhedra in the blood cells, the blood picture did not change as the disease progressed (Drake and McEwen, 1959). Harpaz et al. (1965) observed pathologic changes in the GRs and PLs of *Thaumetopoea wilkinsoni.* Morris (1962) found that most blood cells were heavily infected in larvae of *Lambdina fiscellaria somniaria.* Fewer blood cells were observed in the hemolymph 8–9 days postinfection, death occurring on day 9. During nucleopolyhedrosis of *Pseudaletia,* GRs initially increased (day 4) and then decreased, while SPs and PRs increased in percentages (Wittig, 1968). The THC decreased drastically during infection. On the other hand, Komarek and Breindl (1924) reported that the number of circulating hemocytes in *Lymantria monocha* increased before infected hemocytes were seen. In the cranefly,

Tipula paludosa, the THC increased greatly during nucleopolyhedrosis infection (Xeros and Smith, 1954).

In *Galleria* larvae infected by nucleopolyhedrosis virus no change in the hemocytes was noted until day 10, when the THC decreased (Shapiro, 1967). Fewer round PLs and fusiform PLs and significantly more GRs were observed in virus-infected insects than in controls (Shapiro, 1968b). Other changes observed were loss of ADs, increase in SPs, and increase of degenerating cells. No differences were found in the incidence of cells in mitosis. Only one change appeared to be peculiar to nucleopolyhedrosis: the significant increase in the number and percentage of SPs, both normal and abnormal (Shapiro, 1966).

In *Heliothis zea,* the THC decreased drastically when larvae were exposed to a high virus concentration. No decrease was noted in controls or in larvae exposed to a low virus dose. The number of SPs, which would normally decrease, increased in virus-infected insects (Shapiro et al., 1969). "Results of this study and other reported data reinforce the opinion that qualitative or quantitative changes in hemocytes are nonspecific responses caused by the presence of foreign substances and/or stress conditions" (Shapiro et al., 1969).

During granulosis infection of *Dendrolimus sibiricus,* GRs, PRs, and PLs initially increased. As infection progressed, however, the number of circulating hemocytes drastically decreased (Luk'yanchikov, 1964). During the first 6 days of granulosis in *Pseudaletia,* the GRs greatly increased, while the PLs decreased (Wittig, 1965a).

16.7. Summary

It is difficult to offer a concise and precise summary of hemocyte population changes in insects. Clearly, a vast amount of data has been accumulated, but much of the work offers only a glimpse of hemocyte changes (Arnold, 1974). Many workers have dealt with changes in either total hemocyte count or differential hemocyte count, but not both. Fewer investigations have involved blood volume determinations. For the blood picture of a given insect to be as complete as possible, all three parameters should be studied.

There is no doubt that insect hemocytes do respond to internal changes during development (e.g., at ecdysis) and to external changes (e.g., starvation, wounding, parasitism, including disease). Some workers have shown that the blood picture is a very accurate mirror of developmental changes; others have felt that hemocytes do not appear to respond to drastic conditions. Are both these conclusions correct? Are they attributable to inherent differences, even though the same insect species was used? If so, what is the case when different species are used? Were the rearing conditions the same? Were the hematological techniques used the same?

Studying insect hemocytes offers diversity and challenge to the hematologist. For example, Arnold (1972), using the same techniques, could recognize some taxa by differences in number and size of hemocytes and by cell types; other taxa could be differentiated with difficulty; still others could not be distinguished at all. What criteria should be used to distinguish cell types within a given species or among different species? The use of hemocytology, for example, should be included in basic studies involving DHCs and THCs. The challenge is before us, and I believe the key to our understanding lies in the designation of criteria for study, the standardization of technique, and possibly the selection of key insects for study.

Acknowledgments

I wish to acknowledge the assistance of Drs. R. Granados and A. Nappi. I especially am grateful to Ms. Michele Brown for her excellent typing and patience in reading the manuscript.

References

Aizawa, K., S. Asahina, and H. Fukumi. 1957. Demonstration of the polyhedral diseases of *Euproctis flava* and *Euproctis pseudoconspersa* (Lepidoptera, Lymantriidae). *Jpn. J. Med. Sci. Biol.* 10:61–4.

Arnold, J. W. 1952a. The haemocytes of the Mediterranean flour moth, *Ephestia kühniella* Zell. (Lepidoptera: Pyralididae). *Can. J. Zool.* 30:352–64.

Arnold, J. W. 1952b. Effects of certain fumigants on haemocytes of the Mediterranean flour moth, *Ephestia kühniella* Zell. (Lepidoptera: Pyralididae). *Can. J. Zool.* 30:365–74.

Arnold, J. W. 1966. An interpretation of the haemocyte complex in a stonefly, *Acroneuria arenosa* (Plecoptera: Perlidae). *Can. Entomol.* 98:394–411.

Arnold, J. W. 1969. Periodicity in the proportion of haemocyte categories in the giant cockroach, *Blaberus giganteus. Can. Entomol.* 101:68–77.

Arnold, J. W. 1972. Haemocytology in insect biosystematics: The prospect. *Can. Entomol.* 104:655–9.

Arnold, J. W. 1974. The hemocytes of insects, pp. 201–54. *In* M. Rockstein (ed.). *The Physiology of Insecta*, Vol. 5, 2nd ed. Academic Press, New York.

Arnold, J. W., and C. F. Hinks. 1976. Haemopoiesis in Lepidoptera. I. The multiplication of circulating haemocytes. *Can. J. Zool.* 54:1003–12.

Arvy, L., M. Gabe, and J. Lhoste. 1948. Contribution à l'étude morphologique du sang de *Chrysomela decemlineata* Say. *Bull. Biol. Fr. Belg.* 82:37–60.

Arvy, L., M. Gabe, and J. Lhoste. 1950. Action de quelques insecticides sur le sang du Doryophore (*Chrysomela decemlineata* Say). *Bull. Soc. Entomol. Fr.* 55:122–7.

Ashhurst, D. E., and A. G. Richards. 1964. Some histochemical observations on the blood cells of the wax moth, *Galleria mellonella* L. *J. Morphol.* 114:247–53.

Babers, F. H. 1938. A septicemia of the southern armyworm caused by *Bacillus cereus. Ann. Entomol. Soc. Amer.* 31:371–3.

Bahadur, J., and J. P. N. Pathak. 1971. Changes in the total hematocyte counts of the bug, *Holys dentata* under certain conditions. *J. Insect Physiol.* 17:329–34.

Beard, R. L. 1945. Studies on the milky disease of Japanese beetle larvae. *Conn. Agric. Exp. Stn. Bull.* 491:505–83.

Beard, R. L. 1949. Physiological effects of induced hemorrhage in Japanese beetle larvae. *J. N.Y. Entomol. Soc.* 57:79–91.

Beck, S. D. 1960. Growth and development of the greater wax moth, *Galleria mellonella* (L.) (Lepidoptera: Galleridae). *Trans. Wis. Acad. Sci. Arts Lett.* 49:397–400.

Benz, G. 1963. A nuclear polyhedrosis of *Malacosoma alpicola* (Staudinger). *J. Insect Pathol.* 5:215–41.

Bergold, G. H. 1952. Demonstration of the polyhedral virus in blood cells of silkworms. *Biochim. Biophys. Acta* 8:397–400.

Bettini, S., D. S. Sarkaria, and R. L. Patton. 1951. Observations on the fate of vertebrate erythrocytes and hemoglobin injected into the blood of the American cockroach (*Periplaneta americana* L.). *Science (Wash., D.C.)* 113:9–10.

Boczkowska, M. 1935. Contribution à le étude de le immunité chez les chenilles de *Galleria mellonella* L. contre les champignons entomophytes. *C. R. Soc. Biol.* 119:39–40.

Bowers, B., and C. M. Williams. 1964. Physiology of insect diapause. XIII. DNA synthesis during the metamorphosis of the cecropia silkworm. *Biol. Bull. (Woods Hole)* 126:205–19.

Brehélin, M., D. Zachary, and J. A. Hoffmann. 1976. Fonctions des granulocytes typiques dans la cicatrisation chez l'orthoptère *Locusta migratoria* L. *J. Microsc. Biol. Cell.* 25:133–6.

Brooks, W. M. 1971. The inflammatory response of the tobacco hornworm, *Manduca sexta*, to infection by the microsporidian *Nosema sphingidis*. *J. Invertebr. Pathol.* 17:87–93.

Cameron, G. R. 1934. Inflammation in the caterpillars of Lepidoptera. *J. Pathol. Bacteriol.* 348:441–66.

Cermakova, A., and A. Samsinakova. 1960. Ueber den Mechanismus des durchdringens des pilzes *Beauveria bassiana* Vuill. in die Larvae von *Leptinotarsa decemlineata* Say. *Cesk. Parasitol.* 7:231–6.

Cherbas, L. 1973. The induction of an injury reaction in cultured haemocytes from saturniid pupae. *J. Insect Physiol.* 19:2011–23.

Chorine, V. 1931. Contribution à l'étude de l'immunité chez les insectes. *Bull. Biol. Fr. Belg.* 65:291–393.

Clark, E. W., and D. S. Chadbourne. 1960. The haemocytes of nondiapause and diapause larvae and pupae of the pink bollworm. *Ann. Entomol. Soc. Amer.* 53:682–5.

Clark, R. M., and W. R. Harvey. 1965. Cellular membrane formation by plasmatocytes of diapausing cecropia pupae. *J. Insect Physiol.* 11:161–75.

Collin, N. 1963. Les hémocytes de la larve de *Melolontha melolontha* L. (Coléoptère Scarabaeidae). *Rev. Pathol. Veg. Entomol. Agric. Fr.* 42:161–7.

Crossley, A. C. S. 1964. An experimental analysis of the origins and physiology of haemocytes in the blue bottle blowfly *Calliphora erythrocephala* (Meig.). *J. Exp. Zool.* 157:375–97.

Davis, R. P., and H. A. Schneiderman. 1960. An autoradiographic study of wound healing in diapausing silkworm pupae. *Anat. Rec.* 137:348.

Day, M. F. 1952. Wound healing in the gut of the cockroach *Periplaneta*. *Aust. J. Sci. Res. (B)* 52:282–9.

Day, M. F., and M. J. Bennetts. 1953. Healing of gut wounds in the mosquito *Aedes aegypti* (L.) and the leafhopper *Orosius argentatus* (Ev.). *Aust. J. Biol. Sci.* 6:580–5.

Drake, E. L., and F. L. McEwen. 1959. Pathology of a nuclear polyhedrosis of the cabbage looper, *Trichoplusia ni* (Hübner). *J. Insect Pathol.* 1:281–93.

Dutky, S. R., and E. L. Gooden. 1952. *Coxiella popilliae*, n.sp., a rickettsia causing blue disease of Japanese beetle larvae. *J. Bacteriol.* 63:743–50.

Elbadry, E. 1964. The effect of gamma irradiation on the hemocyte counts of larvae of the potato tuberworm, *Gnorimoschema operculella* (Zeller). *J. Insect Pathol.* 6:327–30.

Feir, D. 1964. Hemocyte counts of the large milkweed bug, *Oncopeltus fasciatus*. *Nature (Lond.)* 202:1136–7.

Feir, D., and E. McClain. 1968a. Mitotic activity of the circulating hemocytes of the large milkweed bug, *Oncopeltus fasciatus*. *Ann. Entomol. Soc. Amer.* 61:413–16.

Feir, D., and E. McClain. 1968b. Induced changes in the mitotic activity of hemocytes of the large milkweed bug, *Oncopeltus fasciatus*. *Ann. Entomol. Soc. Amer.* 61:416 –21.

Feir, D., and G. M. O'Connor, Jr. 1969. Liquid nitrogen fixation: A new method for hemocyte counts and mitotic indices in tissue sections. *Ann. Entomol. Soc. Amer.* 62:246–9.

Fisher, R. A. 1935. The effect of acetic acid vapor treatment on the blood cell count in the cockroach, *Blatta orientalis* L. *Ann. Entomol. Soc. Amer.* 28:146–53.

Fisher, R. A. 1936. The effect of a few toxic substances upon the total blood cell count in the cockroach, *Blatta orientalis* L. *Ann. Entomol. Soc. Amer.* 29:334–40.

Florkin, M. 1937. Variations de composition du plasma sanguin au cours de la métamorphose du ver à soie. *Arch. Int. Physiol.* 60:17–31.

Gilliam, M., and H. Shimanuki. 1967. *In vitro* phagocytosis of *Nosema apis* spores by honey-bee hemocytes. *J. Invertebr. Pathol.* 9:387–9.

Gupta, A. P., and D. J. Sutherland. 1966. *In vitro* transformations of the insect plasmatocyte in some insects. *J. Insect Physiol.* 12:1369–75.

Gupta, A. P., and D. J. Sutherland. 1968. Effects of sublethal doses of chlordane on the hemocytes and midgut epithelium of *Periplaneta americana*. *Ann. Entomol. Soc. Amer.* 61:910–18.

Harpaz, I., E. Zlotkin, and Y. Ben Shaked. 1965. On the pathology of cytoplasmic and nuclear polyhedrosis of the Cyprus processionary caterpillar, *Thaumetopoea wilkinsoni* Tams. *J. Insect Pathol.* 7:15–21.

Harvey, W. R., and C. M. Williams. 1961. The injury metabolism of the cecropia silkworm. I. Biological amplification of the effects of localized injury. *J. Insect Physiol.* 7:81–99.

Hoffmann, J. A. 1969. Etude de la récupération hémocytaire après hémorragies expérimentales chez l'orthoptère, *Locusta migratoria*. *J. Insect Physiol.* 15:1375–84.

Hollande, A. C. 1909. Contribution à l'étude du sang des Coléoptères. *Arch. Zool. Exp. Gén.* 2:271–94.

Hollande, A. C., and M. Aghar. 1928. La phagocytose et le digestion des bacilles tuberculeux par les leucocytes du sang des chenilles autres que *Galleria mellonella*. *C. R. Soc. Biol.* 99:120–2.

Iwasaki, Y. 1927. Sur quelques phénomènes provoqués chez les chenilles de papillons par l'introduction de corps étrangers. *Arch. Anat. Microsc.* 23:319–46.

Jaques, R. P. 1962. The transmission of nuclear-polyhedrosis virus in laboratory populations of *Trichoplusia ni* (Hübner). *J. Insect Pathol.* 4:433–45.

Jones, J. C. 1950. Cytopathology of the hemocytes of *Tenebrio molitor* Linnaeus (Coleoptera). Ph.D. thesis, Iowa State College, Ames, Iowa.

Jones, J. C. 1954. A study of mealworm hemocytes with phase contrast microscopy. *Ann. Entomol. Soc. Amer.* 47:308–15.

Jones, J. C. 1956. The hemocytes of *Sarcophaga bullata* Parker. *J. Morphol.* 99:233–57.

Jones, J. C. 1957. DDT and the hemocyte picture of the mealworm, *Tenebrio molitor* L. *J. Cell. Comp. Physiol.* 50:423–8.

Jones, J. C. 1962. Current concepts concerning insect hemocytes. *Amer. Zool.* 2:209–46.

Jones, J. C. 1964. Differential hemocyte counts from unfixed last-stage *Galleria mellonella*. *Amer. Zool.* 4:337.

Jones, J. C. 1967a. Changes in the haemocyte picture of *Galleria mellonella* (Linnaeus). *Biol. Bull. (Woods Hole)* 132:211–21.

Jones, J. C. 1967b. Estimated changes within the haemocyte population during the last

518   M. Shapiro

larval and early pupal stages of *Sarcophaga bullata* Parker. *J. Insect Physiol.* 13:645–6.

Jones, J. C. 1967c. Normal differential counts of haemocytes in relation to ecdysis and feeding in *Rhodnius. J. Insect Physiol.* 13:1133–41.

Jones, J. C. 1967d. Effects of repeated haemolymph withdrawals and of ligaturing the head on differential counts of *Rhodnius prolixus* Stal. *J. Insect Physiol.* 13:1351–60.

Jones, J. C., and D. P. Liu. 1961. Total and differential hemocyte counts of *Rhodnius prolixus* Stal. *Bull. Entomol. Soc. Amer.* 7:166.

Jones, J. C., and D. P. Liu. 1968. A quantitative study of mitotic divisions of haemocytes of *Galleria mellonella* larvae. *J. Insect Physiol.* 14:1055–61.

Jones, J. C., and D. P. Liu. 1969. The effects of ligaturing *Galleria mellonella* larvae on total haemocyte counts and on mitotic indices among haemocytes. *J. Insect Physiol.* 15:1703–8.

Jones, J. C., and O. E. Tauber. 1952. Effects of hemorrhage, cauterization, ligation, desiccation, and starvation on hemocytes of mealworm larvae, *Tenebrio molitor* L. *Iowa State Coll. J. Sci.* 26:371–86.

Jones, J. C., and O. E. Tauber. 1954. Abnormal hemocytes in mealworms (*Tenebrio molitor* L.). *Ann. Entomol. Soc. Amer.* 47:428–44.

Kamionek, M., and H. Seryczyńska. 1976. Changes in hemocyte ultrastructure of *Galleria mellonella* L. (Lep., Galeriidae) caterpillars due to the effect of fungi and parasitic nematodes. *Bull. Acad. Pol. Sci. Ser. Sci. Biol.* 24:483–5.

Kitano, H. 1969. On the total hemocyte counts of the larva of the common cabbage butterfly, *Pieris rapae crucivora* Boisduval (Lepidoptera: Pieridae) with reference to parasitization of *Apanteles glomeratus* L. (Hymenoptera: Braconidae). *Kontyû* 37:320–6.

Komarek, J., and V. Breindl. 1924. Die Wipfelkrankheit der Nonne und der Erreger derselben. *Z. Angew. Entomol.* 10:99–162.

Kostritsky, M., M. Toumanoff, and S. Metalnikov. 1924. *Bacterium tumefaciens* chez la chenille de *Galleria mellonella. C. R. Acad. Sci.* 179:225–7.

Krieg, A. 1958. Weitere untersuchungen zur Pathologie der Rickettsiose von *Melolontha* spec. *Z. Naturforsch.* 13b:374–9.

Laigo, F. M., and J. D. Paschke. 1966. Variations in total hemocyte counts as induced by a nosemosis in the cabbage looper, *Trichoplusia ni. J. Invertebr. Pathol.* 8:175–9.

Lea, M. S. 1964. A study of the hemocytes of the silkworm *Hyalophora cecropia.* Ph.D. thesis, Northwestern University, Evanston, Illinois.

Lea, M. S., and L. I. Gilbert. 1961. Cell division in diapausing silkworm pupae. *Amer. Zool.* 1:368–9.

Lea, M. S., and L. I. Gilbert. 1966. The hemocytes of *Hyalophora cecropia* (Lepidoptera). *J. Morphol.* 118:197–216.

Lee, R. M. 1961. The variation of blood volume with age in the desert locust (*Schistocerca gregaria* Forsk.). *J. Insect Physiol.* 6:36–51.

Lepesme, P. 1937. L'action externe des arsénicaux sur le criquet pélerin (*Schistocerca gregaria* Forsk.). *Bull. Soc. Hist. Nat. Afrique Nord.* 28:88–103.

Luk'yanchikov, V. P. 1964. Changes in the hemolymph of *Dendrolimus sibiricus* Tschet. (Lepidoptera, Lasiocampidae) during granulosis. *Entomol. Obozr.* 43:297–300.

Mathur, C. B., and B. N. Soni. 1937. Studies on *Schistocerca gregaria* Forsk. IX. Some observations on the histology of the blood of the desert locust. *Indian J. Agric. Res.* 7:317–25.

Metalnikov, S. 1908. Recherches expérimentales sur les chenilles de *Galleria mellonella. Arch. Zool. Exp.* (*Ser.4*) 8:489–588.

Metalnikov, S. 1922. Les changements des éléments du sang de la chenille (*Galleria mellonella*) pendant l'immunisation. *C. R. Soc. Biol.* 86:350–2.

Metalnikov, S. 1927. *L'infection microbienne et l'immunité chez la mite de abeilles Galleria mellonella.* Pasteur Institute. Masson, Paris.

Metalnikov, S., and V. Chorine. 1929. On the natural and acquired immunity of *Pyrausta nubilalis* Hb. *Int. Corn Borer Invest. Sci. Rep.* 2:22–38.

Metchnikoff, E. 1901. *L'Immunité dans les maladies infectieuses.* Masson, Paris.

Morris, O. N. 1962. Studies on the causative agent and histopathology of a virus disease of the western oak looper. *J. Insect Pathol.* 4:446–53.

Nappi, A. J. 1970. Hemocytes of larvae of *Drosophila euronotus* (Diptera: Drosophilidae). *Ann. Entomol. Soc. Amer.* 63:1217–24.

Nappi, A. J., and J. G. Stoffolano, Jr. 1972a. Haemocytic changes associated with the immune reaction of nematode-infected larvae of *Orthellia caesarion. Parasitology* 65:295–302.

Nappi, A. J., and J. G. Stoffolano, Jr. 1972b. Distribution of haemocytes in larvae of *Musca domestica* and *Musca autumnalis* and possible chemotaxis during parasitization. *J. Insect Physiol.* 18:169–79.

Nappi, A. J., and F. A. Streams. 1969. Haemocyte reactions of *Drosophila melanogaster* to the parasites of *Pseudoeoila mellipes* and *P. bochei. J. Insect Physiol.* 15:1551–66.

Neuwirth, M. 1973. The structure of the hemocytes of *Galleria mellonella* (Lepidoptera). *J. Morphol.* 139:105–24.

Niklas, O. F. 1960. Standortein flüsse und natürliche Feinde als Begrengungsfaktoren von *Melolontha*-Larven populationen, eines Waldgebietes (Forstamt Lorsch, Hessen) (Coleoptera: Scarabaeidae). *Mitt. Biol. Bund. Land. Forst.* 101:1–60.

Nittono, Y. 1960. Studies on the blood cells in the silkworm, *Bombyx mori* L. *Bull. Seric. Exp. Stn. (Tokyo)* 16:171–266.

Ogel, S. 1955. A contribution to the study of blood cells in Orthoptera. *Comm. Facult. Sci. Univ. Ankara (Istanbul)* 4:15–33.

O'Kane, W. C., and L. C. Glover. 1935. Studies of contact insecticides. X. Penetration of arsenic into insects. *New Hampshire Agric. Exp. Stn. Tech. Bull.* 63:1–8.

Ovanesyan, T. T. 1951. On the shape of the cells of the hemolymph of the larvae of the mulberry silkworm in the presence of different physiological conditions of the organism. *Zool. Zh.* 30:86–8.

Paillot, A. 1918. *Perezia legeri* nov. sp., microsporidie nouvelle, parasite des chenilles de *Pieris brassicae. C. R. Soc. Biol.* 8:187–9.

Paillot, A. 1920. Sur la caryocinétose et les réactions similaires chez les vertébrés. *C. R. Soc. Biol.* 83:427–8.

Paillot, A. 1923. Sur une techniqne nouvelle permettant l'étude vitale du sang des insectes. *C. R. Soc. Biol.* 88:1046–8.

Paillot, A. 1924a. Sur l'étiologie et l'épidémiologie de la grasserie du ver à soie. *C. R. Acad. Sci.* 179:229.

Paillot, A. 1924b. Sur une nouvelle maladie des chenilles de *Pieris brassicae* et sur les maladies du noyaux des insectes. *C. R. Acad. Sci.* 179:1353–6.

Paillot, A. 1926. Existence de la grasserie chez les papillons de ver à soie. *C. R. Acad. Agric.* 12:201–4.

Paillot, A. 1933. *L'Infection chez les insectes.* G. Patissier, Trevoux.

Patton, R. L., and R. A. Flint. 1959. The variation in the bloodcell count of *Periplaneta americana* (L.) during a molt. *Ann. Entomol. Soc. Amer.* 52:240–2.

Pilat, M. 1935. The effects of intestinal poisoning on the blood of locusts (*Locusta migratoria*). *Bull. Entomol. Rès.* 26:283–92.

Raina, A. K., and R. A. Bell. 1974. Haemocytes of the pink bollworm, *Pectinophora gossypiella*, during larval development and diapause. *J. Insect Physiol.* 20:2171–80.

Ratcliffe, N. A., and S. J. Gagen. 1976. Cellular defense reactions of insect hemocytes in vivo: Nodule formation and development in *Galleria mellonella* (Lep., Galeriidae) and *Pieris brassicae* larvae (Lep., Pieridae). *J. Invertebr. Pathol.* 28:373–82.

Richardson, C. H., R. C. Burdette, and C. W. Eagleson. 1931. The determination of the blood volume of insect larvae. *Ann. Entomol. Soc. Amer.* 24:503–7.

Ries, E. 1932. Experimentells Symbiosestudien. II. Mycetomtransplantationen. *Z. Wiss. Biol. A* 25:184–234.

Rizki, M. T. M. 1957. Alterations in the haemocyte population of *Drosophila melanogaster. J. Morphol.* 100:437–58.

Rizki, M. T. M. 1962. Experimental analysis of hemocyte morphology in insects. *Amer. Zool.* 2:247–56.

Roan, C. C., H. E. Fernando, and C. W. Kearns. 1950. A radiobiological study of four organic phosphates. *J. Econ. Entomol.* 43:319–25.

Rooseboom, A. 1937. Contribution à l'étude de la cytologie du sang de certains insectes, avec quelques considerations générales. *Arch. Neerl. Zool.* 2:432–559.

Rosenberger, C. R., and J. C. Jones. 1960. Studies on total blood counts of the southern armyworm larva, *Prodenia eridania* (Lepidoptera). *Ann. Entomol. Soc. Amer.* 53:351–5.

Salt, G. 1970. *The Cellular Defence Reactions of Insects.* Cambridge University Press, Cambridge.

Schwartz, P. H., and B. G. Townshend. 1968. Effects of milky disease on hemolymph coagulation on the number of hemocytes in infected larvae of the Japanese beetle. *J. Invertebr. Pathol.* 12:288–93.

Shapiro, M. 1966. Pathologic changes in the blood of the greater wax moth, *Galleria mellonella* (Linnaeus), during the course of starvation and nucleopolyhedrosis. Ph.D. thesis, University of California, Berkeley, California.

Shapiro, M. 1967. Pathologic changes in the blood of the greater wax moth, *Galleria mellonella*, during the course of nucleopolyhedrosis and starvation. I. Total hemocyte count. *J. Invertebr. Pathol.* 9:111–13.

Shapiro, M. 1968a. Changes in the haemocyte population of the wax moth, *Galleria mellonella*, during wound healing. *J. Insect Physiol.* 14:1725–33.

Shapiro, M. 1968b. Pathologic changes in the blood of the greater wax moth, *Galleria mellonella*, during nucleopolyhedrosis and starvation. II. Differential hemocyte count. *J. Invertebr. Pathol.* 10:230–4.

Shapiro, M., R. D. Stock, and C. M. Ignoffo. 1969. Hemocyte changes in larvae of the bollworm, *Heliothis zea*, infected with a nucleopolyhedrosis virus. *J. Invertebr. Pathol.* 14:28–30.

Shrivastava, S. C., and A. G. Richards. 1964. The differentiation of blood cells in the wax moth, *Galleria mellonella. Amer. Zool.* 4:312–13.

Shrivastava, S. C., and A. G. Richards. 1965. An autoradiographic study of the relation between hemocytes and connective tissue in the wax moth, *Galleria mellonella* L. *Biol. Bull. (Woods Hole)* 128:337–45.

Shull, W. E., M. K. Riley, and C. H. Richardson. 1932. Some effects of certain toxic gases on the blood of the cockroach, *Periplaneta orientalis* (L.). *J. Econ. Entomol.* 25:1070–2.

Sirotina, M. I. 1961. Hematological detection of microbiological measures taken against the Colorado beetle. *Dokl. Acad. Nauk SSSR* 140:720–3.

Smirnoff, W. A. 1962. A nuclear polyhedrosis of *Erannis tiliaria* (Harris). *J. Insect Pathol.* 4:393–400.

Smirnoff, W. A. 1964. A nucleopolyhedrosis of *Operophtera bruceata* (Hulst) (Lepidoptera: Geometridae). *J. Insect Pathol.* 6:384–6.

Smith, H. W. 1938. The blood of the cockroach *Periplaneta americana* L.: Cell structure and degeneration, and cell counts. *New Hampshire Agric. Exp. Stn. Tech. Bull.* 71:1–23.

Speare, A. T. 1920. Further studies of *Sorosporella uvella*, a fungous parasite of noctuid larvae. *J. Agric. Res.* 18:399–439.

Stairs, G. R. 1964. Changes in the susceptibility of *Galleria mellonella* (Linnaeus) larvae to nuclear-polyhedrosis virus following blockage of the phagocytes with India ink. *J. Insect Pathol.* 6:373–6.

Steinhaus, E. A. 1948. Polyhedrosis ("wilt disease") of the alfalfa caterpillar. *J. Econ. Entomol.* 41:859–65.

Steinhaus, E. A., and K. M. Hughes. 1949. Two newly described species of microsporidia from the potato tuberworm *Gnorismoschema operculella* (Zeller) (Lepidoptera, Gelichiidae). *J. Parasitol.* 33:67–75.

Stephens, J. M. 1963. Effects of active immunization on total hemocyte counts of larvae of *Galleria mellonella* (Linnaeus). *J. Insect Pathol.* 5:152–6.

St. Julian, G., E. Sharpe, and R. A. Rhodes. 1970. Growth pattern of *Bacillus popilliae* in Japanese beetle larvae. *J. Invertebr. Pathol.* 15:240–6.

Sussman, A. S. 1952. Studies of an insect mycosis. III. Histopathology of an aspergillosis of *Platysamia cecropia* L. *Ann. Entomol. Soc. Amer.* 45:233–45.

Takada, M., and H. Kitano. 1971. Studies on the larval hemocytes in the cabbage white butterfly, *Pieris rapae crucivora* Boisduval, with special reference to hemocyte classification, phagocytic activity and encapsulative capacity. *Kontyû* 39:385–94.

Tateiwa, J. 1928. La formule leucocytaire du sang des chenilles normales et immunisées de *Galleria mellonella*. *Ann. Inst. Pasteur* 42:791–804.

Tauber, O. E. 1935. Studies on insect hemolymph with special reference to some factors influencing mitotically dividing cells. Ph.D. thesis, Iowa State College, Ames, Iowa.

Tauber, O. E. 1937. The effect of ecdysis on the number of mitotically dividing cells in the hemolymph of the insect *Blatta orientalis*. *Ann. Entomol. Soc. Amer.* 30:35–9.

Tauber, O. E. 1940. Mitotic response of roach hemocytes to certain pathogens in the hemolymph. *Ann. Entomol. Soc. Amer.* 33:113–19.

Tauber, O. E., and J. F. Yeager. 1935. On the total hemolymph (blood) counts of insects. I. Orthoptera, Odonata, Hemiptera, and Homoptera. *Ann. Entomol. Soc. Amer.* 28:229–40.

Tauber, O. E., and J. F. Yeager. 1936. On the total hemolymph (blood) cell counts of insects. II. Neuroptera, Coleoptera, Lepidoptera, and Hymenoptera. *Ann. Entomol. Soc. Amer.* 29:112–18.

Taylor, A. 1935. Experimentally induced changes in the cell complex of the blood of *Periplaneta americana*. *Ann. Entomol. Soc. Amer.* 28:135–45.

Toumanoff, C. 1930. L'immunisation et la phagocytose chez les larves d'abeilles. *C. R. Soc. Biol.* 103:969–70.

Toumanoff, C. 1949. Les maladies microbienne et l'immunité naturelle chez les insectes. *Rev. Can. Biol.* 8:343–69.

Toumanoff, C., and M. Lapied. 1950. Action du dichlorphenyltrichlorethane (DDT) sur les chenilles de la fausse teigne des ruches *Galleria mellonella* L. Effet de la temperature. Action sur le sang. *Acad. Agric. Fr.* 36:386–71.

Trehan, K. N., and H. R. Pajni. 1961. Mode of action of insecticides. *Beitr. Entomol.* 11:1–11.

Vago, C. 1959. Etudes cytopathologiques sur la rickettsiose bleue du coléoptère *Melolontha melolontha*. *Arch. Inst. Pasteur Tunis.* 36:585–93.

Vinson, S. B. 1971. Defense reaction and hemocytic changes in *Heliothis virescens* in response to its habitual parasitoid *Cardiochiles nigriceps*. *J. Invertebr. Pathol.* 18:94–100.

Webley, D. P. 1951. Blood cell counts in the African migratory locust (*Locusta migratoria migratorioides* Reiche and Fairmaire). *Proc. R. Entomol. Soc.* (*Lond.*) A 26: 25–37.

Weiser, J. 1956. Studie o mikrosporidich z larev Hmyzu nasich vod. II. *Cesk. Parasitol.* 3:193–202.

Weiser, J. 1961. Die mikrosporidien als parasiten der insekten. *Monogr. Z. Angew Entomol. 17.*

Weiser, J. 1963. Sporozoan infections, pp. 291–334. *In* E. A. Steinhaus (ed.). *Insect Pathology: An Advanced Treatise*, Vol. 2. Academic Press, New York.

Weiser, J., and R. L. Beard. 1959. *Adelina sericesthis* n.sp., a new coccidian parasite of scarabaeid larvae. *J. Insect Pathol. 1:*99–106.

Werner, R. A., and J. C. Jones. 1969. Phagocytic haemocytes in unfixed *Galleria mellonella* larvae. *J. Insect Physiol. 15:*425–37.

Wheeler, R. E. 1962. Changes in hemolymph volume during the moulting cycle of *Periplaneta americana* L. (Orthoptera). *Fed. Proc. 21:*123.

Wheeler, R. E. 1963. Studies on the total hemocyte count and hemolymph volume in *Periplaneta americana* (L.) with special reference to the last moulting cycle. *J. Insect Physiol. 9:*223–35.

Wigglesworth, V. B. 1937. Wound healing in an insect (*Rhodnius prolixus* Hemiptera). *J. Exp. Biol. 14:*364–81.

Wigglesworth, V. B. 1955. The role of the haemocytes in the growth and moulting of an insect, *Rhodnius prolixus* (Hemiptera). *J. Exp. Biol. 32:*649–63.

Wigglesworth, V. B. 1956. The haemocytes and connective tissue formation in an insect, *Rhodnius prolixus* (Hemiptera). *Q. J. Microsc. Sci. 97:*89–98.

Wille, H. 1974. Studies on the haemolymph of *Apis mellifera* L. 5. Relationships between the morphology of the leukocytes and four disease elements. *Mitt. Schweiz. Entomol. Ges. 47:*133–49.

Wittig, G. 1962. The pathology of insect blood cells: A review. *Amer. Zool. 2:*257–73.

Wittig, G. 1965a. A study of the role of blood cells in insect disease. *12th Int. Congr. Entomol. (Lond.) 1964:*743.

Wittig, G. 1965b. Phagocytosis by blood cells in healthy and diseased caterpillars. I. Phagocytosis of *Bacillus thuringiensis* Berliner in *Pseudaletia unipuncta* (Haworth). *J. Invertebr. Pathol. 7:*474–88.

Wittig, G. 1966. Phagocytosis by blood cells in healthy and diseased caterpillars. II. A consideration of the method of making hemocyte counts. *J. Invertebr. Pathol. 8:*461–77.

Wittig, G. 1968. Phagocytosis by blood cells in healthy and diseased caterpillars. III. Some observations concerning virus inclusion bodies. *J. Invertebr. Pathol. 10:*211–29.

Xeros, N., and K. M. Smith. 1954. Further studies on the development of viruses in the cells of insects. *Proc. Int. Conf. Electron Microsc. (Lond.), Viruses. 1954:*259–62.

Yeager, J. F. 1945. The blood picture of the southern armyworm (*Prodenia eridania*). *J. Agric. Res. 71:*1–40.

Yeager, J. F., E. R. McGovran, S. C. Munson, and E. L. Mayer. 1942. Effect of blocking hemocytes with Chinese ink and staining nephrocytes with trypan blue upon the resistance of the cockroach *Periplaneta americana* (L.) to sodium arsenite and nicotine. *Ann. Entomol. Soc. Amer. 35:*23–40.

Yeager, J. F., and S. C. Munson. 1942. Changes induced in the blood cells of the southern armyworm (*Prodenia eridania*) by administrations of poisons. *J. Agric. Res. 64:*307–22.

Yeager, J. F., and S. C. Munson. 1950. Blood volume of the roach *Periplaneta americana* determined by several methods. *Arthropoda 1:*255–65.

Yeager, J. F., and O. E. Tauber. 1932. Determinations of total blood volume in the cockroach, *P. fuliginosa,* with special reference to method. *Ann. Entomol. Soc. Amer. 25:*315–27.

Yeager, J. F., and O. E. Tauber. 1933. On counting mitotically dividing cells in the blood of the cockroach, *Periplaneta orientalis* (Linn.). *Proc. Soc. Exp. Biol. Med. 30:*861–3.

Zapol'skikh, O. V. 1976. Morphological and cytochemical studies of the worker honey-bee haemolymph cells (*Apis mellifera:* Hym., Apidae). *Tsitologiia* 18:956–63.

Zielinska, Z. M., and A. Wroniszewska. 1957. Studies on the biochemistry of the wax moth (*Galleria mellonella*). 16. Weight of tissues and organs during starvation of the larvae. *Acta Biol. Exp.* 17:345–9.

Zotta, G., and A. M. Teodoresco. 1933. Formule leucocytaire de le chenille de *Galleria mellonella* infectés par le *Leptomonas pyrrhocoris*. *C. R. Soc. Biol.* 114:314–16.

IV. Techniques

17 Identification key for hemocyte types in hanging-drop preparations

A. P. GUPTA

Department of Entomology and Economic Zoology, Rutgers University, New Brunswick, New Jersey 08903, U.S.A.

Contents

17.1. Introduction

The identification key presented as Table 17.1 is for the novice who is studying insect–or for that matter any other arthropod–hemocytes for the first time. It is often very frustrating for a beginner to try to identify various types of hemocytes in a hemolymph sample or film under light microscope. This key should be helpful in guiding the beginner to become acquainted with the seven main types of hemocytes described in Chapter 4. Because not all hemocyte types are readily observed in any one species, at all developmental stages, and under all physiological conditions, it is important to examine hemolymph samples from different species at various developmental stages and under different physiological conditions. The method of study is no less important. The key is based on hemocyte observations under a phase-contrast microscope in hanging-drop preparations of fixed or unfixed hemolymph from various insects (Gupta and Sutherland, 1966, 1967; Gupta, 1968, 1969).

17.2. Selected species for examination

Although the key can be used to identify hemocytes from any species, I suggest the following insects be used in the beginning: adult *Blaberus* spp. for typical prohemocytes (PRs), plasmatocytes (PLs), granulocytes (GRs), and spherulocytes (SPs); larvae of *Galleria mellonella* and *Porthetria dispar* for typical PRs, PLs, oenocytoids (OEs), SPs, and adipohemocytes (ADs), particularly in larvae about to molt, which

527

Table 17.1. Identification key for insect hemocyte types

1.	Nucleus compact, large in relation to cell size, centrally located, almost filling the cell; very thin, peripheral layer of homogeneous cytoplasm around the nucleus; cells may be round, oval, or elliptical, but always small (compared with other cells in sample) (Figs. 4.1A, 13.8)	Prohemocyte (PR)
1'.	Nucleus not compact, generally small in relation to cell size, not nearly filling the cell	2
2 (1').	Nucleus with chromatin arranged in cartwheel-like fashion, generally eccentric, oval, and sharply outlined; cytoplasm hyaline, generally scant, may contain some spherical or elongate granular inclusions; cell sometimes with cystlike blebs in process of exocytosis (Figs. 4.2D, 7.5, 10.2C, 13.17) . . . . (Beware! COs may be confused with OEs and with GRs in some insects)	Coagulocyte (CO)
2'.	Nuclear chromatin not arranged in cartwheel-like fashion, nucleus eccentric or central; cytoplasm not hyaline, abundant, homogeneous, and without any plate-, rod-, or needlelike inclusions and filaments	3
3 (2').	Cytoplasm generally agranular or slightly granular; nucleus round or elongate and central, and may or may not appear punctate; cells polymorphic and variable in size in various insects (Figs. 4.1B, 13.10) . . .	Plasmatocyte (PL)
3'.	Cytoplasm generally agranular, thick, and homogeneous with or without several kinds of plate-, rod-, or needlelike inclusions and filaments; "vacuoles" may or may not be present; nucleus generally small, round or elongate, and generally eccentric; cells variable in size and shape, generally lyse quickly in vitro, ejecting material into the hemolymph (Figs. 4.2A,B, 8.14, 10.1E, 13.23)	Oenocytoid (OE)
4 (3').	Cytoplasm distinctly granular	4'
4'.	Cytoplasm prominently and characteristically granular; granules may or may not be numerous; nucleus comparatively small (compared with that in plasmatocyte) and compact, round or elongate, and generally central (Figs. 4.1D, 13.14) . . . . (Note that GRs in lower orders are generally larger than in higher orders.)	Granulocyte (GR)
5 (4').	Granules in cytoplasm considerably enlarged and appear as distinct spherules or droplets	5'
5'.	Spherules nonfringent, generally obscuring the nucleus, number of spherules varying from few to many; nucleus rather small, central or eccentric; cells ovoid or round with variable sizes, usually larger than granulocytes, and may be observed releasing material from spherules into hemolymph by exocytosis (Figs. 4.1C,E, 13.19)	Spherulocyte (SP)
6 (5').	Spherules or droplets refringent owing to presence of lipid; nucleus relatively small (compared with that in plasmatocytes or spherulocytes), round or slightly elongate, central or eccentric, and may or may not appear concave, biconvex, punctate, or lobate; cytoplasm may contain other nonlipid granules (Figs. 4.2C, 10.1C, 10.4 (larger cell)	Adipohemocyte (AD)

is when GRs accumulate lipids and appear as ADs; and larval *Tenebrio molitor* for ADs, particularly after the larvae have been chilled at 5 °C for 20–24 hr.

17.3. Procedure for hanging-drop preparation

1. Take a square coverslip and put a tiny drop of saline-versene (NaCl, 0.9 g; KCl, 0.942 g; CaCl$_2$, 0.082 g; NaHCO$_3$, 0.002 g; distilled water, 100 ml + 2% versene).
2. Cut the tip of the antenna or leg (or proleg) and let a drop of hemolymph flow into the saline-versene drop.
3. Carefully turn the coverslip upside down and place it over the depression of a depression or cavity slide. Seal the sides of the coverslip with petroleum jelly.
4. Examine the hanging-drop preparation under the phase-contrast microscope.
5. You will notice that the hemocytes are more evenly distributed near the periphery than at the center of the hemolymph drop. It will, therefore, be easier to focus them sharply near the periphery.
6. Make a fresh preparation for examination every 8–10 min because hemocytes begin to deteriorate after bleeding.
7. For longer-lasting preparations, you may try hemolymph from an insect that has been heat-fixed in water at 60 °C for about 5 min.

17.4. Summary

A beginner's identification key for seven main hemocyte types (prohemocytes, plasmatocytes, granulocytes, spherulocytes, adipohemocytes, coagulocytes, and oenocytoids) is presented. It is suggested that hanging-drop preparations of hemolymph be used to identify these hemocytes under phase-contrast microscope.

References

Gupta, A. P. 1968. Hemocytes of *Scutigerella immaculata* and the ancestry of Insecta. *Ann. Entomol. Soc. Amer. 61*(4):1028–9.

Gupta, A. P. 1969. Studies of the blood of Meloidae (Coleoptera). I. The haemocytes of *Epicauta cinerea* (Forster), and a synonymy of haemocyte terminologies. *Cytologia 34*(2):300–44.

Gupta, A. P., and D. J. Sutherland. 1966. *In vitro* transformations of the insect plasmatocyte in certain insects. *J. Insect Physiol. 12*:1369–75.

Gupta, A. P., and D. J. Sutherland. 1967. Phase contrast and histochemical studies of spherule cells in cockroaches. *Ann. Entomol. Soc. Amer. 60*(3):557–65.

18   Insect hemocytes under light microscopy: techniques

J. W. ARNOLD AND C. F. HINKS

Biosystematics Research Institute, Research Branch, Agriculture Canada, Ottawa, Ontario K1A 0C6, Canada

Contents

18.1. Introduction

In general, the techniques used in vertebrate hematology must be modified for the study of insect hemocytes. The same principles of technique apply, and the insect hemocytes can be observed in vivo, in vitro, in living culture, and in blood films fixed and stained in a variety of ways. The procedures described below are a small sampling of these techniques.

Each procedure provides a somewhat different view of the cells, and it is often useful to employ more than one method, if possible. Such a combination of methods at best includes an in vivo or in vitro technique as a basis for interpretation of fixed and stained cells. To some extent, all the methods are empirical because of variability in the character of both hemolymph and hemocytes in different species. Most of the following techniques have been found suitable for a wide variety of insects, but there are preferred methods for certain species

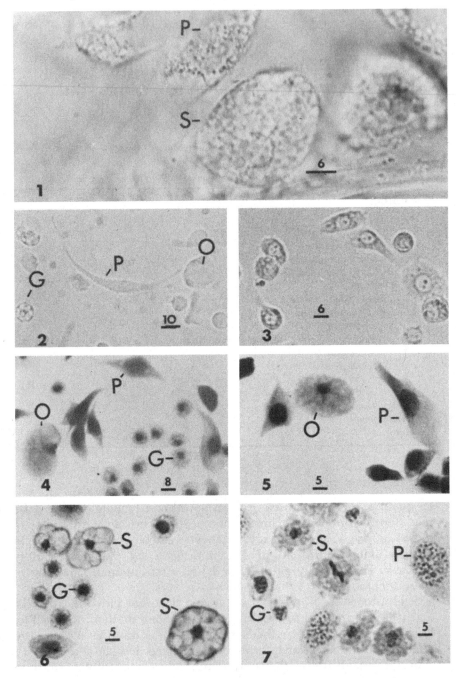

Figs. 18.1–18.7 Hemocytes of insects as seen with various techniques. Magnification as indicated, in microns. **18.1** Hemocytes of *Blaberus giganteus* (Dictyoptera) in vivo, in a wing vein. **18.2.** Hemocytes of *Euxoa declarata* (Lepidoptera) in vitro, as a wet film. **18.3.** Hemocytes of *Malacosoma disstria* (Lepidoptera) in vitro, in subcul-

and for different purposes, and the in vivo procedure is obviously limited to species with transparent areas of the body.

18.2. In vivo procedure

A relatively clear and sometimes excellent view of living hemocytes can be obtained with transmitted light through transparent regions or appendages of certain insects. The procedure is similar to the one long used by vertebrate hematologists to demonstrate living blood cells in thin tissues, such as the ear membrane of the rabbit or the toe membrane of the frog. Simply, it involves immobilization of the insect so that the transparent structure can be manipulated under the compound microscope. Wings are most suitable for this purpose, and for best results they are sandwiched in glycerol or refined immersion oil between thin glass coverslips to reduce diffraction at the cuticle-air interface (Arnold, 1959). The wings of orthopteroid insects are best by far, as the blood circulates freely there in thin veins and sinuses, and the hemocytes are large. Particularly among the large blaberoid cockroaches, the hemocytes can be observed here with great clarity at high magnification (Fig. 18.1), circulating freely in the blood or moving ameboidally in regions out of the main flow (Arnold, 1961). Not all clear-winged insects are so satisfactory, for a variety of reasons (Arnold, 1964), including the small size of the hemocytes (e.g., among Hemiptera), the extreme thickness of the vein walls (e.g., Odonata), the near occlusion of veins by tracheae (e.g., Lepidoptera), the poor circulation of blood in the wings (e.g., Diptera), or the extremely rapid circulation of blood in the wings (e.g., Hymenoptera). Nevertheless, with ingenuity and patience many such insects can be used to gain an impression of the real size and form of the living hemocytes before resorting to techniques that involve injury or sacrifice of the insect or denaturation of the hemocytes. Similarly, with ingenuity and patience other appendages and clear regions of the insect body can serve as windows to the blood. The legs, prothoracic extensions, caudal cerci, and/or respiratory structures of some aquatic insects can be useful, and the dorsum of the abdomen sometimes provides a view of hemocytes in the dorsal sinus or heart. In these cases too, clarity is improved by covering the structure or area with saline (with a trace of wetting agent) or glycerol or immersion oil under a coverslip. Al-

(*continued from facing page*)
ture. **18.4.** Hemocytes of *E. declarata* after rapid Giemsa staining. **18.5.** Hemocytes of *E. declarata* after full Giemsa staining. **18.6.** Hemocytes of *E. declarata* after hematoxylin–eosin–alcian blue staining. Note very selective staining of spherulocytes. **18.7.** Hemocytes of *E. declarata* after acetocarmine staining. Note distinctions between granulocytes and spherulocytes. Note mitosis in the latter. *S* = spherulocyte; *G* = granulocyte; *O* = oenocytoid; *P* = plasmatocyte.

though phase-contrast microscopy can be used here, it is often less effective than standard light microscopy with the condenser manipulated to improve contrast and depth of field (see also Chapter 20).

18.3. In vitro procedure

This procedure includes a variety of techniques that involve the transfer of blood from the insect to glass without or with little exposure to air. There are two common ones: blood under oil and as a wet film under a coverglass. Perhaps the simplest technique is to sever the antenna under mineral oil on a glass slide. Blood issues from the cut end and forms a discrete globule on the slide, protected from the air by the inert oil. Under oil immersion, or at lower magnification with an immersion adapter, the hemocytes can be observed in their normal form as they flow from the antenna and in gradually changing form on the glass surface.

The more common technique, the wet film, involves directly transferring a drop of blood to the microscope slide and covering it immediately with a coverslip ringed with petroleum jelly to exclude air. The hemocytes are seen here clearly (Fig. 18.2), both in suspension for a short time and attached to the glass. They slowly alter their form and size under these conditions and should be examined immediately. It is here that the explosive so-called coagulocytes may be identified best. The view of cells in culture is somewhat similar to this (Arnold and Sohi, 1974), through the flat surface of the standard tissue culture flask (Fig. 18.3), but the typical hemocyte forms are not maintained in subcultures.

The technique for wet films can be used effectively also with treated blood. Heat-fixed blood can be used directly, or living blood can be diluted in fluids that suppress coagulation, such as Turk's diluting fluid (Fisher) or dilute solutions of sequestering agents such as tetrasodium ethylenediaminetetraacetate. Here the form of the hemocytes is retained for longer periods than in wet films of living blood.

18.4. Blood film preparation

Although vertebrate blood can be prepared for microscopy by placing a drop directly on a microscope slide and drawing it out with a coverslip to dry quickly in air, insect blood treated in this way will usually agglutinate before the film can be made or else will show much cellular distortion. For this reason, it is best to fix insect blood before preparing the film. This is accomplished in two ways: by a suitable degree of heat fixation or by dilution of the blood in a fixative that does not immediately cause gelation of the hemolymph. Heat fixation can be followed by other types of fixation if desired, after the film has dried on the slide.

The vertebrate technique, using live blood directly on the slide, can serve in special preparations where cell form is less important than nuclear integrity. In such case, a bead of fresh blood is exposed immediately to acetic acid vapor for rapid fixation. This procedure, as detailed below, is effective for demonstrating mitosis in hemocytes, using nuclear stains.

18.4.1. Fixation and slide preparation

Heat can serve to fix hemocytes very rapidly within the insect, without appreciable change to their shape or size, and at the same time prevent coagulation of the hemolymph. Consequently, after heat fixation, the blood can be withdrawn from the insect and spread easily on a microscope slide without clumping or distortion of the cells. At the same time, the film of blood and its cells adheres evenly to the glass without need for adhesives. Heat fixation is accomplished by plunging the insect into water held at approximately 60 °C for a short period. The temperature can be varied within 5 °C, depending on the size of the insect, and the exposure varied from 1 to 10 min on the same basis. Within these limits, the temperature and time are not critical, but should be regulated for particular species by empirical testing. Slide preparation with heat-fixed blood is a simple process of placing a drop on the slide and touching it with the edge of a coverslip, which is then drawn along the slide with the drop trailing behind. The resulting blood film is air-dried (preferably on a slide warmer at about 32 °C) before staining.

Dilution of living blood directly into a water-soluble chemical fixative also accomplishes rapid fixation of the hemocytes with little coagulation of the hemolymph or cell distortion. It is the preferred method for some insects and for some purposes, especially where a succession of blood samples is required from the same insect. Dilute solutions of formalin (5%) or glutaraldehyde (0.4 M) are recommended. The procedure involves the pooling of some of the fixative solution on the microscope slide, immersing an appendage such as the antenna in it, and severing the appendage so that the blood flows directly into the solution without contacting the air. Blood from other types of wounds can also be dropped from above directly into the fixative so that exposure to air is very brief. In either case the mixture is stirred immediately to prevent clumping of the cells and can be spread on the slide if desired. The mixture is then air-dried completely on the slide and rinsed to remove traces of fixative before staining.

18.4.2. Staining

For most purposes, hemocytes are stained in one of the Romanowsky preparations, which depend on the formation of azure and other oxidation products of methylene blue, usually in combination with eosin (Humason, 1967). With these preparations, variation of the buffer level toward the acid side increases the precision of nuclear staining and decreases cytoplasmic basophilia; the reverse increases the amount of blue in various elements. One can, therefore, alter the effect of the stain for different purposes. We find that the Giemsa preparation is most reliable and use it almost exclusively. It can be used directly for rapid staining or with differentiation for more elegant results (see also Chapter 20).

Rapid Giemsa staining (Fig. 18.4)

Solution of Giemsa (Fisher): 1 drop of concentrate per milliliter distilled water in a Stender dish. Place air-dried slides directly in the solution or after 1 min in absolute methyl alcohol. Inspect slides for depth of staining after 3 min (max. 5 min). Rinse in distilled water for 1 min. Blot-dry using Kodak lens-cleaning tissue. Mount permanent slides in Canada balsam; temporary slides in glycerol or immersion oil.

Full Giemsa staining (Fig. 18.5)

Immerse air-dried films of heat-fixed hemolymph in Giemsa solution (1 drop of concentrate per milliliter distilled water) for 20 min to 2 hr. Rinse in distilled water; then immerse briefly in distilled water to which a few drops of lithium carbonate have been added (to differentiate red-staining structures). Rinse in distilled water, then immerse briefly in distilled water to which a few drops of dilute hydrochloric acid have been added (to differentiate blue-staining structures). Rinse in distilled water and examine. Repeat differentiation if staining is too dense. Blot dry using Kodak lens-cleaning tissue. Mount in Canada balsam.

Comments. Cells are better differentiated than when rapid Giemsa method is used; excellent for photography.

Hematoxylin–eosin–alcian blue staining (Fig. 18.6)

Immerse air-dried films of heat-fixed hemolymph in 15%acetic acid in methanol for 20 min. Hydrate through graded alcohols to distilled water. Immerse in 1% alcian blue 8GX in 0.1 N HCl for 20 min. Rinse in distilled water and immerse in Harris's hematoxylin for 20 min. Rinse in distilled water and differentiate in acid alcohol; then blue in Scott's solution. Examine and repeat differentiation if necessary. Dehydrate to 90% alcohol and immerse in 1% eosin in 90% alcohol for 3 min. Rinse off excess stain in 90% alcohol, transfer to absolute alcohol, then to xylene; mount in Canada balsam.

Comments. Mitotic cells are easier to identify than in Giemsa-stained preparations. Spherulocytes are very distinct, retaining pale blue, whereas the cytoplasm of all other cell types stains pink to pinkish purple.

Acetocarmine staining (Fig. 18.7)
Express small bead of fresh hemolymph from a CO_2-narcotized insect onto the center of a round coverglass and invert over a vial of glacial acetic acid for 90 min. Reverse coverglass with hemolymph uppermost in a solid watch glass and cover with acetocarmine (Humason, 1967) for 90 min. Blot off excess stain carefully and squash the preparation gently onto a microscope slide in a drop of Venetian turpentine.

Comments. Cell types are identifiable; mitotic cells very clear in all cell types; excellent for photography.

Acridine orange staining
Express a small drop of fresh hemolymph from a CO_2-narcotized insect onto a quartz microscope slide. Mix with an equal volume of acridine orange solution, 0.1 mg/ml in 0.9% NaCl. Place coverglass over hemolymph preparation and immediately examine with fluorescence microscope.

Comments. The inclusions of spherulocytes are rapidly and specifically stained to give an intense orange fluorescence. Nuclei of all hemocytes give a yellowish green fluoroscence.

18.5. Summary
Techniques for light microscopy of insect hemocytes are modifications of those used in vertebrate hematology and include in vivo and in vitro procedures as well as methods of fixation and staining. The in vivo procedures utilize transparent regions of the insect body, preferably the wings sandwiched in an inert solution under glass. The in vitro procedures include the examination of blood issuing from a wound under oil, of the cells in culture, or more commonly of preparations of covered wet films of living blood ringed with inert oil to exclude air. Preparation of insect blood for standard histological staining should be preceded by the killing of the insect in hot water to fix the cells and prevent hemolymph coagulation or by dilution of the living

blood directly in a selected fixative solution that does not cause gelation of the hemolymph. Giemsa is probably the best Romanowsky stain for general viewing; other stains, such as hematoxylin–eosin–alcian blue or acetocarmine, are more selective for cell types and for mitoses, respectively. Acridine orange is a very selective stain for spherulocytes.

References

Arnold, J. W. 1959. Observations on living hemocytes in wing veins of the cockroach *Blaberus giganteus* (L.) (Orthoptera: Blattidae). *Ann. Entomol. Soc. Amer.* 52:229–36.

Arnold, J. W. 1961. Further observations on amoeboid haemocytes in *Blaberus giganteus* (L.) (Orthoptera: Blattidae). *Can. J. Zool.* 39:755–66.

Arnold, J. W. 1964. Blood circulation in insect wings. *Mem. Entomol. Soc. Can.* 38:1–48.

Arnold, J. W., and S. S. Sohi. 1974. Hemocytes of *Malacosoma disstria* Hübner (Lepidoptera: Lasiocampidae): Morphology of the cells in fresh blood and after cultivation *in vitro*. *Can. J. Zool.* 52:481–5.

Humason, G. L. 1967. *Animal Tissue Techniques*. Freeman, San Francisco.

19 Techniques for total and differential hemocyte counts and blood volume, and mitotic index determinations

M. SHAPIRO

Gypsy Moth Methods Development Laboratory, U.S. Department of Agriculture, Otis Air Force Base, Massachusetts 02542, U.S.A.

Contents

19.1. Introduction

The three most common measurements made to describe the blood picture of a given insect at a given time or from one time to another are total hemocyte count (THC), differential hemocyte count (DHC), and blood volume (BV). The purpose of this chapter is to examine the methods used to obtain these values and their reliability.

19.2. Total hemocyte count

The first study of THCs in insects was made by Tauber and Yeager (1934). In 1935, they studied Orthoptera, Odonata, Hemiptera, and Homoptera. A year later, these same authors extended their study to include Neuroptera, Coleoptera, Lepidoptera, and Hymenoptera. The insects were heat-fixed (60 °C for 5–10 min) and bled from a proleg, after which the blood sample was diluted with physiological saline. The THC (i.e., the number of circulating hemocytes per cubic millimeter) was determined by the method employed for mammalian blood counts. This work represents an outstanding contribution and has served as a model for subsequent investigations.

Tauber-Yeager (1935) fluid (NaCl, 4.65 g; KCl, 0.15 g; $CaCl_2$,

0.11 g; gentian violet, 0.005 g; and 0.125 ml acetic acid/100 ml) was used also by Fisher (1935), Smith (1938), and Shapiro (1967, 1968). Other physiological saline solutions were utilized by Rosenberger and Jones (1960), Collin (1963), and Gupta and Sutherland (1968). Acetic acid, a component of the Tauber-Yeager fluid, was used for *Galleria* (Stephens, 1963; Shapiro, 1966; Jones, 1967a) and for *Heliothis* (Shapiro et al., 1969; Vinson, 1971). Turk's solution (1–2% glacial acetic acid, slightly colored with gentian violet) was used for *Pectinophora* (Clark and Chadbourne, 1960) and for *Euxoa* (Arnold and Hinks, 1976). Patton and Flint (1959) reported that Turk's solution did not prevent coagulation in hemolymph samples of *Periplaneta*. Versene (1–2% tetrasodium EDTA) was superior to Turk's solution and oxalate and was routinely used. Wittig (1966), studying phagocytosis in *Pseudaletia* larvae, also found versene (2% plus a trace of methylene blue) to be superior to glacial acetic acid. Formalin (10% in 0.85% NaCl) has also been employed as a diluting fluid (Jones, 1956). Physiological saline did not prevent cell agglutination, but the addition of acetic acid (1%) or formaldehyde reduced clumping.

In making total counts from *Pieris*, the first and second drops of hemolymph were utilized (Kitano, 1969). The first drop of hemolymph in unfixed and unfed *Rhodnius* contained more hemocytes than the second drop (Jones, 1962). The number of hemocytes was also reduced from three successive drops of *Bombyx* hemolymph (Matsumoto and Sakurai, 1956). On the other hand, Wittig (1966) found no significant differences in DHCs taken from the first and second drops of *Pseudaletia* hemolymph. Jones (1956) used the first drop of *Sarcophaga* hemolymph for DHCs. The rest of the hemolymph was placed on a second slide, and a portion was used for THCs. Some 3 or 4 drops of *Prodenia* hemolymph were allowed to flow on a glass slide. A portion of the blood was drawn into a Thoma white blood cell pipette, diluted, and counted (Rosenberger and Jones, 1960). After the hemolymph was diluted in a pipette, the first 3 or 4 drops were discarded (Jones, 1967a; Shapiro, 1967, 1968).

In many instances, hemolymph was drawn into a Thoma white blood cell pipette, diluted, and counted in a hemacytometer. The hemolymph dilution ranged from 1:20 (Rosenberger and Jones, 1960; Wittig, 1966) to 1:50 (Clark and Chadbourne, 1960; Shapiro, 1967) to 1:100 (Fisher, 1935; Gupta and Sutherland, 1968). Wittig (1966) adjusted the dilution of hemolymph so that a suspension contained between 800 and 1,600 cells/mm^2 area counted. This adjustment could not be made when the amount of blood available or the number of hemocytes per cubic millimeter was low.

Fisher (1935) found that the standard white cell pipette required too much hemolymph from *Periplaneta*. A micropipette was made that required only 1.7 μl and a final dilution of 1:44. Patton and Flint (1959) also used a special micropipette for *Periplaneta* in which 1 μl of hemolymph could be drawn and diluted 1:100.

THCs are counted in a standard hemacytometer according to the formula (Jones, 1962):

$$\frac{\text{hemocytes in } x \text{ 1-mm squares} \times \text{dilution} \times \text{depth of chamber}}{\text{number of 1-mm squares counted}}$$

Kitano (1969) counted the number of hemocytes in the smallest square (0.00025 mm²), counted 80 squares, and multiplied by a factor of 4,000 to give THC. Fisher (1935) counted hemocytes from three of the four white cell squares in each of the two chambers. Wittig (1966) calculated the THC per cubic millimeter by counting cells in four 1-mm² areas in each of the two chambers. Hemocytes from five 1-mm² squares (the four corner and central squares) were counted by Rosenberger and Jones (1960), Jones (1967a), Shapiro (1966, 1967), and Gupta and Sutherland (1968). White cells in all nine 1-mm² squares were counted by Clark and Chadbourne (1960).

In counting the hemocytes, a variation is to be expected. But when the distribution of the cell count was uneven and clumping was observed, the counts were discarded (Rosenberger and Jones, 1960; Shapiro, 1967; Kitano, 1969). Stephens (1963) questioned whether it was reasonable to discard counts when the means of the two chambers differed greatly. Wittig (1966) felt that such a rejection was justified if it was done on the basis of percent of the mean instead of the number of cells per volume, "for the weight of a number is different for high and low means."

19.3. Differential hemocyte count

In general, the method used by Shapiro (1966) may be considered typical. Larvae were submerged in a hot-water bath (56–58 °C for 1–2 min), and a proleg on the sixth abdominal segment was cut with fine scissors. The hemolymph was allowed to fall on a clean, grease-free microscope slide, and a smear was made in the conventional manner, by drawing a second slide across the first one at a 45 ° angle. The smear was allowed to air-dry and was stained by a modified Pappenheim-panoptic method (Pappenheim, 1914). The details of the method are as follows: (1) flood air-dried smear with May-Grünwald solution and allow to remain on slide for 3 min; (2) add distilled water so that a layer is formed and allow to stand for 2 min; (3) discard the May-Grünwald-water layer; (4) flood slide with Giemsa solution (1 part concentrated Giemsa to 40 parts distilled water) and allow to dry; (5) wash slide in running tap water and allow to dry.

The smear is examined under oil immersion, and 200 cells per slide are differentiated. By this method of staining, azurophilic material appears purple red; chromatin, reddish violet; and basic protoplasm, blue (Pappenheim, 1914).

Using the Yeager (1945) classification system as modified by Jones (1959), the following types of hemocytes were counted: prohemocytes (PRs), plasmatocytes (PLs), granulocytes (GRs), adipohemocytes (ADs), spherulocytes (SPs), podocytes (POs), and oenocytoids (OEs). Degenerating cells and cells in mitosis were also counted. In addition, hemocytes were found that could not be placed in the preceding classes with certainty; these cells were designated as unclassified cells. Jones (1962) recommended that a minimum of five insects of a given stage and physiological status be used. Whenever possible, a minimum of 200 cells should be classified per insect.

In *Rhodnius*, 100 hemocytes were classified from each of five insects as either PRs, PLs, GRs, or OEs. Mitotically dividing cells were also counted (Jones, 1967b). Vinson (1971) examined stained blood films from a minimum of five *Heliothis* larvae per time period. A minimum of 150 cells per larva was classified. A minimum of 200 and a maximum of 5,000 cells were counted in *Tenebrio* larvae (Jones and Tauber, 1954). Jones (1967a) counted 200–1,000 cells in each *Galleria* larva. Whenever possible, 200 cells were counted in *Sarcophaga* (Jones, 1956), in *Periplaneta* (Gupta and Sutherland, 1968), and in *Drosophila* (Nappi and Streams, 1969). Arnold and Hinks (1976) examined 200 hemocytes per smear of the noctuid, *Euxoa*. Five smears were examined per instar.

From 15 to 20 sections of each face fly larva, *Orthellia*, were examined, and a minimum of 100 cells was counted per section (Nappi and Stoffolano, 1972). Wittig (1966) counted 350–400 hemocytes for each DHC from *Pseudaletia*. Yeager (1945) attempted to count at least 400 cells in each DHC from *Prodenia*, but could not from young larvae, old pupae, and adults. Differential counts from *Anagasta* larvae were obtained by classifying 500 cells per smear. At least 10 larvae from each larval stage were used (Arnold, 1952a). In a subsequent study on the effects of fumigants on the hemocytes of *Anagasta*, Arnold (1952b) made DHCs from 20 larvae per time period; 500 cells were also counted in blood smears of *Blaberus* (Arnold, 1969).

19.4. Blood volume

BV is a little used but important value. It has been reported that an inverse relationship exists between the THC and BV in *Bombyx* (Nittono, 1960), in adults of *Locusta* (Webley, 1951), and in last-stage nymphs and adults of *Periplaneta* (Wheeler, 1962, 1963). In addition,

Wheeler (1963) found that the absolute number of circulating hemocytes per cubic millimeter, obtained by multiplying the BV and the THC, was relatively constant, notwithstanding changes in both the BV and the THC.

In a few instances, BVs were determined by weighing insects, removing as much hemolymph as possible, and reweighing the insects (Richardson et al., 1931; Arnold and Hinks, 1976). This exsanguination method appears crude, as more sophisticated methods are available. Smith (1938) employed the cell dilution method of Yeager and Tauber (1932). The following formula was used:

$$V_0 = \frac{d \cdot c_1}{co - c_1} + a;$$

where V_0 = total blood volume; d = amount of dilution fluid in cubic millimeters; co = original cell count per cubic millimeter; c_1 = diluted blood cell count per cubic millimeter; and a = volume of blood drawn in making the original count.

A dye solution method (Yeager and Munson, 1950) was used for *Sarcophaga* (Jones, 1956), *Tenebrio* (Jones, 1957), and *Galleria* (Shapiro, 1966; Jones, 1967a). An 0.2% amaranth red dye in 0.85% NaCl was injected into *Sarcophaga* larvae and pupae, 5% of body weight (Jones, 1956). *Galleria* larvae were injected with 10 μl of 1% amaranth red in saline per gram body weight. The dye was allowed to circulate within the hemocoel for 3–5 min. Then hemolymph was drawn, and the intensity of color was compared to a series of standards. Hemolymph volume percent were converted into microliters (Jones, 1967a).

Shapiro (1966) investigated the BV of *Galleria* larvae. The dye method employed was essentially that used by Yeager and Munson (1950) and modified by Lee (1961) with further modifications by Martignoni and Milstead (pers. comm.). Each larva was weighed and injected with a 1% aqueous amaranth solution. The volume injected was equal to 5% of the insect's body weight. After the amaranth was injected, the larva was placed in a shell vial (1.5 × 6.4 cm), and the dye was allowed to circulate within the hemocoel. After 10 min, a proleg on the sixth abdominal segment was cut, and the blood was collected in a capillary tube (Kimax No. 34500) that had been flooded with pure nitrogen to retard melanization. The blunt end of the tube was sealed on an alcohol burner, and the tube was refrigerated (4 °C) for several minutes and centrifuged (3,100 rpm for 10 min). The tube was recooled, and the portion of the tube containing sedimented hemocytes was cut off and discarded. The plasma was drawn into a disposable Drummond micropipette (10-μl capacity) and diluted in 1 ml Aronsson's buffer (0.995 M). This solution and the standard (Aronsson's buffer) were placed in separate 1.0-ml cuvettes (10-mm light path) and the relative absorbances were determined with a Beckman DB spectrophotometer at 515 mm, the wavelength at which the maximum absorbance of amaranth occurred.

Previously, the absorbances of known concentrations of amaranth had been determined, and, in accordance with Beer's law, the absorbance was proportional to the concentration. The absorbance value of the test sample was plotted against the concentrations of known samples, and the concentration of the test sample was thus obtained. Once the concentration of amaranth in the test sample had been determined, the blood volume was calculated by the following formula:

$$V = \frac{d(c' - c'')1}{(c'')}$$

where V = blood volume in microliters; d = volume of dye injected in microliters; c' = original concentration in percent; c'' = concentration of dye after circulation in percent. In order to obtain the blood volume, V is divided by the body weight of the larva.

19.5. Reliability as influenced by internal and external factors

Reliability of data is influenced by two factors, acting separately or in concert: (1) internal factors, which are related to real differences between insects, and/or (2) external factors, caused by problems in techniques. These two areas will be discussed and assessed.

19.5.1. Internal factors

Some of the variations between similar insects have been shown to be related to real factors in physiology (Arnold, 1974). Patton and Flint (1959) observed that blood cell counts of *Periplaneta* varied significantly intraspecifically and with the stage of development. The mean cell count was 50,400/mm³ with a range of 21,000–99,000 and a standard deviation of 21,000. The authors concluded that a norm must be established for each insect before the THC can be employed as a measure of the physiological state. Collin (1963) showed that the variability of the total count in *Melolontha* was about 30%. He felt that the great variability in the relative proportions of hemocytes (DHC) made it difficult to use that measure as an index of the physiological or pathological state of the insect. Jones (1967b) found the great variability of unfixed DHCs of *Rhodnius* was not attributable to great variations between sequential 100-cell counts per sample or to great differences between the sexes.

Although total counts from insects were variable, the range obtained was "quite comparable to the range obtained by other investigators from mammals" (Tauber and Yeager, 1935). Yeager (1945) tested the reliability of the DHC by examining smears from 100 insects and determining the larval stages of the donors. Some 76% of the smears were identified correctly as to larval stage. Moreover, 82% of these positives were identified correctly to within 48 hr of the true age. In *Galleria*, the ratio of round to fusiform PLs could be utilized as an index of larval weight of insects within the rearing colony (Shapiro, 1966).

Insects used for hematological studies should be as uniform as possible; of the same stage, instar, age, size (Wittig, 1966), and possibly sex. In last-stage armyworms, even among an apparently homogeneous group, individual THCs ranged from 75% to 140% of the mean of the group. The mean of a treated group is related to the average count from a group of untreated insects not only at a given time, but also during a particular time frame in insect development, when cell population changes may occur (Wittig, 1966).

Because of inherent variations, replications must be made. In *Pseudaletia*, more replications of THCs were required than of DHCs, because of greater variabilities in the former. Severe treatments, which might lead to overt disease or death, would be expected to result in hematological changes (Wittig, 1966). Jones and Tauber (1954) noted that normal or high THCs among mealworms had little significance, but greatly reduced THCs accompanied by abnormal hemocytes signified morbidity and eventual mortality. Rosenberger and Jones (1960) believed that the THC in *Prodenia* was not a good indication of health, as normal counts were obtained in both healthy and unhealthy insects. When an insect can recover from a given treatment or condition, however, changes in the blood picture will probably be slight and well within the range of variations found among untreated insects (Wittig, 1966).

19.5.2. External factors

In general, the external factors may be (1) inherent problems of the method itself and/or (2) failure of the investigator to utilize a given method correctly. It is quite possible that two methods, both used to obtain the same data for a given parameter, may have differences in sensitivity. For example, exsanguination of insects for BV measurements is less precise than the dye dilution method. Heat fixation would be useless for studying coagulation in insect hemolymph, as coagulocytes can be recognized only in unfixed preparations. The choice of a given method is dependent upon the types of data required. Differences in total counts between heat-fixed and unfixed hemolymph from stoneflies were felt to be attributable primarily to technical difficulties (Arnold, 1966).

When a given technique is selected, it must be followed precisely.

Even small differences in sampling and counting procedures could lead to large differences in the data produced. The "technique should not be varied in its details during a test, unless it has been established that such changes do not affect the result" (Wittig, 1966). Because it is often difficult to obtain consistent results owing to variations within the insect population, it becomes important not to add more variability to the system. It would be beneficial to the field of insect hematology if workers using the same species of test insect would use the same techniques, so that variations attributable to different techniques could be minimized.

19.6. Summary

Techniques are available to investigate changes in hemocyte populations regarding cell numbers (THCs), cell types (DHCs), and blood volumes. Moreover, these parameters can be combined to determine the absolute numbers of hemocytes within an insect at a given time or through time. Unfortunately, many studies have involved a single parameter. The reliability of these techniques must be critically evaluated so that only the best available ones are used in hematological studies. We must answer the question: Does heat fixation preserve the hemocyte population in situ or does it artifactually cause large numbers of hemocytes to enter the circulation. Perhaps, each method should be defined as to its best usage, so that a mosaic of techniques might be utilized to best advantage.

Once a method is selected for use, it must be followed precisely. Variations in a method might produce variations in results. Assuming that a given method is carried out well, more attention must be given to the test insect. The population should be defined as to age, stage, size, and sex in order to minimize inherent variations. If periodicity is a problem as far as reproducibility of data is concerned, then tests must be carried out at the same time. Is periodicity of hemocyte populations a problem? The growth and development of the insects should be optimal under laboratory conditions. The use of semisynthetic artificial diets and improved rearing techniques should improve the synchrony of growth and minimize the variability of food materials.

References

Arnold, J. W. 1952a. The haemocytes of the Mediterranean flour moth, *Ephestia kühniella* Zell. (Lepidoptera: Pyralididae). *Can. J. Zool.* 30:352–64.

Arnold, J. W. 1952b. Effects of certain fumigants on haemocytes of the Mediterranean flour moth, *Ephestia kühniella* Zell. (Lepidoptera: Pyralididae). *Can. J. Zool.* 30:365–74.

Arnold, J. W. 1966. An interpretation of the haemocyte complex in a stonefly, *Acroneuria arenosa* (Plecoptera: Perlidae). *Can. Entomol.* 98:394–411.

Arnold, J. W. 1969. Periodicity in the proportion of haemocyte categories in the giant cockroach, *Blaberus giganteus*. *Can. Entomol. 101*:68–77.

Arnold, J. W. 1974. The hemocytes of insects. In M. Rockstein (ed.). *The Physiology of Insecta*, 2nd ed. Academic Press, New York.

Arnold, J. W., and C. F. Hinks. 1976. Haemopoiesis in Lepidoptera. I. The multiplication of circulating haemocytes. *Can. J. Zool. 54*:1003–12.

Clark, E. W., and D. S. Chadbourne. 1960. The haemocytes of nondiapause and diapause larvae and pupae of the pink bollworm. *Ann. Entomol. Soc. Amer. 53*:682–5.

Collin, N. 1963. Les Hémocytes de la larve de *Melolontha melolontha* L. (Coleoptère Scarabaeidae). *Rev. Pathol. Veg. Entomol. Agric. Fr. 42*:161–7.

Fisher, R. A. 1935. The effect of acetic acid vapor treatment on the blood cell count in the cockroach, *Blatta orientalis* L. *Ann. Entomol. Soc. Amer. 28*:146–53.

Gupta, A. P., and D. J. Sutherland. 1968. Effects of sublethal doses of chlordane on the hemocytes and midgut epithelium of *Periplaneta americana*. *Ann. Entomol. Soc. Amer. 61*:910–18.

Jones, J. C. 1956. The hemocytes of *Sarcophaga bullata* Parker. *J. Morphol. 99*:233–57.

Jones, J. C. 1957. DDT and the hemocyte picture of the mealworm, *Tenebrio molitor* L. *J. Cell. Comp. Physiol. 50*:423–8.

Jones, J. C. 1959. A phase contrast study of the blood cells in *Prodenia* larvae (Order Lepidoptera). *Q. J. Microsc. Sci. 100*:17–23.

Jones, J. C. 1962. Current concepts concerning insect hemocytes. *Amer. Zool. 2*:209–46.

Jones, J. C. 1967a. Changes in the hemocyte picture of *Galleria mellonella* (Linnaeus). *Biol. Bull. (Woods Hole) 132*:211–21.

Jones, J. C. 1967b. Normal differential counts of haemocytes in relation to ecdysis and feeding in *Rhodnius*. *J. Insect Physiol. 13*:1133–41.

Jones, J. C., and O. E. Tauber. 1954. Abnormal hemocytes in mealworms (*Tenebrio molitor* L.). *Ann. Entomol. Soc. Amer. 47*:428–44.

Kitano, H. 1969. On the total hemocyte counts of the larva of the common cabbage butterfly, *Pieris rapae crucivora* Boisduval (Lepidoptera: Pieridae) with reference to the parasitization of *Apanteles glomeratus* L. (Hymenoptera: Braconidae). *Kontyû 37*:320–6.

Lee, R. M. 1961. The variation of blood volume with age in the desert locust (*Schistocerca gregaria* Forsk.). *J. Insect Physiol. 6*:36–51.

Matsumoto, T., and M. Sakurai. 1956. On the density of haemocytes in the blood bled from a wound in *Bombyx mori* L., *J. Seric. Sci. Jpn. 25*:147–8. (In Japanese, English summary.)

Nappi, A. J., and J. G. Stoffolano, Jr. 1972. Haemocytic changes associated with the immune reaction of hematode-infected larvae of *Orthellia caesarion*. *Parasitology 65*:295–302.

Nappi, A. J., and F. A. Streams. 1969. Haemocyte reactions of *Drosophila melanogaster* to the parasites *Pseudocoila mellipes* and *P. bochei*. *J. Insect Physiol. 15*:1551–66.

Nittono, Y. 1960. Studies on the blood cells in the silkworm, *Bombyx mori* L. *Bull Seric. Exp. Stn. (Tokyo) 16*:171–266.

Pappenheim, A. 1914. *Clinical Examination of the Blood and Its Technique*. Wright, Bristol.

Patton, R. L., and R. A. Flint. 1959. The variation in the blood cell count of *Periplaneta americana* (L.) during a molt. *Ann. Entomol. Soc. Amer. 52*:240–2.

Richardson, C. H., R. C. Burdette, and C. W. Eagleson. 1931. The determination of the blood volume of insect larvae. *Ann. Entomol. Soc. Amer. 24*:503–7.

Rosenberger, C. R., and J. C. Jones. 1960. Studies on total blood counts of the southern armyworm larva, *Prodenia eridania* (Lepidoptera). *Ann. Entomol. Soc. Amer. 53*:351–5.

Shapiro, M. 1966. Pathologic changes in the blood of the greater wax moth. *Galleria mellonella* (Linnaeus), during the course of starvation and nucleopolyhedrosis. Ph.D. thesis, University of California. Berkeley, California.

Shapiro, M. 1967. Pathologic changes in the blood of the greater wax moth, *Galleria mellonella*, during the course of nucleopolyhedrosis and starvation. I. Total hemocyte count. *J. Invertebr. Pathol.* 9:111–13.

Shapiro, M. 1968. Pathologic changes in the blood of the greater wax moth. *Galleria mellonella*, during nucleopolyhedrosis and starvation. II. Differential hemocyte count. *J. Invertebr. Pathol.* 10:230–4.

Shapiro, M., R. D. Stock, and C. M. Ignoffo. 1969. Hemocyte changes in larvae of the bollworm, *Heliothis zea*, infected with a nucleopolyhedrosis virus. *J. Invertebr. Pathol.* 14:28–30.

Smith, H. W. 1938. The blood of the cockroach *Periplaneta americana* L.: Cell structure and degeneration, and cell counts. *New Hampshire Agric. Exp. Stn. Tech. Bull.* 71:1–23.

Stephens, J. M. 1963. Effects of active immunization on total hemocyte counts of larvae of *Galleria mellonella* (Linnaeus). *J. Insect Pathol.* 5:152–6.

Tauber, O. E., and J. F. Yeager. 1934. On the total blood (hemolymph) cell count of the field cricket, *Gryllus assimilis pennsylvanicus* Burm. *Iowa State Coll. J. Sci.* 9:13–24.

Tauber, O. E. and J. F. Yeager. 1935. On total hemolymph (blood) counts of insects. I. Orthoptera, Odonata, Hemiptera, and Homoptera. *Ann. Entomol. Soc. Amer.* 28:229–40.

Tauber, O. E., and J. F. Yeager. 1936. On the total hemolymph (blood) all counts of insects. II. Neuroptera, Coleoptera, Lepidoptera, and Hymenoptera. *Ann. Entomol. Soc. Amer.* 29:112–18.

Vinson, S. B. 1971. Defense reaction and hemocytic changes in *Heliothis virescens* in response to its habitual parasitoid *Cardiochiles nigriceps. J. Invertebr. Pathol.* 18:94–100.

Webley, D. P. 1951. Blood cell counts in the African migratory locust (*Locusta migratoria migratorioides* Reiche and Fairmaire). *Proc. R. Entomol. Soc. (Lond.)* A26:25–37.

Wheeler, R. E. 1962. Changes in hemolymph volume during the moulting cycle of *Periplaneta americana* L. (Orthoptera). *Fed. Proc.* 21:123.

Wheeler, R. E. 1963. Studies on total hemocyte count and hemolymph volume in *Periplaneta americana* (L.) with special reference to the last moulting cycle. *J. Insect Physiol.* 9:223–35.

Wittig, G. 1966. Phagocytosis by blood cells in healthy and diseased caterpillars. II. A consideration of the method of making hemocyte counts. *J. Invertebr. Pathol.* 8:461–77.

Yeager, J. F. 1945. The blood picture of the southern armyworm (*Prodenia eridania*). *J. Agric. Res.* 71:1–40.

Yeager, J. F., and S. C. Munson. 1950. Blood volume of the roach *Periplaneta americana* determined by several methods. *Arthropoda* 1:255–65.

Yeager, J. F., and O. E. Tauber. 1932. Determinations of total blood volume in the cockroach, *P. fuliginosa*, with special reference to method. *Ann. Entomol. Soc. Amer.* 25:315–27.

20   Hemocyte techniques: advantages and disadvantages

M. SHAPIRO

Gypsy Moth Methods Development Laboratory, U.S. Department of
Agriculture, Otis Air Force Base, Massachusetts 02542, U.S.A.

Contents

20.1. Introduction

Insect hemocytes may be studied in several ways: (1) examination of
hemocytes within the living insect; (2) examination of unfixed hemolymph, stained or unstained; (3) examination of fixed hemolymph,
stained or unstained (Arnold, 1974); (4) examination of sections, fixed
or unfixed. It is the intent of this chapter to examine each method and
to analyze its advantages and disadvantages.

20.2. Examination of insect hemocytes

20.2.1. Hemocytes in vivo

The ameboid motion of hemocytes of the giant cockroach, *Blaberus
giganteus,* the course of blood circulation within the mature embryo,
as well as the morphology of hemocytes and blood circulation in insect wings were studied within living insects by Arnold (1959a,b,
1960, 1961, 1964, 1966) and by Arnold and Salkeld (1967). Circulation
of blood was studied in insects belonging to 14 orders. Much valuable
information was obtained, but the method has limited use and has not

549

found widespread advocacy. Jones and Tauber (1954) attempted to differential counts of hemocytes (DHCs) in the wing membrane of the mealworm, *Tenebrio molitor.* Counts could not be made because the cells could not be clearly seen under high magnification.

Clark and Harvey (1965) studied cellular membrane formation in diapausing *Cecropia* pupae by utilizing a modified telobiotic preparation. In this method, pupae were attached to a Plexiglas chamber, so that living hemocytes could be observed directly using phase-contrast optics. Attachment of plasmatocytes (PLs) and migration of cells were also studied. As the hemocyte population in the blood chamber increased, the proportion of fusiform PLs increased. These cells became rounded, however, when they made contact with the hemacytometer surfaces.

Spherical PLs undergo transformation to a flattened disklike lamellocytes (LAs) at about the time of puparium formation in *Drosophila* (Rizki, 1962). These transformations were studied in an ingenious series of experiments. In method A, a capillary tube was attached between the anterior ends of two larvae. In method B, a capillary tube was implanted into the hemocoel of a larva, while the other end was sealed. Morphological transformation of the hemocytes could be observed in methods A and B. In method C, the capillary tube was sealed at both ends after it was filled with hemolymph. This method was useful because a given hemocyte could be observed from various angles. Despite the value of the methods described for *Cecropia* and *Drosophila,* they have not been widely used. Their main disadvantage seems to be the manual skill required to utilize them efficiently.

20.2.2. Unfixed hemolymph

Many workers have examined insect blood cells in unfixed preparations, using phase-contrast microscopy. Living cells can be examined and identified with relative ease. Jones (1954, 1956) readily identified unfixed blood cells from *Tenebrio* and *Sarcophaga,* respectively. The classes of cells described in *Prodenia* from stained smears (Yeager, 1945) were recognized in unfixed, unstained preparations (Jones, 1959). PLs and plasmatocyte-like cells either retained their form, rounded up, spread, or degenerated. Podocytes (POs), spherulocytes (SPs), and adipohemocytes (ADs) were stable for at least 1 hr, and oenocytoids (OEs) were quite unstable and often became rounded. In these cells, nuclear extrusion was observed.

Hemocytes of *Tenebrio* underwent transformations in unfixed coverslip films. Within 15–30 sec of bleeding, the cells rounded up. In hanging-drop preparations, some hemocytes remained fusiform for at least 15 min (Jones and Tauber, 1954). The authors considered that phase-contrast microscopy revealed morphological details that were obscured by heat fixation. They also felt that unfixed drops were not as useful as heat-fixed preparations for the study of abnormal cells.

PLs and coagulocytes (COs) were easily distinguished in the hemolymph of adult *Periplaneta americana* (Jones, 1957). In both *Tenebrio* and *Periplaneta*, COs tended to disintegrate and release cytoplasmic granules. McLaughlin and Allen (1965) described the types of hemocytes in the boll weevil, *Anthonomus grandis*. Coverslip preparations remained in satisfactory conditions for 12–17 hr. Gupta and Sutherland (1967) studied SPs in seven species of cockroaches in a physiological saline solution. Spherules were liberated from the SPs and appeared to grow and resemble young PLs. Hemocytes of saturniid pupae (*Samia cynthia, Antheraea polyphemus*) and *Hyalophora cecropia* larvae survived for several days in hanging-drop cultures. When hemolymph was removed from a small slit in the cuticle, the PLs became ameboid and adhesive. In some cases clumps were formed; other cultures spread. When hemolymph was collected from the anterior half of the pupa as it flowed down, the PLs retained their original form without clumping or spreading (Cherbas, 1973).

Quantitative changes in blood cell populations were followed in *Sarcophaga* (Jones, 1956), in *Prodenia* (Rosenberger and Jones, 1960), and in *Pseudaletia* (Wittig, 1965). The study on *Prodenia* is particularly interesting, as it illustrates a common problem. THCs from unfixed larvae ranged from 9,000/mm³ to 29,000/mm³ ($\bar{x}$ 17,185); counts from heat-fixed larvae ranged from 17,000/mm³ to 45,000/mm³ ($\bar{x}$ 27,734). This difference in counts will be discussed later.

Unfixed hemolymph has been used to study mitosis. As prophase could not be determined, dividing cells were classed as being in metaphase, anaphase, or telophase (Jones and Liu, 1968). The time required for a cell to complete the mitotic cycle from metaphase was estimated as 30 min (Clark and Harvey, 1965).

DHCs of *Tenebrio* (Jones, 1954) and *Prodenia* (Wittig, 1965) could be made using phase optics. Jones (1954) identified OEs, PLs, and COs. Less variability was observed in counts from unfixed than from heat-fixed larvae. The PL/CO ratio in unfixed preparations was significantly different from that in heat-fixed preparations. Unfixed preparations could be used for 1 hr without undergoing degenerative changes (Wittig, 1965). PRs, PLs, GRs, COs, SPs, OEs, ADs, POs, and vermiform cells (VEs) could be distinguished easily. Exploded cells (= COs) and degenerating cells were not counted in DHCs.

Unfixed hemolymph has been utilized for histochemical studies of hemocytes (Arnold, 1952; Jones, 1956; Nittono, 1960; Ashhurst and Richards, 1964; Gupta and Sutherland, 1967; Arnold and Salkeld, 1967; and Zapol'skikh, 1976) in *Anagasta, Sarcophaga, Bombyx, Galleria*, seven species of cockroaches, *Blaberus*, and *Apis*, respectively, as well as for ultrastructural studies (Hoffmann, 1966; Hoffmann et al.,

1968; Bearwald and Boush, 1970, 1971; Lai-Fook, 1973; Neuwirth, 1973).

The use of phase-contrast microscopy has greatly aided the study of insect hemocytes because possible artifacts produced by classic methods of fixation and staining are avoided. Studies of coagulation are feasible only by the use of phase-contrast optics because the COs are recognized only in unfixed preparations (Grégoire, 1951, 1974).

20.2.3. Fixed hemolymph

Chemically fixed hemolymph

P. americana hemocytes were fixed in either Bouin's or Danafo's solution, followed by staining (Taylor, 1935), for studies on cell division. Arnold and Salkeld (1967) compared the effectiveness of several preservatives on the hemocytes of *Blaberus giganteus*. The preservation had to meet certain standards: (1) to fix the hemocytes without resulting in changes in form or shape, (2) to prevent agglutination of the hemolymph, and (3) to be easily removed following fixation. Based on these criteria, the authors eliminated most of the standard fixatives and considered anticoagulants, blood-diluting solutions (usually dilute acetic acid), or dilute aqueous fixatives. Anticoagulants tested included ammonium–potassium oxalate, sodium fluoride–thymol, disodium EDTA (versene), and dipotassium EDTA. Versene (5% in physiological saline) plus 5% formalin was useful. Cell size and shape were preserved, but cytoplasmic granules were not fixed well. Turk's diluting fluid caused distortions as it dried. Formalin alone, at 10% for adult blood and 5% for the blood of small insects, was the best preservative. Hemocytes of the boll weevil, *Anthonomus*, were fixed in 40% formaldehyde or absolute methanol (McLaughlin and Allen, 1965). Hemocytes of the large milkweed bug, *Oncopeltus fasciatus*, were fixed in ethanol (Feir and McClain, 1968); hemocytic changes in the tobacco budworm, *Heliothis virescens*, in response to the parasitoid *Cardiochiles nigrices* were examined following fixation in methanol and subsequent staining (Vinson, 1971).

In most cases, the acetic acid vapor technique (Shull and Rice, 1933) was used by Arnold to fix hemocytes within the insects. *Galleria* hemocytes were primarily fixed with formaldehyde-saline and were then stained (Ashhurst and Richards, 1964). Hemocytes from the antennae of cockroaches were collected in a solution of saline-versene, fixed, and stained for the presence or absence of carbohydrates, lipids, proteins, and nucleic acids (Gupta and Sutherland, 1967).

Heat-fixed hemolymph

Heat fixation has been employed by submersing the insect(s) into hot water. The method is considered satisfactory when (1) hemolymph is easily obtained, (2) hemolymph does not agglutinate or precipitate, (3) cells do not agglutinate (Jones, 1962), and (4) cells are not changed in size, shape, or form. Tauber and Yeager (1935) used the glacial acetic acid vapor treatment or heat fixation (60 °C for 10 min) as measures against coagulation. Rosenberger and Jones (1960) preferred heat fixation to the acetic acid vapor treatment because it was effective, simple to use, and did not cause excessive regurgitation. In phytophagous insects, regurgitation and/or defecation may be excessive following acetic acid vapor treatment (Jones, 1962).

Yeager (1945) employed heat fixation (60 °C for 5–10 min) in his classic study on the blood picture of the southern armyworm. *Prodenia eridania*. This temperature was chosen to prevent changes in form, so that the hemocytes would closely resemble those in situ. In addition to preserving the form of the hemocytes, heat fixation can retard or inhibit coagulation and melanization. In general, temperatures have varied from 55 to 60 °C for 1–10 min. This method has been used to study the hemocytes of the lepidopterans *Anagasta* (Arnold and Hinks, 1976), *Galleria* (Shapiro, 1966, 1967, 1968a,b; Jones, 1967; Jones and Liu, 1969), *Hyalophora cecropia* (Lea and Gilbert, 1966), *Pectinophora* (Clark and Chadbourne, 1960), *Pieris* (Kitano, 1969), and *Prodenia* (Yeager, 1945; Rosenberger and Jones, 1960); the coleopteran *Tenebrio* (Jones, 1950, 1957), the dipteran, *Sarcophaga* (Jones, 1956); the plecopteran, *Acroneuria* (Arnold, 1966); and for general use (Salinger, 1963).

Despite the use of heat fixation, Jones (1959) correctly noted that artifacts of fixation may occur (e.g., loss of spherules from SPs, enlargement of granules in GRs into ill-defined vacuoles). The author also noted that heat fixation preserved the form and structures of other hemocytes, if only for a short time.

20.2.4. Sectioned larvae or other stages

Instead of removing hemocytes from insects for study, whole insects can be fixed and sectioned. Histological techniques were employed to study hemopoietic organs of several lepidopterous larvae (Arvy, 1953) and to investigate host-parasite relationships involving *Pieris* and the braconid parasite, *Apanteles glomeratus* (Kitano, 1969). In the latter study, parasitized larvae were punctured and tissues were fixed with Carnoy's solution. Following fixation, dehydration, and sectioning, tissues were stained. Hemocytic changes in larvae of the fly, *Orthellia caesarion*, parasitized by the nematode, *Heterotylenchus autumnalis*, were studied after fixation and staining (Nappi and Stoffolano, 1971). Lai-Fook and Neuwirth (1972) investigated methods of fixation and concluded that some methods altered the shape, size, and inclusions in hemocytes.

The mitotic index (MI) and total hemocyte count (THC) of *Oncopeltus* were determined after fixation in liquid nitrogen (Feir and O'Connor, 1969). The tissues were frozen rapidly without any loss of hemolymph. Following fixation, sections were made and stained. Counts were made of all hemocytes in the hemocoe, and MIs were taken. Although the method was quite time-consuming and would not be used routinely, the authors felt it was the most accurate method for determining the number of hemocytes within an insect. Nappi and Stoffolano (1971) compared liquid nitrogen fixation of *Musca* larvae with the standard Carnoy B or Kahle's solution. From 70 to 90% of nitrogen-fixed larvae did not remain intact after thawing. After larvae were wrapped in cotton and then placed in liquid nitrogen, mechanical abrasion of the larvae against the vial was prevented. From these studies, the authors concluded that both methods of fixation were similar, utilizing larvae of *M. domestica* and *M. autumnalis*.

20.2.5. Staining of hemocytes

In general, most workers have utilized stained preparations for their studies on insect hemocytes, from unfixed or fixed materials. Metalnikov (1922, 1927), in his work on *Galleria*, relied on May-Grünwald and Pappenheim stains. *Carausius* hemocytes were stained best with May-Grünwald and a Giemsa or Pappenheim counterstain (Rooseboom, 1937). Rooseboom observed that the thickness of the smear influenced not only the color of the cells, but also their dimensions. Cells were more spread and were larger as the smears became thinner. The Pappenheim method combined the merits of Giemsa and May-Grünwald stains, without any of the disadvantages of the latter, such as poor definition of nuclear details and lack of azurophilia (Pappenheim, 1914). Shapiro (1966, 1967, 1968a,b), also working with *Galleria*, utilized May-Grünwald-Giemsa staining, with satisfactory results.

Smith (1938) used several stains in studying the blood of the American cockroach, *P. americana*. Wright's blood stain and methylene blue were inferior to Pianese's methylene blue–eosin and Giemsa. Pianese's methylene blue–eosin, although producing consistent results, was inferior to Giemsa's in differentiating cellular structures. Giemsa's stain was also used by Arnold (1952), Jones and Tauber (1954), Jones (1956), McLaughlin and Allen (1965), Feir and McClain (1968), Nappi and Streams (1969), Vinson (1971), and Arnold and Hinks (1976). Arnold (1952) concluded that Giemsa was the best and most consistent stain for *Anagasta* hemocytes; Crossley (1964) stated that Giemsa was unsatisfactory for *Calliphora* hemocytes, as all types of hemocytes stained blue purple. Wright's blood stain has also been utilized, with good results (Yeager, 1945; Jones and Tauber, 1954;

Clark and Chadbourne, 1960; Salinger, 1963). Hematoxylin stains, in combination with various counterstains, were also successfully employed (Speare, 1920; Taylor, 1935; Smith, 1938; Feir and O'Connor, 1969; Kitano, 1969; Nappi and Stoffolano, 1971).

20.2.6. Problems associated with unfixed hemolymph

Unfixed hemolymph from *Prodenia*, like that from many insects, turns dark brown or black on contact with air. This phenomenon, melanization, occurs when unfixed, fresh hemolymph is drawn from the insect. The injury metabolism of diapausing *Cecropia* pupae was studied by Harvey and Williams (1961). When the integument was excised, phenylthiourea (PTU) crystals were added to the wound to inhibit tyrosinase. Clark and Harvey (1965), studying cellular membrane formation in *Cecropia*, added a 1:1 solution of streptomycin sulfate and PTU into the hemocoels of the insects in order to inhibit both bacterial multiplication and melanization. Cherbas (1973) studied the injury reaction in cultured hemocytes from saturniids. Hemolymph collected was mixed with penicillin (1%) and thiourea (1 M).

In 1933, Yeager and Knight observed the formation of a granular precipitate in the hemolymph of the hemipteran, *Belostoma fluminea*. Precipitation occurred especially in the immediate vicinity of certain cells. The involvement of a class of hemocytes was confirmed by Grégoire and Florkin (1950a,b). The hemocytes were called coagulocytes (= COs). Smith (1938) observed that hemocytes taken from *Periplaneta* tended to round up, rupture, and form a clot. As COs do not appear to settle as rapidly as PLs in thick, unfixed hemolymph films, distorted DHCs may be obtained if "counts are made only at the upper surfaces of such drops" (Jones, 1954). The COs are recognized only in wet films, where they erupt and cause a granular precipitation in the surrounding plasma (Jones, 1962; Grégoire, 1974).

In the American cockroach, coagulation involves blood cells and not the plasma (Smith, 1938). Smith employed acetic acid vapor treatment (Fisher, 1935) to prevent coagulation of the hemolymph, especially in the dilution pipette. In this method, the insect is exposed to acetic acid vapors for 20–40 min (Jones, 1962). Tauber and Yeager (1935) found no significant difference in THCs whether acetic acid vapor treatment (Fisher, 1935) or rapid dilution of hemolymph was employed. Fisher (1935) showed that THCs, after acetic acid vapor treatment (30 min), were about twice as high as those from untreated oriental cockroaches, *Blatta orientalis*. The low counts in untreated insects were primarily attributable to cell clumping. Some of the hemocytes adhered to the wound site and were unavailable for counting. The acetic acid vapor technique was also used by Tauber and Yeager (1935) and Arnold (1952) to inhibit coagulation. Rosenberger and Jones (1960) compared methods of fixation for *Prodenia* blood. Total counts from unfixed hemolymph varied from 9,000/mm^3 to 29,000/mm^3 ($\bar{x}$ 17,185); counts from larvae exposed to acetic acid vapor (45 min at 35 °C) ranged from 19,000/mm^3 to 39,000/mm^3 ($\bar{x}$ 27,471).

Dilution of unfixed hemolymph with physiological saline did not prevent cell agglutination, but the addition of acetic acid (1%) or formaldehyde (4%) to the diluent retarded coagulation (Rosenberger and Jones, 1960). Neither Turk's solution nor oxalate inhibited coagulation of American cockroach hemolymph. A 1–2% solution of versene was successful, however (Patton and Flint, 1959). The hemolymph of the European cockchafer, *Melolontha melolontha*, coagulates upon contact with air. Coagulation is prevented if hemolymph is diluted with an injection of Ringer's solution into the insect without contact with the air (Collin, 1963). Tauber and Yeager's (1935) diluting fluid contained a dilute concentration (0.125%) of acetic acid. Versene (2%) was more effective, however, as a diluting fluid than dilute acetic acid in preventing clumping of *Pseudaletia* hemocytes (Wittig, 1966).

20.2.7. Problems associated with heat-fixed hemolymph

If blood smears are improperly fixed, cell distortions, cell clumping, and nuclear degeneration occur (Jones and Tauber, 1954). In smears from unfixed *Tenebrio* hemolymph, however, cells rapidly agglutinated and COs degenerated. The authors concluded that the most accurate picture of hemocytes in situ was obtained using heat fixation, despite the possible production of artifacts. Some morphological details of hemocytes were obscured by heat fixation (Jones and Tauber, 1954). Later, Jones (1957) concluded that heat fixation reduced the ease of distinguishing between PLs and COs in the mealworm. Although the THCs for heat-fixed DDT-treated and control *Tenebrio* larvae were the same, cells identified as PLs were found only in the DDT-treated group (Jones, 1957). In a prior study, Jones (1954) showed that the number of cells in unfixed larvae ($\bar{x}$ 19,600/mm^3) was significantly less than after heat fixation ($\bar{x}$ 41,700/mm^3). It was not determined whether hemocytes in situ or sessile hemocytes driven into circulation were responsible for the higher counts.

The THCs from unfixed *Galleria* larvae were more variable and significantly lower than those from heat-fixed larvae (Jones, 1967). Stephens (1963), on the other hand, stated that heat fixation did not alter the cell counts. Jones and Liu (1969) found that unfixed counts from head-ligated *Galleria* larvae were lower than those from heat-fixed larvae. The counts from heat-fixed larvae were comparable to those of normal larvae (Jones, 1967). The authors indicated that low counts from unfixed larvae may have been caused by increased clumping and, hence, unavailability of circulating hemocytes.

Counts from heat-fixed *Prodenia* larvae were higher ($\bar{x}$ 27,734/mm³) than those from unfixed larvae ($\bar{x}$ 17,185/mm³) (Rosenberger and Jones, 1960). If larvae were not fixed, about 38% of their blood cells were unavailable because of clumping. Clumping was prevented by immersing larvae at 55 or 66 °C for 1 min, and heat-fixed larvae were used for most studies. Following heat fixation, THCs from *Rhodnius* (Jones and Liu, 1961) and from the stonefly, *Acroneuria* (Arnold, 1966), were significantly higher than those from unfixed specimens. Heat fixation at 45, 50, or 55 °C did not cause sessile hemocytes from *Aedes* larvae to enter the circulation (Jones, 1958).

The differences in numbers of circulating hemocytes among unfixed and heat-fixed insects are "probably due to technical difficulties, but the release of sedentary hemocytes during fixation and manipulation cannot be discounted" (Jones, 1958). This statement concerning heat fixation is as valid now as 20 years ago in that we do not know whether heat fixation preserves all the circulating hemocytes or whether it causes attached, sessile cells to enter the circulation.

20.3. Summary

Several methods are available for studying insect hemocytes: (1) examination within the living insect; (2) examination of unfixed hemolymph after removal from the insect; (3) examination of fixed hemolymph after removal from the insect; (4) examination of sectioned material. Each method has its advocates and its critics, for each method has its strengths and limitations. These methods must be examined critically for optimal usage. Which method should be utilized for what type(s) of study? Despite the possible creation of artifacts by fixation, both chemical and heat, are these methods of value for the study of hemocytes? Is heat fixation itself responsible for apparent differences in THCs between fixed and unfixed cells? Or are these differences attributable to the nonavailability of unfixed hemolymph because of clumping and adhesion? Or both? Problems of methodology (i.e., the best method to use for a given purpose) must be resolved among investigations so that the field of insect hematology may be further advanced.

References

Arnold, J. W. 1952. The haemocytes of the Mediterranean flour moth, *Ephestia kühniella* Zell. (Lepidoptera: Pyralididae). *Can. J. Zool.* 30:352–64.

Arnold, J. W. 1959a. Observations on living haemocytes in wing veins of the cockroach *Blaberus giganteus* (L.) (Orthoptera: Blattidae). *Ann. Entomol. Soc. Amer.* 52:229–36.

Arnold, J. W. 1959b. Observations on amoeboid motion of living haemocytes in the wing veins of *Blaberus giganteus* (L.) (Orthoptera: Blattidae). *Can. J. Zool.* 37:372–5.

Arnold, J. W. 1960. The course of blood circulation in mature embryos of the cockroach *Blaberus giganteus* (L.) (Orthoptera: Blattidae). *Can. J. Zool.* 38:1027–35.

Arnold, J. W. 1961. Further observations on amoeboid haemocytes in *Blaberus giganteus* (L.) (Orthoptera: Blattidae). *Can. J. Zool.* 39:755–66.

Arnold, J. W. 1964. Blood circulation in insect wings. *Mem. Entomol. Soc. Can.* 38:1–48.

Arnold, J. W. 1966. An interpretation of the haemocyte complex in a stonefly, *Acroneuria arenosa* (Plecoptera: Perlidae). *Can. Entomol.* 98:394–411.

Arnold, J. W. 1974. The hemocytes of insects, pp. 201–54. *In* M. Rockstein (ed.). *The Physiology of Insecta*, Vol. 5, 2nd ed. Academic Press, New York.

Arnold, J. W., and C. F. Hinks. 1976. Haemopoiesis in Lepidoptera. I. The multiplication of circulating haemocytes. *Can. J. Zool.* 54:1003–12.

Arnold, J. W., and E. H. Salkeld. 1967. Morphology of the haemocytes of the giant cockroach, *Blaberus giganteus*, with histochemical tests. *Can. Entomol.* 99:1138–45.

Arvy, L. 1953. Données histologiques sur la leucopöiése chez quelques Lepidoptères. *Bull. Soc. Zool. Fr.* 78:45–59.

Ashhurst, D. E., and A. G. Richards. 1964. Some histochemical observations on the blood cells of the wax moth, *Galleria mellonella* L. *J. Morphol.* 114:247–53.

Baerwald, R. J., and G. M. Boush. 1970. Fine structure of the hemocytes of *Periplaneta americana* (Orthoptera: Blattidae) with particular reference to marginal bundles. *J. Ultrastruct. Res.* 31:151–61.

Baerwald, R. J., and G. M. Boush. 1971. Vinblastine-induced disruption of microtubules in cockroach hemocytes. *Tissue Cell* 3(2):251–60.

Cherbas, L. 1973. The induction of an injury reaction in cultured haemocytes from saturniid pupae. *J. Insect Physiol.* 19:2011–23.

Clark, E. W., and D. S. Chadbourne. 1960. The haemocytes of nondiapause and diapause larvae and pupae of the pink bollworm. *Ann. Entomol. Soc. Amer.* 53:682–5.

Clark, R. M., and W. R. Harvey. 1965. Cellular membrane formation by plasmatocytes of diapausing cecropia pupae. *J. Insect Physiol.* 11:161–75.

Collin, N. 1963. Les hémocytes de la larvae de *Melolontha melolontha* L. (Coléoptère Scarabaeidae). *Rev. Pathol. Veg. Entomol. Agric.* 42:161–7.

Crossley, A. C. S. 1964. An experimental analysis of the origins and physiology of haemocytes in the blue bottle blowfly *Calliphora erythrocephala* (Meig.). *J. Exp. Zool.* 157:375–97.

Feir, D., and E. McClain. 1968. Mitotic activity of the circulating hemocytes of the large milkweed bug, *Oncopeltus fasciatus*. *Ann. Entomol. Soc. Amer.* 61:413–16.

Feir, D., and G. M. O'Connor. 1969. Liquid nitrogen fixation: A new method for hemocyte counts and mitotic indices in tissue sections. *Ann. Entomol. Soc. Amer.* 62:146–9.

Fisher, R. A. 1935. The effect of acetic acid vapor treatment on the blood cell count in the cockroach, *Blatta orientalis* L. *Ann. Entomol. Soc. Amer.* 28:146–53.

Grégiore, Ch. 1951. Blood coagulation in arthopods. II. Phase contrast microscopic observations on hemolymph coagulation of sixty-one species of insects. *Blood* 6:1173–98.

Grégoire, Ch. 1974. Hemolymph coagulation pp. 309–60. *In* M. Rockstein (ed.). *The Physiology of Insecta*, Vol. 5, 2nd ed. Academic Press, New York.

Grégoire, Ch., and M. Florkin. 1950a. Etude au microscope à contraste de phase du coagulocyte, du nuage granulaire et de la coagulation plasmatique dans le sang des insectes. *Experientia* 6:297–8.

Grégoire, Ch., and M. Florkin. 1950b. Blood coagulation in arthopods. I. The coagulation of insect blood, as studied with the phase contrast microscope. *Physiol. Comp. Oecol.* 2:126–39.

Gupta, A. P., and D. J. Sutherland. 1967. Phase contrast and histochemical studies of spherule cells in cockroaches (Dictyoptera). *Ann. Entomol. Soc. Amer. 60:* 557–65.

Harvey, W. R., and C. M. Williams. 1961. The injury metabolism of the cecropia silkworm. I. Biological amplification of the effects of localized injury. *J. Insect Physiol.* 7:81–99.

Hoffmann, J. A. 1966. Etude ultrastructurale de deux hémocytes à granules de *Locusta migratoria* (Orthoptère). *C. R. Acad. Sci. Paris 263:*521–6.

Hoffmann, J. A., M. E. Stoekel, A. Porte, and P. Joly. 1968. Ultrastructure des hémocytes de *Locusta migratoria*. *C. R. Acad. Sci. Paris 266:*503–7.

Jones, J. C. 1950. Cytopathology of the hemocytes of *Tenebrio molitor* Linnaeus (Coleoptera). Ph.D. thesis, Iowa State College, Ames, Iowa.

Jones, J. C. 1954. A study of mealworm hemocytes with phase contrast microscopy. *Ann. Entomol. Soc. Amer.* 47:308–15.

Jones, J. C. 1956. The hemocytes of *Sarcophaga bullata* Parker. *J. Morphol.* 99:233–57.

Jones, J. C. 1957. DDT and the hemocyte picture of the mealworm, *Tenebrio molitor* L. *J. Cell. Comp. Physiol.* 50:423–8.

Jones, J. C. 1958. Heat-fixation and the blood cells of *Aedes aegypti* larvae. *Anat. Rec.* 132:461.

Jones, J. C. 1959. A phase contrast study of the blood cells in *Prodenia* larvae (Order Lepidoptera). *Q. J. Microsc. Sci.* 100:17–23.

Jones, J. C. 1962. Current concepts concerning insect hemocytes. *Amer. Zool.* 2:209–46.

Jones, J. C. 1967. Changes in the hemocyte picture of *Galleria mellonella* (Linnaeus). *Biol. Bull. (Woods Hole)* 132:211–21.

Jones, J. C., and D. P. Liu. 1961. Total and differential hemocyte counts of *Rhodnius prolixus* Stal. *Bull. Entomol. Soc. Amer.* 7:166.

Jones, J. C., and D. P. Liu. 1968. A quantitative study of mitotic divisions of haemocytes of *Galleria mellonella* larvae. *J. Insect Physiol.* 14:1055–61.

Jones, J. C., and D. P. Liu. 1969. The effects of ligaturing *Galleria mellonella* larvae on total haemocyte counts and on mitotic indices among haemocytes. *J. Insect Physiol.* 15:1703–8.

Jones, J. C., and O. E. Tauber. 1954. Abnormal hemocytes in mealworms (*Tenebrio molitor* L.). *Ann. Entomol. Soc. Amer.* 47:428–44.

Kitano, H. 1969. On the total hemocyte counts of the larva of the common cabbage butterfly, *Pieris rapae crucivora* Boisduval (Lepidoptera: Pieridae) with reference to the parasitization of *Apanteles glomeratus* L. (Hymenoptera: Branconidae). *Kontyû* 37:320–6.

Lai-Fook, J. 1973. The structure of the haemocytes of *Calpodes ethlius* (Lepidoptera). *J. Morphol.* 139:79–104.

Lai-Fook, J., and M. Neuwirth. 1972. The importance of methods of fixation in the study of insect blood cells. *Can. J. Zool.* 50:1011–13.

Lea, M. S., and L. I. Gilbert. 1966. The hemocytes of *Hyalophora cecropia* (Lepidoptera). *J. Morphol.* 118:197–216.

McLaughlin, R. E., and G. Allen. 1965. Description of hemocytes and the coagulation process in the boll weevil, *Anthonomus grandis* Boheman (Curculionidae). *Biol. Bull. (Woods Hole)* 128:112–24.

Metalnikov, S. 1922. Les changements des éléments du sang de la chenille (*Galleria mellonella*) pendant l'immunisation. *C. R. Soc. Biol.* 86:350–2.

Metalnikov, S. 1927. *L'Infection microbienne et l'immunité chez la mite de abeilles Galleria mellonella.* Masson, Paris.

Nappi, A. J., and J. G. Stoffolano, Jr. 1971. *Heterotylenchus autumnalis:* Hemocyte reactions and capsule formation in the host, *Musca domestica. Exp. Parasitol.* 29:116.

Nappi, A. J., and F. A. Streams. 1969. Haemocyte reactions of *Drosophila melanogaster* to the parasites *Pseudocoila mellipes* and *P. bochei. J. Insect Physiol.* 15:1551–66.

Neuwirth, M. 1973. The structure of the hemocytes of *Galleria mellonella* (Lepidoptera). *J. Morphol.* 139:105–24.

Nittono, Y. 1960. Studies on the blood cells in the silkworm, *Bombyx mori* L. *Bull. Seric. Exp. Stn. (Tokyo)* 16:171–266.

Pappenheim, A. 1914. *Clinical Examination of the Blood and Its Technique.* Wright, Bristol.

Patton, R. L., and R. A. Flint. 1959. The variation in the blood cell count of *Periplaneta americana* (L.) during a molt. *Ann. Entomol. Soc. Amer.* 52:240–2.

Rizki, M. T. M. 1962. Experimental analysis of hemocyte morphology in insects. *Amer. Zool.* 2:247–56.

Rooseboom, M. 1937. Contribution à l'étude de la cytologie du sang de certains insectes, avec quelques considerations générales. *Arch. Neerl. Zool.* 2:432–559.

Rosenberger, C. R., and J. C. Jones. 1960. Studies on total blood counts of the southern armyworm larva, *Prodenia eridania* (Lepidoptera). *Ann. Entomol. Soc. Amer.* 53:351–5.

Salinger, R. 1963. Looking at insect blood. *Turtox News* 41:218–20.

Shapiro, M. 1966. Pathologic changes in the blood of the greater wax moth, *Galleria mellonella* (Linnaeus), during the course of starvation and nucleopolyhedrosis. Ph.D. thesis, University of California, Berkeley, California.

Shapiro, M. 1967. Pathologic changes in the blood of the greater wax moth, *Galleria mellonella*, during the course of nucleopolyhedrosis and starvation. I. Total hemocyte count. *J. Invertebr. Pathol.* 9:111–13.

Shapiro, M. 1968a. Changes in the haemocyte population of the wax moth, *Galleria mellonella*, during wound healing. *J. Insect Physiol.* 14:1725–33.

Shapiro, M. 1968b. Pathologic changes in the blood of the greater wax moth, *Galleria mellonella*, during nucleopolyhedrosis and starvation. II. Differential hemocyte count. *J. Invertebr. Pathol.* 10:230–4.

Shull, W. E., and P. L. Rice. 1933. A method for temporary inhibition of coagulation in the blood of insects. *J. Econ. Entomol.* 26:1083–9.

Smith, H. W. 1938. The blood of the cockroach *Periplaneta americana* L.: Cell structure and degeneration, and cell counts. *New Hampshire Agric. Exp. Stn. Tech. Bull.* 71:1–23.

Speare, A. T. 1920. Further studies of *Sorosporella uvella*, a fungous parasite of noctuid larvae. *J. Agric. Res.* 18:399–439.

Stephens, J. M. 1963. Effects of active immunization on total hemocyte counts of larvae of *Galleria mellonella* (Linnaeus). *J. Insect Pathol.* 5:152–6.

Tauber, O. E., and J. F. Yeager. 1935. On total hemolymph (blood) counts of insects. I. Orthoptera, Odonata, Hemiptera, and Homoptera. *Ann. Entomol. Soc. Amer.* 28:229–40.

Taylor, A. 1935. Experimentally induced changes in the cell complex of the blood of *Periplaneta americana. Ann. Entomol. Soc. Amer.* 28:135–45.

Vinson, S. B. 1971. Defense reaction and hemocytic changes in *Heliothis virescens* in response to its habitual parasitoid *Cardiochiles nigriceps. J. Invertebr. Pathol.* 18:94–100.

Wittig, G. 1965. Phagocytosis by blood cells in healthy and diseased caterpillars. I. Phagocytosis of *Bacillus thuringiensis* Berliner in *Pseudaletia unipuncta* (Haworth). *J. Invertebr. Pathol.* 7:474–88.

Wittig, G. 1966. Phagocytosis by blood cells in healthy and diseased caterpillars. II. A consideration of the method of making hemocyte counts. *J. Invertebr. Pathol.* 8:461–77.

Yeager, J. F. 1945. The blood picture of the southern armyworm (*Prodenia eridania*). *J. Agric. Res.* 71:1–40.

Yeager, J. F., and H. H. Knight. 1933. Microscopic observations on blood coagulation in several species of insects. *Ann. Entomol. Soc. Amer.* 26:591–602.

Yeager, J. F., and O. E. Tauber. 1933. On counting mitotically dividing cells in the blood of the cockroach, *Periplaneta orientalis* (Linn.). *Proc. Soc. Exp. Biol. Med.* 30:861–3.

Zapol'skikh, O. V. 1976. Morphological and cytochemical studies of the worker honeybee haemolymph cells (*Apis mellifera:* Hym., Apidae). *Tsitologiia* 18:956–63.

21 Light, transmission, and scanning electron microscopic techniques for insect hemocytes

R. J. BAERWALD

Department of Biological Sciences, University of New Orleans, Lakefront, New Orleans, Louisiana 70122, U.S.A.

Contents

21.1. introduction

Many technical advances have been made and new techniques developed in recent years in the field of microscopy, especially electron microscopy (EM), resulting in a tremendous expansion of our knowledge of cell substructure and function. Examples are freeze-cleave technique and improved methods for observing cells with scanning electron microscopy (SEM). Unfortunately, these new techniques have often been underutilized or sometimes ignored in the study of hemocyte structure and function. In this chapter I call attention to some of these new techniques and their essential features that should be applicable to future research on hemocytes, especially membrane studies. A comprehensive review of theory and practice of these new methods is beyond the scope of this book, and the reader is referred to the text and list of references for some of the many excellent works that cover these topics in depth.

21.2. Techniques for observing live hemocyte activities

Insect hemocytes are notorious for their ability to lyse and coagulate soon after withdrawal from the host insect. However, it is still possible to view live hemocyte activities in fresh slide preparations (Baerwald and Boush, 1971) and in tissue culture (see Chapter 9). Under ideal optical conditions, when a thin polished glass slide, ultrathin coverslip, and thin (25 μm) tissue spacers are used, a surprising amount of cell detail can be observed and photographed (Baerwald and Boush, 1971) in those hemocytes that attach to the glass and spread out completely (Fig. 21.1). Motile hemocytes do not reveal nearly as much detail as sessile hemocytes that spread completely.

Phase-contrast optics and Nomarski phase-contrast optics are the best optical methods for viewing the activities of live hemocytes in vitro, such as phagocytosis, ameboid-type migrations, and cytoplasmic granule release under various experimental conditions. Carl Zeiss (Germany) offers one of the best phase-contrast and Nomarski phase-contrast systems on the market. In addition, the Zeiss planapochromat phase objectives, 40× oil (NA 1.0) and 100× oil (NA 1.4), offer a totally flat field and resolution close to the limit of light optical systems. The 40× oil objective offers special potential because it has a much longer working distance that the 100× oil objective. This allows the researcher to work with special glass chambers, such as Rose chambers, which often have substantially thicker plastic or glass covers. This feature is especially useful for time-lapse studies of cultured hemocytes. Carl Zeiss has available a long-working-distance phase-contrast condenser that can also be used with the optically superior Rose chambers, obviating the need for an inverted microscope. For observations of hemocytes cultured in tubes or Petri plates, Zeiss offers an excellent inverted light microscope that can utilize the complete line of Zeiss objectives and condensers. Wild Herbrug also manufactures an outstanding inverted light microscope.

Development of a tissue culture preparation that would allow capsule formation, coupled with the above techniques, would allow observation of live cell activity with good resolution and under controlled experimental conditions. Such experiments, coupled with

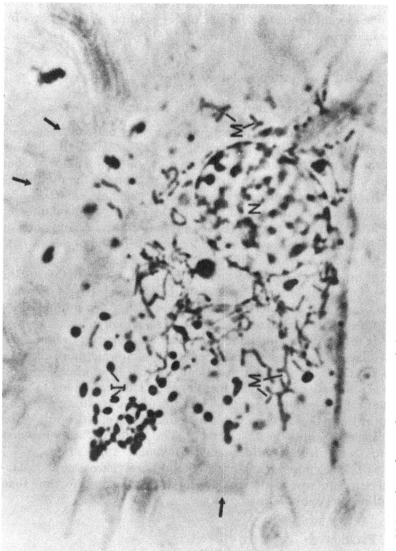

Fig. 21.1. High-resolution phase-contrast light microscopy is capable of producing micrographs with a surprising amount of cellular detail of hemocytes that have spread out and attached themselves to glass surfaces in vitro. Mitochondria (*M*), cytoplasmic inclusion (*I*), nucleus (*N*), and outer cell boundary (arrows) can be seen. × 4,000. (From Baerwald, 1971)

electron microscopic examination of the same capsules produced in vitro, would contribute to a further understanding of the factors involved in capsule formation in vivo.

21.3. Standard thin-section techniques for electron microscopy

21.3.1. Difficulties associated with fixing hemocytes

Electron microscopists consider it important that most fresh tissues, but not all (Hayat, 1970), be fixed without delay to avoid cell autolysis after removal from the host animal. The problem of cell autolysis can be especially acute with hemocytes, perhaps more so than with any other tissue. Some hemocytes may undergo changes, detectable by EM, within seconds after exposure to air. Oozing insect blood directly into a fast-penetrating fixative (Baerwald and Boush, 1970) is one of the most straightforward ways to reduce the likelihood of hemocytes' undergoing partial cytoplasmic changes before fixation. Karnofsky's fixative (Karnofsky, 1967) offers an interesting advantage in this regard because it contains formaldehyde as well as glutaraldehyde. Formaldehyde, the smallest-molecular-weight and fastest-penetrating aldehyde, presumably stabilizes the hemocytes initially. Fixation then proceeds further with glutaraldehyde, which is superior to formaldehyde as fixative, but is thought to penetrate somewhat more slowly.

The fixation, dehydration, and embedding schedule described below is routinely used in our laboratory and has been found to yield consistent high-quality results. The procedure is straightforward enough so that a person completely new to EM can obtain a surprising level of success after the first attempt. The textbook *Principles and Techniques of Electron Microscopy: Biological Applications,* Vol I (Hayat, 1970) is a popular and fairly comprehensive treatment of standard electron microscope techniques for those who require more than the outline presented here.

21.3.2. Fixation

Although Karnofsky's fixative offers a special advantage in preserving hemocytes, we have found glutaraldehyde (first introduced by Sabatini et al., 1963), buffered with sodium cacodylate, to be an excellent all-around primary fixative for a wide variety of insect as well as mammalian tissues. Purified (dialdehyde) glutaraldehyde ($> 0\%$ in H_2O) is available from a variety of sources (see Appendix) and is preferred because the dialdehyde is generally considered to be a better fixative than other oligomers of the chemical (Hayat, 1970). Adjustment of the glutaraldehyde concentration (2–6%) and/or molarity of the buffer (0.05–0.2 M), depending on the tissue, usually results in a nearly isosmolar fixative, which should prevent undue shrinking or swelling of the tissue. Sodium cacodylate, as a buffer, is convenient for preparation, handling, refrigeration, and storage of buffer solutions. Microorganisms and fungi will not grow in stock cacodylate buffer solutions, as they do in phosphate buffers, because cacodylate contains arsenic. In addition, the sodium salt of cacodylate is much easier to handle and store in the laboratory than veronal or collidine buffers. However, because the chemical contains arsenic, the powder should be handled under a hood and contact by inhalation or skin avoided (Weakley, 1977). Although certain tissues continue to be notoriously difficult to fix and may require special fixatives such as acrolein, purified glutaraldehyde with or without formaldehyde and buffered with cacodylate has proved an excellent all-around fixative for a wide variety of insect tissues, including blood cells.

21.3.3. Postfixation

After aldehyde fixation, hemocytes are postfixed for 1 hr with osmium tetroxide (OsO_4) at room temperature in snap cap vials (1% OsO_4 in 0.1 M cacodylate buffer). We routinely maintain frozen stocks of 2% OsO_4 in distilled H_2O in our laboratory. Refreezing and rethawing OsO_4 solutions do not affect the quality of the fixative. However, freezing effectively eliminates reduction of the oxide to metal in solution, substantially reduces osmium fumes during storage in the refrigerator, and eliminates much waste of the comparatively expensive OsO_4.

21.3.4. Dehydration

We have found acetone to be a convient and relatively nontoxic dehydration agent. Since Epon is fairly soluble in acetone, the propylene oxide steps during infiltration can be avoided. After OsO_4 fixation, the material is exposed for 10 min each to 30%, 70%, and 95% acetone and three separate 10-min changes of 100% acetone. Our acetone stock is maintained with Linde molecular sieve pellets type 4-A (Fullam, New York) to take up any water absorbed from the air by the hygroscopic pure acetone.

21.3.5. Infiltration and embedding

A modification of Luft's (1961) original technique is routinely used in our laboratory where the epoxy/anhydride ratio is changed from 0.7 to 0.6 to improve the cutting quality of the plastic (Hayat, 1970). To further ensure the cutting quality of the plastic (Epon 812), several precautions are taken in procuring, storing, mixing, and polymerization of the plastic. Plastic batches may vary in their cutting quality from distributor to distributor. We have found Epon 812 supplied by Electron Microscopy Sciences (see Appendix) to be both economical and of high cutting quality. One mixing convenience Epon users enjoy (Araldite users also) is that the epoxy equivalent, which can change with time, is labeled on the bottles.

Stock mixtures of Epon-anhydride and Epon-NMA (mixtures A & B, respectively) are prepared according to Luft's (1961) suggestions and mixed thoroughly by vigorous shaking at room temperature, where mixtures A and B have a relatively low viscosity. After the air bubbles have escaped from the mixtures, the air in the bottles is replaced with Freon and the bottles are tightly stoppered and carefully sealed with Parafilm before refrigeration. Refrigerated stock solutions are usable for a least 6 months when stored in this fashion. Bottles are warmed for at least 1 hr before opening to prevent water from condensing on the cold plastic. Final plastic batches (10 ml) are mixed thoroughly with a geared Teflon stirring device (Fig. 21.2) obtainable through the various EM retail suppliers (see Appendix). A gradual, thorough, and extensive infiltration is accomplished with a commercial Infiltron (Fig. 21.2), using a 50:50 mixture of the final embedding mix and 100% water-free acetone. The pieces are gently and contin-

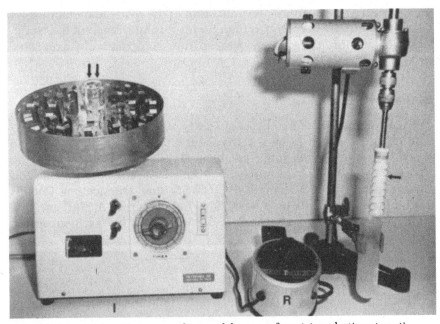

Fig. 21.2. Two devices that are used in our laboratory for mixing plastic automatically and that contribute to a block of high cutting quality. Infiltron (*I*) is used overnight for infiltration of specimen with plastic; rheostat (*R*) operated Teflon stir bar (arrow) ensures even mixing of plastic prior to embedding.

uously swirled overnight in the device, while the acetone slowly evaporates at room temperature. The next day the pieces are embedded in fresh plastic, either in flat embedding molds or in Beem capsules. The plastic is slowly polymerized for 2 days at 37 °C and 2 days at 60 °C to ensure even polymerization of the block. Extraction is rare using the above procedures, but if evident, ethanol can be substituted for acetone.

21.4. Membrane-enhancing staining techniques

21.4.1. En bloc staining with uranyl acetate

A variety of useful heavy-metal staining techniques for EM are available for enhancing membrane substructure. These methods often increase the normally limited usefulness of thin-section techniques, particularly when intercellular junctions and other membrane substructural details are being studied.

Perhaps the most widely used general membrane-enhancing staining method is en bloc staining with uranyl acetate. After fixation in the usual way, the material may be exposed to 2% uranyl acetate dissolved in 10%, 30%, or 70% alcohol or acetone for 20 min. A 30-min soak in saturated aqueous uranyl acetate immediately before dehydration also works quite well as a general membrane-enhancing stain for thin-section studies (Forbes and Sperelakis, 1977). Similarly, all hemocyte membranes were enhanced with a 2-hr soak in 0.5% uranyl acetate solution in veronal acetate buffer (Baerwald and Boush, 1970). This extra staining of the block results in an enhanced density of the two electron-dense laminae of the unit membrane and generally makes all thin-sectioned cell membranes more apparent. For maximum effect, sections are stained with uranyl acetate and lead citrate, although the use of uranyl acetate at this stage is considered optional.

21.4.2. The lanthanum method

Lanthanum continues to be a useful external marker for a wide variety of EM thin-section studied involving cell membrane modifications such as intercellular junctions, caveole, or t-tubule invaginations of the plasma membrane in muscle tissue. Lanthanum is an electron-dense heavy metal that ordinarily does not penetrate or stain cell membrane, but will penetrate even the smallest space between two cells, such as the 20-Å extracellular gap of gap junctions (McNutt and Weinstein, 1973; Shimono and Clementi, 1977). As a result, when oblique, unstained sections of lanthanum-impregnated intercellular junctions are examined, various regular electron-dense arrays are often seen. These regular arrays indicate that protein complexes are present, sometimes arranged in hexagonal arrays, spanning the extra-cellular gap associated with gap or septate junctional complexes. Al-though this technique has been used often to study mammalian gap junctions and, to some extent, septate junctions, it has not been often used to study insect gap junctions. This method and other related techniques could be used to good advantage to study hemocyte gap junctions and the extent of extracellular space remaining after capsule formation.

21.4.3. Ruthenium red

Ruthenium red generally shares with lanthanum a reputation for marking extracellular structures, such as the cell coat, for EM studies (Wanson et al., 1977). Unlike lanthanum, ruthenium red has been shown to penetrate cells under certain conditions (Howell, 1974; Al-lenspach and Babiars, 1975). This staining method also greatly en-hances membrane details in a wide variety of cells (Mariani Colombo and Rascio, 1977) and should not be overlooked as a useful special-ized staining procedure in high-resolution studies of hemocyte mem-brane substructure using thin-section techniques. In at least one study (Rowley and Ratcliffe, 1976), ruthenium red was not effective as a stain for hemocyte cytoplasmic contents extruded during coagulation of *Galleria mellonella* hemolymph. Nevertheless, it is felt that this stain and others of its type offer excellent potential for studying hemo-cyte membrane, cell surfaces, or cell junctions under a variety of ex-perimental conditions.

21.4.4. Tannic acid

The use of tannic acid as a stain mordant is a comparatively new ap-proach for enhancing membrane and other cell structural components for EM. It was first used to view microtubule protofilaments in situ (Tilney et al., 1973), but is now gaining popularity as a specialized marker for cell membranes (Bonilla, 1977). The tannic acid method of stain enhancing apparently works as an external membrane stain for cells that have retained membrane integrity, but produces an unusual in situ negative stain effect for microtubules (Tilney et al., 1973), mi-crofilaments, and F-actin (LaFountain et al., 1977) in cells that have lost it. If tannic acid is used before osmium fixation, the plasma mem-brane is enhanced; if used after osmium fixation, internal membranes are enhanced as well (Wagner, 1976). The unusual accumulation of electron-dense metals at membranes, which this procedure produces, often results in the detection of additional structural detail. For exam-ple, tannic acid has served to reveal additional structural details in desmosomes of thin-sectioned newt epidermis (Kelly and Shienvold,

1976). This technique should be considered along with other stain-enhancing techniques for future fine structural studies of hemocytes and hemocyte capsules.

21.5. Negative staining techniques

Negative staining techniques continue to be some of the best approaches to high resolution for ultrastructural substructure analysis of isolated membrane components. Unusually high quality and resolution are often achieved by this technique, with useful resolutions near 15 Å point-to-point (e.g., in isolated mouse liver gap junction: Caspar et al., 1977).

Many excellent reviews and technical chapters have been written on the negative staining technique (e.g., Haschemeyer and Meyers, 1972). Elaborate procedures for support film preparation, solution preparation, and application are often described. We have found, however, that the following technique serves well for a variety of applications and has the advantage of being quite straightforward. We use 400-mesh grids coated with 2% freshly prepared collodion in water-free amyl acetate. We prefer to avoid carbon coating by operating our microscope (Philips 200) with low electron emission levels, double condenser, short exposure times (0.5–1 sec), and at 80 kV. Collodion films are very delicate and break easily, even with the above precautions. However, we feel that staining, picture quality, and contrast are generally superior to those obtained using carbon-coated grids.

We have also found uranyl acetate (2%, freshly prepared in double-distilled water) to be an easier stain to work with than phosphotungstic acid (PTA), generally yielding photomicrographs with contrast superior to PTA. Although certain materials may require unusual handling, simple mixing of a droplet of stain with a droplet of material in suspension (for 30 sec) on a collodion-coated grid and then drawing off the excess with filter paper have generally proved successful in our laboratory. It is anticipated that the above procedure would be applicable to high-resolution studies of isolated membrane fractions and cell components from a wide variety of hemocytes.

21.6. The freeze-fracture technique

21.6.1. Introduction

The freeze-fracture or freeze-cleave method has become extremely popular among cell biologists in recent years, largely because of its utility for studying membrane substructure. This technique is unique in several respects. First, the method splits or cleaves biological membranes near or at the hydrophobic interior of the membrane, generating two, often large, novel fracture faces of half membranes. Second, the method is a nonaveraging approach to membrane substructure analysis, with resolutions comparable to those achieved with standard electron microscope techniques.

A comprehensive review of freeze-fracture methods is beyond the scope of this book and is probably not necessary, as several excellent reviews of this type are available (Koehler, 1972; Bullivant, 1973; Benedetti and Favard, 1973; Zingsheim and Plattner, 1976). In addition, a superb atlas of freeze-etch ultrastructure is available in large format, often with equivalent thin-sectioned material on the facing page (Orci and Perrelet, 1975). Finally, the reader is referred to Chapter 6 of this book for further comments on interpretation of freeze-cleave micrographs. The purpose of this section is to relate a basic technique that has worked well for us on hemocytes (Teigler and Baerwald, 1972; Baerwald, 1973, 1974, 1975) as well as other types of tissues. It is hoped that the procedure outlined below will serve as a guide for other hemocyte researchers contemplating use of this technique.

Because hemocytes are normally found monodispersed and free in circulation in a liquid hemolymph, special handling of these cells is required before freeze fracturing can be carried out. Some sort of clot or pellet must first be prepared. Depending on the insect, clots can often be formed by oozing hemolymph from a clipped antenna or appendage directly into a glutaraldehyde fixative; this is the most desirable approach. Varying the temperature and/or concentrations of the fixative slightly may facilitate clot formation. Alternatively, hemocytes may be pelleted by gentle centrifugation in the presence of the fixative and the pellet cut into pieces (1–2 mm) suitable for freeze fracturing. For this method, any sort of loose clot or pellet that can be handled long enough for freezing is suitable. Once frozen, the material remains stable through the rest of the procedure.

21.6.2. Simplified step-by-step procedure

A basic method for *routine* freeze fracturing has now been established and is, with few modifications, widely accepted. It is, therefore, worth outlining. These procedures have been used with few modifications for a wide variety of cells, including hemocytes:

1. Small pieces of tissue (1–2 mm) are quickly fixed with isotonic phosphate- or cacodylate-buffered glutaraldehyde in a manner similar to the first fixation steps for routine EM. (Glutaraldehyde stabilizes the material in preparation for glycerol exposure.)
2. After glutarldehyde fixation, the tissues are exposed to 30% buffered

glycerol for 30–60 min. This serves as a necessary antifreeze so that large ice crystals are avoided during the freezing step.

3. The material is then quickly plunged into Freon 12, chilled to a liquid state ($-150\,°C$) with liquid nitrogen. Rapid freezing, on the order of 10,000 °C/min, occurs under these conditions, avoiding ice crystal formation (McNutt and Weinstein, 1973).

4. The frozen material is quickly inserted into some sort of freeze-fracture device, care being taken to avoid partial thawing at this stage. A typical freeze-fracture device includes some sort of high-vacuum system, bell jar, liquid nitrogen apparatus for cooling the specimen stage and fracturing knife, and electrode assemblies for the shadowing of carbon and platinum. A diagram of the Balzers freeze-fracture device (bell jar assembly) is shown in Fig. 21.3. It is currently the most popular and, unfortunately,

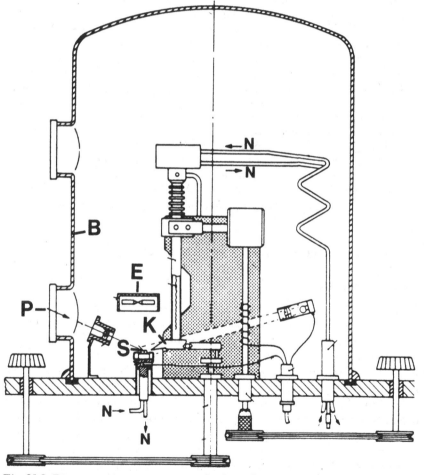

Fig. 21.3. Diagrammatic representation of Balzers freeze-fracture device. Note vacuum bell jar (*B*) with portholes (*P*) surrounding specimen stage (*S*), knife (*K*), and electrodes (*E*). Liquid nitrogen (*N*) chills both knife and specimen stage. (Redrawn from Bullivant, 1973)

most expensive device on the market. However, recent technical improvements in other makes indicate that devices from other manufacturers (Denton Vacuum, Inc., Cherry Hill, New Jersey 08003; Edwards High Vacuum, Inc., Grand Island, New York 14072) may also gain popularity and be available at a fraction of the cost.

5. After a reasonably high vacuum (10^{-6} mm Hg) has been established, the material is carefully "warmed" to -100 °C or -115 °C in preparation for the fracturing step.

6. Fracturing the frozen tissue is accomplished by a knife chilled to liquid nitrogen temperature (-196 °C) attached to a rotary microtome apparatus (in the case of the Balzers freeze-fracture device) or a hinge device (in a recently modified version of the Denton machine). Hinge devices can also be utilized in the Balzers machine. The frozen material is split by fracture planes traveling out ahead of the knife analogous to an icepick splitting a large block of ice. The fracture planes preferentially follow areas within the frozen block of tissue where electrostatic binding forces are weakest. Fortunately for the membranologist, this happen to be along the hydrophobic interior of biological membranes.

7. A replica is next prepared of the fractured surface of the frozen tissue. Carbon is first evaporated from the top to hold the replica together, and then platinum is evaporated from the side (45 °) to give three-dimensional shadowing detail and contrast to the replica.

8. The tissue is then removed from the freeze-fracture device and allowed to thaw. The replica is separated from the material, typically using sodium hypochlorite (bleach) to dissolve the tissue. The replica remains floating on the surface of the bleach solution and then is cleaned with several distilled water rinses. Finally, the replica is transferred to a grid and can be inserted directly into an EM without further preparation.

21.7. Scanning electron microscope techniques

21.7.1. Introduction

Although scanning electron microscopy (SEM) of biological structures (teeth, bone, hair) was done as far back as the early 1960s (Boyde and Stewart, 1962), it was not until quite recently that useful SEM studies of soft tissues and cells were done (Porter et al., 1972). Many excellent SEM atlases are generally available that include discussions on technique, application, and theory (e.g., Kessel and Shih, 1974). In this discussion, I call attention to recent technical improvements in preparation of soft tissue and cells for SEM that should generally be useful for future SEM studies of hemocytes under a variety of experimental conditions. The reader should refer to more exhaustive accounts of SEM methods, such as those listed above, for complete background on this technique, which is currently growing in popularity among cell biologists.

21.7.2. Critical-point drying

SEM and transmission electron microscopy (TEM) require specimen dehydration before insertion into the instruments because vacuums of 10^{-4}–10^{-6} Torr are necessary for normal operation of the instruments. However, unlike TEM the mode of operation of SEM (a narrow, scanning electron beam) allows the operator to view extensive surface areas of cells and tissues with a great depth of field. Under special conditions and with state-of-the-art instrumentation, useful 50–70-Å resolution, point-to-point, is obtainable.

Faithful preservation of the total true surface along with the smallest details of soft tissues and cells is technically difficult because of surface tension forces that are destructive to the delicate plasma membrane and underlying cytoplasm in the liquid phase during dehydration at normal atmospheric pressures. There is little doubt that critical-point drying (Anderson, 1951) represents the single most important technical advance dealing with the dehydration problems mentioned above. The details of this method and the general preparation of soft tissues and cells for SEM have been reviewed by Kessel and Shih (1975) and Porter et al. (1972). The important feature of the critical-point technique is that dehydration is accomplished at a critical temperature and pressure point (1,072 psi and 31 °C for CO_2). Under these conditions, liquid CO_2 will pass to the gaseous phase with a minimum of distortional forces on the drying cells. Freon 13 or nitrous oxide can be substituted (Porter et al., 1972), although the original method using CO_2 seems to be currently popular. Recently, it has been shown that undesirable drying effects, although drastically reduced, are still present (Kistler and Kellenberger, 1977). However, there is little question that the critical-point technique greatly improves the appearance of cells in the SEM compared with standard freeze-dry or freeze-substitution techniques. It is anticipated that this technique could be used to resolve new surface features of hemocytes under a variety of conditions.

21.7.3. Improved heavy-metal evaporating techniques

After careful dehydration of the cells, coating of the specimens with heavy metals is required. A gold-palladium alloy or gold alone is usually evaporated onto the specimen using standard bell jar vacuum evaporating devices, with various interchangeable attachments that allow rotation of the specimen in several planes to ensure an even coating over the surface. Coating of the specimen with gold-palladium produces a fine-grain coating capable of emitting secondary electrons after being struck by the scanning beam at energies and quantities optimal for the scintillator-photomultiplier collector system in the SEM, and avoids surface charging.

Recent refinements of the basic shadowing techniques have somewhat improved shadowing reproducibility and inherent resolution of the specimen, and reduced the possibility of heat damage to the specimen during metal evaporation. Some recent reports (Miller et al., 1977; Kirschner et al., 1977) have described the use of a quartz crystal monitoring device during evaporation. This device, which is based on a readout system of crystal vibration frequency changes during evaporation, enables the investigator to apply the proper thickness of metal onto the specimen in a more reproducible manner from specimen to specimen. Optimum metal thickness is especially important for high-resolution SEM ($\simeq 50$ Å) studies of cells.

Sputter-coater devices are now available from many companies and are currently gaining wide acceptance among SEM users. They have the advantage over the conventional bell jar vacuum evaporator of speed, reducing damage to the specimen from heat during metal evaporation, and of being generally less expensive and more compact than conventional evaporators.

21.7.4. Instrumentation

Newer SEM models generally offer significant advantages over older models with respect to maintenance and accessory adaptability of the instrument on a retrofit basis. Most instruments offer goniometer stages of various descriptions and often extremely large specimen ports for special use. Columns have been redesigned to facilitate cleaning and alignment procedures. Most new SEMs incorporate solid-state electronics, often modular, with various built-in test circuits to facilitate repair. Operator panels have often been redesigned to make routine operation easier.

Even though most SEM electronics are now in modular solid-state form, good technical service from the parent company is still deemed critical. A researcher should consider the availability of technical service personnel in his locality very carefully before deciding on a particular make. The following is a partial list of some parent companies currently manufacturing SEMs: Cambridge, United Kingdom; Phillips Electronic Instruments, Inc., The Netherlands; JEOL, Japan; ETEC Corporation, United States; AMR Corporation, United States.

21.8. Summary

Light, transmission, and scanning electron microscope techniques have been briefly reviewed with emphasis on those procedures that show special potential for further studies on hemocyte movement, cell surfaces, and membrane substructural studies. Light microscopic techniques such as phase contrast and Nomarski phase contrast have been emphasized as the best approaches for viewing live hemocyte activities in vitro. Successful standard electron microscope techniques have been reviewed, along with specialized heavy-metal membrane staining techniques and simplified negative staining methods applicable to future high-resolution studies of isolated hemocyte. A step-by-step freeze-fracture procedure, which has been successful for hemocytes, has been presented. Finally SEM procedures have been briefly reviewed, focusing on recent developments such as the critical-point technique, sputter coating, and instrumentation that show special potential for future studies of hemocyte surfaces under a variety of experimental conditions.

Appendix

Most of the distributors listed below will supply free complete catalogs and price lists on request. Most have supplies, parts, and accessory equipment for both transmission electron microscopy and scanning electron microscopy.

LKB Instruments, 4840 Rugby Avenue, Washington, D.C. 20014
Polysciences, Inc., P.O. Box 4, Warrington, Pennsylvania 18976
Ted Pella, Inc., P.O. Box 510, Tustin, California 92680
Electron Microscopy Sciences, Fort Washington, Pennsylvania 19034
Ladd Research Industries, Inc., P.O. Box 901, Burlington, Vermont 05401
Ernest F. Fullam, Inc., P.O. Box 444, Schenectady, New York 12301

References

Allenspach, A. L., and B. S. Babiars. 1975. Intramitochondrial binding of ruthemium red in degenerating chondroblasts. *J. Ultrastruct. Res. 51*:348–53.

Anderson, T. 1951. Techniques for the preservation of three-dimensional structure in preparing specimens for the electron microscope. *Trans. N.Y. Acad. Sci. Ser. III 13*:130–4.

Baerwald, R. J. 1973. Freeze-fracture studies of insect hemocyte gap junctions and associated cell membranes. *J. Cell Biol. 59*:27a.

Baerwald, R. J. 1974. Freeze-fracture studies on the cockroach hemocyte membrane complex: Symmetric and asymmetric membrane particle distribution. *Cell Tissue Res. 151*:383–94.

Baerwald, R. J. 1975. Inverted gap and other cell junctions in cockroach hemocyte capsules: A thin section and freeze fracture study. *Tissue Cell 7*(3):575–85.

Baerwald, R. J., and G. M. Boush. 1970. Fine structure of the hemocytes of *Periplaneta americana* (Orthoptera: Blattidae) with particular reference to marginal bundles. *J. Ulstrastruct. Res. 31*:151–61.

Baerwald, R. J., and G. M. Boush. 1971. Time-lapse photographic studies of cockroach hemocyte migrations *in vitro. Exp. Cell Res. 63*:208–13.

Benedetti, E. L., and P. Favard (eds.). 1973. *Freeze-Etching Techniques and Applications.* Société Française de Microscopie Electronique, Paris.

Bonilla, E. 1977. Staining of transverse tubular system of skeletal muscle by tannic acid–glutaraldehyde fixation. *J. Ultrastruct. Res. 58:*162–5.

Boyde, A., and A. Stewart. 1962. SEM of the surface of developing mammalian dental enamel. *J. Ultrastruct. Res. 7:*159–72.

Bullivant, S. 1973. Freeze-etching and freeze-fracturing, pp. 67–112. *In* J. K. Koehler (ed.). *Advanced Techniques in Biological Electron Microscopy.* Springer-Verlag, New York.

Caspar, D. L. D., D. A. Goodenough, L. Makowski, and W. C. Phillips. 1977. Gap junction structures. I. Correlated electron microscopy and X-ray diffraction. *J. Cell Biol. 74:*605–28.

Forbes, M. S., and N. Sperelakis. 1977. Myocardial couplings: Their structural variations in the mouse. *J. Ultrastruct. Res. 58:*50–65.

Haschemeyer, R., and R. Meyers. 1972. Negative staining, pp. 101–47. *In* M. A. Hayat (ed.). *Principles and Techniques of Electron Microscopy,* Vol. 2. Van Nostrand Reinhold, New York.

Hayat, M. 1970. *Principles and Techniques of Electron Microscopy: Biological Applications,* Vol. 1. Van Nostrand Reinhold, New York.

Howell, J. N. 1974. Intracellular binding of ruthenium red in frog skeletal muscle. *J. Cell. Biol. 62:*242–7.

Karnovsky, M. 1967. A formaldehyde-glutaraldehyde fixative of high osmolarity for use in electron microscopy. *J. Cell Biol. 27:*137a.

Kelly, D. E., and F. Shienvold. 1976. The desmosome: Fine structural studies with freeze-fracture replication and tannic acid staining of sectioned epidermis. *Cell Tissue Res. 172:*309–23.

Kessel, R. G., and C. Shih. 1974. *Scanning Electron Microscopy in Biology: A Student's Atlas on Biological Organization.* Springer-Verlag, New York.

Kirschner, R., M. Rusli, and T. Martin. 1977. Characterization of the nuclear envelope, pore complexes and dense luminae of mouse liver nuclei by high resolution SEM. *J. Cell Biol. 73:*118–32.

Kistler, J., and E. Kellenberger. 1977. Collapse phenomena in freeze-drying. *J. Ultrastruct. Res. 59:*70–6.

Koehler, J. K. 1972. The freeze-etching technique, pp. 53–98. *In* M. A. Hayat (ed.). *Principles and Techniques of Electron Microscopy: Biological Applications,* Vol. 2. Van Nostrand Reinhold, New York.

LaFountain, J. R., C. Zobel, H. Thomas, and C. Galbreath. 1977. Fixation and staining F-actin and microfilaments using tannic acid. *J. Ultrastruct. Res. 58:*78–86.

Luft, J. H. 1961. Improvements in epoxy resin embedding methods. *J. Biophys. Biochem. Cytol. 9:*409.

Mariani Colombo, P., and N. Rascio. 1977. Ruthenium red staining for electron microscopy of plant material. *J. Ultrastruct. Res. 60:*135–9.

McNutt, N. S., and R. S. Weinstein. 1973. Membrane ultrastructure at mammalian intercellular junctions. *Prog. Biophys. Mol. Biol. 26:*45–101.

Miller, S. C., E. Hay, and J. Codington. 1977. Ultrastructural and histochemical differences in cell surface properties of strain-specific and non strain-specific T # 3 adenocarcinoma cells. *J. Cell Biol. 72:*511–29.

Orci, L., and A. Perrelet. 1975. *Freeze-Etch Histology: A Comparison Between Thin Sections and Freeze-Etch Replicas.* Springer-Verlag, New York.

Porter, K., D. Kelley, and P. Andrews. 1972. The preparations of cultured cells and soft tissues for scanning electron microscopy. pp. 1–19. *In Proceedings, Fifth Annual Stereoscan Scanning Electron Microscopy Colloquium.* Kent Cambridge Scientific, Morton Grove, Illinois.

Rowley, A., and N. Ratcliffe. 1976. The granular cells of *Galleria mellonella* during clotting and phagocytic reactions *in vitro. Tissue Cell 8:*437–46.

Sabatini, D. D., K. Bensch, and R. Barnett. 1963. Cytochemistry and electron microscopy: The preservation of cellular ultrastructure and enzymatic activity by aldehyde fixation. *J. Cell Biol.* 17:19–58.

Shimono, M., and F. Clementi. 1977. Intercellular junctions of oral epithelium. II. Ultrastructural changes in rat buccal epithelium induced by trypsin digestion. *J. Ultrastruct. Res.* 59:101–12.

Teigler, D. J., and R. J. Baerwald. 1972. A freeze-etch study of clustered nuclear pores. *Tissue Cell* 4(3):447–56.

Tilney, L. G., J. Bryan, D. Bush, K. Fujiwara, M. Mooseker, D. B. Murphy, and D. H. Snyder. 1973. Microtubules: Evidence for 13 protofilaments. *J. Cell Biol.* 59:267–73.

Wagner, R. C. 1976. The effect of tannic acid on electron images of capillary endothelial cell membranes. *J. Ultrastruct. Res.* 57:132–9.

Wanson, J., P. Drochmans, R. Mosselmans, and M. Ronveaux. 1977. Adult rat hepatocytes in primary monolayer culture. *J. Cell Biol.* 74:858–77.

Weakley, B. S. 1977. How dangerous is sodium cacodylate? *J. Microsc.* 109(2):249–51.

Zingsheim, A., and H. Plattner. 1976. Electron microscopic methods in membrane biology, pp. 56–94. *In* E. D. Korn (ed.). *Methods in Membrane Biology*, Vol. 7. Plenum Press, New York.

22 Histochemical methods for hemocytes

DOREEN E. ASHHURST

Department of Structural Biology St. George's Hospital Medical School, London, SW17 0RE, U.K.

Contents

22.1. Introduction

Histochemical techniques are designed to give information about the chemical composition of tissues. There is a vast range of techniques that can be used for the location and identification of nucleic acids, carbohydrates, lipids, proteins, enzymes, and some inorganic ions. The subject is complex, and there are many pitfalls for the inexpert. For many biologists, to do some histochemistry is merely to follow the procedures, as outlined, at the end of a histochemistry textbook–it all seems so simple! But many procedures are inadequately described; often, for example, no source is given for the vital reagent that it is essential to purchase from a particular supplier, and the interpretation of a positive result may be described so categorically as to be misleading or only half true.

The techniques that have been most frequently used on smears of insect hemocytes are primarily those for localizing carbohydrates and lipids, as the main purpose has been the identification of the chemical nature of the many granules and droplets that can be seen in certain types of hemocytes in histological preparations or by phase-contrast microscopy. For this reason, I shall concentrate on these methods in this chapter.

22.2. Fixation

An important prerequisite for the majority of histochemical tests is the fixation of the tissue. In general, only proteins are completely fixed, i.e., rendered insoluble and resistant to subsequent procedures, by the commonly used fixatives. Carbohydrates are not fixed, but are retained in the tissues by virtue of the enmeshing proteins. Lipids, in general, are not fixed, and most are still readily soluble in alcohols and other organic solvents.

The choice of fixative is important. Fixative mixtures, such as Bouin's and Zenker's fluids, should be avoided, except where specifically mentioned in the method. Formaldehyde is the fixative of choice for most purposes because it fixes the proteins efficiently by cross linking the polypeptide chains. It is usually used as a 4% aqueous solution buffered to approximately neutral pH; this can be done with a phosphate buffer or by adding calcium acetate to make a 2% solution in the fixative.

Hemocytes are usually prepared for light microscopic investigations in the form of smears, and the hemocytes are fixed in situ on the slide. The smear is very thin, and hence very little time is required for the fixative to penetrate the thickness of the smear. As a result, only very short fixation times are required. Too-long fixation can actually result in the removal of material from tissues. Fixation times between 10 min and 1 hr are adequate for blood smears. The smear is then washed in running water for 30 min or longer to remove excess fixative before the histochemical test is performed.

22.3. Tests for carbohydrates

It is possible, using well-tried techniques, to characterize many classes of carbohydrate-containing substances. It is not appropriate here to discuss the general classification of carbohydrates, but it is probably useful to mention the types that might be found in hemocytes. The polysaccharide glycogen might be present in the cytoplasm. Many hemocytes contain numerous cytoplasmic granules composed of glycoprotein. These glycoproteins are sometimes acidic owing to attached sulfate or sialic acid residues. Glycosaminoglycans, such as hyaluronic acid and chondroitin sulfate, would not be expected to be present because they are components of the connective tissues and are not usually found intracellularly.

A carbohydrate should be characterized by the site at which it is found and by whether it is a neutral glycoprotein or mucin, a sulfo- or sialomucin, or a glycosaminoglycan. These are defined, first, on the presence or absence of (1) *vic*-glycol groups, (2) uronic acid, (3) sulfate ester groups, (4) sialic acid carboxyls. Further subdivision is possible on criteria such as (1) affinity for basic dyes (e.g., azure A and toluidine blue), (2) affinity for alcian blue, (3) lability toward testicular hyaluronidase, (4) lability toward *Vibrio cholerae* neuraminidase. Thus, using a battery of histochemical tests, it is possible to distinguish, for example, small differences in the mucins secreted by different glands.

The presence of *vic*-glycol groups can be detected by using the periodic acid–Schiff (PAS) test devised by MCManus (1946). Periodic acid is an oxidizing agent for glycols and it breaks the C—C bonds where they occur as 1–2 glycol groups. The resulting aldehyde

$$\text{(hexose ring with OH, H, H, OH substituents)} + HIO_4 \longrightarrow \text{(ring with H, H and two C=O aldehyde groups)}$$

groups can be visualized using colorless Schiff's reagent, which attaches to aldehydes and, as a result of electron reorganization in the product thus formed, assumes a bright magenta color. All carbohydrates, except glycosaminoglycans, give a positive reaction with the *standard PAS test*, i.e., after a 10-min period of oxidation by 0.5% periodic acid. The important stage in this test is the periodic acid oxidation; the length of time in Schiff's reagent is not significant. The uronic acid-containing components of the glycosaminoglycans do possess *vic*-glycol groups, but owing to steric hindrance, these are oxidized to aldehydes only after periodic acid oxidation of several hours (Scott and Dorling, 1969); thus, with the standard PAS test, they are negative. PAS positivity merely indicates that a carbohydrate is present; it does not reveal anything about the nature of the carbohydrate or whether it is attached to protein or lipid. Thus, two very dissimilar substances can be PAS-positive and adjacent in a tissue section. A positive PAS reaction due to the presence of glycogen can be confirmed by digesting the tissue with diastase to remove the glycogen before performing the PAS test.

The most reliable methods currently available for the detection of anionic groups on carbohydrates are based on the dye alcian blue. Other basic dyes, such as azure A and toluidine blue, may be used, but they are much more capricious, the staining is less intense and is not stable after dehydration, so that they are not so useful for studies of hemocytes. In most instances, anionic groups are situated along the length of the molecule, and such anionic substances may be termed polyanions. Naturally occurring polyanions are the phosphates, particularly of the nucleic acids, the carboxyls of uronic and sialic acids, and the sulfate ester groups on various carbohydrates and lipids. Alcian blue is a basic dye and hence may be electrostatically bound by any of the acidic groups in the tissue. It is a copper phthalocyanin dye with a central chromophore, and its four positive charges are around the periphery of the molecule (Fig. 22.1) (Scott, 1970). The shape of the molecule is important in determining its affinities; for primarily steric reasons, it is bound more readily by the polyanions of glycosaminoglycans than those of RNA and DNA. By varying the pH and salt concentration of the dye bath, it is possible to distinguish the phosphate, carboxyl, and sulfate ester groups.

Alcian blue is most commonly used at pH 2.6 (in 3% acetic acid) (see Appendix), and under these conditions it binds to the phosphate groups of nucleic acids and proteins and to the carboxyl and sulfate ester groups of glycoproteins and glycosaminoglycans. At pH 1.0 (in 0.1 N HCl) the binding is restricted to the sulfate groups because the phosphate and carboxyl groups are no longer dissociated (Quintarelli et al., 1964). Dye binding by highly sulfated substances may be more intense at this pH because the complexes formed between the polyanions and the polycations of the proteins are split by the chloride ions in the dye solution, so releasing many polyanions to bind the dye. The chloride ions also compete for the cations of the dye; when no other anions are present, the four cations of the dye molecule may react with four charges on the polyanion. When other anions are present, theoretically only one cation of the dye molecule may be bound by the polyanion, the other three being bound by the competing anions in the dye solution. Such competition will result in more dye molecules being bound by the polyanion, and hence an enhancement of staining will occur (Scott et al., 1964). If a simple electrolyte such as MgCl$_2$ is added to the dye solution (see Appendix), the electrolyte cation com-

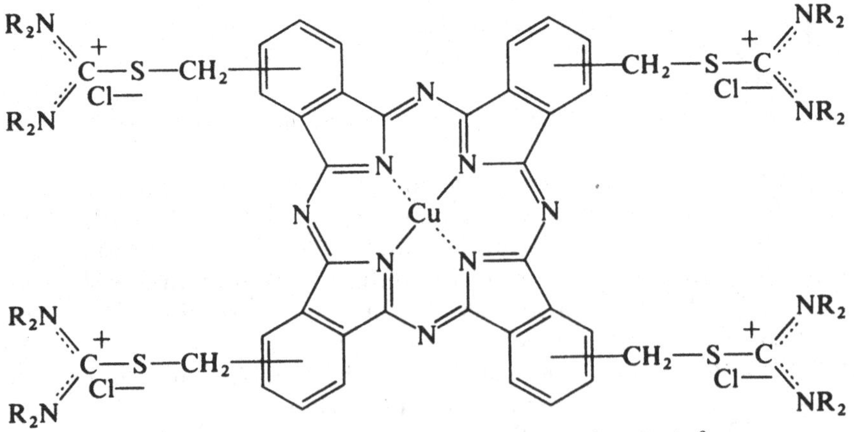

Figure 22.1. Structure of Alcian Blue. Dyue exists in several isomeric forms.

petes with the dye cation for the polyanion, and the amount of dye bound depends on the competitive affinity of the dye and cation for the polyanion. The critical electrolyte concentration (CEC) is the concentration of added cation required to displace the dye from the substrate, or put another way, the concentration required to prevent the dye from binding to the substrate, and this is characteristic for each polyanion. The phosphates of nucleic acids and the carboxyls of hyaluronic acid and sialic acid bind alcian blue in the presence of less than 0.2 M $MgCl_2$, chondroitin sulfates and dermatan sulfate bind alcian blue in the presence of up to 0.6 M $MgCl_2$, and heparin binds up to 0.9 M $MgCl_2$, while the highly sulfated keratan sulfate still binds alcian blue in the presence of 1.0 M $MgCl_2$ (Scott and Dorling, 1965). The sulfomucins have variable CECs, but all bind alcian blue in the presence of concentrations of $MgCl_2$ greater than 0.2 M. Thus alcian blue used under specific conditions can be used to distinguish carbohydrates with carboxyl and sulfate groups.

From the foregoing account, it is obvious that the conditions under which alcian blue is used must always be fully described.

Another point to be borne in mind when using alcian blue is that staining times should be in the order of 18 hr to allow equilibrium between the dye in solution and in the tissue to occur. Because electrostatic dye binding is, by its nature, reversible, great care must be taken after the slide is removed from the dye bath. For example, if a slide is taken from a dye bath containing 1.0 M $MgCl_2$ and plunged into distilled water, the dye immediately goes into the water and the tissue is now bathed with the dye under different salt conditions. This can result in dye being bound at this later stage; indeed, this was found to occur when locust hemocyte smears were treated in this way (Costin, 1975). It is, therefore, essential that when the CEC experiment is performed, the slides are washed in solutions at the same pH and containing the same amount of $MgCl_2$ as the dye bath (Mowry and Scott, 1967).

These procedures, together with the location of the stained material, will indicate the nature of the carbohydrates in a tissue or cell. Further confirmation can be obtained by enzyme digestion tests (see Appendix). The commercially available testicular hyaluronidases will remove hyaluronic acid, chondroitin sulfates, and probably dermatan sulfate. *Vibrio cholerae* neuraminidase will remove labile sialic acid residues; some sialic acid residues are neuraminidase-resistant. It should be remembered when interpreting the results of enzyme extractions that commercial samples of enzymes may be impure and may contain nonspecific proteases.

Another way in which confirmation can be obtained is by using various blocking procedures. Methylation for 4–5 hr at 37 °C blocks carboxyl groups but has no effect on sulfate ester groups, so only the sulfomucins and sulfated glycosaminoglycans bind alcian blue after this procedure. Methylation for 3–4 hr at 60 °C blocks carboxyl groups and removes sulfate esters, so eliminating all dye binding (Sorvari and Stoward, 1970). The reactivity of the blocked carboxyls can be restored by subsequent saponification with KOH (Sorvari and Stoward, 1971). In this situation, the sialomucins will regain their ability to bind alcian blue, but the sulfomucins will remain unstained. Saponification may also render resistant sialic acid residues labile to neuraminidase digestion (Culling et al., 1974).

The acidic groups may be further distinguished by using diamine methods (Spicer, 1965) (see Appendix). The diamine solution is made up of a mixture of *meta-* and *para*-dimethylphenelene diamines to which either a low or higher concentration of ferric chloride is added to provide the low-iron diamine (LID) and high-iron diamine (HID) tests. Salt complexes are formed between the cationic oxidation product of the diamine mixtures and the polyanions. With the LID mixture the dye is bound by both carboxyls of sialic acid and hyaluronic acid, and by sulfate ester groups; with the HID mixture, only the sulfate ester groups of glycosaminoglycans and sulfomucins bind the dye.

The use of aldehyde fuchsin and colloidal iron will not be discussed. It is often difficult to obtain a satisfactory solution of aldehyde fuchsin owing to variation in dye samples. Colloidal iron is somewhat capricious, and the conditions under which it may be bound by various acidic groups have not been worked out as thoroughly as has been done for alcian blue.

There is often great heterogeneity among the granules within one hemocyte (Costin, 1975). In such instances, many granules may be reactive with, for example, the PAS test and bind alcian blue, but when the HID test is performed, very few may stain black. Such situations may be sorted out by performing two histochemical reactions in sequence. Some possibilities and the results are listed in Table 22.1.

After the series of tests described here has been performed, it should be possible to distinguish the following types of granules containing carbohydrates:

1. Neutral glycoprotein: PAS +
2. Glycoprotein with sialic acid residues: PAS+; AB pH 2.6+; AB pH 5.7 with 0.2 M $MgCl_2$ –; neuraminidase-labile; LID +; HID –
3. As 2, but sialic acid residues neuraminidase-resistant
4. Glycoprotein with sialic acid and sulfate ester groups: PAS +; AB pH 2.6 +; AB pH 1.0 +; AB pH 5.7 with 0.2 M $MgCl_2$ +; neuraminidase-labile; LID +; HID +
5. As 4, but sialic acid residues neuraminidase-resistant
6. Glycoprotein with sulfate groups: PAS +; AB pH 2.6 +; AB pH 1.0 +; AB pH 5.7 with 0.2 M $MgCl_2$ +; neuraminidase-resistant; LID +; HID + ·

Table 22.1. Some test sequences that enable neutral, sialo- and sulfomucins, and glycosaminoglycans (GAGs) to be distinguished.

Test[a] sequence		Results	
1	2	Substance	Color
AB pH 2.5	PAS	Neutral mucins	Magenta
		Acidic mucins	Purple
		GAGs	Blue
AB pH 1.0	PAS	Neutral mucins	Magenta
		Sialomucins	Magenta
		Sulfomucins	Purple
		Hyaluronic acid	Unstained
		Sulfated GAGs	Blue
AB pH 5.7 with 0.05 M MgCl₂	PAS	Neutral mucins	Magenta
		Acidic mucins	Purple
		GAGs	Blue
LID	AB pH 2.5	Sulfomucins	Black[b]
		Some sialomucins	Black
		Sulfated GAGs	Black
		Some sialomucins	Blue
		Hyaluronic acid	Blue
HID	AB pH 2.5	Sulfomucins	Black
		Sulfated GAGs	Black
		Sialomucins	Blue
		Hyaluronic acid	Blue
HID	PAS	Neutral mucins	Magenta
		Sialomucins	Magenta
		Hyaluronic acid	Unstained
		Sulfomucins	Black
		Sulfated GAGs	Black

[a] AB = alcian blue; PAS = periodic acid–Schiff; LID = low-iron diamine; HID = high-iron diamine.
[b] The black color produced in the LID and HID tests may not appear very dense and in some instances may have a distinct red tinge.

22.4. Tests for lipids

Some hemocytes contain lipid droplets, and to verify their occurrence it is necessary to perform tests for lipids. Lipids are not adequately fixed by any fixatives, except osmium tetroxide and chromic acid, and are thus held in the tissues by virtue of the enmeshing proteins. They are extracted by ethanols and other organic solvents, so that dehydration must be avoided. Osmium tetroxide fixation cannot be used routinely because it is deposited as lower oxides in the lipids and other tests cannot then be performed. Formaldehyde with added calcium is usually used in practice. The formaldehyde fixes the proteins, and the calcium ions form coacervates with the lipids, rendering them less soluble. Formaldehyde saline should be avoided because sodium ions enhance the solubility of some lipids (Bayliss-High, 1977). Their retention can also be further increased by subsequent postchromation of the tissue (Baker, 1946). Fixation times vary with the test to be performed (Bayliss-High, 1977); fixation of 1 hr should be sufficient for a smear of hemocytes.

Tests for lipids can be classified as histochemical or histophysical (Adams, 1965). The lipids can be divided into two main groups: hydrophobic and hydrophilic. The hydrophobic lipids include the unsaturated triglycerides, fatty acids, and cholesterol; they tend to form globules, i.e., droplets within the aqueous cytoplasm. The hydrophilic lipids often form bimolecular layers, i.e., membranes and other structural components of the cells, and include the phospholipids, cerebrosides, and sulfatides.

The dyes most often used to stain lipids are the Sudan dyes, Sudan black B, red Sudan IV, and oil red O. These dyes are fat-soluble and are absorbed by the hydrophobic globules of triglycerides, fatty acids, etc. Sudan black also stains phospholipids, but this is due to a salt linkage with the phosphatide group. Nile blue is another lipid stain, but in solution there are two components, a red oxazone, which is fat-soluble and hence stains the hydrophobic lipids, and a blue oxazine, which stains phospholipids, again by a salt linkage.

Phospholipids can also be detected by Baker's (1946) acid hematein test (see Appendix). The tissues are first postchromed with potassium dichromate, which is thought to be reduced by the unsaturated groups of the fatty acids of the phospholipids to a lower oxide, and this mordants the reaction of the phospholipids with the hematoxylin in acid solution. Although this is based on an old method, Baker's modification, which introduced precise timing for each procedure, is considered to be specific for the phospholipids lecithin and sphingomyelin (Adams, 1965).

There are many other tests for lipids, but their applicability to studies of hemocytes is not obvious. Some employ the enzyme lipase; great care must be exercised in the interpretation of any results produced by this enzyme because commercial samples may contain various proteases and other contaminants that can give rise to false results. These other methods for lipids are discussed by Adams (1965) and Pearse (1968).

Some lipids are "masked" by protein and cannot be detected without prior treatment. Wigglesworth (1971) proposed sodium hypochlorite as an agent for unmasking lipids, and he found that the lipids unmasked in this way were then stainable by Sudan black B. The lipids involved are primarily the structural lipids, such as those found in mitochondria, and they probably bind Sudan black by salt linkage.

The applicability of this method to hemocytes is not obvious, as the hydrophobe lipids that form the globules do not require unmasking.

22.5. Tests for nucleic acids

The standard histochemical test for DNA, the Feulgen reaction, is so well established and documented that it does not require further discussion here. The tests for RNA, however, are not so well proven.

RNA is basophilic, i.e., it is negatively charged owing to the phosphate groups along the length of the molecule, and hence it binds basic dyes electrostatically. The tests for RNA are, in fact, tests for basophilia because they employ basic dyes. The other strongly basophilic tissue constituents are the glycosaminoglycans and the sialo- and sulfomucins. Hence all these substances will bind the same dyes (see above, under "Tests for Carbohydrates").

It was mentioned earlier that alcian blue is, for steric reasons, a good dye for polyanionic carbohydrates, but not for nucleic acids. Other dyes, for steric reasons, are good for nucleic acids and not for carbohydrates. Pyronine is one such dye; it is a much smaller molecule than alcian blue and appears able to penetrate the helix of RNA more readily. A detailed comparison of the staining of glycosaminoglycans and RNA by alcian blue, pyronine, and methyl green is given by Scott and Willett (1966). Pyronine, in Brachet's original method for RNA, is used in conjunction with methyl green. The latter dye has a greater affinity for DNA, and hence the RNA binds the pyronine and is bright pink, while the nucleus binds the methyl green alone to give a green color, or both dyes to give a purple color. That the coloration by pyronine is due to the presence of RNA is confirmed by prior digestion of the cells with ribonuclease. This control should always be done. The spherules found in one type of insect hemocyte contain an acidic glycoprotein that binds pyronine strongly (Ashhurst and Richards, 1964; Lai-Fook, 1973; Neuwirth, 1973), and unless enzymatic digestion is performed, misinterpretation of the result can occur.

Commercial samples of methyl green contain methyl violet, which has to be extracted with chloroform before the dye can be used. This can be avoided by using malachite green (Baker and Williams, 1965) (see Appendix).

Other basic dyes, such as toluidine blue at pH 4.2, have been used to detect RNA, but these have no obvious advantage over pyronine.

22.6. Tests for proteins

There are many good histochemical tests for specific groups, configurations, and amino acids in proteins. These include tests for the amino groups of the terminal amino acids, terminal or side chain carboxyls, disulfide groups of cystine, the sulfhydryl groups of cysteine, the guanidinyl groups of arginine, the p-substituted phenol configuration of tyrosine, and the indole configuration of tryptophan.

Because proteins are ubiquitous throughout the cell, and because each contains many amino acids, there is rarely a high concentration of any one amino acid. Thus, the result of tests such as the Millon reaction for tyrosine or the Sakaguchi reaction for arginine tends to be an overall faint coloration of the tissue. These tests are often referred to or regarded as "tests for proteins" simply because the amino acid they detect is found in all proteins; such tests are obviously of limited value. The dimethylaminobenzaldehyde (DMAB) reaction for tryptophan gives a faint blue coloration in most tissues, but a strongly positive blue coloration is found in the zymogen granules of the cells of the mammalian exocrine pancreas, the parotid gland, and the peptic cells of the gastric pits, i.e., the proteolytic enzymes produced by these cells contain much tryptophan; Such intense coloration is, however, rare. Protein tests have not so far been very useful in the study of hemocytes for the reasons just discussed. Descriptions of the various tests for different amino acids, etc., are given by Adams (1965) and Pearse (1968).

22.7. Tests for enzymes

In order to perform their various roles, hemocytes must possess many enzymes. For example, the lysosomes of the phagocytic blood cells would be expected to contain hydrolytic enzymes active at acid pH, such as the acid phosphatases. In fact, acid phosphatase has been localized in the lysosomes of granulocytes of *Calpodes ethlius* and *Galleria mellonella* (Lai-Fook, 1973; Neuwirth, 1973) and the phagocytic hemocytes of *Calliphora erythrocephala* (Crossley, 1968).

Many pitfalls await the investigator of hemocyte enzymes, and it is pertinent at this point to discuss the work that has been done on human leucocytes. Many enzymes, such as alkaline phosphatase, lactate dehydrogenase, and nonspecific esterases, have been localized, but experience has shown that great care must be taken in the interpretation of the results. Each blood cell contains only very small amounts of enzyme, and so as much as possible must be retained during the test procedures. The critical stage in any test for an enzyme is the incubation of the test tissue in a medium containing a substrate at appropriate pH and temperature on which the enzyme can react. It is during the incubation period that loss of enzyme can occur, partly as a result of the preceding treatment of the cells.

In the preparation of a smear of blood cells, the shearing forces flatten the cells and can lead to some disruption of the membranes. Furthermore, even if the smears are dried well before immersion in the incubation mixture, cell loss and disintegration can occur. This may be prevented by fixation, but fixation itself leads to a loss of enzyme activity owing to changes in the permeability of the membranes and to inactivation of the enzymes. To reduce these effects, studies of human leucocytes have been done on unfixed and minimally fixed smears, which are incubated in a substrate mixture containing polyvinylpyrrolidone, agar, gelatin, or Ficoll (Stuart et al., 1975); this enables unfixed cells to remain on the slide throughout the incubation period. Alternatively, the cells may be taken through the test in suspension and then transferred to the slide by cytocentrifugation (Stuart et al., 1975). It is obvious that before meaningful studies of the enzymes of insect hemocytes can be done, attention will have to be paid to the methods used for preparing the smears.

A large number of methods for specific enzymes are described by Pearse (1968) and other authors, and it is superfluous to describe them here. The reliability of most of these methods is very dependent on the substrate and subsequent capture agent used; in many instances, it is essential to purchase these from a specific company. At all times the reaction should be controlled, not only against the same tissue under different conditions, but also against a tissue, which is known to give a positive reaction. False-negative reactions are as common as false-positive reactions in enzyme histochemistry.

22.8. Discussion

The foregoing account of histochemical methods for hemocytes is not intended to be exhaustive; it is an attempt to acquaint the reader with some of the problems that beset the "cookery book" histochemist. An example from the entomological literature that may prove to be a misinterpretation of histochemical results is the suggestion that the chorion of the eggshell contains acid mucopolysaccharide. The chorions of the eggshells of *Drosophila* and *Nasonia vitripennis* bind alcian blue (presumably at pH 2.6, as the method of Steedman was used) (King and Koch, 1963; King et al., 1968), and it was assumed that this result indicates the presence of acidic carbohydrates. Later, the chorions of *Acheta domestica* and *Gryllus mitratus* were found to contain a phosphoprotein; the serine residues have phosphate groups attached to form phosphoserine (Furneaux, 1970; Kawasaki et al., 1971). The eggshell of *Bombyx mori* differs, however, in that it contains little phosphorus (Kawasaki et al., 1971), so there is obviously diversity in the eggshell proteins of different species. But until the eggshells of *Drosophila* and *Nasonia* are shown not to contain a phosphoprotein, it remains a distinct possibility that the alcian blue binding may be due to phosphate groups and not to a polyanionic polysaccharide.

The Appendix contains the methods for some tests that are not adequately described in the textbooks or that include some modifications for hemocytes. It is pertinent to mention here that when writing papers, authors should give either full details of the method they used; or an original reference. Many authors omit these essential details and so make detailed appraisal of their results by others difficult or, in some instances, impossible.

At the present time, few histochemical reactions are quantitative, and it would be wrong to suggest that any of the tests described here can give an indication of the amount of the reactive substance present in the hemocyte. Such information will have to await the development of new techniques.

The aim of any histochemical study of hemocytes is not only to characterize the different types of hemocytes and their granules, but also to provide information about the function of the cells and their inclusions. So far, our knowledge of hemocyte function is rudimentary; the cells are involved in the encapsulation of foreign bodies, the phagocytosis of degenerating tissues at metamorphosis and in wound healing, etc., but apart from the obvious function of the lysosomes, we know nothing of the function of the other inclusions, such as the lipid droplets and carbohydrate-containing granules. Further elucidation of the nature of the particulate inclusions should provide evidence of their possible functions.

Thus, while histochemical tests properly understood and used can be of immense value in the characterization of the hemocytes, the unwary face many pitfalls that can completely invalidate their work.

Appendix

Histochemical methods for carbohydrates

1. Fixation

For the methods discribed here, fixation of hemocyte smears in 4% formaldehyde in 2% calcium acetate, which is at approximately neutral pH, for 30–60 min is recommended. The fixed smear is washed in running water for 2–3 hr before the chosen test is performed.

2. Alcian blue

Solutions
Alcian blue 8GX 300 should be used.
1   0.1% or 1.0% alcian blue in 0.1 N HCl, pH 1.0.
2   0.1% or 1.0% alcian blue in 3% acetic acid, pH 2.6.
The solutions should be filtered before use. They keep for only 3–4 days.

Procedure
1   Stain smears for 18 hr at room temperature in alcian blue solutions.
2   pH 1.0 – Blot dry, dehydrate in two changes of absolute ethanol, clear in
        xylene, and mount.
    pH 2.6 – Wash for 5 min in distilled water, dehydrate rapidly, clear in xy-
        lene, and mount.

Result
pH 1.0: Only sulfated mucosubstances are alcianophilic, i.e., sulfated glycosa-
        minoglycans and sulfomucins.
pH 2.6: All polyanions can bind the dye. These include:
        1   Phosphates of nucleic acids, e.g., nuclear staining
        2   Carboxyls of glycosaminoglycans and sialic acid residues on gly-
            coproteins
        3   Sulfate ester groups of glycosaminoglycans and sulfomucins, but
            to lesser extent (see text).

3. Alcian blue at critical electrolyte concentrations

(Scott and Dorling, 1965; Mowry and Scott, 1967)

Solutions
0.01–0.1% alcian blue 8GX 300 in 0.05 M acetate buffer, pH 5.7.
    To this dye solution, $MgCl_2$ is added to give a range of concentrations as fol-
lows: 0.05 M, 0.1 M, 0.2 M, 0.4 M, 0.6 M, 0.8 M, 1.0 M, and intermediate mo-
larities if required. The dye solutions should be filtered before use. Solutions
keep only for 2–3 days.

Procedure
1   Take sections to water.
2   Stain in alcian blue solutions for 18 hr.
3   Wash in three 5-min rinses of buffer containing the appropriate molarity of
    $MgCl_2$.
4   Dehydrate rapidly, clear in xylene, and mount.

Results

0.05 M–0.1 M $MgCl_2$	Phosphates	DNA, RNA
	Carboxyls	Sialic acid, hyaluronic acid, other glycosaminoglycans

	Sulfate esters	Sulfomucins, glycosaminoglycans other than hyaluronic acid
0.2 M	Phosphates	Very weakly stained
	Sulfate esters	Sulfomucins, sulfated glycosaminoglycans
0.4 M	Sulfate esters	Some sulfomucins, chondroitin sulfate, dermatan sulfate, keratan sulfate, heparin
0.6 M	High-density sulfate esters	Some sulfomucins, keratan sulfate, heparin
0.8 M	As 0.6 M	
1.0 M	Very high density sulfate esters	Keratan sulfate, possibly some sulfomucins

4. Low-iron diamine and high-iron diamine tests

(Spicer, 1965; Stoward, 1967)

Solutions

1 Low-iron diamine: 30 mg of N,N-dimethyl-m-phenylenediamine-$(HCl)_2$ and 5 mg N,N-dimethyl-p-phenylenediamine-HCl are dissolved in 50 ml distilled water. To this solution add 0.5 ml 40% wt/vol ferric chloride solution.
2 High-iron diamine: 120 mg of *meta*-diamine and 20 mg *para*-diamine are dissolved in 50 ml distilled water. To this solution add 1.4 ml 40% wt/vol ferric chloride solution.

The solutions must be freshly made before use.

Note: The concentration of the ferric chloride solution is given as 10% in Spicer's (1965) paper and in some textbooks; this is stated wrongly (Stoward, 1967).

Procedure

1 Stain smears in diamine solutions for 18 hr at room temperature.
2 Take smears straight to 95% ethanol (two changes), followed by two changes of absolute ethanol, clear in xylene, and mount.

Result

1 Low-iron diamine: Mucosubstances with carboxyls and sulfate groups stain gray to reddish—purple to black.
2 High-iron diamine: Only sulfated mucosubstances stain reddish—purple to black.

5. Neuraminidase digestion

(Stoward, 1967)

Solution

Commercially obtained *Vibrio cholerae* neuraminidase solution is added to 0.05 M acetate buffer, pH 5.5, containing 1% sodium chloride and 0.1% calcium chloride to make a final enzyme activity concentration of 100 units/ml.

Procedure

1 Mark smear slides "N" and "NC."
2 Put slides flat on damp filter paper in Petri dish. Flood "N" with enzyme solution, "NC" with buffer.
3 Incubate at 39 °C for 18 hr.
4 Wash for 5 min in running water, rinse in distilled water.
5 Stain smear with alcian blue, pH 2.6 or pH 5.7, with 0.05 M or 0.1 M $MgCl_2$.

Result

Neuraminidase-labile sialic residues are removed.

6. Hyaluronidase digestion

(Spicer et al., 1967)

Solution

0.05% solution of testicular hyaluronidase in 0.1 M phosphate buffer, pH 5.5 (activity and animal origin of enzyme vary).

Procedure

1 Mark smear slides "H" and "HC."
2 Put slides in Petri dish on damp filter paper. Flood "H" with enzyme solution, "HC" with buffer.
3 Incubate for 2 hr at 37 °C.
4 Wash for 5 min in running water, rinse in distilled water.
5 Stain smears with alcian blue, pH 2.6, pH 1.0, or pH 5.7, with appropriate molarity of $MgCl_2$.

Results

Testicular hyaluronidase removes hyaluronic acid and chondroitin sulfates. It probably also removes dermatan sulfate.

Histochemical methods for lipids

1. Acid hematein test for phospholipids

(Baker, 1946)

The method of fixation (i.e., steps 1–4) described here is recommended for smears to be stained with Sudan black B, also.

Solutions
1 Formol calcium fixative
     10 ml   40% formaldehyde
     10 ml   10% aqueous calcium chloride (anh.)
     80 ml   distilled water
2 Dichromate-calcium
      5 g   potassium dichromate
      1 g   calcium chloride (anh.)
     100 ml   distilled water
3 Acid hematein
     0.05 g   hematoxylin
      48 ml   distilled water
       1 ml   1% aqueous sodium iodate
     Heat until just boils. Cool, add 1 ml glacial acetic acid. Use at once.
4 Borax ferricyanide
     0.25 g   potassium ferricyanide
     0.25 g   borax 10 H_2O
     100 ml   distilled water
     Keep in dark.

Procedure
1 Fix smear in formol-calcium for 2 hr.
2 Transfer to dichromate-calcium for 3 hr at room temperature.
3 Transfer to dichromate-calcium for 19 hr at 60 °C.
4 Wash smear in distilled water.
5 Transfer sections to acid hematein solution for 5 hr at 37 °C.
6 Rinse in distilled water.
7 Transfer to borax-ferricyanide solution at 37 °C for 18 hr
8 Wash in water.
9 Mount in Farrant's medium or glycerine jelly.

Results
Phospholipids blue or blue black (ignore feeble reactions); cytoplasm yellowish.

Histochemical methods for ribonucleic acid

1. Pyronine–malachite green

(Baker and Williams, 1965)
     The advantage claimed for this method is that malachite green is a stable substance, not requiring extraction with chloroform.

Stock solutions required
1 Acetate buffer at pH 4.8
       acetic acid, 0.2 N   81 ml
       sodium acetate, o.2 M   119 ml

2 Pyronine Y 4% aqueous
3 Malachite green, 0.3% aqueous
4 Sodium bicarbonate, 0.025% aqueous

Pyronine/malachite green solution
Acetate buffer at pH 4.8   50 ml
Pyronine Y solution   40 ml
Malachite green solution   10 ml

Note: Different specimens of the dyes, obtained from the same manufacturer, do not appear to be quite identical in their effects, and it may be found desirable to change the proportions of the pyronine Y and malachite green solutions from 40:10 to 40.5:9.5 or 41:9.

Procedure
1 Fix smears in Zenker's fluid for 10–60 min. Wash in running water for 30–60 min.
2 Flood slide with pyronine–malachite green solution for 20 min.
3 Blot with filter paper.
4 Dip momentarily in sodium bicarbonate solution.
5 Dip momentarily in 95% ethanol.
6 Absolute ethanol, two lots, 2 min exactly altogether.
7 Clear in xylene and mount.

Note: It is important that the 95% and absolute ethanols used for dehydration are perfectly clean.

Result
Areas of basophilia stain bright pink to pink red.
Nuclei stain blue green to purple.

To confirm that the basophilia is attributable to RNA, ribonuclease digestions should be done.

Acknowledgment

I am most grateful to Dr. P. J. Stoward for his critical reading of and comments on this paper.

References

Adams, C. W. M. 1965. *Neurohistochemistry.* Elsevier, Amsterdam.
Ashhurst, D. E., and A. G. Richards. 1964. Some histochemical observations on the blood cells of the wax-moth, *Galleria mellonella* L. *J. Morphol. 114*:247–54.
Baker, J. R. 1946. The histochemical recognition of lipine. *Q. J. Microsc. Sci. 87*:441–70.
Baker, J. R., and E. G. M. Williams. 1965. The use of Methyl Green as a histochemical reagent. *Q. J. Microsc. Sci. 106*:3–13.
Bayliss-High, O. 1977. Lipids, pp. 168–85. *In* J. D. Bancroft and A. Stevens (eds.). *Theory and Practice of Histological Techniques.* Churchill Livingstone, London.
Costin, N. M. 1975. Histochemical observations of the haemocytes of *Locusta migratoria. Histochem. J. 7*:21–43.

Crossley, A. C. 1968. The fine structure and mechanism of breakdown of larval interseg-mental muscles in the blowfly *Calliphora erythrocephala*. *J. Insect Physiol.* *14*:1389–1407.

Culling, C. F. A., P. E. Reid, M. G. Clay, and W. L. Dunn. 1974. The histochemical demonstration of O-acylated sialic acid in gastrointestinal mucins. *J. Histochem. Cytochem.* *22*:826–31.

Furneaux, P. J. S. 1970. O-phosphoserine as a hydrolysis product and amino acid analy-sis of new laid eggs of the house cricket, *Acheta domesticus* L. *Biochim. Biophys. Acta 215*:52–6.

Kawasaki, H., H. Sato, and M. Suzuki. 1971. Structural proteins in the egg-shell of the oriental garden cricket, *Gryllus mitratus*. *Biochem. J. 125*:495–505.

King, P. E., J. G. Richards, and M. J. W. Copland. 1968. The structure of the chorion and its possible significance during oviposition in *Nasonia vitripennis* (Walker) (Hy-menoptera: Pteromalidae) and other chalcids. *Proc. R. Entomol. Soc. Lond.* (A) *43*:13–20.

King, R. C., and E. A. Koch. 1963. Studies on the ovarian follicle cells of *Drosophila* *Q. J. Microsc. Sci. 104*:297–320.

Lai-Fook, J. 1973. The structure of the haemocytes of *Calpodes ethlius* (Lepidoptera). *J. Morphol. 139*:79–104.

McManus, J. F. A. 1946. Histochemical demonstration of mucin after periodic acid. *Na-ture (Lond.) 158*:202.

Mowry, R. W., and J. E. Scott. 1967. Observations on the basophilia of amyloids. *Histo-chemie 10*:8–32.

Neuwirth, M. 1973. The structure of the hemocytes of *Galleria mellonella* (Lepidop-tera). *J. Morphol. 139*:105–24.

Pearse, A. G. E. 1968. *Histochemistry, Theoretical and Applied,* Vol. 1, 3rd ed. Chur-chill, London.

Quintarelli, G., J. E. Scott, and M. C. Dellovo. 1964. The chemical and histochemical properties of Alcian Blue. III. Chemical blocking and unblocking. *Histochemie 4*: 99–112.

Scott, J. E. 1970. Histochemistry of Alcian Blue. I. Metachromasia of Alcian Blue, Astra-blau and other cationic phthalocyanin dyes. *Histochemie 21*:277–85.

Scott, J. E., and J. Dorling. 1965. Differential staining of acid glycosaminoglycans (mu-copolysaccharides) by Alcian Blue in salt solutions. *Histochemie 5*:221–33.

Scott, J. E., and J. Dorling. 1969. Periodate oxidation of acid polysaccharides. III. A PAS method for chondroitin sulphates and other glycosamino-glycuronans. *Histochemie 19*:295–301.

Scott, J. E., G. Quintarelli, and M. C. Dellovo. 1964. The chemical and histochemical properties of Alcian Blue. 1. The mechanism of Alcian Blue staining. *Histochemie 4*:73–85.

Scott, J. E., and I. H. Willett. 1966. Binding of cationic dyes to nucleic acids and other biological polyanions. *Nature (Lond.) 209*:985–7.

Sorvari, T. E., and P. J. Stoward. 1970. Some investigations of the mechanism of so-called "methylation" reactions used in mucosubstance histochemistry. I. "Methy-lation" with methyl iodide, diazomethane, and various organic solvents containing either hydrogen chloride or thionyl chloride. *Histochemie 24*:106–13.

Sorvari, T. E., and P. J. Stoward. 1971. Saponification of methylated mucosubstances at low temperatures. *Stain Technol. 46*:49–52.

Spicer, S. S. 1965. Diamine methods for differentiating mucosubstances histochemi-cally. *J. Histochem. Cytochem. 13*:211–34.

Spicer, S. S., R. G. Horn, and T. J. Leppi. 1967. Histochemistry of connective tissue mucopolysaccharides, pp. 251–303. *In* B. M. Wagner and D. E. Smith (eds.). *The Connective Tissue.* Williams & Wilkins, Baltimore.

Stoward, P. J. 1967. The histochemical properties of some periodate-reactive mucosubstances of the pregnant Syrian hamster before and after methylation with methanolic thionyl chloride. *J. R. Microsc. Soc.* 87:77–103.

Stuart, J., P. A. Gordon, and T. R. Lee. 1975. Enzyme cytochemistry of blood and marrow cells. *Histochem. J.* 7:471–87.

Wigglesworth, V. B. 1971. Bound lipid in the tissues of mammal and insect: a new histochemical method. *J. Cell Sci.* 8:709–25.

Taxonomic index

600

Subject index